This textbook provides a clear, coherent introduction to standard cosmology, tracing a line from early-universe physics to the present-day cosmic web and galaxy evolution, with consistent links between observations, statistical methods, and simulations. The dialogue format of the book makes assumptions and reasoning steps explicit for the reader. Well suited for advanced students and researchers as a rigorous primer that provides the tools to assess contested issues.

—Anastasiia Lazutkina
PhD student at the University of Wuppertal

As would be expected from a collaboration of renowned authors, "Spinning the Cosmic Web" is an authoritative and up to date tract on the current view of almost all aspects of cosmology. If that were all there was to it, it would be among the top books on anyone's list. But it has a novel and important feature that raises it to another level: it directly addresses and discusses the questions that need to be asked about our so-called "Standard Model of Cosmology".

The way in which this is done is through conversations involving three people. This trialogue runs through the material of every chapter. In effect the authors are addressing the questions, uncertainties, and doubts that the reader may have, and other ideas that might be explored. This sheds an important light on how research in cosmology takes place, how questions are raised and what steps can be taken to resolve the issues. This makes the book an excellent basis for teaching the subject: the questions and answers are posed on behalf of the reader.

The three people in this trialogue are researchers in cosmology: a fusty professor, a pre-tenured post-doc and an ebullient graduate student. Each has their own personal biases that are exposed and commented on by the other two. This mode of presentation is unusual though not new, and in this case, it is beautifully done. It also goes some way towards addressing the modern way of electronic teaching in which everything is presented on-screen and face to face discussion is a rarity. It certainly makes the book more entertaining than a dry textbook.

The trialogues are the discussion that the students might be having. This is particularly important at a time like the present when cosmology is being thrown into chaos: we have more questions than answers. "Spinning the Cosmic Web" presents the questions and problems in a lucid and very readable way.

—Bernard Jones
Professor Emeritus at University of Groningen

This excellent and timely book describes the development of the theory of structure formation in the Universe. It includes the discovery of the Cosmic Web, a description of the wide range of physical processes at play and the key astronomical observations that together provide a comprehensive understanding of the formation and evolution of galaxies in the Cosmic Web. Einasto's seven decades of active research in this area adds an unusually broad perspective. The dialogs in the classical style clarify key concepts and are fun to read.

—Tim de Zeeuw
Professor Emeritus at Leiden University Director General of European Southern Observatory 2007–2017

Jaan Einasto is one of the most admired and respected astronomers in the world. He and his group are specially acclaimed for their work mapping galaxies and analysing how they are grouped into clusters. Such work led in the 1970s to general acceptance of the remarkable conclusion that most of the gravitating stuff that binds clusters together isn't the stars and gas we observe, but some still-mysterious "dark matter".

This authoritative and clearly written book brings the story up to date — recounting all that's been learnt from the availability of more powerful telescopes (on the ground and in space). We now have a better understanding of the galaxies themselves, the role of dark matter in their formation, and how the cosmos itself has evolved from a hot, dense beginning about 13.8 billion years ago.

We should be grateful to Jaan Einasto for the clear and comprehensive book, which deserves wide readership.

—Lord Martin Rees
Plumian Professor at Cambridge University 1973–1991 Astronomer Royal 1995–2025

SPINNING
THE
COSMIC WEB

SPINNING
THE
COSMIC WEB

Jaan Einasto
Tartu Observatory, Estonia

Gert Hütsi
National Institute of Chemical Physics and Biophysics, Estonia

Istvan Szapudi
University of Hawaii, USA

Peeter Tenjes
Tartu Observatory, Estonia

World Scientific

NEW JERSEY · LONDON · SINGAPORE · BEIJING · SHANGHAI · HONG KONG · TAIPEI · CHENNAI · TOKYO

Published by

World Scientific Publishing Co. Pte. Ltd.

5 Toh Tuck Link, Singapore 596224

USA office: 27 Warren Street, Suite 401-402, Hackensack, NJ 07601

UK office: 57 Shelton Street, Covent Garden, London WC2H 9HE

Library of Congress Cataloging-in-Publication Data
Names: Einasto, Jaan author | Hütsi, Gert author | Szapudi, Istvan author | Tenjes, P. author
Title: Spinning the cosmic web / authors Jaan Einasto, Tartu Observatory, Estonia,
    Gert Hütsi, National Institute of Chemical Physics and Biophysics, Estonia,
    Istvan Szapudi, University of Hawaii, USA, Peeter Tenjes, Tartu Observatory, Estonia.
Description: New Jersey : World Scientific, [2026 Spinning the Cosmic Web] |
    Includes bibliographical references and index.
Identifiers: LCCN 2025013901 | ISBN 9789819813469 hardcover |
    ISBN 9789819813476 ebook for institutions | ISBN 9789819813483 ebook for individuals
Subjects: LCSH: Cosmology--Textbooks | Cosmology--Mathematical models--Textbooks |
    LCGFT: Textbooks
Classification: LCC QB981 .E38 2025
LC record available at https://lccn.loc.gov/2025013901

British Library Cataloguing-in-Publication Data
A catalogue record for this book is available from the British Library.

For any available supplementary material, please visit
https://www.worldscientific.com/worldscibooks/10.1142/14322#t=suppl

Desk Editors: Nambirajan Karuppiah/Carmen Teo Bin Jie

Typeset by Stallion Press
Email: enquiries@stallionpress.com

Preface

The origin of this book was a request from World Scientific Publishing to write a textbook on recent development in cosmology, including dark matter and dark energy as well as the cosmic web. There exist historical reviews on astrophysics and cosmology by Longair (2006), Sanders (2010), Gott (2016), and Thompson (2021). Introduction to cosmology was described by Ryden (2017), intergalactic medium by Ryden & Pogge (2021), chemical evolution of galaxies by Pagel (1997), cosmological physics by Peacock (1999), and galaxy formation and evolution by Ellis (2022) and Longair (2023). The development of modern cosmology was described in a series of excellent monographs by Peebles *et al.* (2009), and Peebles (2020, 2022a, 2022c, 2024, 2025), As this overview shows, cosmology and physics of galaxies were discussed as separate scientific fields.

Our expertise is essentially empirical in discussing observations and comparing observations with results of numerical simulations of the evolution of the cosmic web and galaxies. We decided to discuss the formation and evolution of galaxies as part of the cosmology, since galaxies form in the environment — the cosmic web. *Spinning the Cosmic Web* is a textbook on modern cosmology with an observational focus using only elementary calculus and statistics. Unlike previous titles that focus on linear cosmology at an undergraduate level, this work will extend to non-linear phenomena, particularly emphasising the cosmic network of clusters connected with filaments and voids separated by walls. For the first time, the intertwined processes on various scales, from the largest scales to the structure and emergence of the cosmic web and galaxies and their populations, will get their due. The observational approach facilitates the often neglected formation and evolution of galaxies in dark matter haloes, forming the cosmic web, crucial to understanding the interconnection between the observed galaxy distribution and the underlying dark matter structure — the bias.

After a short introduction to the contemporary cosmological paradigm in historical context, we discuss the current observations of the Universe and its constituents. The book is divided into 10 chapters. The book's first half focuses on cosmic web, while the second half deals with galaxies and their evolution within the cosmic web. It is manifested by the understanding that dark matter and dark energy are important components of matter–energy content of the Universe and that galaxies are not distributed randomly but form the cosmic web.

We introduced in most chapters discussions in a dialogue form (like Platon, etc.). This would distinguish us from most physics books, although it takes up time and it would be quite unconventional. Nevertheless, we could use that format to keep the readers mind on cosmology. The dialogues serve a dual purpose of elucidating different sides of issues and providing some lightness among the dense science paragraphs. While this is unusual for textbooks, we were particularly inspired by the book, *Gödel, Escher, Bach* (Hofstadter, 1999) (among others), to add dialogues.

There are three actors in the dialogues: Callistus, an old-fashioned professor with a penchant for alternative models and Latin proverbs; Rhea, a soon-to-be-tenured assistant professor who champions the concordance model and also prefers MOND over GR and is secretly Swiftie; finally, Ariel, a socially conscious grad student, who has life and big decisions in front of them. Throughout the book, we follow this team working through their differences of opinions on cosmic matters and beyond while building and reinforcing their bond as scientists and humans.

Authors thank Prof. Bernard Jones, Mark Neyrinck and John Learned for suggestions made after reading preliminary version of the book. We thank Ph. Andre, Daniel Anglés-Alcázar, Teresa Antoja, Doris Arzoumanian, Peter Behroosi, Nate Bastian, Bahar Bidaran, Lucas Bignore, Bruno Bingelli, Tom Broadhurst, Gustavo Bruzual, Fernndo Buitrago, Adam Burrows, Michele Cappellari, Markus Cautun, Renye Cen, Paul Clark, Douglas Clowe, Christopher

Conselice, Benjamin Davis, Avishai Dekel, Alexandra Dupuy, George Efstathiou, Maret Einasto, Frank Eisenhauer, Brian Fields, Steven Finkelstein, Wendy Freedman, Karl Gebhardt, Margaret Geller, Reinhard Genzel, Orly Gnat, Richard Gott, Alister W. Graham, Yuichi Harikane, William Harris, Tilman Hartwig, Timothy Heckman, Amina Helmi, Patrick Hennebelle, Yehuda Hoffman, Philip Hopkins, Icko Iben, Kohei Inayoushi, Anna Ivleva, Jason Kalirai, Rahul Kannan, Igor Karachentsev, Dušan Kereš, Anatoly Klypin, Eiichiro Komatsu, John Kormendy, Pavel Kroupa, Diederik Krujssen, Tongyan Lin, Marshall McCall, Ted Mackereth, Dougal Mackey, Jorryt Matthee, Lamiya Mowia, Dylan Nelson, Jerry Ostriker, Jim Peebles, Saul Perlmutter, Brant Robertson, Aaron Romanowsky, Friedrich Röpke, Arnab Sarkar, Sergei Shandarin, Prateek Sharma, Benjamin Seidel, Jenny Sorce, Daniel Stark, Margherita Talia, Elmo Tempel, Laird Thompson, Brent Tully, Jason Tumlinson, Aurelien Valade, Stephen Walker, and Ned Wright for figures from their publications. Figures are reproduced by permission of the AAS, Monthly Notices of the Royal Astronomical Society, and EDPsciences. Permission is granted for the purpose of reuse in "Spinning the Cosmic Web", and does not extend to any other forms of distribution or reproduction beyond what is customary for the journal's dissemination practices. Many thanks also to the desk editors of the book Carmen Teo Bin Jie and Nambirajan Karuppaiah from World Scientific Publishing.

About the Authors

Jaan Einasto was born in 1929 in Tartu, Estonia. He received a Ph.D. equivalent from the Tartu University in 1955 and a senior research doctorate in 1972. Einasto joined Tartu Observatory as a research associate in 1952, served as the head of the Department of Physics of Galaxies and Cosmology (1976–1997). Jaan Einasto is a winner of Gruber Cosmology Prize and Viktor Ambartsumian International Prize, author of Dark Matter and Cosmic Web Story (two editions).

István Szapudi is a theoretical cosmologist at the Institute for Astronomy, University of Hawaii, who explores the mysteries of the universe using statistics, simulations, and observations. He studies how the patterns of temperature and density fluctuations in the early and present universe reveal the nature of dark matter, dark energy, and inflation. He is also interested in the anomalous correlation between the cosmic microwave background and the distribution of galaxies, which challenges the standard cosmological model. His paper that discovered the largest void in the universe, the likely cause of the Cold Spot on the cosmic microwave background, was featured on the cover of *Scientific American* in 2015. His recent work with renowned philosopher Silvia De Bianchi on the achronotopic interpretation of quantum mechanics resolves many mysteries of the standard Copenhagen interpretation. He recently proposed a tethered sun shield near the L1 Lagrange point to mitigate the effects of climate change. He also invented a slowly rotating model of the universe that solves the Hubble tension. He has published on various topics beyond cosmology, including applied mathematics, computer science and machine learning, general relativity, geoengineering, philosophy of science, economics, and epidemiology.

Peeter Tenjes was born in 1955 in Türi, Estonia. He studied theoretical physics at the University of Tartu. After graduating from the university, P. Tenjes started to work as a teaching assistant in the Department of Physics. Soon, he joined the Tartu Observatory, the Department of Physics of Galaxies and Cosmology as a research associate and senior research associate. P. Tenjes defended the degree of Doctor of Astronomy in 1993. From 1997 to 2017, he started working at the Institute of Theoretical Physics at the University of Tartu as an Associate Professor and then as a Professor. His lectures included several courses in theoretical physics, mathematical physics and astrophysics. Scientific interests of P. Tenjes include galactic structure, galactic dynamics and evolution of galaxies in galactic groups and clusters.

Gert Hütsi is a senior researcher in the High-Energy and Computational Physics group at NICPB in Tallinn. He completed his PhD in 2006 at the Max Planck Institute for Astrophysics (MPA), where his work involved one of the earliest measurements of baryon acoustic oscillations in the large-scale matter distribution and their use as a geometric probe of the Universe's expansion history. After his PhD, he held postdoctoral positions at University College London, Tartu Observatory and at MPA. His research covers range of topics in cosmology and astrophysics, including the formation and evolution of large-scale structure, cosmic diffuse radiation fields, gravitational waves, black holes, and astroparticle physics.

Contents

Preface   vii

About the Authors   xi

1. **Introduction**   1

    1.1   Scope of the book   1
    1.2   Structure of the book   3

2. **Formation of the Modern Cosmological
    Paradigm**   7

    2.1   Classical cosmological paradigm   8
          2.1.1   Theory of general relativity   12
          2.1.2   The expansion and age of the Universe . .   12
          2.1.3   The nature of spiral nebulae   15
          2.1.4   Distribution of galaxies on sky   16
          2.1.5   Formation and evolution of stars   18
    2.2   Dynamical and physical evolution of galaxies . . .   22
          2.2.1   Models of spatial structure of galaxies . . .   23
          2.2.2   Hydrodynamical models of galaxies   29
          2.2.3   Dynamical evolution of galaxies   36
          2.2.4   Physical evolution of galaxies   39
          2.2.5   Summary of the classical cosmological
                  paradigm   46
    2.3   New cosmological paradigm   47
          2.3.1   Dark matter   48
          2.3.2   The cosmic web   57

2.3.3 Comparison with simulations 62
2.3.4 Inflation 68
2.4 Cosmological parameters 71
2.4.1 Critical dialogues in cosmology 71
2.4.2 Dark energy and the accelerating
Universe 72
2.4.3 Expansion rate and the age of the
Universe 74
2.4.4 Big Bang nucleosynthesis and the amount
of baryonic matter 78
2.4.5 Matter–energy densities of various cosmic
populations and other cosmological
parameters 79
2.4.6 New cosmology paradigm is ready: What
next? 83

3. Inflationary Concordance Cosmology   **85**

3.1 The homogeneous universe 85
3.1.1 Redshift 88
3.1.2 The Hubble parameter and constant 90
3.1.3 The Friedmann equation 91
3.1.4 Energy conservation and acceleration . . . 93
3.1.5 Equation of state 94
3.1.6 Multi-component universes 96
3.1.7 Flatness problem 97
3.1.8 Distances 98
3.1.9 Local Universe 99
3.2 Inflation and Big Bang 102
3.2.1 Inflation 103
3.3 Thermal history 109
3.3.1 Concise timeline 112
3.3.2 Cosmological baryogenesis 113
3.3.3 Nucleosynthesis 113
3.4 The cosmic microwave background 117
3.4.1 The surface of last scattering 117
3.4.2 CMB anisotropies 119
3.5 Gravitational instability 124
3.6 The power spectrum as a statistical probe 129
3.7 Indicator power spectra 132

4. **Formation and Evolution of the Cosmic Web**   **137**

   4.1   Numerical simulations 138
       4.1.1   Initial conditions and simulation
              methods 140
       4.1.2   Simulations with cosmological constant . . 146
       4.1.3   Simulating galaxy samples 149
   4.2   Smoothing density fields 150
       4.2.1   Smoothing density fields with constant
              smoothing length 150
       4.2.2   Defining cosmic web populations 151
   4.3   Tracing cosmic web populations 153
       4.3.1   Adaptive smoothing methods 154
       4.3.2   Cosmic web populations 157
   4.4   Density fields of galaxies and matter 168
       4.4.1   The luminosity density field 168
       4.4.2   The evolution of density perturbations
              of various scales 174

5. **Integrated Properties of the Cosmic Web**   **177**

   5.1   Percolation properties of galaxies and matter . . . 177
       5.1.1   Extended percolation analysis 178
       5.1.2   Evolution of percolation functions 180
       5.1.3   Percolation functions of DM and SDSS
              clusters and voids 182
   5.2   Correlation function 186
       5.2.1   Calculation of correlation function and its
              derivatives 188
       5.2.2   Correlation functions of galaxies and
              matter 189
       5.2.3   Fractal properties of the cosmic web 192
       5.2.4   Relation between 2D and 3D correlation
              functions 196
       5.2.5   Cosmological implications 203

6. **Evolution of the Cosmic Web**   **207**

   6.1   Evolution of the shape of the density field 207
       6.1.1   Mathematical and cosmological estimates
              of the shape of the density field 210

6.1.2 The growth of structures of various
scales 212

6.1.3 Evolution of the skewness and kurtosis of
DM density fields 215

6.1.4 Cosmological implications 218

6.2 Biasing phenomenon 221

6.2.1 Statistical biasing 223

6.2.2 Physical biasing 224

6.2.3 The influence of a homogeneous
population in voids 227

6.2.4 Evolution of the luminosity function
of galaxies 229

6.2.5 Evolution of CFs of simulation
galaxies and particle density-limited
DM samples 231

6.2.6 Evolution of bias functions of simulated
galaxies and DM particles 232

6.2.7 Evolution of bias parameters of simulated
galaxies and DM particles 234

6.2.8 Bias parameter of faintest galaxies and the
fraction of matter in the clustered
population 237

6.2.9 Cosmological implications 238

7. Physical Processes in Galaxies   245

7.1 Formation and evolution of galaxies in dark
matter haloes 247

7.1.1 Physical conditions before and after the
recombination 247

7.1.2 First stars 250

7.1.3 First galaxies 254

7.1.4 Observations 258

7.1.5 Reionisation 261

7.2 Star formation 263

7.2.1 Thermal instability 265

7.2.2 Molecular clouds 267

7.2.3 Initial mass function 271

7.3 Gas inflows and outflows in galaxies 275
    7.3.1 Gas in and around galaxies: Current
        situation 275
    7.3.2 Basic physical characteristics of the gas . . 277
    7.3.3 Gas accretion 285
    7.3.4 Gas inflows and outflows 290
7.4 Supernovae explosions and chemical evolution
    of galaxies 295
    7.4.1 Supernovae explosions 295
    7.4.2 Merger of neutron stars 299
    7.4.3 Production of chemical elements 301
    7.4.4 Chemical evolution of galaxies 304
7.5 Supermassive black holes in galactic centres 312
    7.5.1 Direct observations of the supermassive
        black holes in the centres of galaxies 312
    7.5.2 Statistical properties of SMBHs 317
    7.5.3 A pathway from black holes to SMBHs . . 323
7.6 Observing the galaxy evolution with redshift 327
    7.6.1 Historical notes 327
    7.6.2 A time machine of distant galaxies 328
    7.6.3 Properties of young high-redshift
        galaxies 331

8. **Structure and Evolution of Galaxies in the
   Cosmic Web**   **341**

8.1 Formation and evolution of galactic populations . . 341
    8.1.1 Dynamical processes in the evolving
        Universe 342
    8.1.2 Dynamical processes in galactic
        population 346
    8.1.3 Evolution of galactic disks and stellar
        haloes 348
    8.1.4 Evolution of galactic bulges 354
    8.1.5 Formation and evolution of globular
        clusters 359
    8.1.6 Galactic luminosity functions in various
        environments 369
    8.1.7 Evolution of galactic luminosity
        functions 373

8.2   Structure and evolution of local galaxies 374
    8.2.1   Structure and evolution of Milky Way . . . 376
    8.2.2   Structure and evolution of Andromeda
         galaxy M31 380
    8.2.3   Structure and evolution of galaxies in the
         Local Universe 391
8.3   Structure and evolution of clusters of galaxies . . . 398
    8.3.1   Structure and evolution of the Virgo
         cluster and the Local Supercluster 398
    8.3.2   Structure and evolution of colliding
         clusters 406
    8.3.3   Structure and evolution of the galaxy
         clusters of various richness 408
8.4   Structure and evolution of superclusters of
    galaxies 412
    8.4.1   Structure of clusters in supercluster
         filaments 412
    8.4.2   Structure and evolution of the A2142,
         Corona Borealis and BOSS Great Wall
         superclusters 413

9.   Structure and Evolution of the Ensemble of Superclusters and Voids   417

9.1   Structure and evolution of the
    Local Universe 417
    9.1.1   Magellanic stream and polar ring
         galaxies 418
    9.1.2   Structure and evolution of the Local
         Group 420
    9.1.3   A Council of Giants 422
    9.1.4   Constrained simulations 422
    9.1.5   Evolution of basins of dynamical
         influence 426
    9.1.6   Structure and future evolution of the Local
         Universe 429
9.2   Some geometrical properties of the cosmic web . . 433
    9.2.1   Regularity of the cosmic web 434
    9.2.2   Supercluster walls 437

9.3 Structure and evolution of the ensemble of
superclusters and voids 439
 9.3.1 Structure and evolution of the ensemble of
superclusters from density field 442
 9.3.2 Structure and evolution of ensemble of
superclusters and voids from velocity
data 448
 9.3.3 Method and first results 449
 9.3.4 Structure and evolution of the ensemble of
basins of attractions and repulsions 450
9.4 Concluding remarks 452

10. ΛCDM and Beyond   **457**

10.1 ΛCDM 458
 10.1.1 Radiation 461
 10.1.2 Matter 462
 10.1.3 Vacuum energy 464
 10.1.4 Minimal set of ΛCDM parameters 468
 10.1.5 Problems with ΛCDM 469
10.2 Dark energy 473
10.3 Dark matter 478
10.4 Concluding comments 487

Epilogue   489

References   495

Index   531

Chapter 1

Introduction

1.1 Scope of the book

Cosmology has emerged during the last century as the natural science of the Universe as a whole. The physics of its ingredients, such as stars and galaxies, has initially developed on a parallel but somewhat independent track. The convergence of these disciplines and the emergence of observational cosmology during the 1950s resulted in an explosive and exponential development, as can be seen from the number of papers published on the topic (Fig. 1.1). It appears that in the last decade, cosmology has matured, and despite the abundance of data from unprecedented giant surveys such as SDSS, Planck, eBOSS, and most recently Euclid, the number of papers stabilised.

The revolution of integrating mathematical, physical, and observational branches of cosmology started in the middle of the 20th century. In the 1960s, Robert Dicke in Princeton, Dennis Sciama in Cambridge, and Yakov Zeldovich in Moscow created new centres of theoretical cosmology to study processes in the Universe as physical ones rather than purely mathematical. Their theoretical progress relied heavily on observations facilitated by the emergence of novel techniques and facilities. In the late 20th century, several 10-m class telescopes and numerous space observatories further accelerated the rapid growth of observational cosmology. Thus, the observational cornerstones of modern cosmology emerged, such as the accelerating expansion of the Universe attributed to dark energy (DE) and large redshift surveys of galaxies, which in turn led to the discovery of the cosmic web, studying the early phases of galaxy evolution, and

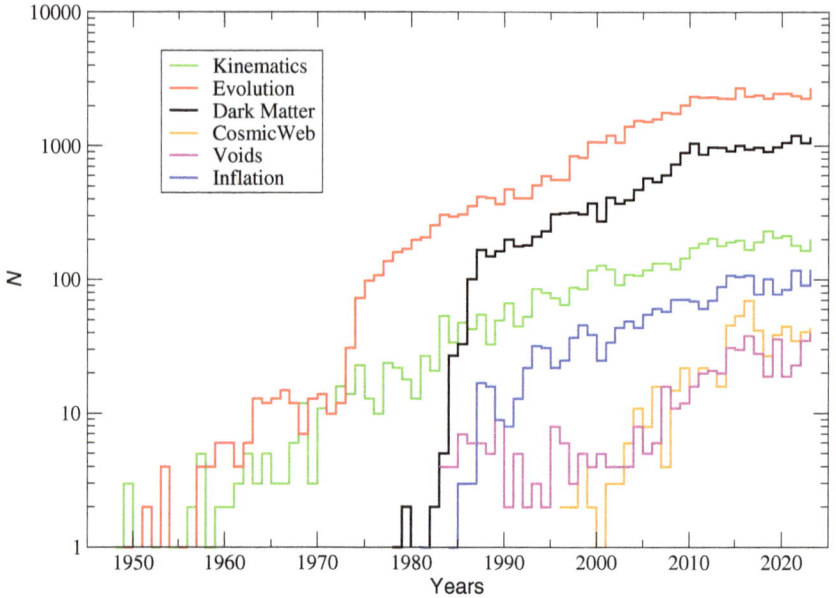

Fig. 1.1. The number of yearly published papers on various topics in cosmology (Einasto, 2024).

recognising the necessity of an inflationary period in the evolution of the Universe. The rapid growth rebalanced various fields of cosmology. Galaxy evolution dominates in the number of papers, followed by dark matter (DM) and inflation (cf. Fig. 1.1). The cosmology community accepted initially left-field concepts, such as DM, cosmic web, and inflation.

Galaxies are born and evolve in a cosmic environment — the cosmic web influenced by the presence of DM. Thus, the formation and evolution of galaxies must be treated simultaneously with the evolution of the cosmic web. For the first time, this introductory book treats this complex problem primarily from an observational point of view complemented by numerical simulations and theoretical analysis. The trinity of albeit incomplete and complex, observational data, simulations, and theory provides the clearest picture of galaxy evolution in the cosmic web attainable at present. We concentrate on the global properties of the web and discuss the evolution of the components of the web — galaxies and systems of galaxies in the context of their environment.

1.2   Structure of the book

The book consists of 10 chapters in four conceptual units. The first unit contains the introductory chapters (Chapters 1 and 2), where we discuss the formation of the modern cosmological paradigm. The second unit focuses on traditional cosmology (Chapters 3–6). In the third conceptual unit, we discuss the structure, formation, and evolution of galaxies in the cosmic web (Chapters 7–9). The last unit (Chapter 10) is devoted to open problems in cosmology.

In Chapter 2, we describe the formation of cosmology as a physical science in the 20th century. In the first half of this period, the theoretical and observational foundation of cosmology was created. These pioneering studies included the theoretical foundation of the expanding universe by Einstein (1916), Friedmann (1922, 1924) and Lemaître (1927), which is the first pillar of modern cosmology. The second pillar is the discovery of the expansion of the Universe and the determination of the expansion speed by Hubble (1929a) and Hubble & Humason (1931). The third pillar is the understanding that most nebula are external galaxies, as shown by Öpik (1922) and Hubble (1925, 1926, 1929b).

In parallel, the structure and dynamics of galaxies were studied, which forms the fourth pillar of cosmology. The theoretical foundation of this development was created by Eddington (1914) and Jeans (1922). The observational basis of the understanding of galaxies was the detection of various stellar populations with different kinematics in our Milky Way by Lindblad (1927) and Oort (1927, 1928).

Galaxies consist of stars. Understanding the structure and evolution of stars is needed to understand the evolution of galaxies, which is the fifth pillar of cosmology. This includes the model of the formation of our Galaxy by Eggen *et al.* (1962), which shows that the Galaxy contracted during its formation. The theory of the physical evolution of galaxies on the basis of stellar evolution tracks was suggested by Tinsley (1968).

The sixth pillar is the distribution of galaxies, based on Lick galaxy counts, reduced by Seldner *et al.* (1977) and described quantitatively by Peebles (1980), using basically the correlation function of galaxies. According to this analysis, galaxies and clusters of galaxies are distributed randomly.

In the 1970s, cosmology was in a crisis. Some observational data could not be explained within the traditional framework, such as the mass paradox in clusters of galaxies (Zwicky, 1933). A new cosmological paradigm emerged, which includes the presence of DM in the Universe, its web-like structure, and the suggestion of the inflation period in the early evolution of the Universe. The last part of Chapter 2 describes the observational evidence for the presence of DE, the accelerating Universe and cosmological parameters.

At the end of the 20th century, the modern cosmology paradigm has been matured. All of its matter–energy constituents were known, and the accuracy of essential cosmological parameters was already about 1%. The new 21th century has added many details to our picture of the Universe and its components. The rest of our book is devoted to the description and analysis of these details.

The second conceptual unit describes classical cosmology in Chapters 3–6. Chapter 3 is devoted to the theoretical basis of the cosmology. This includes the discussion of redshifts as distance indicators, the Hubble parameter as expansion speed, the Friedmann equations, the equations of energy conservation and acceleration, and the description of the inflationary model of the Universe, cosmological density parameters to baryonic matter, DM, and DE. Further, we discuss cosmological nucleosynthesis and the cosmic microwave background (CMB) radiation. Finally, we examine the gravitational instability of the primordial medium.

In Chapter 4, we describe numerical simulations, which allow us to follow the formation and evolution of the Universe. Integrated properties of the cosmic web are described in Chapter 5. This includes the description of the percolation analysis and the analysis of the correlation function and the related fractal function. We also discuss the relationship between 2D and 3D correlation functions. In Chapter 6, we discuss the evolution of the cosmic web, including its shape as described by skewness and kurtosis parameters. The last part of Chapter 6 is devoted to analysing the biasing phenomenon — the relation between the distributions of galaxies and DM.

The third conceptual unit is devoted to the structure, formation and evolution of galaxies in the cosmic web. It starts with Chapter 7, where we discuss physical processes in galaxies. This includes physical conditions in the Universe before and after the recombination epoch, the formation of first stars and galaxies, gas inflows and outflows in galaxies, supernova explosions and chemical evolution of galaxies. We also describe supermassive black holes in galactic centres and young galaxies.

Chapter 8 further discusses the structure and evolution of galaxies in the cosmic web. It starts with a discussion of the formation and evolution of galactic populations. Next, we discuss the structure and evolution of the nearest galaxies, the Milky Way, Andromeda galaxy M31. It follows the discussion of the structure and evolution of clusters and superclusters of galaxies: the Virgo Cluster, the Local Supercluster, and several rich superclusters. Chapter 9 discusses the structure and evolution of the ensemble of superclusters and voids. It starts by describing the structure and evolution of the Local Universe using constrained simulations. Thereafter, we discuss the structure and evolution of superclusters and voids using spatial and velocity data.

The last conceptual unit describes the standard cosmology model in Chapter 10. Here, we discuss the searches to determine the nature of DM and DE. In the Epilogue, we raise some open questions: the birth of the Universe and our Universe as one realisation of the multiverse. The book ends with the bibliography and index.

Chapter 2

Formation of the Modern Cosmological Paradigm

Until the 20th century, cosmology was mostly a philosophical and metaphysical discipline because very little was known about the actual global structure of the Universe and the nature of the various astronomical objects within it. In the beginning of the 20th century, most astronomers believed that our Milky Way system is the principal constituent of the Universe. Sir Arthur Eddington (1914) wrote his famous book, *Stellar Movements and the Structure of the Universe*, where he identified the Milky Way with the whole Universe. The presence of other stellar systems similar to the Milky Way was discussed, but there existed no proof for this. At this time, most astronomers accepted the view that the whole stellar universe is very old, of the order of 10^{14} years. This age estimate was based on the observation that stellar orbits in our Milky Way system are well mixed and relaxed. The relaxation time of this process by star–star encounters is very long, of the order mentioned above, and this estimate was taken as the possible age of the Universe.

In this chapter, we describe the classical cosmological paradigm and its observational basis. Next, we discuss new observational data which led to the formation of the modern cosmological paradigm: the discovery of dark matter (DM), the cosmic web, inflation and dark energy (DE).

2.1 Classical cosmological paradigm

[*A dark office full of books. Prof. Callistus is lost in his thoughts. His mentees, Ariel, a grad student, and Rhea, an assistant professor, sit across from him. Waiting. Finally, he scribbles some notes into a notepad and looks up.*]

CALLISTUS: Do you ever wonder why the night sky is dark?

ARIEL: Duh? The sun is down? Are we here to talk about trivialities or to penetrate the secrets of the universe?
[*Rhea is eager to jump in.*]

RHEA: Think about the stars! Shouldn't they brighten up the night sky?

ARIEL: They do, but the stars are much further away than the Sun. Therefore, starlight is extremely dim, except for a few stars closest to us.

CALLISTUS: Very good. How many stars are there? Imagine the universe filled with stars.

RHEA: Think about the Sun. It's a typical star.

ARIEL: Okay, if our Sun is typical, and the universe is filled with Sun-like stars, in a shell at a distance r, the number of stars would grow with r^2, while their brightness would decrease as $1/r^2$, thus –

RHEA: The total brightness from each shell should be constant.

CALLISTUS: Right. Looks like the infinite shells would add up to infinite brightness.

RHEA: Stars by the pocketful.

ARIEL: Of course, when we look at a star, it blocks the light from the stars behind it. It pains me to come to the obviously wrong conclusion that the night sky should be uniformly bright. If our sun is typical, should the night be as bright as looking into the sun? Crazy.

CALLISTUS: It's called –

RHEA: Olbers' paradox!

CALLISTUS: The topic of our session today.
[*He looks out of the window. The sun is about to set.*]

CALLISTUS: Well, the night sky is still dark. Now what?

RHEA: One of our assumptions must be wrong. At least.

ARIEL: I know the universe is not filled uniformly with stars. Our Galaxy has an edge, and the number of stars drops.

RHEA: Cold. You could repeat the argument, replacing stars with galaxies and quasars.

ARIEL: That's at least not as crazy. The night sky should be as bright as the nearest galaxy. But it doesn't sound like the solution... The universe is expanding, so the starlight reaching us from a distance will be shifted towards the red, towards less energy, through the Doppler effect. Faraway shells will contribute less and less.

RHEA: Warm.

CALLISTUS: Right, you're onto something. What about time?

ARIEL: About time? A silly time-travel movie. I can enjoy such fluff sometimes, but –

RHEA: How long does it take for light to reach us?

ARIEL: The time for light to reach us is equal to the distance divided by the speed of light...

CALLISTUS: And?

ARIEL: Big Bang Theory! [*sings*] Our whole universe was in a hot, dense state. Then nearly 14 billion years ago, expansion started, wait –

RHEA: More like 13.7 billion years ago. Light reaches us from a finite sphere corresponding to the age of the universe, which is also the size of the universe.

CALLISTUS: Our particle horizon. The whole universe might be bigger. Personally, I'd bet on it being bigger... we can imagine the universe outside our horizon, especially if we are philosophers, but as physicists, we cannot observe anything outside of it.

ARIEL: What about tachyons? [*Rhea looks at her watch.*]

RHEA: What about the Hot Tub Time Machine?

ARIEL: Now that's a fun movie.

RHEA: Exactly. A movie. Now, let's get back to reality and summarise our combined mentoring session that certainly saved time for Professor Callistus... but do I need to be here for that?

CALLISTUS: Absolutely, your contribution is crucial, and you gain valuable teaching experience.

[*She rolls her eyes.*]

RHEA: Should we continue or split these sessions?

CALLISTUS: Admit you enjoyed it. I did.

ARIEL: O.M.G. Tension. Fun. You guys should start a TikTok. It would go viral instantly.

CALLISTUS: What's tik tok? It's millennial slang for... time?

ARIEL: Naahh, I'm Gen Z. I'll explain TikTok later.

RHEA: Let's keep it on topic.

ARIEL: I'd say let's continue these sessions.

RHEA: I mean, Olbers' paradox.

ARIEL: Today, I learned that the night sky is dark because the universe expands and has a finite age. Each far-away shell within the horizon contributes less and less due to redshift.

CALLISTUS: Yet, if you look at this incredible JWST image, it's almost a wallpaper of beautiful galaxies filling the image. Olbers' dream. I just got this data last night.

[*Shows off an incredible JWST image Fig. 7.12 on his screen.*]

ARIEL: Awesome!

RHEA: Actually, I downloaded it. <u>Our data.</u>

CALLISTUS: You sure did. I thank you. As PI.

RHEA: I wrote the proposal.

CALLISTUS: Cedo maiori.[1] More importantly, Ariel, do you want to help with data reduction?

ARIEL: Uhhmm. Maybe some. Other. Time. Tik. Tok.

[*Ariel exits.*]

CALLISTUS: Right. My secret mission was to get them interested in helping us.

RHEA: Why did you start with Olbers' paradox? The cosmological principle is more fundamental.

CALLISTUS: OP is hands-on. The CP is so complicated. Even defining it clearly eludes me.

RHEA: The universe is homogeneous and isotropic. How about that?

CALLISTUS: Yes, but nothing is homogeneous or isotropic around us. Look at this image: galaxies are distributed, forming the cosmic web; there are clusters, walls, filaments, voids –

RHEA: Well, the cosmic microwave background is pretty isotropic.

CALLISTUS: If we disregard fluctuations.

RHEA: OK, the universe is *statistically* homogeneous and isotropic. On large scales.

CALLISTUS: Each void or cluster is a huge fluctuation. And how large a scale should we consider?

RHEA: Most of us would say 100 mega-parsec even larger scales. I admit that there will be small fluctuations even on those scales as well.

CALLISTUS: Right. How do I explain all these caveats to a student? We need a specific model, like your favourite concordance ΛCDM, to calculate the expected fluctuations and compare them with the observations to test homogeneity and isotropy. Even experts disagree on the details.

RHEA: What if we follow Copernicus? It helps us to recognise that we're not in a special place in the universe.

CALLISTUS: True, but we don't need to be in a special place if we live in an inhomogeneous universe. Like one of the general relativistic Bianchi models... I agree, though; the cosmological principle grew out of the Copernican idea, even if it's not identical.

RHEA: And if we bring up statistical isotropy and homogeneity, we must mention non-Gaussianity.

CALLISTUS: You see? We could conclude from a simple Gaussian model that the universe is anisotropic and find later that it is consistent with statistical isotropy under a non-Gaussian model with a larger variance.

---

[1] I yield to a greater person.

RHEA: We can mistake anisotropy for non-Gaussianity and vice versa. On second thoughts, these ideas might be a bit too intricate to begin with.

CALLISTUS: But I also agree that the cosmological principle is a fundamental assumption we must mention early on. Even if we cannot define it precisely without more advanced concepts.

RHEA: We'll cover those later. Like GR and the concordance ΛCDM model. From the beginning, it's important to emphasise that the cosmological principle is an *assumption* that we need to test continuously.

CALLISTUS: That's a great plan. We introduce it at an intuitive level and get back to it later when the students have had a taste of the cosmic web, non-Gaussianity, general relativity, etc. You can work out the details.

RHEA: What?

CALLISTUS: Right. Since you raised the issue, it's only fair that I let you solve it.

RHEA: Yes, of course. I get it. Now, let's get to our JWST data... I know I have to reduce it.

CALLISTUS: Because you love to reduce data and you're good at it... I wish Ariel would help... but I will. Before we start, I just realised we haven't discussed the perfect cosmological principle that has been refuted –

RHEA: Let's get to work! Now!
           [*Callistus sighs and sits down next to Rhea. They start working.*]

From 1920s to 1970s, six important observational data and theoretical ideas changed our understanding of the Universe and galaxies, which form pillars of the classical cosmological paradigm. The first pillar is the formulation of the general relativity theory and its extension — the theory of the Universe. The second pillar is the discovery of the expansion of the Universe and the determination of the distance–velocity or the Hubble diagram of galaxies The third pillar is the determination of the nature of spiral nebula. The fourth pillar is the determination of the structure and evolution of stars. The fifth pillar is the determination of basic data on the structure and evolution of the Milky Way and other galaxies. The sixth pillar is the determination of the distribution of galaxies on the sky, which is almost random. In this section, we discuss these pillars in more detail.

2.1.1   *Theory of general relativity*

The first major pillar of the classical cosmological paradigm is the formulation of the theory of general relativity by Einstein (1916), and its extension, the theory of the expanding universe. Based on this theory, de Sitter (1917) suggested a model with cosmological constant Λ but with zero or negligible matter density. A few years later, Friedmann (1922, 1924) and Lemaître (1927) discovered solutions to Einstein's equations that contained realistic amount of matter. Friedmann (1924) found three solutions for the cosmic evolution: one with ever-accelerating expansion, one periodic scenario with evolution from and back to zero radius (the oscillating universe), and the third, where initially the universe is decelerating due to gravity, but after some time, the expansion accelerates due to the influence of the cosmological constant. Einstein & de Sitter (1932) also proposed a model with the critical cosmological density.

2.1.2   *The expansion and age of the Universe*

The second major pillar of the cosmology is the observational discovery that the Universe is expanding, which allows one to find the velocity–distance relation of galaxies and clusters of galaxies. These data yield information on the distribution of galaxies in the depth of the cosmic space.

In the 1920s, radial velocities of some tens of galaxies were measured, and almost all of them showed a shift of spectral lines to the red part of the spectrum i.e. lines were redshifted. The larger the shift, the fainter the galaxies, and soon, the hypothesis was made that the whole Universe is expanding, the expansion velocity being proportional to the distance to the galaxy.

The first steps of this discovery were made by Wirtz (1922, 1924) who found that redshifts and distances of galaxies are related. Wirtz (1924) suggests a clear relationship of this phenomenon with the de Sitter (1917) cosmological model. Lundmark (1924) discussed the curvature of the space–time in the de Sitter universe and the relationship between distances and redshifts. Lemaître (1927) presented his idea of an expanding Universe, derived the velocity–distance relation and provided the first observational estimate of the constant of proportionality in this law. The expansion of the Universe was confirmed independently by Hubble (1929a). Lemaitre proposed that the Universe expanded from an initial point, which he called the "Primeval

Atom". Presently, the theory is known as the Big Bang Theory. Hoyle (1948) and Bondi & Gold (1948) preferred a different theory of the origin of the Universe, called the steady state theory, where matter is continuously created and the mean density of matter remains constant. According to this theory, the Universe has no beginning and will have no end.

The expanding Universe can be described by two fundamental constants: the mean expansion rate of space, measured by the Hubble constant, and the mean density of the Universe, expressed in units of the critical cosmological density. The critical density is the amount of matter/energy required to make the Universe spatially flat. A flat Universe has no curvature. If the density is less than the critical density, then the Universe will expand forever according to the classical picture. In the opposite case, the density is greater than the critical one, and gravity is strong enough to make the Universe collapse back, the so-called "Big Crunch".

Very large efforts have been made to measure the value of the Hubble constant. The distance of galaxies can be derived directly for objects with known absolute magnitude, such as cepheids. This relationship can be calibrated using similar variable stars in our Galaxy. First measurements by Lundmark and Hubble yielded a value of about 500 km/s per megaparsec (Mpc). The first major correction to this value came in the 1950s when Walter Baade (1951) discovered that there are two types of cepheids. Some of them belong to Population II, which dominate in galactic haloes and have a different luminosity–period relation. Also, it was found that stars in the most distant galaxies, observed by Hubble, were actually star clusters.

A special role in these efforts was played by the 200-inch Hale telescope in the Mount-Palomar observatory. However, different authors obtained rather different values for the Hubble constant. Sandage (1961), Sandage & Tammann (1976) and Sandage *et al.* (2006) found for the Hubble constant a value $H_0 = 50 \pm 7 \, \mathrm{km \, s^{-1} \, Mpc^{-1}}$, de Vaucouleurs (1978, 1982) preferred a value $H_0 = 95 \pm 10 \, \mathrm{km \, s^{-1} \, Mpc^{-1}}$, and van den Bergh (1992, 1994) got $H_0 = 76 \pm 9 \, \mathrm{km \, s^{-1} \, Mpc^{-1}}$. The presently accepted value was found later when new observational possibilities were available, including Hubble Space Telescope and satellite observations of the microwave background radiation.

RHEA: Can you believe that some don't like Big Bang Theory?

ARIEL: I know. It's my favourite show.

RHEA: No! I'm talking about believers in the steady state theory.
      [*Prof. Callistus enters.*]

CALLISTUS: Apologies for being late. Faculty meeting. Boooring... but I see that you're discussing one of my favourite non-working theories.

RHEA: Unbelievable.

ARIEL: I can see the difficulty with a TV show titled *Steady State Theory*. So, I'm assuming it's not a show.

RHEA: The universe expands, and as such, its density and temperature decrease. In the past, it was hotter and denser. You can extrapolate back to the point of infinite density –

ARIEL: The Big Bang!

CALLISTUS: The Hot Big Bang. Some would say the Big Bang starts only after inflation when the universe reheats.

RHEA: That's a 10^{-32} second detail.

CALLISTUS: A conceptual difference, but go on.

RHEA: Steady state theory states, no pun intended, that the density and temperature of the universe are constant.

ARIEL: That can't be unless... matter is created from nothing!

CALLISTUS: It's not as crazy as it seems. If the universe is homogeneous and isotropic in space <u>and</u> time, it adheres to the perfect cosmological principle–

RHEA: and hydrogen atoms appear out of nothing.

CALLISTUS: Just a little. According to Bondi & Gold (1948), "at most one particle of proton mass per litre and the rate 10^9 years"! Could you observe that?

RHEA: I give you that. But the cosmic microwave background –

CALLISTUS: Right. There are variations, such as quasi-steady state, and cyclical universe...

RHEA: And they are all refuted by several observations. Like it or not, the universe... they are a-changing. In time and space.

CALLISTUS: Right. In our field, observations always beat beauty. But it's intriguing to imagine creation little by little, almost unnoticed, instead of BOOM!

RHEA: Big Bang!

ARIEL: So, in the modern Big Bang Theory, the universe started with inflation, reheated, which can be considered to be the Big Bang. What about the future? How will our universe end?

CALLISTUS: While we can observe the past of the universe, we can only speculate about its future. It looks like we happen to live in a time when dark energy overtakes dark matter.

RHEA: The future's bright and dazzling. Not. The universe will expand exponentially, and astronomers in the extremely far future, tens of billions of years from now, will not see any extragalactic objects.

ARIEL: The universe starts with inflation and ends with inflation. That's pretty.

CALLISTUS: That is the Big Rip, I see that as the most likely as well. The Big Crunch, the opposite scenario, where the universe collapses, is unlikely.

RHEA: The end is probably just a lonely galaxy where stars fade away and cool down. Big Rip with a Heat Death or Big Chill.

ARIEL: That's lovely.

CALLISTUS: Could be worse. We could have vacuum decay at any moment, and that phase transition could sweep through the whole universe. Causa mortis.[2]

ARIEL: Saying it in Latin doesn't help. Not only must I worry about climate change, but the whole universe can disappear from under my feet.

CALLISTUS: You would not have any feet left. The Higgs vacuum might be metastable, so it could start anywhere, anytime. But don't worry about it, it's unlikely to happen before your PhD defence.

RHEA: And if it does, you cannot do anything about it.

ARIEL: Thanks. Sounds like I should keep my focus on global warming...

2.1.3 *The nature of spiral nebulae*

In the early years of the 20th century, a hot topic was the nature of spiral nebulae: Are they gaseous objects within the Milky Way system or distant worlds similar in structure to our Galaxy? On 26 April 1920, the Great Debate between astronomers Harlow Shapley and Heber Curtis was held in the Smithsonian Museum of Natural History on the nature of spiral nebulae and the size of the Universe. Arguments in favour of both concepts were serious and it was difficult to decide who was right.

This problem interested Ernst Öpik (1921). Rotation velocity measurements near the centre of M31 had been published by Pease (1918), and Öpik developed a method to estimate the distances to M31 from relative velocities within them. Öpik assumed that the acceleration at a distance r from the centre is equal to the gravitational acceleration due to the mass inside the sphere of radius r and that stars on the disk move approximately with circular velocities

---

[2]Cause of death.

V_c. In this case, the gravitational acceleration at the distance from the centre r is

$$\frac{GM}{r^2} = \frac{V_c^2}{r} = \frac{V_c^2}{\theta D}. \tag{2.1}$$

Öpik expressed the distance r through the angular distance θ and distance to the Andromeda, D, $r = \theta D$. The energy flux f from the M31 is

$$f = \frac{L}{4\pi D^2}. \tag{2.2}$$

Combining these equations, Öpik found the equation for the distance,

$$D = \frac{V_c^2 \theta}{4\pi G f} \times \frac{L}{M}. \tag{2.3}$$

Öpik assumed that the mass-to-luminosity (M/L) ratio of M31 is the same as this ratio within our Galaxy, which can be estimated from the stellar luminosity function in the solar neighbourhood. From these data, he obtained the distance of 785 kiloparsecs (kpc) (Öpik, 1921). A few years later, he made a new estimate (Öpik, 1922), which gives for the distance of the Andromeda nebula 440 kpc. This means that M31 is not within the Milky Way and must be an external independent system.

Astronomical community was not prepared to understand Öpik's method, which combined information from different fields: the measured rotation speed V_c, angular distance θ and flux f, and the estimated M/L ratio. But astronomers trusted the Hubble (1925, 1926, 1929b) estimate, found from cepheids in spiral nebulae NGC 6822, M33 and M31, using the 100-inch telescope of the Mount Wilson Observatory. Hubble's observations confirmed the large distance and the extragalactic nature of spiral nebulae. The existence of the world of galaxies was accepted by the astronomical community.

2.1.4 *Distribution of galaxies on sky*

The distribution of galaxies on the sky forms the sixth major pillar of the cosmology. In the first half of the 20th century, many attempts have been made to count galaxies in relatively small areas of sky.

Shapley & Ames (1932) started a systematic survey of northern and southern galaxies using photographic plates made with the 24-inch telescopes at Harvard and South Africa with a limiting $m = 13$ magnitude. Shapley found almost all galaxies catalogued 100 years ago by William Herschel listed in their NGC catalogue. The Local or Virgo cloud of galaxies was the most strong concentration of galaxies.

A different search strategy was applied by Edwin Hubble (1934). He used Mont Wilson 60- and 100-inch telescopes and counted galaxies in small fields for a limiting magnitude 20.0 separated by $5°$ on sky in declination and right ascension (near equator). Each field contained in the mean about 100 galaxies. Hubble concluded on the basis of these counts that galaxies are located randomly in space and that the *mean* spatial density of galaxies is approximately independent of the distance and the direction on the sky, as expected from the general cosmological principle.

The first deep catalogue of galaxies, covering the whole northern hemisphere, was made in the Lick Observatory with the 20-inch Carnegie astrograph by Shane & Wirtanen (1967). The published catalogue contains count of galaxies in declination from $\delta = -20°$ to $\delta = +90°$ in $1° \times 1°$ cells. Actual count were made in $10' \times 10'$ cells. Seldner *et al.* (1977) used these actual counts and corrected count for various errors and plate sensitivity differences. The final map of galaxies in the northern galactic hemisphere $b \geq 40°$ is shown in Fig. 2.1. Several well-known clusters of galaxies are seen on the map. For example, the Coma cluster appears near the centre of the map. The general impression is that field galaxies are distributed approximately randomly.

Soneira & Peebles (1978) developed a non-dynamical computer model universe to match the character of the galaxy distribution in the Lick survey. The model assigns 'galaxy' positions in a three-dimensional clustering hierarchy, fixes absolute magnitudes, and projects angular positions of objects brighter than $m = 18.9$ onto sky of an imaginary observer. This procedure yields a galaxy map that can be compared with that of the Lick survey, see Fig. 2.2. Both the real Lick map and the computer-generated map can be used to calculate two-point angular correlation functions. Authors show that both correlation functions are practically identical.

The conclusion from these studies, based on the apparent (two-dimensional) distribution of galaxies and clusters on the sky and

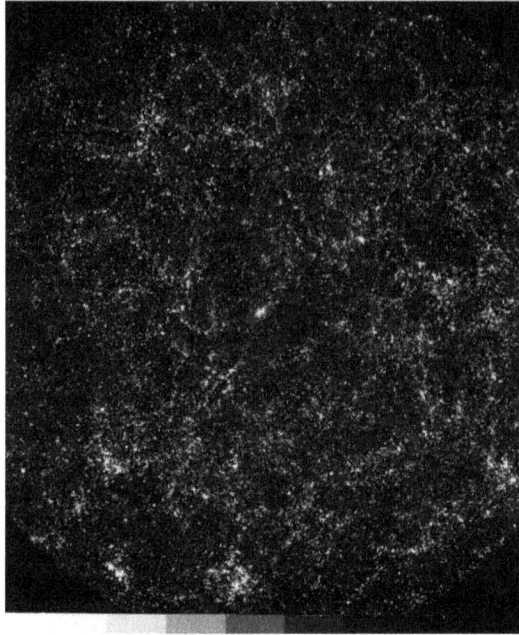

Fig. 2.1. Map of Lick survey galaxies in the northern galactic hemisphere brighter than $m_B \leq 18.9$ and north of galactic latitude $b \geq 40°$ (Soneira & Peebles, 1978).

the velocity–distance relation, confirmed the picture suggested by Kiang (1967) and de Vaucouleurs (1970) that galaxies and clusters of galaxies are hierarchically clustered. The concept of superclusters was introduced by de Vaucouleurs (1953, 1958) and Abell (1958). de Vaucouleurs (1970) emphasised that Abell's rich clusters are themselves clustered not only on the characteristic scale of superclusters, but on larger scales, i.e. indefinite clustering of galaxies. However, this hierarchy does not continue to very large scales as this contradicts observations, which show that on very large scales, the distribution is homogeneous. A theoretical explanation of the hierarchical clustering scenario was given by Peebles & Yu (1970).

2.1.5 *Formation and evolution of stars*

The evolution of galaxies and the Universe as a whole depends on the evolution of stars. The understanding of sources of stellar energy and the path of the evolution of stars are important elements of the classical cosmological paradigm.

Fig. 2.2. Simulated map of galaxies imitating the 2D distribution of Lick galaxies (Soneira & Peebles, 1978).

In the early years of the 20th century, astronomers adopted the Russell hypothesis on stellar evolution: stars are born as red giants, they contract to form blue giants, and then cool and move along the dwarf branch (main sequence) towards red dwarfs. The dominating sources of energy according to Russell are gravitation and radioactive decay.

Stellar evolution was a topic of Ernst Öpik's interest. O- and A-type main sequence stars have 10–30 solar masses, whereas the masses of red dwarfs are only a fraction of the solar mass. Öpik concluded that, if the Russell hypothesis is correct, then the stellar evolution should be accompanied with mass loss. If mass loss occurs in double stars, the distance between components must increase from blue to red double stars of the main sequence, the expected increase is approximately 20 times. To check this scenario, Öpik (1923, 1924) studied double stars and found that, contrary to the expectation, the mean distance between components of double stars decreases about two times when moving from blue to red main sequence stars.

Another fact contrary to the Russell hypothesis comes from geological data which indicate that the mean temperature on the Earth's surface has been almost constant during its whole geological history. If the Sun evolved according to the Russell hypothesis, its luminosity must decrease along the main sequence by a factor of 1000, and it is impossible to avoid similar changes of the temperature on Earth.

The conclusion from these calculations was as follows: the Hertzsprung–Russell diagram (HRD) is not an evolutionary diagram, but a diagram of various initial conditions in mass and chemical composition. The energy production per unit mass of blue giants is much higher than that of red dwarfs, thus the energy production must depend on physical conditions in the star. In main sequence stars, the luminosity per unit mass is proportional to the ninth power of the mass (Öpik, 1923). The basic physical parameter which changes among the main sequence stars of different mass is the temperature, thus a similar dependence must also be valid for the temperature. Since the temperature rises inwards, the energy source of stars must be located near the centre (Öpik, 1922). Based on these arguments, Öpik (1922) concluded that stars obtained their energy from nuclear reactions. This argument was confirmed by Eddington (1924, 1926). The gravitation source of energy is effective only in the first stage of the evolution during the contraction towards the main sequence of the HRD.

The basic nuclear process is the burning of hydrogen to helium, as found by Öpik (1938) and Bethe (1939). Since the energy source is located in the centre of the star, this leads to convection similar to the formation of convection in boiling water in a kettle heated from below. Due to convection, the active matter is continuously replaced. Calculations suggested that the star is convective only in the central parts (Öpik, 1938). In this case, its structure is a composite one — the convective core is surrounded by a radiative envelope. There exists no mixing of stellar matter between the core and the envelope. This evolutionary stage is the longest, and during this stage, star is located in the main sequence. When all hydrogen in the convective core is exhausted, the core contracts and heats up, and the envelope expands — the star becomes a red giant. As the temperature in the inner region of the star increases, nuclear reactions demanding higher temperature start. Thus, a red giant looks like an onion: an outer shell of hydrogen burning, while in the core and in inner sheets, heavier elements burn to form carbon, neon, oxygen, silicon, etc.

After the hydrogen burning in the core, first helium is burning and the star lies in the horizontal branch. Near the end of the star evolution in the very central zone, where the temperature is high enough, iron is produced. But this is the end of energy releasing nuclear reactions. All previous nuclear reactions produce energy, but when iron fuses into heavier elements, it absorbs energy out of the reaction, slowing it down. The central region of the star no longer produces energy and its gravity pulls outer layers inwards. Evolutionary tracks of stars in the HR diagram depend on the mass of the star, see Fig. 2.3. During the quiet evolution, low-mass stars slowly eject

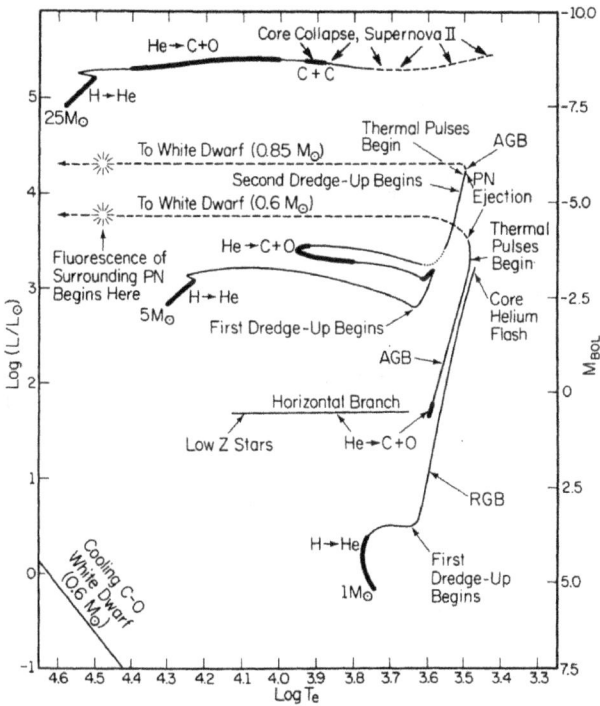

Fig. 2.3. Tracks in the HR diagram of theoretical model stars of low (1 $M_\odot$), intermediate (5 $M_\odot$), and high (25 $M_\odot$) mass. Nuclear burning on a long timescale occurs along the heavy portions of each track. The places where first and second dredge-up episodes occur are indicated, as are the places along the asymptotic giant branch (AGB) where thermal pulses begin. The third dredge-up process occurs during the thermal pulse phase, and it is here that one may expect the formation of carbon stars and ZrO-rich stars. The luminosity where a given track turns off from the AGB is a conjecture based on comparison with the observations (Iben, 1991).

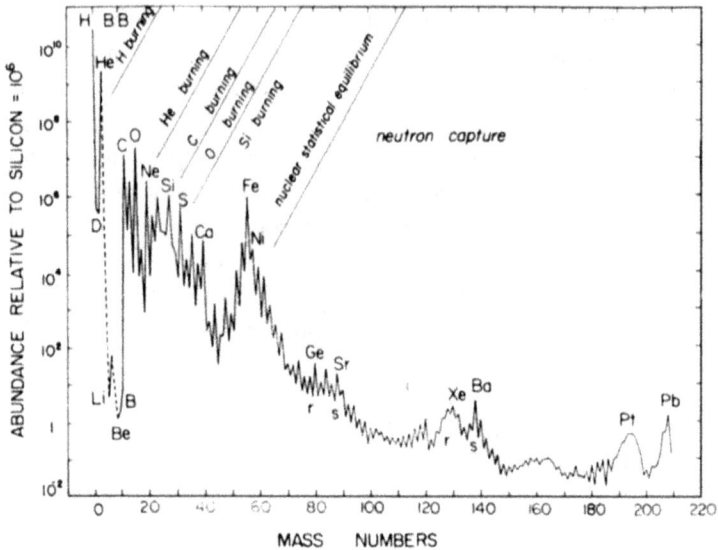

Fig. 2.4. The local galactic abundance of nuclear species normalised to 10^6 ^{28}Su atoms (Pagel, 1997).

its atmospheres via stellar wind. Higher-mass stars collapse quickly and explode as supernovas.

During supernova explosions, large amounts of energy are released and the temperature rises, thus in a short timescale, all elements heavier than iron are synthesised. During the explosion, outer layers are expelled and enrich the interstellar matter with heavier elements. Thus, the next generations of stars form from a medium which already contains elements heavier than hydrogen, helium and lithium. Expanding shock waves from a supernova explosion can trigger the formation of new stars. In this way, all chemical elements form and an equilibrium distribution forms, see Fig. 2.4.

Öpik's theory was fully accepted only after the publication of independent analyses by Schwarzschild *et al.* (1953), Schwarzschild & Spitzer (1953), Hoyle & Schwarzschild (1955) and Burbidge *et al.* (1957).

2.2 Dynamical and physical evolution of galaxies

During the first half of the 20th century, data on the structure of our Galaxy accumulated rapidly. The rotation of the Galaxy was discovered (Oort, 1927; Strömberg, 1924) and also the position of the Sun far from the Galactic centre (Oort, 1928). Also, it was found that the Galaxy consists of many stellar populations with different kinematical, spatial and physical properties (spectral class), such as young main-sequence stars in the galactic disk, old globular clusters, RR Lyrae variables (Lindblad, 1927; Strömberg, 1924). These populations form a near continuum from the very lowest to the very highest space velocities. These populations allow us to reconstruct the galactic past because the time required for stars in the galactic system to exchange their energies and momenta is very long compared with the age of the galaxy.

In this section, we discuss first the spatial structure and dynamical models of regular galaxies. Next, we describe hydrodynamical models of galaxies. Finally, we discuss models of dynamical and physical evolution of galaxies. The theory of structure and dynamics of galaxies was formed among others by Grigori Kuzmin (Kuzmin, 2022) and is summarised by Binney & Tremaine (2008).

2.2.1   *Models of spatial structure of galaxies*

Several problems must be solved to model galaxies' structure in more detail. First, which functions are to be used to describe galaxies? These functions need to be the ones that can be compared to observations, or they should be important in calculating some other functions. From observations, we can measure the surface brightness, the rotation curve, the velocity dispersion curves, and the stellar content.

The second problem arises: What connection formulae are to be used to calculate all the functions describing a galaxy? The form of these functions depends on the shape of galaxies. When we look for the structure of regular galaxies, the natural assumption is that disk galaxies are flattened systems with rotational symmetry and symmetry with respect to the galaxy's plane. Elliptical galaxies can be handled in a similar way as biaxial ellipsoidal figures. Although, in principle, elliptical galaxies can have a triaxial form, this introduces two additional free parameters (one axial ratio and one viewing angle) and let this remain for the future.

The third problem is connected with the choice of the principal descriptive function. In observations, we see the surface brightness; should we start from the surface density? Surface density is not a good function to start from. In some sense, the surface density is not a physical quantity, but the spatial density is. The spatial density distribution determines the gravitational potential, accelerations, and motions of stars, i.e. all the behaviour of a galaxy. Surface density is only a projection, and it is used only due to our limited observational abilities. Thus, the most important descriptive function is the spatial density. All other functions describing a galaxy can be calculated thereafter from the spatial density as a rule via integrals.

Now, we may ask the final question: Can we formulate some simple physical criteria that the principal description function we selected must satisfy, or in another way, how can we choose between various analytical expressions for the density distribution? It is reasonable to accept that some physically motivated criteria for the density law are needed. It seems natural that the following criteria must be fulfilled (see Einasto, 1969):

- The spatial density must be non-negative and finite.
- Some statistical moments of the spatial density must be finite as well, at least moments which define the central gravitation potential, the mass and the effective radius of the system.
- At large distances from the centre, the density must smoothly approach zero value.
- The description functions have no breaks.
- The acceptable formula for the spatial density must be flexible enough to describe various galaxies. Still, on the other hand, the number of free parameters must be kept as small as possible.

Let us assume a galaxy or a galaxy population is of a regular form with the equidensity surfaces being ellipsoids of rotation with a constant axial ratio ϵ. In this case, it is convenient to use the usual cylindrical coordinates (R, θ, z). In the meridional plane, the coordinates are (R, z), and symmetry is assumed to be the z-axis. The experience of modelling has shown that the spatial density distribution of galactic populations can be expressed by a generalised exponential model (Einasto, 1965):

$$\rho(a) = \rho_0 \exp\left(-(a/a_c)^{1/N}\right), \qquad (2.4)$$

where ρ_0 is the central density, a is the semimajor axis of the equidensity ellipsoid with $a^2 = R^2 + z^2/\epsilon^2$, a_c is a characteristic radius, and N is a structural parameter, which allows one to vary the shape of the density profile. Despite the scale factors, it has only one free parameter, the power N. The cases $N = 1/2$, $N = 1$ and $N = 4$ correspond to the Gaussian, conventional exponential and the de Vaucouleurs (1948) models, respectively. This model allows a natural extrapolation of the density for large distances from the galaxy's centre and fits observed density profiles of known galactic populations very well. This density profile satisfies the physical criteria mentioned above.

Let us stop for a moment at the meaning of the characteristic radius. Obviously, it should characterise the real extent of the system. In statistics, there are two main kinds of means: the usual mean and the harmonic mean. In the present case, one needs to calculate the "mean" radius with the weights of the mass of an ellipsoidal shell $d\mu = 4\pi\epsilon\rho(a)a^2 da$. According to the criteria given in textbooks of mathematical statistics, in this case, we need to use the harmonic mean radius defined as

$$a_0^{-1} = M^{-1} \sum_i \mu_i a_i^{-1}.$$

(In the case of continuous density distribution, the sum should be replaced with the integration of course.) Indeed, the harmonic mean characterises a system's real extent well independent of the value of N. Using the harmonic mean radius a_0, the previous formula (2.4) transforms to

$$\rho(a) = \rho_0 \exp\left(-(a/ka_0)^{1/N}\right), \tag{2.5}$$

where the central density is $\rho_0 = hM/(4\pi\epsilon a_0^3)$. Here, two normalising parameters were introduced due to the use of a_0, h and k. They depend on N. The details to calculate them are given in Tenjes *et al.* (1994) and Tamm *et al.* (2012, see appendixes).

To construct a physical model of a galaxy, one needs to fix which are the basic quantities, describing the spatial and kinematical structure of the galaxy and its subsystems of different ages and compositions (stellar populations). As such parameters, one can adopt the mass of the subsystem M, its characteristic radius a_0, the axial ratio of equidensity ellipsoids ϵ, and the structural parameter N determining the degree of concentration of the mass to the centre of the system. As morphological parameters, we can consider colour and mass-to-light ratio of subsystems.

From observations, we have the surface brightness distribution, the rotational curve (or even the velocity field) and the velocity dispersion curve (or the velocity dispersion field). Knowing the kinematics (e.g., the rotation law) of a stellar system makes it possible to obtain the mass distribution of that system.

Again, let us suppose that the isodensity surfaces of the galaxy are ellipsoids of revolution with constant axial ratio ϵ and the angle between the sky plane and the symmetry plane of the galaxy is i. In this case, the surface density distribution is

$$P(A) = \frac{1}{2\pi E} \int_A^\infty \frac{\mu(a)da}{a\sqrt{a^2 - A^2}},$$

where A is the semimajor axis of the projected isodensity ellipsoid with the apparent axial ratio E:

$$E^2 = \cos^2 i + \epsilon^2 \sin^2 i$$

and

$$\mu(a) = 4\pi\epsilon\rho(a)\,a^2$$

is the mass function (the mass of a spheroidal sheet per unit interval of a). The surface brightness distribution of a galaxy allows us to calculate all the density distribution parameters, a_0, ϵ, N and the total luminosity of the galaxy L. Here, we assumed that the same density distribution law can describe a galaxy's mass density and luminosity density.

Galaxies consist of subsystems of different ages, compositions and shapes (e.g., the bulge, the disk, the stellar halo, etc.). We can assume that the density of subsystems can be approximated by ellipsoids of revolution with the similar exponential law (2.5) but with a different set of parameters.

Now, let us turn to dynamical modelling. Using known formulae for the gravitational attraction of an infinitely thin ellipsoidal layer, it can be shown (see, e.g., Binney & Tremaine, 2008, Section 2.5.2) that in the equatorial plane of a galaxy ($z = 0$), the circular velocity of the whole system can be found by summing contributions of individual subsystems to V_c^2:

$$V_c^2(R) = G \sum_i \int_0^R \frac{\mu_i(a)da}{\sqrt{R^2 - e_i^2 a^2}}. \tag{2.6}$$

Here,

$$e = \sqrt{1 - \epsilon^2}$$

is the eccentricity and G is the gravitational constant.

When calculating the mass distribution of galaxies from their rotational curves, rotational velocities are often identified with circular velocities. One should remember that this is only an approximation valid when the velocity dispersions are much smaller than rotational velocities. In a general case, the hydrodynamical modelling should be used. We describe this in the following section.

Since the 1950s, models of mass distribution in galaxies, including our Galaxy, have been constructed using the method described above. For external galaxies, the distribution of light in populations can be used to find the shape of the population. The critical step in model calculations of external galaxies is finding the mass-to-light ratio of populations, M/L, to convert the light distribution to the spatial density distribution. For this purpose, several methods have been applied.

First, we can calculate the M/L ratios of the components simply by fitting the light distribution and the mass distribution and handling M/Ls of components as free parameters. We can do it since different populations dominate in various regions of the galaxy, so the M/Ls do not couple with each other. The second possibility is to use spectroscopic observations of the stellar content of populations together with the chemical evolution models. Unfortunately, at least one additional free parameter, the age of a component, is added here. The third possibility is to use velocity dispersions of star clusters of similar physical properties (colour, age). It is well known that stars form in various star-formation clouds, which later evolve to stellar associations and clusters. Later associations and clusters are dissolved by various perturbations. Probably all field stars of galaxies formed in just such a way.

Different data and methods can lead to rather different values of M/Ls, and it is not clear which data can be trusted. Thus, a need exists to bring all estimated M/L values to a coherent system. There exist indirect methods to put constraints to the possible limits of M/Ls of components. It is clear that the luminosity of a population depends on its chemical composition and age. To understand how

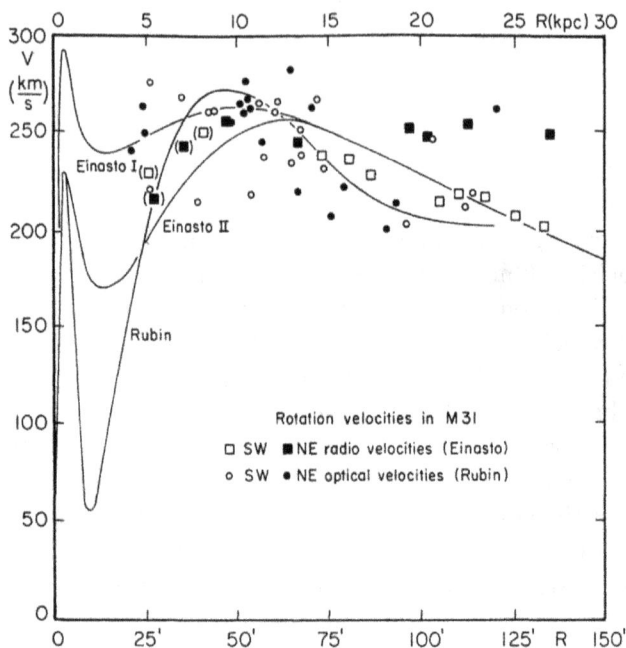

Fig. 2.5. The rotation curve of M31. Observed rotation data from optical and radio measurements according to a summary by Rubin & Ford (1970) are shown by various symbols. Solid curves show rotation according to the model by Rubin & Ford (1970), and two variants of the model by Einasto (1972). In this model, mass-to-light ratios f_V of nucleus, bulge, halo, thick disk and flat disk populations are in variant I, 42, 4.2, 4.1, 10.1, 3.2, and in variant II, 42, 2.7, 0.07, 14.7, 6.5 (Einasto, 1972).

M/L depends on the composition and the age of the population, their evolution models are needed, as discussed in the following.

Einasto (1972) calculated a model of the Andromeda galaxy, M31, using five populations: nucleus, bulge, stellar halo, thick disk and flat disk. The rotation curve for this model is shown in Fig. 2.5 with two variants of mass-to-light ratios of populations. For comparison, the model by Rubin & Ford (1970) is also shown. The figure shows that at a large distance from the centre, observed rotation velocities remain almost constant, whereas model curves suggest decreasing rotation velocities.

Subsequent data indicated the presence of almost flat rotation curves for almost all galaxies, and it was increasingly difficult to accept models of galaxies with known stellar populations. The main conclusion of the galactic model calculations is that something is wrong in the classical picture of the matter distribution of the galaxies. Later, this discrepancy turns out to be the beginning of the change in the whole cosmological paradigm.

2.2.2 *Hydrodynamical models of galaxies*

Let us describe the kinematics and dynamics of a stellar system in more detail. These models are often called hydrodynamical models because the equations are in some sense similar to ordinary hydrodynamical equations. In the case of stellar systems, they are the Jeans equations.

A general starting point here is the collisionless Boltzmann equation. The main function in this equation is the phase-space density f. The phase-space is the space of coordinates and velocities, i.e. in general, it is 6D space with three coordinates $\vec{x}$ and three velocities $\vec{v}$. Despite the 6D form, the phase-space density is like an ordinary mass density. In addition, the phase space-density can be an explicit function of time. Thus, $f = f(\vec{x}, \vec{v}, t)$. We assume the stellar system to be collisionless — a rather good (but not always) approximation. In principle, the Boltzmann equation can be written in a very concise form

$$\frac{Df}{Dt} = 0.$$

Here, D/Dt is the so-called Stokes operator, being a "full" time derivative, i.e. the time derivative, by taking into account that the phase-space coordinates are also functions of time:

$$\frac{D}{Dt} \equiv \frac{\partial}{\partial t} + \sum_{i=1}^{3} \left(\frac{dx_i}{dt} \frac{\partial}{\partial x_i} + \frac{dv_i}{dt} \frac{\partial}{\partial v_i} \right).$$

If the acceleration is caused by the potential Φ, then according to Newton's law,

$$\frac{dv_i}{dt} = \frac{\partial \Phi}{\partial x_i}$$

(and of course $v_i = dx_i/dt$). Now, we have the Boltzmann equation

$$\frac{\partial f}{\partial t} + \sum_{i=1}^{3} \left(v_i \frac{\partial f}{\partial x_i} + \frac{\partial \Phi}{\partial x_i} \frac{\partial f}{\partial v_i} \right) = 0.$$

The fact that the right side of the Boltzmann equation is zero means that we have a collisionless equation. In the case of a collisional system on the right side, there is a function characterising collisions.

Integrating the phase-space density over all velocities, we have the ordinary spatial mass density

$$\rho(\vec{x}, t) \equiv \int f(\vec{x}, \vec{v}, t) d^3 v.$$

This can be handled as a definition of the spatial mass density.

The Jeans equations result from the Boltzmann equation by multiplying the equation with v_i and integrating thereafter over all velocities. We do not do it here; a reader can look, e.g., the textbook by Binney & Tremaine (2008). Only one point we note here. An important concept is the introduction of velocity dispersions. When we multiply the Boltzmann equation with v_i and integrate over velocities, first, averages will be introduced:

$$\overline{v_j v_i} \equiv \frac{1}{\rho} \int f(\vec{x}, \vec{v}, t) v_j v_i d^3 v,$$

and thereafter, the velocity dispersion tensor[3]

$$\sigma_{ij}^2 \equiv \overline{(v_i - \bar{b}_i)(v_j - \bar{v}_j)} = \overline{(v_i v_j - v_i \bar{v}_j - \bar{v}_i v_j + \bar{v}_i \bar{v}_j)} = \overline{v_i v_j} - \bar{v}_i \bar{v}_j.$$

This means that the term $\overline{v_i v_j}$ divides into two parts: the "flowing" motion $\bar{v}_i \bar{v}_j$ and the random motion σ_{ij}^2.

After doing all this, the result is

$$\rho \frac{\partial \bar{v}_j}{\partial t} + \rho \bar{v}_i \frac{\partial \bar{v}_j}{\partial x_i} = -\rho \frac{\partial \Phi}{\partial x_j} - \frac{\partial (\rho \sigma_{ij}^2)}{\partial x_i}. \tag{2.7}$$

Here, $i, j = 1, 2, 3$ and summation is over repeating indices. It gives us three equations called the Jeans equations. The equation is similar to the Euler equation in fluid mechanics, and often, it is called the stellar hydrodynamical equation.

As we intend to model stellar systems in cylindrical galactocentric coordinates (R, θ, z), the Jeans equations should be written in cylindrical coordinates. We do not derive it here. Next, in these equations,

---

[3]It can be shown that it is a symmetric tensor.

additional simplifying assumptions are usually used: it is assumed that the stellar system we intend to model is stationary, has symmetry with respect to the z-axis, and there are no systematic motions in the R- and z-directions. In this case, the terms with $\partial/\partial t$ and $\partial/\partial\theta$ in Eq. (2.7) are zero, and $\bar{v}_R = \bar{v}_z = 0$. In addition, the terms $\overline{v_R v_\theta} = \overline{v_z v_\theta} = 0$ vanish and the second equation in Eq. (2.7) turns to identity. Only two Jeans equations remain.

Hydrodynamical models of galaxies connect spatial densities and velocities of galactic populations to a coherent system. Velocities of stars of various populations depend on two factors: on the gravitational field of the galaxy as a whole (including all populations) and on the structure of the particular population. Structural and kinematical properties of individual populations are determined by the conditions during their formation and further evolution.

In the following, we describe the hydrodynamical model developed by Einasto (1972). The Jeans equations in cylindrical coordinates for a stellar system within assumptions, referred above, can be written as

$$\frac{1}{R}\left(\sigma_R^2 - \sigma_\theta^2\right) + \frac{1}{\rho}\frac{\partial}{\partial R}\left(\rho\sigma_R^2\right) + \frac{1}{\rho}\frac{\partial}{\partial z}\left[\rho\gamma\left(\sigma_R^2 - \sigma_z^2\right)\right] - \frac{V_\theta^2}{R} = \frac{\partial\Phi}{\partial R},$$
(2.8)

$$\frac{1}{R}\gamma\left(\sigma_R^2 - \sigma_z^2\right) + \frac{1}{\rho}\frac{\partial}{\partial R}\left[\rho\gamma\left(\sigma_R^2 - \sigma_z^2\right)\right] + \frac{1}{\rho}\frac{\partial}{\partial z}\left(\rho\sigma_z^2\right) = \frac{\partial\Phi}{\partial z}.$$
(2.9)

In these equations, V_θ is the rotational velocity of the population ($\bar{v}_\theta$ in our earlier designations), and

$$\gamma = \frac{1}{2}\tan 2\alpha,$$
(2.10)

where α is the inclination angle of the major axis of the velocity ellipsoid of the population with respect to the galactic symmetry plane.[4]

---

[4]We mentioned earlier that σ_{ij}^2 is a symmetric tensor, i.e. it can be transformed with a suitable coordinate change into a diagonal form; α is just the inclination angle of this coordinate system to the (R, z) coordinates.

Calculating the necessary derivatives, Eqs. (2.8) and (2.9) can be written as

$$V_\theta^2 - p\sigma_R^2 = -R\frac{\partial \Phi}{\partial R} = V_c^2 \tag{2.11}$$

and

$$\frac{1}{\rho}\frac{\partial(\rho\sigma_z^2)}{\partial z} + q\frac{\sigma_z^2}{R} = \frac{\partial \Phi}{\partial z}, \tag{2.12}$$

where

$$p = (1 - k_\theta) + G_R\{\rho\} + G_R\{\sigma_R^2\} + \frac{R}{z}\gamma(1 - k_z)$$
$$\times [G_z\{\rho\} + G_z\{\gamma\} + G_z\{1 - k_z\}] \tag{2.13}$$

and

$$q = \gamma\left(\frac{1}{k_z} - 1\right)[1 + G_R\{\rho\} + G_R\{\gamma\} + G_R\{\sigma_R^2 - \sigma_z^2\}]. \tag{2.14}$$

In these equations,

$$k_\theta = \frac{\sigma_\theta^2}{\sigma_R^2}, \quad k_z = \frac{\sigma_z^2}{\sigma_R^2} \tag{2.15}$$

are ratios of velocity dispersions and $G\{\ \}$ is the logarithmic derivative, e.g.,

$$G_R\{\rho\} = G\{\rho(R)\} = \frac{\partial \ln \rho}{\partial \ln R}. \tag{2.16}$$

In these equations, on the left side denote the kinematical and spatial data on populations, and those on the right side denote the gravitational acceleration of the whole galaxy.

Equations (2.11) and (2.12) include five unknown kinematical functions: σ_z, V_θ, k_θ, k_z, γ. To calculate these functions, we have only two equations at present, thus the system of hydrodynamical equations is not closed. To solve the problem, one needs to have three additional independent relations between these unknown functions. It is convenient to give these additional relations for k_θ, k_z, and γ, which determine the shape and the orientation of the velocity ellipsoid. In this case, Eq. (2.12) allows us to calculate the dispersion σ_z, giving the scale of the velocity dispersion, and Eq. (2.11) allows us to calculate the centroid velocity V_θ, giving the shift of the velocity ellipsoid with respect to the local standard of rest.

We see that the calculation of the hydrodynamical model of a galaxy reduces to the problem of finding equations for auxiliary kinematical functions k_θ, k_z, and γ.

The Poisson equation in cylindrical coordinates can be written as (Kuzmin, 1952)

$$4\pi\, G\rho = -\frac{\partial^2 \Phi}{\partial z^2} - \frac{\partial^2 \Phi}{\partial R^2} - \frac{1}{R}\frac{\partial \Phi}{\partial R}, \tag{2.17}$$

where Φ is the acceleration potential, ρ is the dynamical density, and G is the gravitational constant. To find the dynamical density of the galaxy by this method, it is thus necessary to know the acceleration potential Φ.

Kuzmin (1952) analysed the distribution of flat population stars in the Galaxy. He found that the vertical distribution of these stars is approximately normal (Gaussian) and can be characterised by the parameter

$$C = \frac{\sigma_z}{\zeta}, \tag{2.18}$$

where σ_z is the velocity dispersion of stars in vertical direction and ζ is the variance of z. This parameter is related to the vertical gradient of Φ:

$$C^2 = -\partial^2 \Phi / \partial z^2. \tag{2.19}$$

This is the main term in the dynamical density formula (2.17). The remaining terms are less significant and can be expressed for the planar subsystems through the galactic rotation constants A and B. Since the planar subsystems rotate practically with a circular velocity, we can express the Galactic rotation constant in this case as follows:

$$A = \frac{1}{2}\left(\frac{V_z}{R} - \frac{dV_z}{dR}\right)$$

and

$$B = -\frac{1}{2}\left(\frac{V_z}{R} + \frac{dV_c}{dR}\right).$$

Given that $V_c^2 = -R\,\partial\Phi/\partial R$, we find for the sum of the second and third terms of formula (2.17) the expression

$$\frac{\partial^2\Phi}{\partial z^2} + \frac{1}{R}\frac{\partial\Phi}{\partial R} = -2\frac{dV_c}{dR}\frac{V_c}{R} = 2(A^2 - B^2). \tag{2.20}$$

The Poisson equation for the dynamical density takes the form:

$$4\pi G\rho = C^2 - 2(A^2 - B^2). \tag{2.21}$$

In the case of $z = 0$, the velocity ellipsoid is not tilted with respect to the galactic plane, and $\gamma = 0$, but $\partial\gamma/\partial z \neq 0$. Thus, Eq. (2.13) for the parameter p will have the form

$$p = (1 - k_\theta) + n_R(1 - k_z) + G_R\{\rho\} + G_R\{\sigma_R^2\}, \tag{2.22}$$

where

$$n_R = R\left(\frac{\partial\gamma}{\partial z}\right)_{z=0}. \tag{2.23}$$

On the basis of his theory of the third integral of motion of stars as a quasi-integral, Kuzmin (1953) demonstrated that

$$n_R = -\frac{1}{4}G_R\{C_c^2\}\left[1 + \frac{B_c(A_c - B_c)}{C_c^2}\right]^{-1}, \tag{2.24}$$

where A_c, B_c, C_c are the Oort–Kuzmin dynamical parameters for flat populations. Using the Poisson equation in this form, and taking into account that $A_c, B_c \ll C_c$, Kuzmin (1953) derived an approximate formula

$$n_R = -\frac{1}{4}G_R\{\rho_t\}. \tag{2.25}$$

The theory of irregular forces gives us the following relation between the velocity dispersion ratios (Kuzmin, 2022):

$$k_z^{-1} = 1 + k_\theta^{-1} \tag{2.26}$$

and

$$\frac{1}{\sigma_z^2} = \frac{1}{\sigma_\theta^2} + \frac{1}{\sigma_R^2} \tag{2.27}$$

found from the theory of irregular forces for the case $z \ll R$.

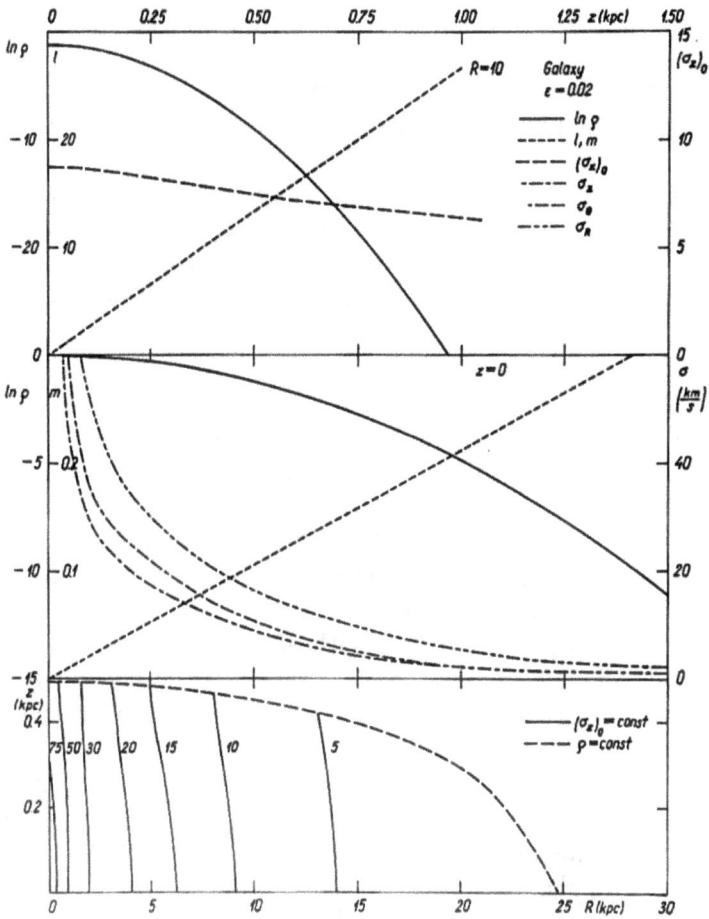

Fig. 2.6. Descriptive functions for the test population of flattening $\epsilon = 0.02$.

Einasto (1972) extended this method to the more general case $z \neq 0$ and calculated hydrodynamical models for six populations of the Galaxy with axial ratios $0.02 \leq \epsilon \leq 0.8$ and for the Andromeda galaxy M31 for $\epsilon = 0.02, 0.08, 0.30$. Descriptive functions for the flat population with $\epsilon = 0.02$ are shown in Fig. 2.6. The figure has three panels.

In the upper panel, we show with solid line the density logarithm, $\ln \rho$, with dotted line the vertical density gradient, $l = -\partial \log \rho / \partial z$, and with dashed line the velocity dispersion $(\sigma_z)_0$ in Jeans approximation. Velocity dispersion $(\sigma_z)_0$ was calculated within the range

$0 \le z \le z_u$, where z_u is the outer vertical limit of the population. All functions are plotted as functions of z at $R = 10$ kpc, the adopted distance of the Sun from Galactic centre. The vertical scale z is shown at the top of the panel, the density scale on the left border, and the velocity dispersion scale on the right border.

Central panel of the figure shows with solid line the logarithm of the density, ρ, with dotted line the radial density gradient, $m = -\partial \log \rho / \partial R$, and with various dot-dashed lines velocity dispersions σ_R, σ_θ, σ_z. These data are given as functions of R on the plane of the Galaxy, $z = 0$. The radial distance scale R is shown at the lower border of the figure, the density scale is on the left vertical border, and the velocity dispersion scale is on the right border. Densities are given in units of the central density and gradients l and m in kpc^{-1}, dispersions in km/s. Vertical velocity dispersion σ_z was calculated from the second hydrodynamical equation (2.12), using for k_z Kuzmin equation (2.26) from the theory of irregular forces. Radial velocity dispersion σ_R was found from definition equation (2.15) from σ_z and k_z, taking into account the Kuzmin equation (2.27).

In the bottom panel, we show with dashed line isoline, $\rho = $ const. and with solid lines isolines $(\sigma_z)_0 = $ const., in the plane of R, z-coordinates, shown at the bottom and the left border of the panel. Velocity dispersion σ_z was calculated within ranges $0 \le R \le 30$ kpc and $0 \le z \le z_u$, where z_u is the outer vertical limit of the population. We see that in the disk population lines $(\sigma_z)_0 = $ const. are vertical, i.e. the velocity dispersion at given radial distance is constant. Available observational data support this picture. In halo populations lines $(\sigma_z)_0 = $ const. are similar to lines of constant density.

2.2.3 *Dynamical evolution of galaxies*

There is a possibility to reconstruct the evolution of galaxies by studying the spatial structure and kinematics of stars of various ages in the Galaxy. The relationship between the velocity dispersion and the rotational speed of populations was established by Strömberg (1924). The greater the (negative) galactocentric rotational velocity $-V_\theta$, the larger the velocity dispersion of a population of stars. Figure 2.7 presents the Strömberg diagram for stars of various galactic populations. Populations with normal chemical composition are marked with dots, metal-poor populations with open circles. The

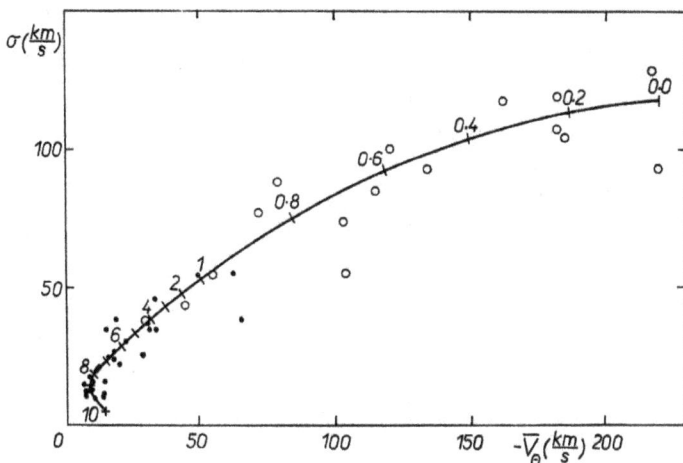

Fig. 2.7. Strömberg diagram for galactic populations. In the horizontal axis, we show the heliocentric centroid velocity of the population in the direction of the Galactic rotation; in the vertical axis, we plot the mean velocity dispersion $\sigma = \sqrt{\frac{1}{3}(\sigma_R^2 + \sigma_\theta^2 + \sigma_z^2)}$. Open circles are for metal-poor populations, dots for populations with normal metal abundance. The numbers give the birth dates in 10^9 years starting from the formation of the oldest populations, assuming for the age of the Galaxy 10^{10} years (Einasto, 1972).

mean dependence between rotation velocity and mean velocity dispersion is approximated by a solid line. Numbers give the birth dates in billion years, starting from the formation of oldest populations, assuming for the age of the Galaxy 10 billion years. Ages of populations were estimated from evolutionary tracks.

Figure 2.7 shows that there exist deviations from this relationship: stars with very small velocity dispersion and interstellar gas have rotational speeds, smaller than stellar populations with velocity dispersion of the order of 15 km/s. This effect can be explained by the hypothesis that interstellar gas rotates with a velocity lower than the circular velocity. If this is the case, then young stars just "fall" in the direction of the Galactic centre after their birth. A similar effect is observed in the vertical direction. Radio observation suggests that the shell of interstellar gas does not coincide exactly with the plane of the Galaxy and has a wave form. In the solar neighbourhood, these waves have an amplitude of about 50 pc. After formation, stars are free from electromagnetic forces and fall towards the Galactic plane.

Very young stars are still located close to their places of origin. The frequency of such oscillations, measured by Kuzmin parameter C, does not depend on the amplitude (if it is small). Thus, young stars should oscillate around the plane of the Galaxy. Such oscillations were analysed by Jõeveer (1972, 1974) to estimate the dynamical density of matter in the solar vicinity.

Kinematical data on disk and halo stars can be used to reconstruct the dynamical evolution of the Galaxy. Eggen *et al.* (1962) used data on old stars in our Galaxy to reconstruct the galactic history. Their arguments are based on the following assumptions: (i) Stellar populations of forming stars get spatial and kinematic characteristics of gas clouds from which they formed. (ii) Spatial and kinematical characteristics of stellar populations do not change with time or change very slowly. In particular, the angular momentum of stellar populations and eccentricity of their orbits do not change even if the gravitational field of the whole Galaxy changes. (iii) Mass and total angular momentum of the whole Galaxy are constant.

Eggen *et al.* (1962) developed a dynamical model of the Galaxy, calculated the gravitational potential in the galactic plane, and used this potential to find orbits of old stars. The stars of the largest ultraviolet excess are moving in highly elliptical orbits, whereas stars with little or no excess move in nearly circular orbits. These correlations require that the oldest stars were formed out of gas falling towards the galactic centre in the radial direction and collapsing from the halo onto the plane. The collapse was very rapid, and only a few times 10^8 years were required for the gas to attain circular orbits, where gravitational attraction is balanced by centrifugal acceleration. The initial contraction must have begun near the time of formation of the first stars some 10^{10} years ago.

In the early phase of newly formed galaxies, their gravitational field changes rapidly. This leads to violent relaxation of galaxies mostly in their inner regions. Lynden-Bell (1967) developed the statistical mechanics of violent relaxation in stellar systems. Numerical experiments of the early evolution of galaxies confirmed that the statistical method by Lynden-Ball describes well the observed energy distribution of stellar systems. These results suggest that during the formation of elliptical galaxies and cores of spiral galaxies, the turbulence played an important role.

2.2.4 *Physical evolution of galaxies*

The change of physical characteristics of galaxies with time is caused by the dynamical evolution of galaxies (redistribution of mass) and by the change of physical characteristics of stars due to stellar evolution. Advances in the understanding of stellar evolution permit us to follow the physical evolution of galaxies and build models of the evolution of stellar populations and gas in galaxies, and pioneering models were calculated by Tinsley (1968) and Einasto (1972). Models of the evolution of stellar populations and galaxies are based on stellar evolution tracks, star formation rates (SFR) (as a function of time), and the initial mass function (IMF).

2.2.4.1 *Star formation rate*

We define the local SFR as the time derivation of the density of stars. Following Schmidt (1959), we assume that the star formation rate is proportional to the density of gas in power S:

$$R_l = \frac{d\rho_s}{dt} = \gamma \rho_g^S, \tag{2.28}$$

where ρ_s and ρ_g are star and gas densities, respectively, and γ and S are constants. We assume that the full matter density in a volume element, $\rho = \rho_s + \rho_g$, does not depend on time. In this case, after integration of Eq. (2.28), we get

$$\rho_g = \rho[1 + (S-1)\tau]^{\frac{-1}{S-1}}, \tag{2.29}$$

where

$$\tau = t/K, \tag{2.30}$$

and for the characteristic time K, we have

$$K = (\gamma \rho^{S-1})^{-1}. \tag{2.31}$$

In case $S = 1$, we get

$$\rho_g = \rho\, e^{-\tau} \tag{2.32}$$

and

$$K = \gamma^{-1}. \tag{2.33}$$

Observational data on the density and mass of stars and gas suggest that $S = 2$. In this case, the characteristic time is $K = (\gamma \rho)^{-1}$.

2.2.4.2 *Initial mass function*

A critical step in the calculation of the model of the physical evolution of galaxies is the determination of the initial mass function (IMF). The initial distribution of stars according to their mass and luminosity and stellar evolution was discussed by Salpeter (1955) and Schmidt (1959). The number of stars of mass M, $F(M)$, formed in unit time interval per cubic parsec, can be expressed by the following equation:

$$F(M) = a \times M^{-n}, \tag{2.34}$$

where a and $n = 2.35$ are constants. The total mass of forming stars is equal to

$$\int_{M_0}^{M_u} F(M) M \, \mathrm{d}M. \tag{2.35}$$

The mass-to-light ratio M_i/L_i of the population i depends critically on the lower mass limit of the IMF, M_0. We cannot use for the minimal mass a value $M_0 = 0$, since the integral in Eq. (2.35) is not converging in this case. Results do not depend on the exact form of the function $F(M)$ for small M. For this reason, we can take M_0 as the effective lower limit of mass of forming stars and take $F(M) = 0$, if $M < M_0$. Blue supergiant stars are unstable for pulsations, cross the instability zone very fast, and do not lose their mass during this period very much. For this reason, Einasto (1972) accepted as the upper limit of forming stars $100 \, M_\odot$.

The choice of parameters M_0, M_u and n is crucial in modelling the physical evolution of galaxies. Earlier, it was assumed that these parameters are constants. However, in this case, it is impossible to explain differences in mass-to-light ratios of globular clusters ($M/L_V \approx 1$), dwarf galaxies ($M/L_V \approx 10$), and cores of giant elliptical galaxies ($M/L_V \approx 100$), all having approximately similar ages. In calculations of the evolution of the Galaxy, Einasto (1972) accepted for stellar populations with a normal metal content a lower star formation limit $M_0 = 0.03 \, M_\odot$, for metal-rich populations a limit $M_0 = 0.001 \, M_\odot$, and for metal-poor populations a limit $M_0 = 0.1 \, M_\odot$. Most limits are lower than the lowest masses needed to start hydrogen burning in stars, $M^* = 0.08 \, M_\odot$.

Differences in the parameters of star formation function can probably be explained by differences in the chemical composition of old stellar populations of galaxies. The fraction of heavy chemical elements is different in globular clusters and in old open clusters, both having approximately equal ages. Heavy elements are synthesised in stars. Rapid enrichment of interstellar gas by heavy elements is done by massive stars with fast evolution. The large variability of chemical compositions of stars of different old populations suggests that in the early period of the evolution, various populations had different fractions of massive stars, much higher than the present fraction of massive stars, depending on the density of the gas.

2.2.4.3 *Star evolution tracks*

In the early 1970s, stellar evolution tracks were available for a large range of stellar masses from 0.05 to 60 solar masses and for several values of the metal content, normal stars with $Z = 0.02$. Evolution tracks can be checked using colour-magnitude (HR) diagrams of star clusters, and for selected clusters, HR diagrams are shown in Fig. 2.8.

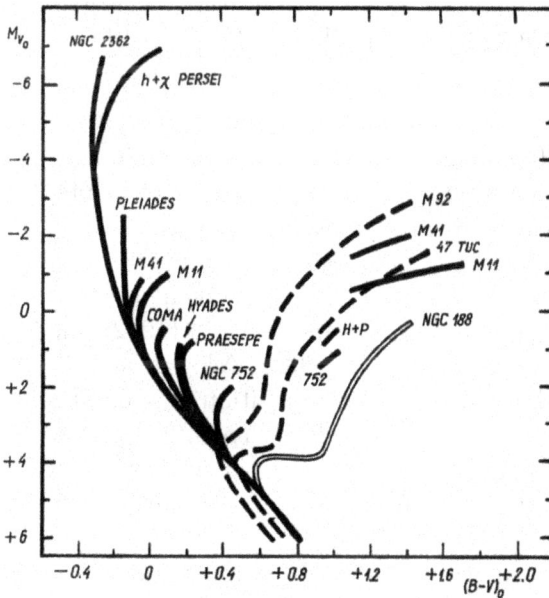

Fig. 2.8. Colour-magnitude diagrams for selected star clusters (Einasto, 1972).

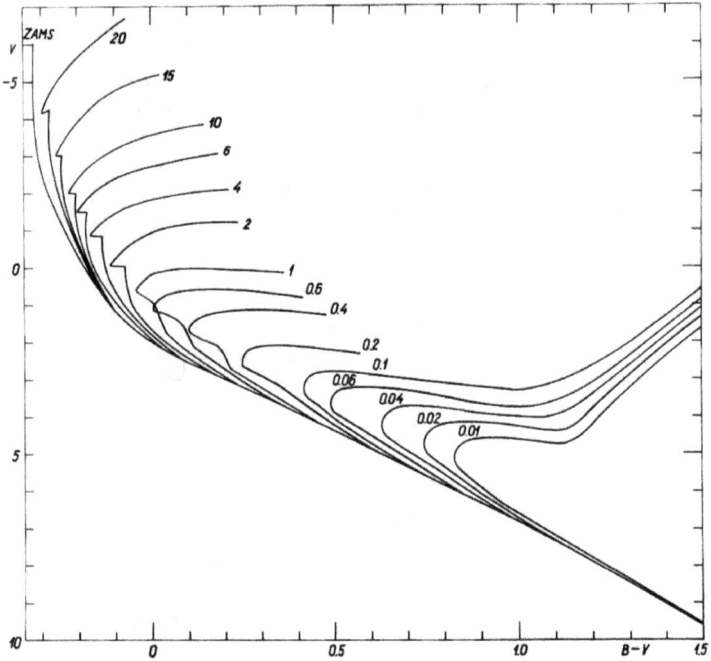

Fig. 2.9. Isochrones of stars of normal chemical composition $Z = 0.02$. Masses of stars in solar mass units are labelled 20, 15, 10, ..., 0.01 (Einasto, 1972).

Isochrones of stars of normal metal content $Z = 0.02$ for stars of masses from 0.01 to 20 solar masses are shown in Fig. 2.9. To get evolution tracks for a population, rich in heavy elements, Einasto (1972) added to composition $Z = 0.02$ tracks the following corrections: $\Delta \log t = -0.12$, $\Delta \log T_e = -0.10$, $\Delta \log L = -0.20$. We attribute these corrections to stars of heavy element content $Z = 0.08$. To get tracks for metal deficit stars, Einasto (1972) added to $Z = 0.02$ tracks corrections: $\Delta \log t = -0.22$, $\Delta \log L = 0.25$. We attributed these corrections to stars of composition $Z = 0.001$.

This dataset allows through interpolation to find luminosity — temperature diagrams for stellar populations of an arbitrary age, star mass and composition. Model populations can be found for a large range of epochs from 0.01 to 20 billion years. Luminosity functions, colours and mass-to-luminosity ratios can be calculated in the UBVRIJKL colour system. Results of these calculations are described in the following section.

2.2.4.4 *Evolution of galactic populations*

We show in the following figures some results of calculations of the evolution of stellar populations using star formation function parameter $S = 2$ and the characteristic time K. Galaxies consist of populations with various star formation parameters, thus the evolution of real galaxies is a superposition of several populations with different values of K and lower star mass formation limit M_0, depending on the density of the gas and the nature of the population.

An important property of the population is the luminosity function and its dependence on the age of the population, as shown in Fig. 2.10 in a large range of absolute luminosities from very low magnitude $M_B = 40$ to a high magnitude $M_B = -12$. The figure shows that in very young populations of age less than one billion years, the luminosity function has two isolated regions. The bright region is formed by stars in the main sequence and giants and the lower

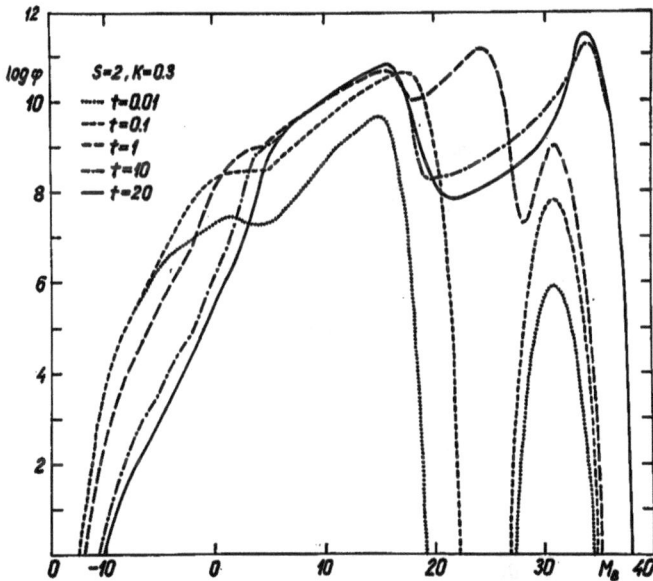

Fig. 2.10. The dependence of the luminosity function of galaxies in B system on the age of the galaxy t in billion years. As argument, we use the absolute magnitude in B system. Star formation function parameters are taken as follows: $S = 2$, $K = 0.3 \times 10^9$ years. The mass of the galaxy is $M = 10^{11}\, M_\odot$ (Einasto, 1972).

region with stars around the absolute magnitude $M \approx 31$. At the age $t = 1$ billion years, both regions join. In very young star systems, stars of high luminosity are dominating, there also exist pulsars — supernova remnants. In young systems, low mass stars are still in the stage of gravitational contraction, and their contribution to the luminosity of the system is surprisingly high. With the increasing age of the system, the fraction of low luminosity stars increases due to the increase in the number of evolved stars.

The evolution of the energy distribution in the spectra of model populations is shown in Fig. 2.11. Very young populations are very bright in ultraviolet region of the spectrum. In populations of the age $t = 1$ billion years and older energy distributions are rather similar, only the level of the curve decreases as the total luminosity of the population decreases.

The dependence of the mass-to-light function on chemical composition and star formation rate parameters is shown in Fig. 2.12. The general diapason of the function $f = M/L$ of old populations of different compositions allows us to explain the differences of this function for globular clusters and elliptical galaxies of various mass.

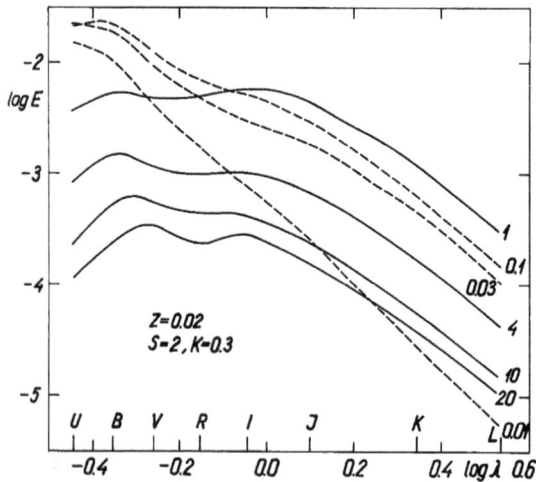

Fig. 2.11. The energy distribution in spectra of a model galaxy of normal metal content, $Z = 0.02$, and mass $10^{11}\,M_\odot$ at different ages, given in billion years. Wavelength is shown in microns. Also, we show effective wavelengths of UBVIJK colours. Star formation function parameters are taken as follows: $S = 2$, $K = 0.3 \times 10^9$ years. The mass of the galaxy is $M = 10^{11}\,M_\odot$ (Einasto, 1972).

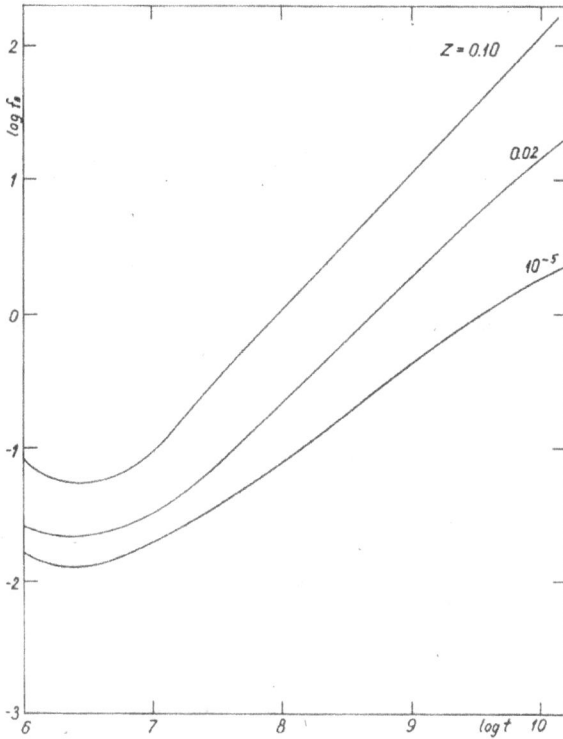

Fig. 2.12. Mass-to-light ratio of galaxies, $f_B = M/L_B$, as a function of the age t, for extremal values of the metal content Z (Einasto, 1972).

To explain these variations for different populations, parameters of star formation function and respective evolutionary tracks must be changed in reasonable limits.

The critical point in model construction is the determination of mass-to-light ratios for individual populations. There exist several possibilities to find these parameters from observations. For the nucleus and the core, this ratio can be determined from observations by two methods: from spectrophotometric data and from the virial theorem. For the halo, we can use the value for globular clusters determined from velocity dispersions. For disk and bulge, we can use the value for open clusters, found from velocity dispersion, from the rotation velocity at distance from the centre where disk or bulge dominates and from calculations of the physical evolution of populations.

Mass-to-light ratios $f_B = M/L_B$ of galactic populations are formed during the evolution of stars and are incorporated in dynamical models of galaxies. M/L_B ratios depend on the age and the chemical content of populations and are fixed by the minimal mass of stars in the star formation function, M_0, see Eq. (2.35). Einasto (1972) accepted for stellar populations of various chemical compositions different lower limits of star formations, as discussed above. Using these lower mass limits, Einasto (1972) got for old metal-poor populations $M/L_B \leq 3$, for old intermediate populations $M/L_B \leq 10$, and for old extremely metal-rich populations $M/L_B \leq 100$, see Fig. 2.12. New data by Faber & Jackson (1976) suggested that velocity dispersions of the nuclei of elliptical galaxies are much lower than accepted so far, which leads to a considerable decrease in mass-to-light ratios of elliptical galaxies. This suggests that corrections are needed to previous galaxy evolution models. This can be done by changing the lower mass limit of forming stars and using for all populations identical lower mass limits, $M_0 \approx 0.1$ $M_\odot$, which yields lower M/L values for all populations.

2.2.5 *Summary of the classical cosmological paradigm*

The classical cosmological paradigm can be described shortly as follows:

- The Universe formed as a result of the Big Bang about 10–15 billion years ago. Big Bang itself was considered as a mathematical singularity.
- The Universe is presently expanding with a rate which corresponds to the Hubble parameter in the range of 50–100 km/s/Mpc.
- The expansion of the Universe is spatially very uniform.
- The mean density of the Universe is about 0.04 of the critical cosmological density. This density concerns only baryonic matter, but this was recognised later.
- Galaxies are distributed in space almost randomly. About 10% of galaxies form clusters and superclusters of galaxies.
- During the early phase of the evolution of the Universe, only light chemical elements formed via the Big Bang nucleosynthesis; all heavier elements formed inside stars by stellar and supernova

nucleosynthesis. Nucleosynthesis is the basic source of energy in stars.

- Galaxies consist of populations of various ages, compositions, kinematics, and spatial structures. HR diagrams of homogeneous populations (as star clusters) can be used to determine the ages of populations and to reconstruct the history of the formation and evolution of galaxies.

Almost all observations known in the early 1970s supported this paradigm. General properties of the Universe can be described by two parameters, the Hubble constant and the deceleration parameter, which depends on the mean density of the Universe (Sandage, 1961). The foundation of this paradigm is still valid today. There were only a few clouds on the horizon which did not fit into this paradigm. One of these facts was the inability to present observed rotation curves with known stellar populations. A related difficulty was the mass paradox in clusters of galaxies found by Zwicky (1933). A theoretical problem was that the Universe has expanded enormously, thus even small deviations from the critical density should increase during this time. How is it possible that the present density is smaller than the critical one, but only by a factor of the order of 10?

From these small clouds and new observational data, a new cosmological paradigm formed. Most elements of the classical picture are valid today, similar to classical physics, whose basic elements remain unaltered. New data show that on top of the classical picture, new elements must be added. But there was a long way to go. In the following sections, we discuss some of these steps which led to the formation of the new paradigm in cosmology.

2.3 New cosmological paradigm

From 1970s to 1990s, four ideas transformed the cosmology into one of the most rapidly developing branches of physics. The first idea was the detection that in addition to the ordinary matter, there exists dark matter (DM), which dominates large-scale gravitational forces and consists of a new (and still unidentified) weakly interacting elementary particle. The second idea was the detection that galaxies are not distributed randomly in space, but form the cosmic web, which evolves very slowly and thus contains information on the very early Universe. The third idea is "cosmic inflation" — the proposal that the Universe grew exponentially after the Big Bang, driven by the vacuum energy density of an effective scalar field that rolls slowly from a false to the true vacuum. Quantum fluctuations in this "inflaton" field are blown up to macroscopic scales and eventually produce galaxies and the cosmic web. The fourth idea is the detection of dark energy and accelerating Universe.

In this section, we discuss how, step by step, these four developments changed our understanding of the structure and evolution of the Universe and its constituents — galaxies and systems of galaxies.

2.3.1 *Dark matter*

RHEA: I've just been to the saddest conference in my life.

CALLISTUS: The ski conference you were raving about?

ARIEL: Ski conference? I feel a sudden urge to become a cosmologist.

RHEA: It just means talks in the morning, skiing, talking science in the ski lift, more talk talks at night. This is exhausting, you know?

ARIEL: Someone has to do it. And I'm so ready.

RHEA: The few hours of skiing we got were great though. But the talks were all about dark matter detection. Or non-detection. One brilliant scientist after another describes a brilliant experiment that detects...

CALLISTUS: Nothing. But if they do, it's a Nobel.

RHEA: Long story short, I survived. Never again.

ARIEL: Feel free to call on me next time.

CALLISTUS: Rhea always picks on me for my light predisposition towards alternative theories, but when it comes to dark matter, she has an unorthodox taste.

RHEA: Not seriously, but I do like MOND. Modified Newtonian Gravity.

ARIEL: What's wrong with general relativity?

CALLISTUS: The theory that rules them all!

RHEA: And cannot be quantised. It needs all these fanciful things, like dark matter and dark energy, to explain the world. Dark matter? We have candidates, but we don't know what it is. Dark energy? With negative pressure that vaguely behaves like vacuum energy, except 10^{-120} times smaller? We don't even have a clue what it is or even a faint hope for direct detection. Not to mention, the inflaton, a hypothesised scalar field? All these baroque words combine under the grammar of ignorance.

CALLISTUS: Also, they represent the interesting areas of science where you can make a real contribution.

ARIEL: What's so great about MOND?

RHEA: It's a simple and intuitive modification of the Newtonian law of gravity for low accelerations.

CALLISTUS: Around 10^{-20} m/s^2, so really low.

RHEA: Are you ready for it? MOND explains galaxy rotation curves with baryonic, i.e ordinary, matter only. We don't have to assume a form of matter we don't know. Some large ellipticals are more consistent with MOND than with ΛCDM cosmology.

CALLISTUS: I hate to throw cold water on your fire, but what about the Bullet Cluster?

ARIEL: I've read that its visible and gravitational masses are shifted with respect to each other. How do we know that?

CALLISTUS: The background light is slightly curved when passing near a large mass. The mass behaves like a lens. We call this phenomenon gravitational lensing. From the lensing distortions of the background light, we can reconstruct the mass profile of the lens. In the case of the Bullet Cluster, the reconstructed mass profile is not centred on the visible mass. MOND cannot explain this. Locus desperatus.[5]

RHEA: Or at least it has a hard time explaining it. I still have a soft spot for it.

CALLISTUS: And you should. Especially if you were to find an extension that is consistent with Einstein's general relativity. I would be more open to that.

ARIEL: I guess until then we have to live with dark matter.

RHEA: And hope somebody finally detects it!

CALLISTUS: Dum spiro, spero.[6]

2.3.1.1 *The discovery of dark matter*

The first hint to difficulties in masses of systems of galaxies came from Zwicky (1933, 1937). He measured redshifts of galaxies in the Coma cluster and found that the velocities of individual galaxies with respect to the cluster mean velocity are much larger than those expected from the estimated total mass of the cluster calculated from masses of individual galaxies. The only way to hold the cluster from rapid expansion is to assume that the cluster contains huge quantities of some invisible dark matter. According to his estimate, the amount of dark matter in this cluster exceeds the total mass of cluster galaxies at least 10-fold, probably even more. However, at this time, astronomers were interested in the structure and evolution of stars, and Zwicky's work seemed to be remote and uninteresting.

Slowly more dynamical data on clusters and groups of galaxies were collected. The possible presence of dark matter in groups

---

[5] A hopeless passage.

[6] While I breathe, I hope (Cicero).

of galaxies was suggested by Karachentsev (1966). To explain high velocities of cluster galaxies, Ambartsumian (1961) suggested the idea that clusters are recently formed and are now expanding. However, van den Bergh (1961b) drew attention to the fact that the dominating population in elliptical galaxies is the bulge consisting of old stars, indicating that cluster galaxies are old. It is very difficult to imagine how old cluster galaxies could form an unstable and expanding system.

A completely new and innovative approach in the study of masses of systems of galaxies was applied by Kahn & Woltjer (1959). The authors paid attention to the fact that most galaxies have positive redshifts as a result of the expansion of the Universe; only the Andromeda galaxy M31 has a negative redshift of about 120 km/s, directed towards our Galaxy. This fact can be explained if both galaxies, M31 and our Galaxy, form a physical system. The negative radial velocity indicates that these galaxies have already passed the apogalacticon of their relative orbit and are presently approaching each other. From the approaching velocity, the mutual distance, and the time since passing the perigalacticon (taken equal to the present age of the Universe), the authors calculated the total mass of the double system. They found that $M_{tot} \geq 1.8 \times 10^{12} \, M_{\odot}$. The conventional masses of the Galaxy and M31 were estimated to be of the order of $2 \times 10^{11} \, M_{\odot}$. In other words, the authors found evidence for the presence of additional mass in the Local Group of galaxies. The authors suggested that the extra mass is probably in the form of hot gas of temperature about $5 \times 10^5 \, K$.

Galaxies have extended stellar haloes with approximately exponential light distribution, and it is expected that at large distances beyond the stellar halo, the rotation velocity follows Keplerian orbits. In contrast, measured rotation curves are flat at large galactocentric distances. First hints to the presence of flat rotation curves of galaxies come from Babcock (1939), Oort (1940), Roberts (1966), Rubin & Ford (1970) and Freeman (1970).

Independent evidence for the presence of difficulties in masses of galaxies showed that it is impossible to represent the mass distribution of galaxies with known stellar populations, as discussed above (Einasto, 1972). A theoretical difficulty in models of galaxies was found by Ostriker & Peebles (1973), who showed that flattened galaxies are unstable and must be stabilised by a massive thick population.

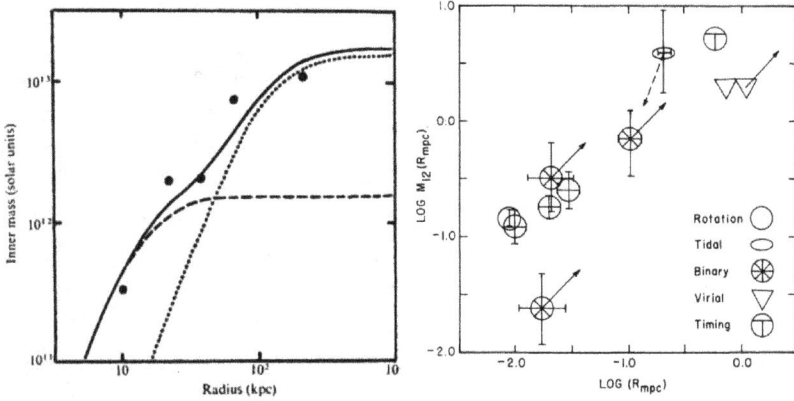

Fig. 2.13. Left: The mean internal mass $M(R)$ as a function of the radius R from the main galaxy in 105 pairs of galaxies (dots). Dashed line shows the contribution of visible populations, dotted line the contribution of the dark corona, solid line the total distribution (Einasto *et al.*, 1974a). Right: Mass (in units 10^{12} M$_\odot$) of local giant spiral galaxies within a distance (R in h^{-1} Mpc) of their centres, as determined by various methods (Ostriker *et al.*, 1974).

To estimate the extent of possible massive haloes of galaxies, Einasto *et al.* (1974a) used satellites of giant galaxies as test bodies to find the extent of the massive population. Authors were able to find the mean internal mass as a function of galactocentric radius in 105 pairs of galaxies, separately for visible populations and for the extended dark corona, see Fig. 2.13. This analysis suggested that all giant galaxies have massive coronae, exceeding the mass and the radius of known populations about 10-fold. Ostriker *et al.* (1974) obtained similar results using similar arguments, as shown in the right panel of Fig. 2.13. Both papers suggest that the total cosmological density of the matter in galaxies is about 0.2 of the critical cosmological density.

As noted by de Swart (2024), *usual dark matter histories center on the roles of astronomers Fritz Zwicky and Vera Rubin, those observations are considered as evidence for the existence of dark matter. But facts and observations themselves do not tell a history. To understand the origin of the case for dark matter, we need to know how prior observations made by Zwicky, Rubin, and others were interpreted to be evidence for its existence. In what context were they used to show that the Universe had preponderous amounts of missing matter? Who started to care, and why? That happened independently 50 years ago on both sides of the Iron Curtain by publications of papers by Einasto et al. (1974a) and Ostriker et al. (1974)... The evidence they presented was neither a simple proof nor a single observation, like that of Zwicky or Rubin, but an inference using a combination of different arguments. As Peebles stated when I interviewed him, "What was the best argument? None of them. This is a case of no one argument being compelling, but so many arguments pointing in the same direction." The two papers were exemplars of the nascent field of physical cosmology and its interdisciplinary teamwork and methodology: combining data and arguments from different scales — from stars and galaxies to clusters — to form a consistent physical picture of the cosmos.*

As mentioned by Peebles (2022c), discussions of dark matter by Einasto *et al.* (1974a) and Ostriker *et al.* (1974) formed a remarkable Merton (1961) multiple, when similar new ideas were suggested by two teams independently. Merton (1961) noticed that similar discoveries are often made by scientists working independently of each other, and called these as "multiple independent discoveries". Merton believed that it is multiple discoveries, rather than unique ones, that represent the common pattern in science.

These results were not accepted immediately by the astronomical community. Dark matter problem was discussed in the Third European Astronomical Meeting, which took place on 1–5 July, 1975 in Tbilisi, Georgia. In the dark matter session, the principal discussion was between the supporters of the classical paradigm with conventional mass estimates of galaxies and of the new one with dark matter. Arguments favouring the classical paradigm were presented by Materne & Tammann (1976). Their most serious argument was as follows: *Big Bang nucleosynthesis suggests a low-density Universe with the density parameter $\Omega \approx 0.05$; the smoothness of the Hubble flow also favours a low-density Universe.* If dark matter exists in quantities as suggested by new data, then the arguments by Materne and Tammann must be explained in some other way. The explanation came 30 years later, when the smallness of CMB fluctuations suggested that most of the matter in Universe is non-baryonic dark matter, and the dominating matter–energy component is dark energy. However, scientific problems cannot be solved by dispute, and new observational evidence is needed.

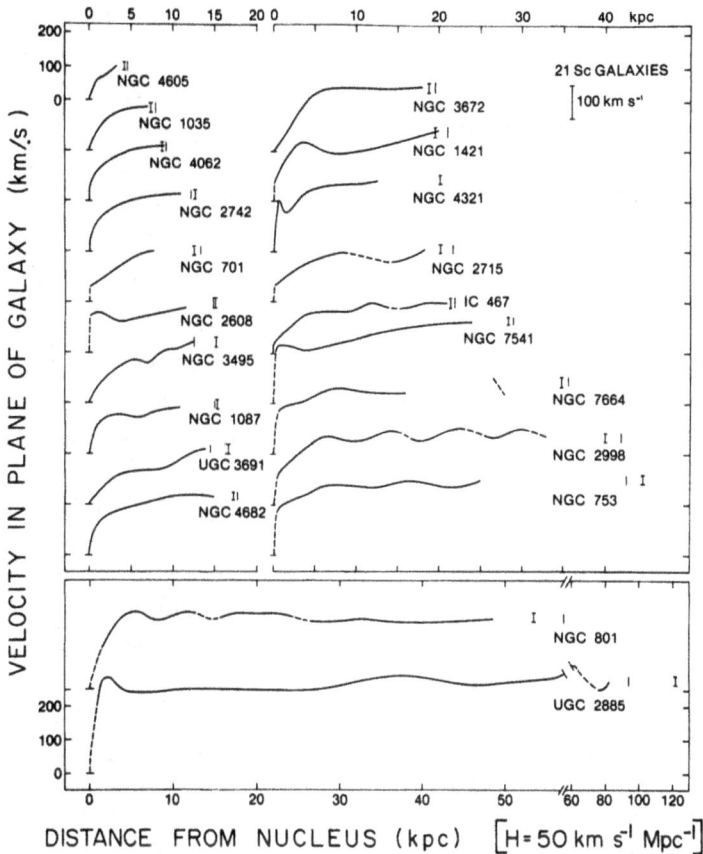

Fig. 2.14. Rotation velocities of 21 Sc galaxies (Rubin *et al.*, 1980).

Additional observational data on rotation curves of galaxies became soon available. Flat rotation curves were found by Bosma (1978) from radio observations of neutral hydrogen. Rubin *et al.* (1980) determined rotation curves for 21 Sc galaxies and showed that they are flat up to about 50 kpc, as seen in Fig. 2.14.

According to new estimates, the total mass density of matter is 20% of the critical cosmological density, thus dark matter is the dominant population in the whole Universe. Opinions about the nature of dark matter were different. Ostriker *et al.* (1974) argued that the very great extent of spiral galaxies can perhaps most plausibly be understood as due to a giant halo of faint stars. Einasto (1972) argued that cold gas (hydrogen) is excluded since it would be observable. Einasto *et al.* (1974a) noticed that dark matter in clusters cannot be explained by hot gas, since X-ray data suggest that the mass of hot gas is not sufficient to stabilise clusters — clusters must be stabilised by dark matter. Arguments against a stellar population are more complex, and we discuss this issue in the following section.

2.3.1.2 *The nature of dark matter*

Einasto (1972) emphasised that rotation of galaxies in outer regions must be influenced by a hypothetical new population. To avoid confusion with the conventional metal poor stellar halo, the new population was called corona. This population should form a large, massive, and an almost spherical population. In particular, for its formation, dissipation is not needed. Different size, shape, mass and dissipation properties suggest a different formation history and nature.

The dark matter problem was also discussed during the IAU General Assembly in Grenoble, 1976, at the Commission 33 Meeting. Arguments for the non-stellar nature of dark coronae were presented by Einasto *et al.* (1976a, 1976b). The analysis of kinematical and physical properties of stellar populations of various ages suggested the presence of a continuous distribution of stellar populations of various age and kinematical characteristics. As kinematical characteristics, the mean velocity dispersion and the heliocentric centroid velocity can be used, as expressed in the Strömberg diagram in Fig. 2.15. The oldest halo populations have the lowest metallicity and M/L-ratio, the highest velocity dispersion, and the largest (negative) heliocentric velocity in Fig. 2.15. The dark population is almost spherical and non-rotating. It has a much larger radius than all known stellar populations. In order to be in equilibrium in the Galactic gravitational potential, these coronal stars must have a high mean velocity dispersion, about $\sigma \approx 200\,\mathrm{km/s}$, much more than all known stellar populations, up to $125\,\mathrm{km/s}$, see Fig. 2.15. There is no place to put this new population into the sequence of known populations, and its estimated velocity dispersion is too high for a conventional stellar population. These kinematical data suggest that if the hypothetical population is of stellar origin, it must be formed much earlier than all known populations, and there exist no intermediate populations. It is known that star formation is not an efficient process: usually in a contracting gas cloud, only about 1% of the mass

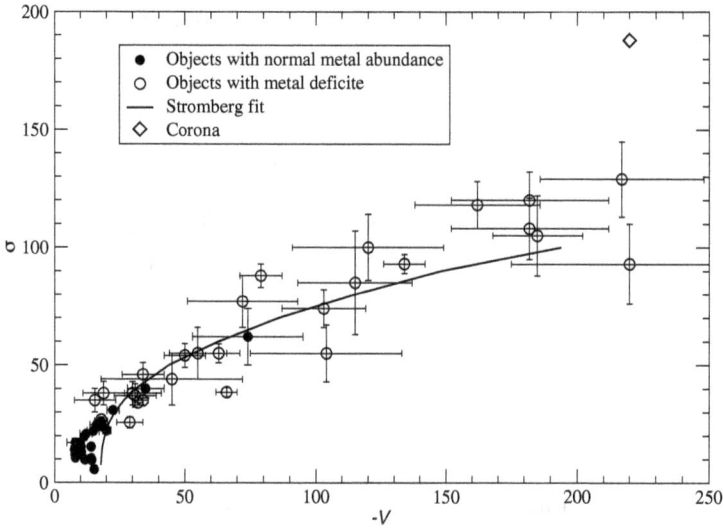

Fig. 2.15. Strömberg diagram for galactic populations. Kinematical data for the corona are taken from the model by Tenjes *et al.* (1994).

is converted to stars. Thus, we have a problem of how to convert, in an early stage of the evolution of the Universe, a high fraction of primordial gas into this population of first-generation stars. All these arguments show that the hypothetical population cannot be of stellar origin.

Already in the 1970s, suggestions were made that some sort of non-baryonic elementary particles, such as massive neutrinos, may serve as candidates for dark matter particles. There were several reasons to search for non-baryonic particles as a dark matter candidate. First of all, no baryonic matter candidate fits the observational data. Second, the total amount of matter is of the order of 0.2–0.3 in units of the critical cosmological density, while the nucleosynthesis constraints suggest that the amount of baryonic matter cannot be higher than about 0.04 of the critical density.

The only known non-baryonic particle was the neutrino, thus it was natural that first neutrinos were considered as dark matter particle candidates. Szalay & Marx (1976) considered neutrinos as dark matter candidates using as argument the total density of matter and the density of known baryonic matter. A similar suggestion was made by Rees (1977). Fluctuations of the relict radiation are smaller than the detection limit, which suggests that the dominant matter component must be non-baryonic. If dark matter is non-baryonic, then this helps to explain the paradox of small temperature fluctuations of the cosmic microwave background radiation (Chernin, 1981). Density perturbations of non-baryonic dark matter already start growing during the radiation-dominated era, whereas the growth of baryonic matter is damped by radiation. If non-baryonic dark matter dominates dynamically, the total density perturbation can have an amplitude of the order of 10^{-3} at the recombination epoch, which is needed for the formation of the observed structure of the Universe. However, the neutrino-dominated or hot dark matter generates almost no fine structure of the Universe, see the following section. Thus, some other solution had to be found.

In the early 1980s, difficulties with baryonic dark matter were well known, and non-baryonic dark matter was seriously considered. Also, difficulties with neutrino-based dark matter were known, thus dissipationless particles heavier than neutrinos were suggested by Blumenthal *et al.* (1982), Bond *et al.* (1982), and Peebles (1982b). Here, hypothetical particles like axions, gravitinos or photinos play the role of dark matter named as cold dark matter (CDM).

Primack & Blumenthal (1984) discussed arguments that dark matter is not baryonic, based on the deuterium abundance, and the absence of small-scale fluctuations in the microwave background radiation. The authors suggested the following classification of elementary particles as dark matter candidates: hot if free streaming erases all but supercluster-scale fluctuations; warm if free streaming erases fluctuations smaller than galaxies; and cold if free streaming is unimportant. Hot particles are light ($\sim 100 \, \mathrm{eV}$) and remain relativistic until just before recombination. Warm particles are 10–100 times heavier and thus become non-relativistic sooner. The formation of galaxies and the large-scale distribution of galaxies in CDM universe was analysed by Blumenthal *et al.* (1984). Now, the CDM scenario was accepted by the cosmology community.

As an alternative to dark matter, Milgrom & Bekenstein (1987) suggested the Modified Newton dynamics (MOND) to explain flat rotation curves of galaxies. However, the body of data from various areas, as well as the inflation model of the early Universe give strong support to the dark matter and dark energy concept.

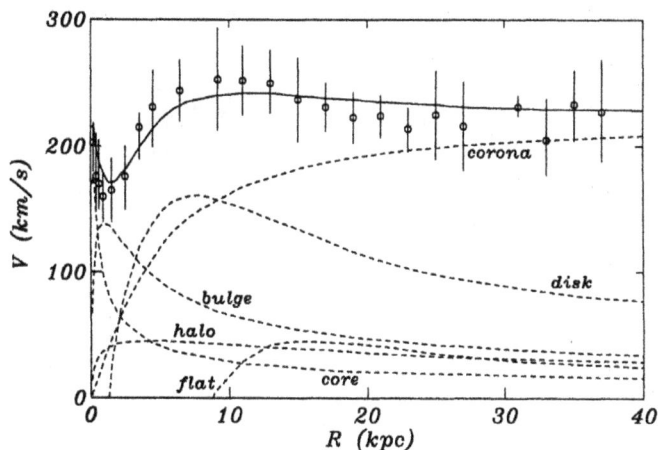

Fig. 2.16. The rotation curve of M31. Open circles: observations; thick curve: best-fit model, dashed lines: rotation curves of components (Tenjes *et al.*, 1994).

Tenjes *et al.* (1994) calculated a model of M31 with dark corona with stellar populations: flat, disk, bulge, core and halo, and dark corona, see Fig. 2.16. The figure shows that this model represents observed rotation data very well.

2.3.2 The cosmic web

According to the classical cosmological world view, based on the study of the two-dimensional distribution of galaxies on the sky, most galaxies belong to the general field, and only a relatively small fraction of galaxies is located in clusters. This means that galaxies are clustered on small scales, but are distributed on large scales more or less randomly. According to the dominant ideas of the early 1970s, the first objects to form are globular cluster sized systems which, by clustering, form larger objects, such as galaxies and clusters of galaxies. A theoretical explanation of such distribution was suggested by Peebles & Yu (1970) with the hierarchical clustering model called the bottom-up scenario.

One of the first hints of difficulties in the classical paradigm came from the study of the distribution of a homogeneous sample of Sc I galaxies by Rubin *et al.* (1973). Rubin compiled an all-sky sample of Sc I and Sc II galaxies in apparent magnitude interval $14.0 \leq m \leq 15.0$. The authors found that the distribution of redshifts of these galaxies is curious. In the Coma cluster area of the sky, redshifts of galaxies cluster around a value of $6400 \, \mathrm{km \, s^{-1}}$, but in the Perseus cluster area redshifts are clustered around a value of $4950 \, \mathrm{km \, s^{-1}}$. Areas of different mean values of redshifts are approximately located in opposite regions of sky, thus Rubin *et al.* suggested that one possible reason for this anisotropy may be a large motion of the Galaxy and the Local Group with respect to the general field of galaxies.

To check the classical paradigm on the distribution of galaxies, four groups started to collect redshifts to find the spatial distribution of galaxies. Results of all four groups were discussed in the IAU Symposium "Large Scale Structure of the Universe" in Tallinn, September 1977. Jõeveer & Einasto (1978) presented the study of the structure of the Perseus–Pisces supercluster and its surroundings and of the global network of superclusters and galaxy chains/filaments. Tifft & Gregory (1978) and Tarenghi *et al.* (1978) reported results of the distribution of galaxies in the Coma and Hercules superclusters, respectively. Tifft in his talk gave an overview of the recent study of the Coma supercluster and its environment by Gregory & Thompson (1978), see Fig. 2.19. These studies were based on redshifts obtained in Kitt Peak National Observatory in Arizona. Tully & Fisher (1978) presented a movie of the Local Supercluster. To obtain a spatial image of the supercluster, he used the simple trick of making the image rotate, which created a three-dimensional illusion. The movie showed that the Local Supercluster consists of a number of chains of galaxies which branch off from the supercluster's central cluster in the Virgo constellation as legs of a spider.

The global network of galaxies can be seen in Figs. 2.17 and 2.18. The first one shows the distribution of "near" Zwicky clusters on the sky in the Perseus region in restricted redshift interval $3{,}500 \leq V_0 < 6{,}500 \, \mathrm{km/s}$, and Fig. 2.18 presents the wedge diagram of galaxies for the $30°$–$45°$ declination zone. Both diagrams show that galaxies of the Perseus-Pisces supercluster form a long chain in which clusters and groups of galaxies are embedded as pearls. An important characteristic of the chain is its thickness — it is very narrow, as thick as the diameter of the clusters. There are no galaxies either

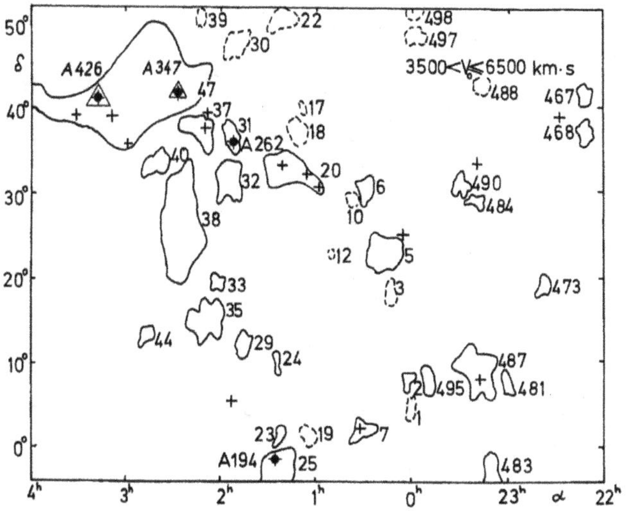

Fig. 2.17. The distribution of Zwicky clusters of the distance class "near" in the Perseus area of the sky in redshift interval $3,500 \leq V_0 < 6,500$ km/s. Abell clusters A426, A347, A262 and Zwicky clusters 37, 31, 20, 10, 6, 5 form the main chain of clusters of the Perseus–Pisces supercluster (Einasto *et al.*, 1980b; Jõeveer *et al.*, 1977).

up or downwards of the chain (as seen from our viewpoint) as well as in front of and behind the chain.

The comparison of adjacent redshift slices and sky views showed that chains of galaxies and systems of galaxies (groups and clusters) form an almost continuous network, as seen in the wedge diagram in Fig. 2.18. Here and there in the network, there are denser regions with more clusters in an aggregate, which can be identified with superclusters. The superclusters are branched: in addition to clusters, there are numerous galaxies which are not randomly scattered but also located as chains or filaments.

Jõeveer *et al.* (1977) and Gregory & Thompson (1978) also estimated the filling factor of the Universe covered by superclusters of galaxies and other filled regions (groups outside superclusters). The data indicated that superclusters fill only about 4% of the total space; the remaining 96% of space forms voids between superclusters. Since voids have been found to exist within superclusters, the filling factor of the Universe with systems of galaxies is even smaller.

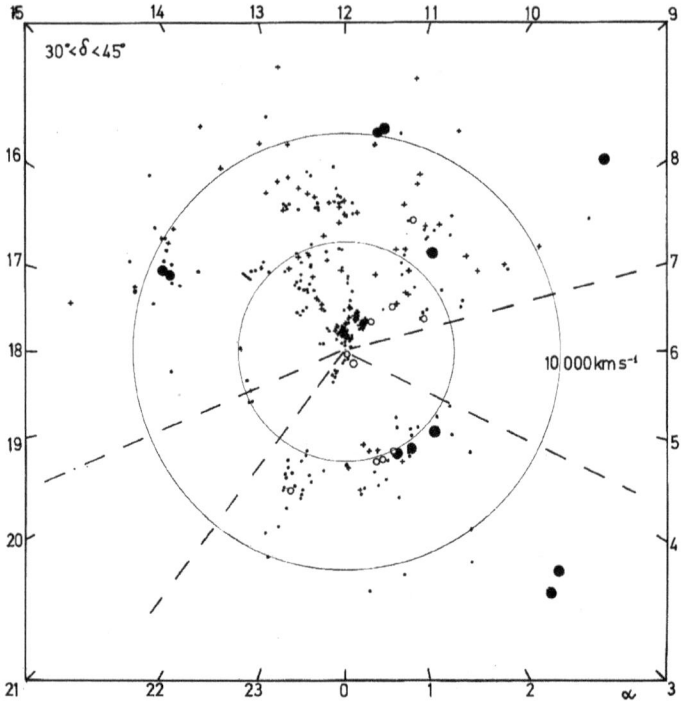

Fig. 2.18. Wedge diagram for the 30°–45° declination zone. Filled circles show rich clusters of galaxies, open circles — groups, dots — galaxies, crosses — Markarian galaxies (Jõeveer & Einasto, 1978).

The principal morphological property of the Universe is the pattern of the large-scale geometry — the presence of a continuous network of clusters, filaments and sheets, forming superclusters, and voids between them. This network has a remarkable fine structure. To that time, superclusters were considered as clusters of clusters, as advocated by Abell (1958). New data showed that superclusters are much richer; they contain, in addition to rich Abell-type clusters, groups and poor clusters galaxies, as well galaxies. But most importantly, cluster and galaxy filaments join superclusters to a single connected network.

The network has a hierarchical character — there exist rich chains of clusters/groups, and poor filaments, containing only galaxies, see wedge diagrams in Fig. 2.19. Clusters and groups as well as galaxies are not isolated elements in space, they are located within filaments.

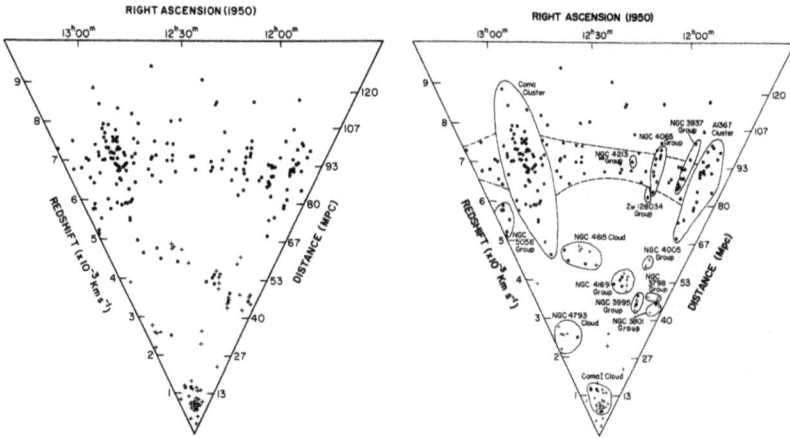

Fig. 2.19. Wedge diagram for all galaxies in survey by Gregory & Thompson (1978). The supercluster is clearly seen at an average redshift of approximately 7000 km/s. The distribution of foreground galaxies is very clumpy. The angular size has been magnified by approximately two times.

Examples of poor filaments are galaxy filaments which join the Local supercluster with other nearby superclusters. Two galaxy filaments at RA about 13^h join Local and Coma superclusters, two filaments at RA about 0^h and 2^h join Local and Perseus–Pisces superclusters, see Fig. 2.18, These filaments contain no clusters and are seen only in wedge diagrams.

A structural detail is important: the main cluster chain of the Perseus–Pisces supercluster contains clusters and groups of various richness called the Perseus chain. Clusters and groups are located at fairly regular intervals from each other. In the sky plane, the Perseus chain forms a smooth curve, and mean redshifts of clusters/groups (and their main galaxies) are almost identical. This means that the Perseus chain is very long and essentially a *one-dimensional* structure. Galaxy filaments, seen in Fig. 2.18, are also essentially one-dimensional structures (Jõeveer *et al.*, 1978; Gregory *et al.*, 1981).

Other morphological properties are the shape of clusters and the morphological types of galaxies. Practically, all clusters of the Perseus chain are elongated along the main ridge of the chain (Jõeveer *et al.*, 1978; Gregory *et al.*, 1981). All main galaxies of clusters in the Perseus chain are supergiant elliptical (cD) galaxies and are elongated along the axis of the chain. These properties mean that there exists a close physical link between cluster main galaxies and their *large-scale* environment. These properties hint to a common origin and evolution, not to random clustering. Additional detail is the morphological type of galaxies: in clusters and filaments, the fraction of elliptical galaxies is high, in sheets of galaxies outside filaments, there are practically no elliptical galaxies (Jõeveer *et al.*, 1978; Gregory *et al.*, 1981).

Jõeveer *et al.* (1977) emphasised that the Perseus cluster of galaxies A426 is peculiar in many respects. It has one of the highest velocity dispersion measured, and it is a strong emitter of X-rays and contains a number of radio galaxies. Its main galaxy, NGC 1275, is a peculiar Seyfert galaxy. This peculiarity is not surprising if we take into account the position of the cluster in the Universe. It is located at the corner of many filaments and sits in a very deep potential well of the Perseus–Pisces supercluster.

Reports on the presence of cosmic web with knots, filaments and voids by Tully & Fisher (1978), Jõeveer *et al.* (1977), Tifft & Gregory (1978) and Gregory & Thompson (1978) at the IAU Symposium in Tallinn 1977 form a remarkable Merton (1961) multiple. As noted by Jim Peebles earlier for the dark matter discovery: "What was the best argument? None of them. This is a case of no one argument being compelling, but so many arguments pointing in the same direction". This statement is also valid for the discovery of cosmic voids and the whole web.

The filamentary character of the galaxy distribution was confirmed by de Lapparent *et al.* (1986) and Geller & Huchra (1989). Now, the non-random distribution of galaxies was taken seriously. Such distribution was actually predicted by the numerical simulations of the evolution of the Universe, which we discuss in the following section.

2.3.3 *Comparison with simulations*

The critical link between the early, nearly uniform Universe and the rich structure seen at the present epoch has been provided by numerical simulation. Here, we discuss how our understanding of the evolution of the structure of the Universe has been improved. This has made use of the increase in the power of modern computers to create

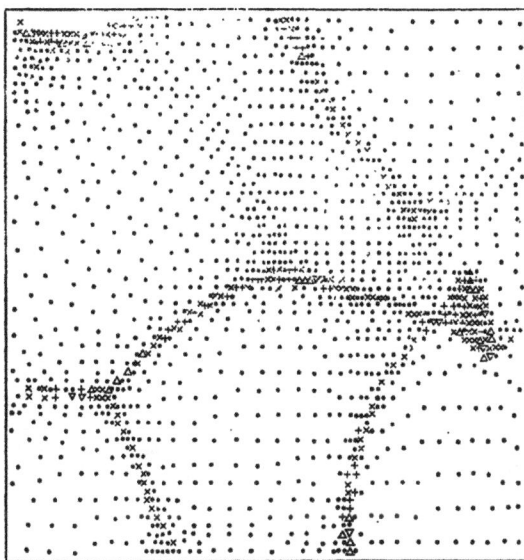

Fig. 2.20. Distribution of particles in simulations (Shandarin 1975, private communication), Doroshkevich *et al.* (1980).

ever more realistic virtual universes: simulations of the growth of cosmic structures from the primordial soup.

Sergei Shandarin obtained the first results of numerical simulations of the evolution of particles according to the theory of gravitational clustering developed by Zeldovich (1970), see Fig. 2.20. In this picture, a system of high- and low-density regions is seen: high-density regions form compact clumps which are joined by filaments, together they form a cellular network which surrounds large underdense regions. This picture is the first indication of what structures could be expected in the Universe according to the Zeldovich model. Earlier versions of the model, as described by Doroshkevich *et al.* (1974), did not predict any regularities in the distribution of galaxies.

The three-dimensional distribution of galaxies and clusters of galaxies was described in the Tallinn symposium only qualitatively. The first quantitative comparison of Peebles and Zeldovich structure formation models was done by Zeldovich *et al.* (1982). Authors investigated properties of the distribution of real galaxies in the Virgo–Coma region, the distribution of particles in a 3D simulation

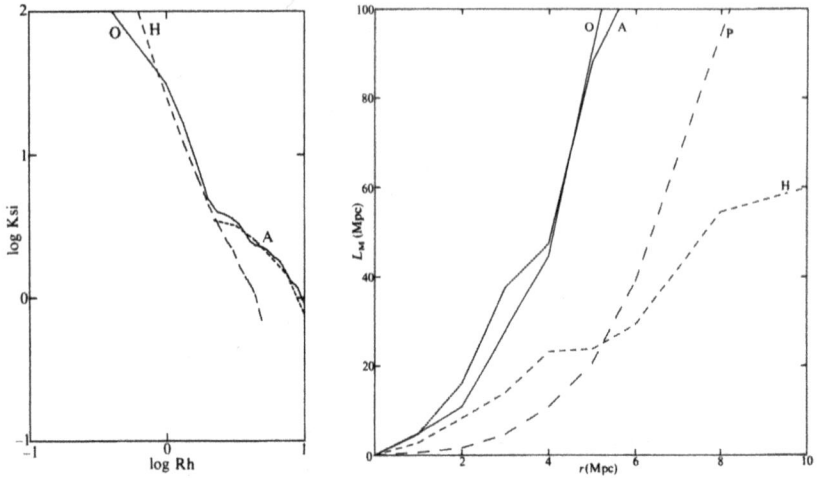

Fig. 2.21. Left: the correlation function of the observed sample O around the Virgo cluster (cube of side 80 Mpc) of the sample generated by the hierarchical clustering model H and the adiabatic model A. Right: the maximal length L_M of connected regions as a function of neighbourhood radius r for four catalogues: O, A, H, and P (Poisson model). All distances are expressed for Hubble constant $h = 0.5$ (Zeldovich *et al.*, 1982).

by Klypin & Shandarin (1983), and the distribution of particles in the model, constructed according to the prescription described by Soneira & Peebles (1978). The two-dimensional view of this model is shown in Fig. 2.2. Also, authors used the Poisson distribution of particles. Three tests were used: the spatial correlation function, percolation and multiplicity tests.

The left panel of Fig. 2.21 shows the spatial correlation functions for three samples. The absolute magnitude-limited observed sample around the Virgo supercluster (using CfA redshift survey, complete up to $m = 14.5$) is designated as O. The sample A was calculated using the simulation of the adiabatic model by Klypin & Shandarin (1983). H is for the sample generated using the method to find a sample of particles in the hierarchical clustering scenario (it is not an actual dynamical simulation). The most important feature of the O and A samples is the presence of a knee in the correlation function and absent in the H model. At small distances, the correlation function is sensitive to the distribution of galaxies/particles at small mutual distances. At such small distances, the majority of galaxies belong to clusters and groups, which have an almost spherical shape. At larger distances, the correlation function feels the presence of filaments of galaxies, which are essentially one-dimensional. Thus, the geometry of the structure changes when we move from small to large mutual distances of galaxies/particles. The hierarchical model has no filaments, thus the correlation function is featureless.

The percolation method allows us to use the length of the largest system as a test. The right panel of Fig. 2.21 shows the maximal lengths of systems of galaxies/particles as a function of the neighbourhood radius r. Here, we compare the behaviour of four samples: the observed sample O, the models A and H, and the Poisson sample P. Galaxies as well as particles in simulations are clustered, this means that at small radii r, the length of the longest system grows with increasing r faster than in the Poisson sample. But at larger radii, the behaviour of samples is different. In the observed sample O as well as in the model sample A, there are filaments joining clusters to a network. These filaments make the formation of longer systems easy, and the length of the longest system grows more rapidly than in the Poisson case. Figure 2.21 shows that samples O and A have almost identical growth. In contrast, in the sample H, at larger r, the growth of the length L with radius r is *slower* than in the Poisson sample. The reason is simple: the density of the field particles of the sample H is lower than in the Poisson sample, since a large fraction of particles are used in clusters. Thus, we see that this test is sensitive to the presence of filaments which join clusters to a connected network.

Next, Zeldovich *et al.* (1982) used the multiplicity function of systems of galaxies/particles found for various neighbourhood radii r. At each radius r, authors counted the number of systems of certain multiplicity (number of galaxies/particles). The frequency of systems of various multiplicities for the neighbourhood radius $r = 5$ Mpc (for Hubble constant $h = 0.5$) is shown in Fig. 2.22. The multiplicity is expressed in a logarithmic scale in powers of 2: 0 corresponds to $2^0 = 1$, i.e. isolated galaxies, 5 to $2^5 = 32$, i.e. medium rich systems, etc. Actually, a histogram of multiplicities is shown; between indices i and $i + 1$, all systems with multiplicities between 2^i and 2^{i+1} are counted.

Figure 2.22 shows that all samples studied have different distributions of multiplicities. The observed sample has approximately

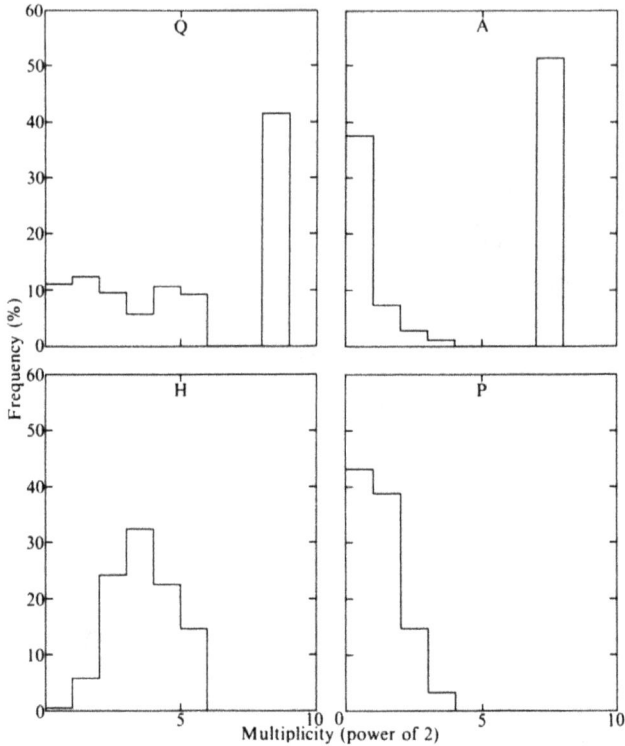

Fig. 2.22. Distribution of galaxies according to the multiplicity of the system. Neighbourhood radius $r = 5$ Mpc (for Hubble constant $h = 0.5$). Multiplicity is expressed in powers of 2 (5 corresponds to $2^5 = 32$). Samples are designated as in Fig. 2.21: O: observed, A: adiabatic model, H: hierarchical clustering model, P: Poisson model (Zeldovich *et al.*, 1982).

an equal fraction of systems of various richness, i.e. there exists a fine structure of systems of galaxies of different richness. The largest fraction of galaxies belongs to one single large system — the Virgo supercluster. The A sample also has one large system, but the distribution of smaller systems is more similar to the Poisson sample distribution, i.e. there are almost no systems of intermediate richness — small-scale filaments. This shows that the A sample, based on the neutrino-dominated Universe model, is in conflict with observations.

One more important difference: in the A sample, there exists a large fraction of isolated particles — these are void particles, forming a smooth low-density population in voids, absent in the O sample. The H model has no superclusters and no population of smoothly distributed isolated void particles; multiplicities have a peak at medium multiplicity level. The difference between A and O samples is an early indication of the biasing in the evolution of the cosmic web.

The analysis by Zeldovich *et al.* (1982) shows that the hierarchical clustering model fails in all tests, and the pancake model fails in the multiplicity test. The distribution of particles shown in Fig. 2.20 was calculated by Sergei Shandarin using only the Zeldovich approximation. The first true N-body simulations (Zeldovich approximation was applied only in the early phase of the evolution) were done by Doroshkevich *et al.* (1980) in two dimensions (2D) (64^2 mesh) and by Klypin & Shandarin (1983) in three dimensions (3D) (32^3 mesh). Both simulations were done using power spectra cut on small scales. At this time, the non-baryonic nature of the dark matter was a serious possibility to consider, and the first natural candidate was massive neutrinos called hot dark matter (HDM). Massive neutrinos move with very high speed, thus density perturbations on small scales are damped. For this reason, simulations of the neutrino-dominated Universe were similar to previous simulations of the baryonic matter, where spectra were also cut on small scales. In all simulations, the formation of a cellular structure of particles was clearly visible, as seen in Fig. 2.20. However, the structure forms too late. Einasto *et al.* (1980a) estimated the epoch of supercluster formation using simple geometrical models (sheets, cylinders and knots). According to this model, the formation epochs of sheets, cylinders and knots are at redshifts 12.7, 5.7 and 2.3, respectively. Observations suggest an early formation of galaxies and rich clusters of galaxies, as discussed by van den Bergh (1961a, 1962). In other words, the conventional neutrino-dominated cosmology is not possible.

Late formation of clusters and the failure of the multiplicity test were basic difficulties of the Zeldovich pancake model based on the assumption that dark matter is neutrino-dominated. In early 1980s, it was clear that the total density of matter is about 0.2 of the critical density and that the majority of the mass is in some form of dark matter. To avoid difficulties with neutrinos as the dark matter candidate, Blumenthal *et al.* (1982), Bond *et al.* (1982), and Peebles (1982a) assumed that dark matter is in the form of weakly interacting particles (cold dark matter).

First simulations of the formation of the structure with both neutrino-dominated and CDM were made by Centrella & Melott (1983) and Melott *et al.* (1983). The last paper is the first one where the advantages of the CDM model were tested using all quantitative tests, as discussed by Zeldovich *et al.* (1982), to analyse the HDM model by Doroshkevich *et al.* (1982). In contrast to the HDM model, in the CDM scenario, the structure formation starts at an early epoch. Superclusters consist of a network of filaments of DM haloes which can be identified with galaxies, groups, and clusters of galaxies similar to the observed distribution of galaxies. Thus, CDM simulations reproduce quite well the observed structure with clusters, filaments and voids, including quantitative characteristics: the correlation function, the percolation or connectivity, and the multiplicity distribution of systems of galaxies (Melott *et al.*, 1983). The Melott *et al.* paper ends with a statement: *"We see here strong support for the structure formation process in an axion-, gravitino-, or photino-dominated Universe. Galaxy formation proceeds from collapse of small-scale perturbations, as in the Hierarchical Clustering theory, but large-scale coherent structure forms as in the Adiabatic theory."*

In the CDM model, the structure forms through clustering of small-scale perturbations, as in the Hierarchical clustering theory by Peebles & Yu (1970). However, the clustering is not a random process but a regular flow of smaller systems towards larger dynamical attractors. This regular flow was confirmed by Tully *et al.* (2014). Large-scale coherent structure forms as suggested by Zeldovich (1970) with the introduction of the pancaking concept. The pancaking concept was generalised by Arnold *et al.* (1982) to the more general concept of Lagrange singularities of various spatial dimensions from knots and filaments to surfaces. The formation of galaxies and large-scale structure in CDM universe was discussed in detail by Blumenthal *et al.* (1984). Further support to the CDM concept was given by Bond *et al.* (1996) by analysing the physical processes in the formation of the cosmic web.

2.3.4 *Inflation*

According to the presently accepted Big Bang model, the Universe started from a singularity. But "singularity" is a mathematical term.

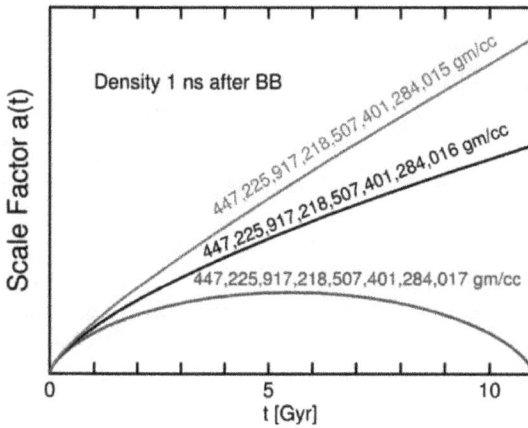

Fig. 2.23. The growth of the density scale factor from 1 ns after the Big Bang (Ned Wright talk, *A New MAP of the Early Universe*, 2003).

Big Bang theory says nothing about the physics of the primordial explosion. The theory of inflation is a physical description of the bang itself. It tries to answer a number of questions which could not be explained in the framework of the classical Big Bang model.

The first problem is the flatness problem. During the evolution, the Universe expands so much that any deviation from exact critical density would increase during the expansion. In order to have an approximately critical density today, it must have critical density in the early epoch, and at the time of nucleosynthesis, with an accuracy of at least 24 decimal places, i.e. it must be very accurately tuned, see Fig. 2.23. Here, we see that if the density changes by one digit in a 24 digit number of the density, the further evolution of the growth of the density scale factor changes radically.

This and some other difficulties of the classical Big Bang theory can be avoided if in the very early phase of the evolution of the Universe there was a period of very rapid expansion by a factor of at least 10^{26}. This rapid expansion is called inflation. The inflation scenario was suggested by Aleksei Starobinsky (1980, 1982, 1985) and independently by Alan Guth (1981). Starobinsky (1980) and Guth (1981) papers contain no mutual references and were completely independent, another Merton (1961) multiplicity.

This inflation model solves the flatness problem. During the inflationary period the Universe is driven very accurately towards the critical mass density. The model also solves the problem of homogeneity. The presently visible Universe has a radius of about 15 billion light years. Since the expansion factor is at least 10^{25} times, the present Universe was before the inflation so small that there was plenty of time for it to come to a uniform temperature. So, in the inflationary model, the uniform temperature was established before the inflation took place, in an extremely small region.

However, the classical version of the inflation model does not answer several fundamental questions: Why does the Universe have parameters as they are? What happened before the inflation started? To answer these questions Linde (1982, 1983, 2002a) suggested a model of chaotic inflation, shortly discussed in the Epilogue.

ARIEL: Professor, your fondness for alternative theories is well known: is there one for inflation?

CALLISTUS: I guess it would be colliding branes?

RHEA: You got him this time. There is no alternative to inflation, it solves four problems –

ARIEL: The flatness problem, the horizon problem, the monopole problem, and provides initial conditions!

CALLISTUS: Very good Ariel! You definitely read your homework assignment. And Rhea is right: colliding branes, obscure higher dimensional surfaces from string theory, are not likely to replace inflationary cosmology any time soon. Not to mention that it's not even that different when you start writing equations. But –

RHEA: Wait! Let me savour this moment. My smile is like I won a contest. [*Callistus laughs.*]

CALLISTUS: I see the new generation as dogmatic about inflation. The concordance model is almost perfect; it is supported by many, if not most, observations. But I came of age at a time when many more models and ideas were floating around –

RHEA: I'm not dogmatic, I just don't waste my time on crazy ideas.

CALLISTUS: Unless it's MOND, which is not a crazy idea. I'm just open-minded. It's ironic that we have an age reversal: I should be the grumpy old conservative guy and you should be –

RHEA: You're grumpy enough.

ARIEL: Ouch. I guess inflation is here to stay.

2.4 Cosmological parameters

2.4.1 *Critical dialogues in cosmology*

In the middle of the 1990s, the situation in cosmology was confusing. The presence of dark matter and cosmic web were announced, and the inflation paradigm was also suggested, but it was not known what the geometry of the Universe is and which of the Friedmann–Robertson–Walker models is correct. To discuss various aspects of cosmology a special conference, "Critical Dialogues in Cosmology", was organised, which took place in Princeton 24–27 June 1996 (Turok, 1997).

According to the dominating ideas, the Universe was expanding and can be characterised by two cosmological parameters: the expansion speed, measured by the Hubble parameter H_0, and the deceleration parameter q_0, which depends on the total matter–energy density of the Universe (Sandage, 1961). As noted by Very Rubin in her concluding dialogue talk, even the first digit of the Hubble parameter was not certain: Sandage and Tammann in their talk preferred a value $H_0 = 55 \pm 10 \, \mathrm{km \, s^{-1} \, Mpc^{-1}}$, whereas Wendy Freedman got from preliminary results of HST observation $H_0 = 73 \pm 4 \, \mathrm{km \, s^{-1} \, Mpc^{-1}}$. The inflation model of the Universe was suggested, which predicts a flat Universe with the total matter–energy density $\Omega_{\mathrm{tot}} = \Omega_{\mathrm{M}} + \Omega_{\mathrm{r}} + \Omega_{\Lambda} = 1$, where $\Omega_{\mathrm{M}} = \Omega_{\mathrm{b}} + \Omega_{\mathrm{DM}}$ is the matter density, which is sum of baryonic density Ω_{b} and DM density Ω_{DM}, Ω_{r} is the density of radiation, and Ω_{Λ} is the density of dark energy or cosmological constant. According to this model, the dominant constituent of the Universe at the present epoch is dark energy, but there exists no direct observational evidence of it.

Scientific problems cannot be solved by discussion, and new observational data are needed. In years after the Critical Dialogues in Cosmology, three important observational advantages took place: (i) the observational confirmation of the presence of dark energy, (ii) the use of HST and later JWST to measure the Hubble parameter more accurately, and (iii) the use of CMB satellites to find cosmological parameters.

These developments changed our view on cosmological parameters. New observational data and theoretical ideas suggested that the zoo of cosmological parameters is much richer than those adopted earlier. We begin our review of these developments by the observational testimony of the presence of dark energy, which confirms the prediction of the inflation model that the present Universe should have a flat geometry.

2.4.2 *Dark energy and the accelerating Universe*

Gunn & Tinsley (1975) showed that new data on the Hubble diagram, combined with constraints on the density of the Universe and the ages of galaxies, suggest that the most plausible cosmological models have a positive cosmological constant, are closed, and will expand for ever.

Observational evidence for the presence of a cosmological term in the mass–energy relation comes from the distant supernova experiments. Two teams, led by Riess *et al.* (1998) (High-Z Supernova Search Team) and Perlmutter *et al.* (1999) (Supernova Cosmology Project), initiated programs to detect distant type Ia supernovae in the early stage of their evolution and to investigate their properties with large telescopes. These supernovae have an almost constant intrinsic brightness (depending slightly on their evolution). By comparing the luminosities and redshifts of nearby and distant supernovae, it is possible to calculate how fast the Universe is expanding at different times, see Fig. 2.24. The supernova observations give strong support to the cosmological model with the Λ term.

The cosmological Λ term is presently interpreted as vacuum or dark energy. Dark energy has two important properties: its density ρ_v is constant, i.e. the density depends neither on time nor on location, and it acts as a repulsive force, i.e. it accelerates the expansion of the Universe. The best-fit confidence region in the $\Omega_M - \Omega_\Lambda$ plane is shown in Fig. 2.25. We see that the flat model with $\Omega_\Lambda = 0$ is excluded by a wide margin and that $\Omega_\Lambda > 0$ is required to reconcile theory with observations.

The first property means that in an expanding Universe in the earlier epoch, the density of matter (ordinary + dark matter) exceeded the density of dark energy, see Fig. 10.1. As the Universe expands, the mean density of matter decreases, and at a certain epoch, the matter density and the absolute value of the dark energy effective gravitating density were equal. This happened at an epoch which corresponds to redshift $z \approx 0.7$. Before this epoch, the gravity of matter decelerated the expansion; after this epoch, the antigravity of dark energy accelerated the expansion. This is a global phenomenon — it happened for the whole Universe at once.

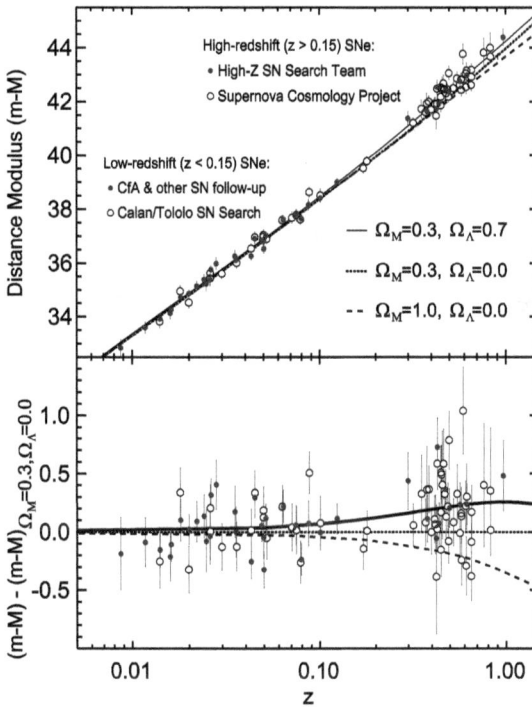

Fig. 2.24. Upper panel (a) shows Hubble diagram for high-redshift type Ia supernovae from the Supernova Cosmology Project and low-redshift type Ia supernovae from the Calan/Tololo Supernova Survey. Panel (b) shows residuals. Solid curves are for theoretical models with $\Omega_M = 0.3$, $\Omega_\Lambda = 0.7$, dotted lines with $\Omega_M = 0.3$, $\Omega_\Lambda = 0.0$, and dashed lines for models with $\Omega_M = 1.0$, $\Omega_\Lambda = 0.0$ (Perlmutter & Schmidt, 2003).

Dark energy also influences the local dynamics of astronomical bodies. The local effect of dark energy on the dynamics of bodies has been studied by Igor Karachentsev, Brent Tully and collaborators. Using the Hubble Space Telescope and large ground-based telescopes, Karachentsev determined accurate distances and redshifts of satellite galaxies in the Local group and several nearby groups of galaxies. This study shows that near the group centre up to distance $R \sim 1.25 \ h^{-1}$Mpc, satellite galaxies have both positive and negative velocities with respect to the group centre; at larger distance, all relative velocities are positive and follow the Hubble flow, see Fig. 2.26. This test demonstrates that dark energy influences both the local and global dynamics of astronomical systems.

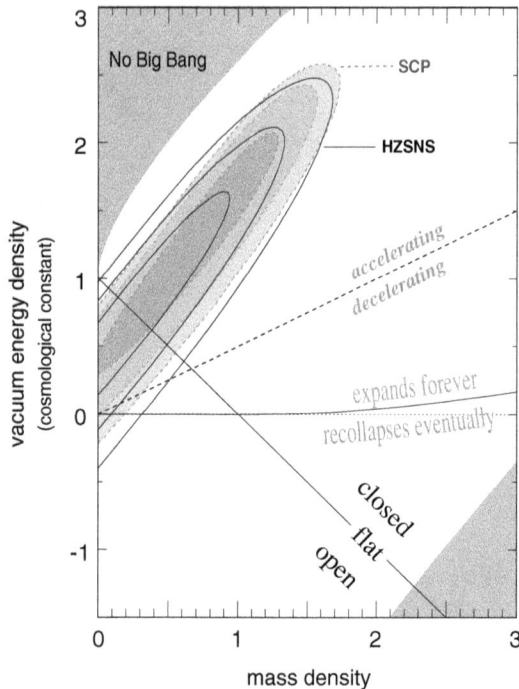

Fig. 2.25. Best-fit confidence regions in the Ω_M–Ω_Λ plane (Perlmutter & Schmidt, 2003).

2.4.3 *Expansion rate and the age of the Universe*

In 1960s–1970s, the efforts to determine the Hubble constant used mainly cepheids as distance indicators. A new possibility to find the Hubble constant is to measure photometric distances using the Tip of the Red Giant Branch (TRGB) distance indicator. The TRGB uses the luminosity of the brightest red-giant-branch stars in a galaxy as a standard candle, see Fig. 2.27. The luminosity of most luminous stars in the infrared I-band is insensitive to composition of other chemical elements and can be used as a distance indicator. For larger distances, the measurements of Type Ia supernovae (SNe Ia) can be used as a standard candle. Recent determination of the velocity–distance relation by Sandage *et al.* (2010) is shown in Fig. 2.28. As argument, we use the distance modulus

$$m - M = 5\log_{10}(d) - 5, \tag{2.36}$$

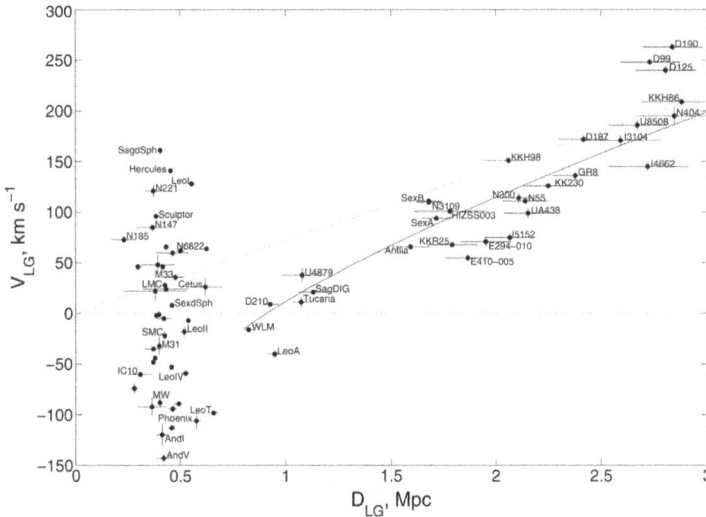

Fig. 2.26. Velocities of nearby galaxies in the Local Group rest frame. The nearest galaxies have a large velocity dispersion. Galaxies beyond 900 kpc are systematically redshifted. The solid curve is an empirical fit to galaxies beyond 900 kpc (Tully, 2015).

where m and M are apparent and absolute magnitudes of galaxies, corrected for interstellar absorption, and d is the distance in parsecs.

An important property of the velocity–distance relation is its scatter — this is very small. This shows that *The local velocity field is as regular, linear, isotropic, and quiet as it can be mapped with the present material. The lack of measurable velocity perturbations, in spite of the observed density inhomogeneities, suggests that the gravitational potential energy is small compared with the kinetic energy of the expansion (provided that there is no high-density, uniform intergalactic medium), and hence that $q_0 < 1/2$* (Sandage & Tammann, 1975). In other words, the structure of the Universe evolves very slowly and if there exist some regularities in the large-scale distribution of galaxies, these regularities must reflect the conditions in the Universe during the formation of galaxies.

The expansion parameter is $q_0 = 1/2 \times \Omega$, where Ω is the mean matter–energy density of the universe in units of the critical density. For an "empty" Milne's model universe with density parameter $\Omega \ll 1$, the age of the universe is $1/H_0$, or 9.7 gigayears (Gyr) for $h = 1$ and 19.4 Gyr for $h = 0.5$. Here, we use the Hubble constant

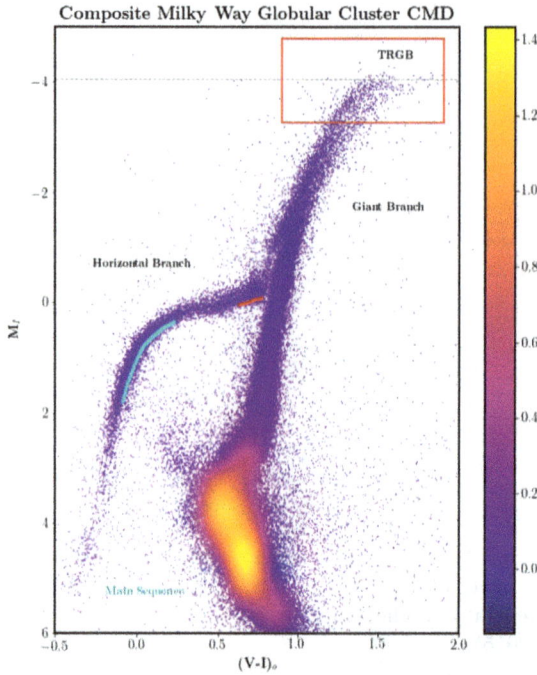

Fig. 2.27. Composite colour magnitude diagram (CMD) M_I versus $(V - I)_0$, based on 46 Galactic globular clusters, colour-coded by the density of the points. The clusters span a range in metallicity of $-2.4\,\mathrm{dex} < [\mathrm{Fe/H}] < -1.0\,\mathrm{dex}$. The horizontal branch, main-sequence turnoff, and giant branch are labelled. The cyan and red lines indicate the blue and red horizontal branch fits (Freedman, 2021).

in dimensionless units h, defined as follows: $H_0 = 100\ h\mathrm{km}^{-1}\mathrm{Mpc}^{-1}$. For a universe with critical density $\Omega = 1$ (Einstein–de Sitter model), the age is $2/3$ of that for the empty universe. For the Hubble constant $h = 0.7$, and an empty universe, as assumed in 1960s, the age of the universe is 13.5 Gyr. We show in Fig. 2.28 the curve for a ΛCDM model with $\Omega_m = 0.3$ and $\Omega_\Lambda = 0.7$, as discussed later in this chapter.

Measuring an accurate value of H_0 was one of the motivating reasons for building the NASA/ESA Hubble Space Telescope (HST). One of the key projects of HST was the determination of H_0. Final results of this study were published by Freedman *et al.* (2001). Authors combined different weighting schemes and found a consistency value $H_0 = 72 \pm 8\,\mathrm{km\,s}^{-1}\,\mathrm{Mpc}^{-1}$.

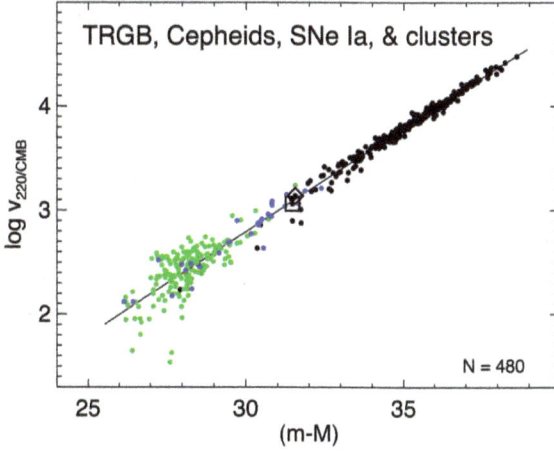

Fig. 2.28. The Hubble diagram of galaxies with TRGB distances (green), cepheid distances (blue) and clusters (black). The curve corresponds to a ΛCDM model with $\Omega_m = 0.3$ and $\Omega_\Lambda = 0.7$ (Sandage *et al.*, 2010).

Planck 2018 measurements yield for the Hubble parameter $H_0 = 67.37 \pm 0.54 \, \text{km s}^{-1} \, \text{Mpc}^{-1}$, and the age of the Universe $t_0 = 13.801 \pm 0.0024 \, \text{Gyr}$ (Planck Collaboration *et al.*, 2020). Earlier WMAP seven year analysis plus BAO and type Ia supernova data yielded slightly different values: $H_0 = 70.2 \pm 1.4 \, \text{km s}^{-1} \, \text{Mpc}^{-1}$, and the age of the Universe $t_0 = 13.77 \pm 0.12 \, \text{Gyr}$ (Komatsu *et al.*, 2011), but the difference is of the order of the error of WMAP data.

Additional information was obtained from the James Webb Space Telescope (JWST) launched in January 2021. Results from JWST on the Hubble parameter were discussed by Freedman *et al.* (2024). Authors observed with the JWST 11 nearby galaxies to find cepheids and stars near the TRGB and JAGB (J-region Asymptotic Giant Branch), see Fig. 2.27. The analysis showed that the data from TRGB and JAGB are very similar, slightly different from cepheid data. Best estimate from JWST from TRGB and JAGB data is $H_0 = 69.85 \pm 1.75(\text{stat}) \pm 1.54(\text{sys}) \, \text{km s}^{-1} \, \text{Mpc}^{-1}$ and from cepheids $H_0 = 72.05 \pm 1.86(\text{stat}) \pm 3.10(\text{sys}) \, \text{km s}^{-1} \, \text{Mpc}^{-1}$. Freedman *et al.* (2024) argued that TRGB and JAGB data can be more trusted. This helps us to remove the Hubble tension — the difference in the Hubble parameter from nearby Universe, as found from galaxies, and distant Universe, as measured from CMB temperature measurements.

2.4.4   *Big Bang nucleosynthesis and the amount of baryonic matter*

According to the Big Bang model, the Universe began in an extremely hot and dense state. For the first second, it was so hot that atomic nuclei could not form — space was filled with a hot soup of protons, neutrons, electrons, photons and other short-lived particles. Occasionally, a proton and a neutron collided and stuck together to form a nucleus of deuterium (a heavy isotope of hydrogen), but at such high temperatures, they were broken immediately by high-energy photons. When the Universe cooled off, these high-energy photons became rare enough that it became possible for deuterium to survive. These deuterium nuclei could keep sticking to more protons and neutrons, forming nuclei of helium and other light elements. This process of element formation is called "Big Bang nucleosynthesis", suggested by Alpher *et al.* (1948). The denser the proton and neutron "gas" is at this time, the more light elements will be formed. As the Universe expands, the density of protons and neutrons decreases. Neutrons are unstable unless they are bound up inside a nucleus. After a few minutes, the free neutrons will be gone and nucleosynthesis will stop. The relationship between the expansion rate of the Universe and the density of protons and neutrons (the baryonic matter density) determines how much of each of these light elements are formed in the early Universe. The baryon to photon ratio controls the light element abundance. The Planck best-fit range is $\Omega_B h^2 = 0.022433 \pm 0.00015$ (Planck Collaboration *et al.*, 2020), which yields abundance predictions shown in Fig. 2.29.

The Big Bang cosmological nucleosynthesis at a temperature of the order of 10^9 K created all the hydrogen and helium and some lithium, leading to primordial mass fractions $X \simeq 0.76$ for hydrogen, $Y \simeq 0.24$ for helium, and $Z = 0$ for all heavier elements, referred in astronomy as 'metals'. First generations of stars, called population III, consisted only of hydrogen and helium, and produced during nuclear reactions heavier elements, expelled to the interstellar space when the stars die. Next generation of stars, called population II, contain certain amount of heavier elements with $10^{-5} \le Z \le 0.02$. Soon stars, called population I, with approximately solar abundance formed with $Z \ge 0.02$.

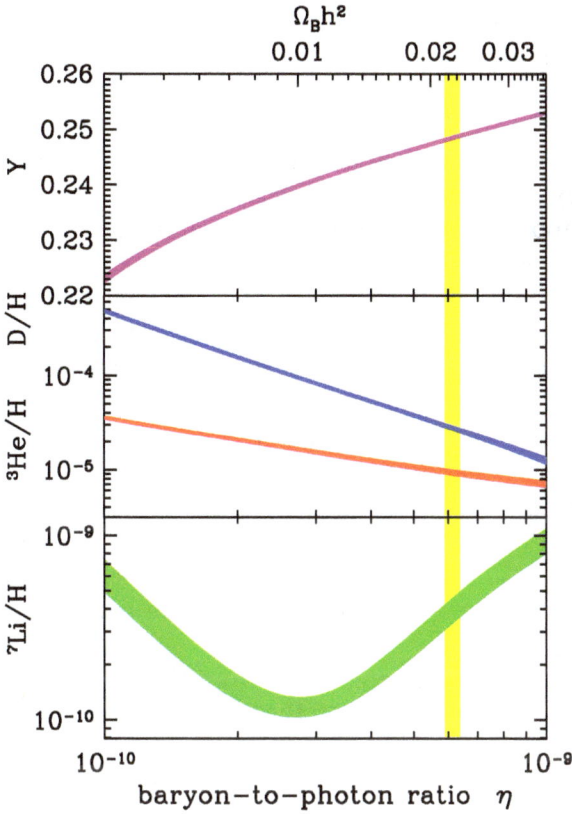

Fig. 2.29. Abundance predictions for standard BBN to helium fraction Y and fractions of deuterium D, helium He and lithium Li to hydrogen H. The WMAP baryon-to-photon range is shown in the vertical (yellow) band (Cyburt *et al.*, 2003).

2.4.5 *Matter–energy densities of various cosmic populations and other cosmological parameters*

The most important advance was the measurement of cosmological parameters with three CMB satellites: the Cosmic Background Explorer (COBE), the NASA operated Wilkinson Microwave Anisotropy Probe (WMAP), and the Planck space observatory operated by the European Space Agency (ESA). COBE satellite was launched in November 1989, WMAP mission in June 2001, and Planck mission in May 2009. These three satellites allowed us to measure the spectrum of the CMB, the shape of angular power spectrum and the polarisation of CMB fluctuations with high precision.

Data from the COBE satellite indicate that the gas in opposite directions of the sky have the same temperature with an accuracy of one part in 100,000 at the recombination epoch. Such high accuracy is possible only if these different regions have communicated. However, this was impossible at the time of recombination, since such communication would be possible only with a speed roughly 100 times the speed of light. In other words, the identical temperature must be achieved much earlier when the Universe was more compact. The homogeneity of the Universe can be explained by the inflation model.

Results of five-year and seven-year WMAP observation were published by Komatsu *et al.* (2009, 2011), and results of Planck 2018 data by Planck Collaboration *et al.* (2020), Planck 2018 temperature power spectrum and polarisation spectrum are shown in Fig. 2.30, as found by Planck Collaboration *et al.* (2020). Both spectra have multiple wiggles. The shape of these spectra depends on cosmological parameters and allows us to determine these parameters with high precision.

Most important parameter found from WMAP and Planck observations is the angular scale of the maximum of the CMB power spectrum. The position of this maximum depends on the total matter–energy density of the Universe. Planck 2018 data in combination with BAO and type Ia supernova data yield for the spatial curvature of the Universe $\Omega_k = 0.0007 \pm 0.0019$ (Planck Collaboration *et al.*, 2020). This means that Planck data did not find any deviations from a spatial flat universe with $\Omega_k = 0$ and $\Omega_{tot} = 1$. For the amount of matter, Planck 2018 data give the following values: baryon density $\Omega_b h^2 = 0.02233 \pm 0.00015$, CDM density $\Omega_c h^2 = 0.1198 \pm 0.0012$, dark energy density $\Omega_\Lambda = 0.6889 \pm 0.0056$ (Planck Collaboration *et al.*, 2020). Earlier analyses from WMAP five- and seven-year data gave very similar results (Komatsu *et al.*, 2009, 2011).

The analysis of temperature power spectra and polarisation spectra of CMB observations yields information for a large set of cosmological parameters. One of such parameters is the fluctuation amplitude parameter, for which Planck 2018 data yield a value $\sigma_8 = 0.8102 \pm 0.0061$ (Planck Collaboration *et al.*, 2020). The other important parameter is the index n_s of the primordial power spectrum of fluctuations. We show in Fig. 2.31 constraints on the index

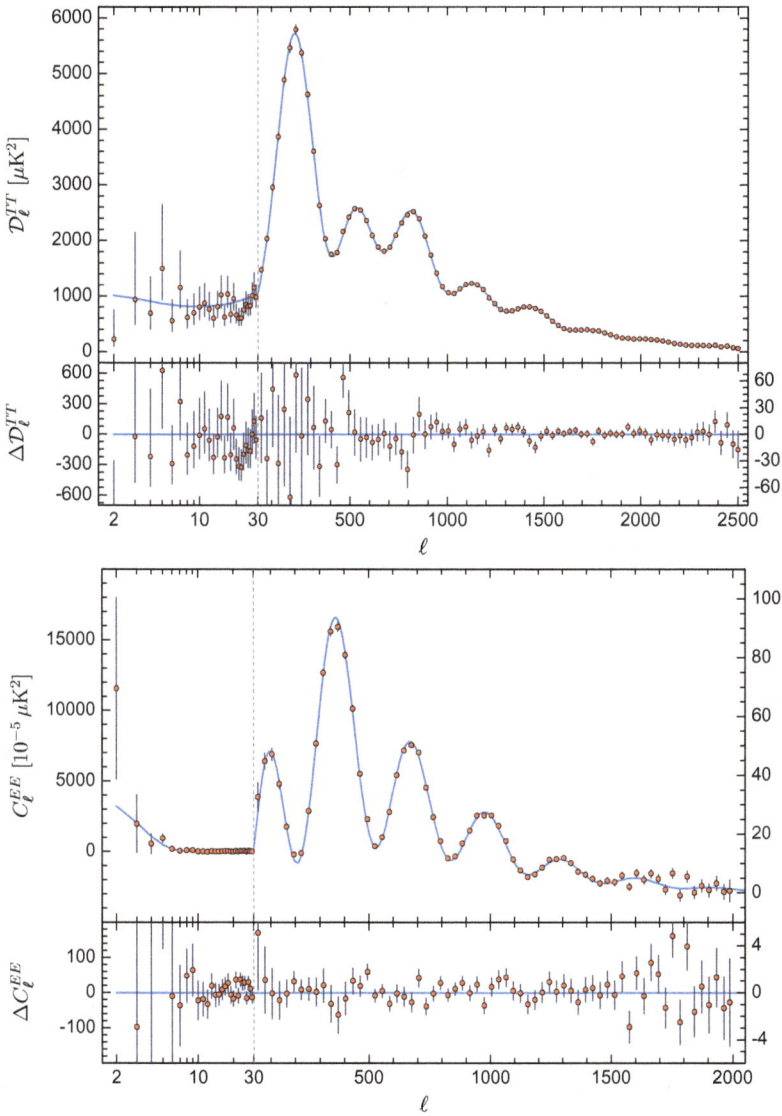

Fig. 2.30. Top: Planck 2018 temperature power spectrum; bottom: Planck 2018 polarisation spectrum. The base ΛCDM theoretical spectrum best fits to the Planck likelihoods are plotted in light blue in the upper panels. Residuals with respect to this model are shown in the lower panels. The error bars show 1σ uncertainties (Planck Collaboration *et al.*, 2020).

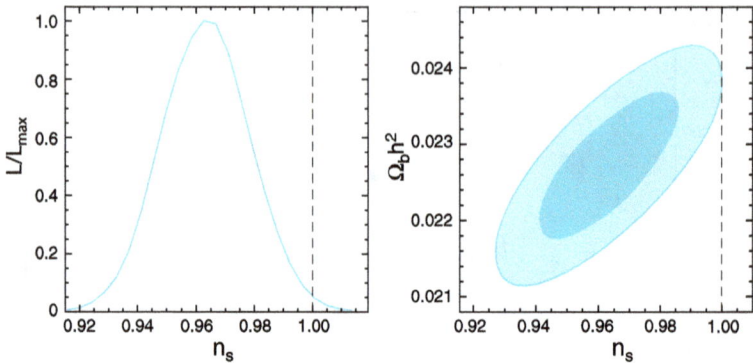

Fig. 2.31. Constraint on the primordial tilt, n_s. Left: One-dimensional marginalised constraint on n_sc from the WMAP-only analysis. Right: Two-dimensional joint marginalised constraint (68% and 95% CL), showing a strong correlation between n_s and $\Omega_b h^2$ (Komatsu *et al.*, 2009).

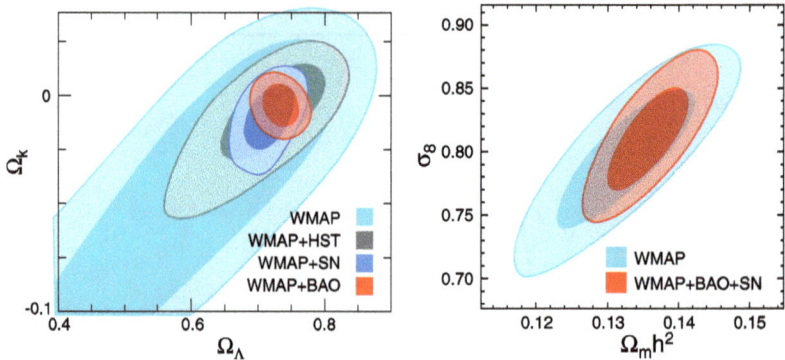

Fig. 2.32. Left: Constraints on parameters Ω_k and Ω_Λ Right: Constraints on parameters σ_8 and $\Omega_m h^2$ (Komatsu *et al.*, 2009).

n_s and the two-dimensional constraint between n_s and baryon density $\Omega_b h^2$. The adopted value for the index is $n_s = 0.9652 \pm 0.0042$ (Planck Collaboration *et al.*, 2020). A deviation from the spectrum with index $n_s = 1.0$ was predicted by Starobinsky (1982, 1985). For the temperature of CMB radiation, Fixsen (2009) found a value $2.72548 \pm 0.00057\,\mathrm{K}$ from WMAP data.

Constraints between parameters Ω_k and Ω_Λ, and between parameters σ_8 and $\Omega_m h^2$, are shown in Fig. 2.32 according to WMAP data by Komatsu *et al.* (2009). For the fluctuation amplitude parameter,

Planck 2018 data give $\sigma_8 = 0.8101 \pm 0.0061$. Of other parameters obtained by CMB observations, we mention the recombination epoch $z_{re} = 7.82 \pm 0.7$.

2.4.6 *New cosmology paradigm is ready: What next?*

During the last 50 years, the astronomy community has been witness to several paradigm changes. The most important paradigm changes are as follows:

- Cosmic microwave background (CMB) radiation was detected, which confirmed the paradigm of hot big bang cosmology.
- Most of the matter in the Universe is dark and consists of weakly interacting non-baryonic particles; the density of dark matter is about 0.25 of the critical cosmological density.
- The early evolution of the Universe includes a period of very rapid expansion or inflation which made the Universe homogeneous and its density equal to the critical density. The latest version of inflation theory suggests an initial chaotic stage within a multiverse.
- The Universe has structure in the form of a cosmic web. The seeds of the cosmic web were created at the very early stages of the evolution and give us information on the properties of the Universe at this epoch.
- Most of the matter/energy content is the dark energy, about 0.7 of the critical density. Dark energy causes an accelerating expansion of the Universe.

Can we say that the modern cosmological paradigm is now complete? Probably not. The modern paradigm explains almost all observational facts, but there are still a few important problems open. As suggested by Peebles (2002), the situation now has some similarity with the situation in physics at the eve of the 20th century. A 100 years ago, almost all experimental data were explained by the classical 19th century physics. But there were some clouds on the horizon — the Michelson experiment which suggested the isotropy of the velocity of light and difficulties in the theory of gases. To explain these clouds, the whole modern 20th century physics evolved: the theory of relativity, quantum physics, and much more.

Chapter 3

Inflationary Concordance Cosmology

We present a concise summary of the standard cosmological model to serve as background for all subsequent chapters. We provide a simplified treatment in a flat, Newtonian framework, since most observations prefer a critical universe. For more details and derivations in the full GR framework and a more extended range of cosmological models, see, e.g., Dodelson & Schmidt (2020), Peacock (1999), and Kolb & Turner (1990).

3.1   The homogeneous universe

CALLISTUS: You're early!

RHEA: I wanted to get here before Ariel to make sure we don't confuse them during the next session. [*Callistus smiles.*]

CALLISTUS: You mean no Einstein–Cartan gravity, no teleparallel gravity, what about –

RHEA: Not even MOND. Well, maybe MOND is OK.

CALLISTUS: I love GR as much as you do, but won't you agree that it's important to be aware of these extensions that are equally capable of describing gravitational phenomena?

RHEA: I call Occam's razor!

CALLISTUS: Raze away, but I find gauge theories easier and better motivated, so I will not be satisfied with simplicity alone.

RHEA: Ninety-nine per cent of cosmologists use GR, and the rest consider simple extensions, like MOND, or modified gravity theories. Only a few –

CALLISTUS: Outliers, like me –

RHEA: No. Well, yes. You are an out-of-the-box thinker, but I meant mathematically inclined people –

ARIEL: What about mathematically inclined people? Aren't they cool?

CALLISTUS: Aren't they?

[*Beat.*]

RHEA: Sometimes, math is a temptation that can become a distraction. The key is to focus on the fundamentals.

CALLISTUS: Which is our topic today. What are the fundamental interactions of nature?

ARIEL: Strong, electromagnetic, weak, and gravity.

RHEA: And you put them in descending order, their coupling strength in this order is $1, 1/137, 10^{-4}$ and 10^{-38}, respectively.

ARIEL: Wow, gravity is so weak if you put it this way.

CALLISTUS: So, which of these forces are important for cosmology?

ARIEL: You can't trick me. The strong and weak forces are short-ranged; therefore, they cannot be important on cosmological scales.

CALLISTUS: The electromagnetic force is supreme. We don't fall through the floor because of them.

RHEA: Very funny. It's strong, but the universe is...

ARIEL: Neutral! Positive and negative charges cancel each other's effect. Therefore, the weakest of all, gravity, dominates on the largest scales!

RHEA: Oh, cause it's gravity.

CALLISTUS: Ubi concordia, ibi victoria.[1] And the theory of gravity is

RHEA: MOND... I know you want me to say Einstein's GR!

CALLISTUS: Expressing the supreme beauty of geometry along with... some extensions that... we talk about some other time.

ARIEL: Space–time is curved, and matter follows geodesics. Matter creates curvature. I took GR.

CALLISTUS: Like you should before you study cosmology.

RHEA: Although we can derive the dynamics of the expanding universe from Newtonian considerations.

CALLISTUS: The correct derivation of the Friedmann equations is from GR. The Newtonian derivation is at best a mnemonic.

RHEA: The Newtonian derivation yields a clear physical idea of curvature: the universe's total energy is proportional to the negative curvature.

ARIEL: That's pretty. There must be something to it.

CALLISTUS: Let's get to work then. [*They start to work out the equations on the blackboard.*]

As inherited from Einstein's special relativity, space–time is four-dimensional with one time and three space coordinates, $x_\mu = (t, \vec{x})$.

---

[1]Where there is unity, there is victory.

Greek letters go through four indices $\mu = 0, 1, 2, 3$, while the spatial part, written as Latin indices, is $x_i = (\vec{x}) = (x_1, x_2, x_3)$. The principal quantity is the metric tensor, a generalisation of the Pythagorean theorem for small differences dx_μ:

$$ds^2 = g_{\mu\nu} dx^\mu dx^\nu, \tag{3.1}$$

where repeated upper and lower indices imply a summation. The original Pythagoras theorem would be $ds^2 = dx_1^2 + dx_2^2 + dx_3^2$ in three-dimensional space, and the generalisation to four-dimensional space–time in special relativity is $ds^2 = -dt^2 + dx_1^2 + dx_2^2 + dx_3^2$. This corresponds to the Minkowski metric

$$g_{\mu\nu} = \begin{pmatrix} -1 & 0 & 0 & 0 \\ 0 & 1 & 0 & 0 \\ 0 & 0 & 1 & 0 \\ 0 & 0 & 0 & 1 \end{pmatrix}, \tag{3.2}$$

where the lower right 3×3 submatrix is the spatial part corresponding to the Pythagoras theorem. The most general mathematical universe can have a more complicated metric, but we are lucky with the existing universe: it appears to be flat, and apart from expansion, it is almost identical to the Minkowsky metric

$$g_{\mu\nu} = \begin{pmatrix} -1 & 0 & 0 & 0 \\ 0 & a^2(t) & 0 & 0 \\ 0 & 0 & a^2(t) & 0 \\ 0 & 0 & 0 & a^2(t) \end{pmatrix}. \tag{3.3}$$

This is the Friedmann–Lemaitre–Robertson–Walker (FLRW) metric of an expanding homogeneous flat universe corresponding to $ds^2 = -dt^2 + a^2(t)(dx_1^2 + dx_2^2 + dx_3^2)$. The physical coordinates expand isotropically. The function $a(t)$, the expansion rate, is the fundamental quantity of classical cosmology. The comoving coordinates are identical to $a = 1$ and have a Minkowski or flat metric. We will see that observations support this metric with extremely high accuracy, but for now, let's accept that this describes our universe.

Sometimes, we write this metric in a spherical coordinate system centred on us, $ds^2 = -dt^2 + a^2(t)(dr^2 + r^2 d\Omega^2)$, where the angular part $d\Omega^2 = d\theta^2 + \sin^2(\theta)d\phi^2$. This metric form explicitly expresses the isotropy of the universe around the observer, while the previous form emphasises homogeneity.

Note that there are relatively simple curved generalisations of the above metric. The universe could be positively curved, analogous to a ball in two dimensions, or negatively curved, like a saddle in two dimensions. Formally, the r^2 before $d\Omega^2$ in the metric generalises with a function $R_0^2 \cos^2(r/R_0)$ or $R_0^2 \cosh^2(r/R_0)$ for closed and open universes, respectively. The former would be a compact universe, favoured by quantum cosmologists, but not the observations. A negatively curved universe is infinite, as is a flat universe. We will not use the metrics of curved universes, since the real universe appears flat, and dealing with the curvature math would be an unnecessary distraction.

3.1.1 Redshift

Given our picture of an expanding universe, waves corresponding to light would stretch as the universe expands. This is the redshift; since stretching always means a longer wavelength. Therefore, light gets redder. If we lived in a contracting universe, we would experience a universal blueshift.

Let's consider a wave with wavelength λ_e emitted at time t_e. We will observe that wave at time t_o stretched to λ_o. Also, let's introduce the time corresponding to the wavelength $\delta t = \lambda/(c = 1)$. For light rays, $ds = 0$ as in special relativity. Then

$$dt^2 = a^2(t)(dx_1^2 + dx_2^2 + dx_3^2) = a^2(t)dr^2, \qquad (3.4)$$

where the second equation is the definition of the comoving distance r. Using the above, we can calculate the comoving distance at the beginning and end of the emission:

$$r = \int_{t_e}^{t_o} dt/a(t) = \int_{t_e+\delta t_e}^{t_o+\delta t_o} dt/a(t). \qquad (3.5)$$

Subtracting $\int_{t_e+\delta t_e}^{t_o} dt/a(t)$ from both sides, we get two infinitesimal integrals that can be approximated as

$$\frac{\delta t_e}{a(t_e)} = \frac{\lambda_e}{a(t_e)} = \frac{\lambda_o}{a(t_o)} = \frac{\delta t_o}{a(t_o)}. \qquad (3.6)$$

Since at present $a(t_0) = 1$ by definition, the redshift characterising the wavelength stretch $z = (\lambda_0 - \lambda_e)/\lambda_e$ relates the expansion rate simply as

$$1 + z = \frac{1}{a(t_e)}. \tag{3.7}$$

The above is the fundamental redshift relation. It confirms our intuition that light rays stretch as the universe expands.

ARIEL: Professor, is cosmological redshift a Doppler shift?[2]

CALLISTUS: There is a lot of confusion about this even in some textbooks.

ARIEL: We argued about this with my friends till the wee hours of the morning.

RHEA: It's a semantic question. We know how to calculate it. Does it matter if we say it's due to the stretching space or the motion of galaxies?

ARIEL: Last night, I reviewed our derivation and I concluded that it is the expansion of space between us and a galaxy that causes the redshift.

RHEA: That's correct.

CALLISTUS: A contrario, the other view is also correct.

ARIEL: Well... I argued that redshifts above $z > 1$ would imply a velocity greater than the speed of light. Nonsense, obviously. There are confirmed JWST redshifts –

RHEA: – above 10! Did you read our paper?

ARIEL: Of course I did.

CALLISTUS: Hold your TikTok. In special relativity or Minkowski space–time, you could say "nonsense, obviously". But we live in a Friedmann universe, or something close to it –

ARIEL: And FLRW coordinate velocities can exceed c.

RHEA: You can read the paper by Bunn and Hogg from 2009[3] if you must.

CALLISTUS: It's fun.

ARIEL: I agree.

CALLISTUS: They interpret redshift as a series of small Doppler shifts in overlapping Minkowskian space–time regions.

ARIEL: I will read it!

RHEA: Duh. Just words. Measure and calculate. That's all you need.

---

[2]This dialogue is dedicated to the late Nick Kaiser, who particularly enjoyed such arguments.

[3]"The kinematic origin of the cosmological redshift," Bunn & Hogg (2009: pp. 688–694).

3.1.2 *The Hubble parameter and constant*

Let's take a galaxy at a distance $x = a(t)r$ at rest locally. For a straightforward interpretation, we can assume it is not too far. Then, we can differentiate according to time to get its recession velocity as a function of distance:

$$v = \dot{x} = \dot{a}r = \frac{\dot{a}}{a}x = H(t)x. \tag{3.8}$$

The velocities, hence redshifts, of galaxies, are proportional to their distance. Edwin Hubble and Vesto Slipher found this observationally, and the quantity $H = \dot{a}/a$ is the Hubble parameter. The Hubble parameter today $H(t = 0) = H_0$ is the Hubble constant. It is a supremely important *cosmological parameter*, distilling cosmological models into a few numbers.

The unit of the Hubble constant is $1/\text{time}$, but traditionally it is expressed as $H_0 = 100h\,\text{km/s/Mpc}$, and $h \simeq 0.71 \pm 0.04$ is the dimensionless Hubble constant. Measurements of h by different groups have tended to disagree with each other over the history of cosmology, then these tensions disappeared with improved observations. We are in another cycle of Hubble tension *between early* ($h \simeq 0.68$) measurements from cosmic microwave background (CMB) and more direct *late* ($h \simeq 0.74$) measurements based on supernovae. The disagreement is at about 5σ statistical significance. Only time will tell if the cause is observational issues or the holy grail of *new physics*.

ARIEL: The units of the Hubble constant are weird.
　　　　[*Writes on the blackboard*]
CALLISTUS: Astronomers are a conservative bunch –
RHEA: Don't get me started...
CALLISTUS: Quid novi?[4] It makes <u>some</u> sense though. When the measurements were much less accurate than today, the little h helped compare simulations and observations, scaling them to the same baseline.
ARIEL: But km/s and Mpc? Why is the km not cancelled by the Mpc? They are both distances.
CALLISTUS: Astronomers measure redshifts cz in km/s as a recession velocity. Dividing with the Hubble constant directly gives the distance in Mpc over h.

---

[4]What's new?

RHEA: Moreover, you can cancel the distance units to get the inverse of the Hubble time, $9.78h^{-1}$Gy. A good rule of thumb to remember is that a year is $\pi \times 10^7$s.

ARIEL: If I plug in $h = 0.7$, I get the age of the universe!

CALLISTUS: Right, and you get the Hubble distance by multiplying the Hubble time with the speed of light. [*Calculating*]

ARIEL: $3000h^{-1}$ Mpc, the radius of the universe! But I don't understand why we still use h when the observations are so precise.

CALLISTUS: It's... still easier to compare with old papers...

RHEA: Nobody reads them. We should change.

CALLISTUS: Some now use h_{70} which renormalises the Hubble constant to 70 km/s/Mpc.

ARIEL: Eww.

CALLISTUS: An abomination.

RHEA: Damn You, Little h^5! Just quoting a paper.

3.1.3 *The Friedmann equation*

According to Eq. (3.3), to understand the expansion of the universe, we need to explore the dynamics of the expansion rate, i.e. find an equation for $a(t)$. In general relativity (GR), we would start with the metric and plug it into Einstein's equation. This is a relatively easy but still tedious task (perhaps with the help of computer algebra). Therefore, we omit it here. More insight can be gained from the Newtonian derivation.

Let's consider a Newtonian homogeneous universe uniformly filled with matter of density $\rho(t)$. Following our metric and the cosmological principle, a homogeneous and isotropic expansion means that the distance between any two points will scale with $a(t)$. Due to homogeneity, we can imagine cutting out a sphere of radius $R(t) = R_0 a(t)$ around an arbitrary point O. In Newtonian theory, the net gravitational net force of the outside world is zero. This follows from the *Shell theorem*, which states that the net gravitational force inside a thin shell is zero. Then if we decompose the outside universe into thin shells, the sum of the zeros from each shell add up to zero inside the sphere of radius $R(t)$.

It is relatively easy to prove the Shell theorem with direct calculation. Still, thinking about the forces on an inside point from a small

---

[5]See Croton (2013) for history and more details.

solid angle Ω in two opposite directions is more illuminating. If the point is at a distance r_1 from a section of the shell corresponding to Ω in one direction and r_2 from the other, the net force on the point will be zero: this follows from the fact that Newton's gravitational law scales as $\propto 1/r_i^2$ and the total shell mass within the solid angle is $\propto r^2$. Note the similarity of the argument with *Olber's* paradox.

The other aspect of the Shell theorem is that the gravitational force outside a thin shell is the same as if the total mass of the shell is concentrated in the centre. This follows from Gauss's law, which states that the integral of the gravitational force perpendicular to a surface is proportional to the mass inside. Applying Gauss's law to any shell larger than the original one readily gives the Shell theorem for the outside.

If the force within our sphere is zero, we can forget about the universe and consider the original homogeneous sphere of radius $R(t)$. A unit point mass placed on the surface of our sphere will experience a force:

$$\ddot{R} = -\frac{GM}{R^2}. \tag{3.9}$$

Since the total mass of our sphere is constant, we can integrate this to get an energy balance equation

$$\frac{1}{2}\dot{R}^2 = \frac{GM}{R} + C, \tag{3.10}$$

where C is a constant, the total energy of our unit mass. As foreshadowed earlier, zero curvature, supported by most observations, means $C = 0$. Finally, we obtain the Friedmann equation by substituting the density $M = \frac{4\pi}{3}\rho R^3$ and cancelling R_0 the arbitrary comoving size of our sphere to obtain

$$H(t)^2 \equiv \left(\frac{\dot{a}}{a}\right)^2 = \frac{8\pi G}{3}\rho(t) - \frac{k}{R_0^2 a^2} \tag{3.11}$$

the Friedmann equation with $-2C = k/(R_0^2 a^2)$ the curvature term. With some foresight, we can choose $k = \pm 1$, in which case R_0 is the comoving curvature radius of the universe. The rigorous derivation of this term is from the curved generalisation of the metric of Eq. (3.3). If $c = 1$ units are not used, $k \to kc^2$.

The Friedmann equation is the foundational equation of cosmological dynamics, a differential equation in terms of the universe's energy density ρ. We can solve it as soon as we know how the energy density scales with the expansion rate.

If we substitute Eq. (3.3) into Einstein's equation, we will get two more equations: one related to energy conservation of the cosmic fluid energy density (and for connoisseurs, the Bianchi identities) and one related to acceleration. The three equations are not independent: we can derive the second equation from thermodynamics and the third from combining the first two.

3.1.4 *Energy conservation and acceleration*

Without resorting to GR, we can obtain the equation corresponding to energy conservation from the thermodynamics of the source density filling up the homogeneous universe. Usually, we assume that the universe's expansion is adiabatic (although some theories drop this assumption, assuming *bulk viscosity*). Therefore, during expansion, the change in entropy, $dS = E + pdV = 0$. From the first law of thermodynamics with $E = \rho V$ (remember $c = 1$!) in a unit volume $V = a^3$,

$$\dot{\rho} + 3\frac{\dot{a}}{a}(\rho + p) = 0. \tag{3.12}$$

We can differentiate Eq. (3.11) and use the above equation to obtain

$$\frac{\ddot{a}}{a} = -\frac{4\pi G}{3}(\rho + 3p). \tag{3.13}$$

This equation differs slightly from our starting Eq. (3.9): from M, we would get only the first term above the missing $3p$ term. If you know relativity, you will recognise that $\rho(t) + 3p$ is the trace of the energy–momentum tensor. The three equations (3.11), (3.12), and (3.13) are correct, and we could have derived them from plugging Eq. (3.3) into Einstein's equation. We can interpret the first two equations as energy conservation, the first balancing the kinetic and potential energy of a test particle and the second involving the local energy density and pressure of a fluid filling up the universe.

3.1.5 *Equation of state*

We have two independent differential equations of (3.11)–(3.13), and three variables ρ, p, and a. To solve the dynamics, we need an additional equation, which is usually the equation of state, $p = p(\rho) = w\rho$. The second equality encompasses almost all interesting cases: non-relativistic matter $w = 0$, ultra-relativistic gas or radiation $w = 1/3$, cosmological constant $w = -1$, and anything in between.

For a low-temperature classical gas, when the particle velocities are $v \ll 1$, we expect the pressure to be negligible from the ideal gas law, therefore $w \simeq 0$. We expect the same for dark matter (DM) that does not interact. Radiation is ultra-relativistic; its equation of state is $p = \rho/3$. Thus, for any ordinary matter, we expect $0 \leq w \leq 1/3$. The sound speed from fluid mechanics is $c_s = \sqrt{w}c$. Therefore, we expect $w < 1$ for any form of matter.

Dark energy (DE) is an exotic form of energy that can have a wider range of ws. The cosmological constant Λ has negative pressure $p = -\rho$. The DE equation of state can change with the expansion rate. A popular parametrisation of the dynamical DE is $w = w_0 + w_a(1 - a)$ with $w_a \neq 0$ weakly supported by the recent DESI results. We will see that, formally, curvature behaves as if it had an equation of state with $w = -1/3$. Finally, a truly exotic form of energy, phantom energy, would have negative kinetic energy with $w < -1$.

Equation (3.12) can be solved analytically for an arbitrary constant w. Substituting $p = w\rho$ and cancelling dt yield

$$\frac{d\rho}{\rho} = -3(1 + w)\frac{da}{a}. \tag{3.14}$$

The solution is

$$\rho =\simeq a^{-3(1+w)}, \tag{3.15}$$

$$\rho =\simeq a^{-3}, \qquad w = 0 \text{ (M)}, \tag{3.16}$$

$$\rho =\simeq a^{-4}, \qquad w = \frac{1}{3}\text{(R)}, \tag{3.17}$$

$$\rho = \text{const.}, \qquad w = -1 \text{ (}\Lambda\text{)}. \tag{3.18}$$

where M, R, and Λ denote matter, radiation, and cosmological constant. If there is more than one component in our universe, each component follows its own law. For instance, any ordinary or dark matter will always scale with one over the volume.

For some time, only one component will dominate the universe. In this case, one can readily insert these results into the Friedmann equation

$$\dot{a}^2 \simeq \rho a^2 \simeq a^{-3w-1} \tag{3.19}$$

to get analytic scaling solutions when $w \neq -1$:

$$a =\simeq t^{2/(3+3w)}, \qquad w \neq 1, \tag{3.20}$$

$$a =\simeq t^{2/3}, \qquad w = 0 \text{ (M)}, \tag{3.21}$$

$$a =\simeq t^{1/2}, \qquad w = \frac{1}{3} \text{ (R)}, \tag{3.22}$$

$$a = \exp\left(H_0(t - t_0)\right), \qquad w = -1 \text{ (}\Lambda\text{)}, \tag{3.23}$$

where the last line for $w = -1$ requires a separate calculation from the equation $\dot{a} = H_0 a$ obtained from $\rho = \text{const.}$

At this point, we can calculate the Hubble constant by assuming $a = (t/t_0)^{2/(3+3w)}$, which is consistent with our conventions that today $a(t_0) = 1$:

$$H_0 = \left(\frac{\dot{a}}{a}\right)_0 = \frac{2}{3 + 3w} t_0^{-1}. \tag{3.24}$$

The age of a universe filled with DM or ordinary matter $w = 0$ during most of its evolution is $t_0 = 2/(3H_0)$. With $h \simeq 0.7$, this would give ages slightly above 9 Gy below the age of our Galaxy's oldest stars. This is an early hint of DE, which changes the universe's expansion history such that its age becomes 13.7 Gy, consistent with observations. To understand that, we need to be able to solve the Friedmann equation for more complex universes with more than one ingredient.

When $w \neq 0$, we can combine the above equations and use (3.24) to show that $\rho = \rho_0(t_0/t)^2$ regardless of w, or more precisely,

$$\rho = \frac{3H_0^2}{8\pi G}\left(\frac{2}{3 + 3w}\right)^2 H_0^2 t^{-2} = \frac{1}{6\pi G(1 + w)^2 t^2} \tag{3.25}$$

3.1.6   *Multi-component universes*

We have only one Universe, and it has more than one ingredient. Typically, one of them, such as radiation, or matter, dominated during long periods. During rare transitions, more than one component was important. Our universe is best described by the ΛCDM model, which posits that today, matter and cosmological constant have similar contributions.

The Friedmann equation (3.11) at present time divided by H_0^2 is

$$1 = \frac{8\pi G}{3H_0^2}\rho_0 - \frac{k}{R_0^2 H_0^2} = \Omega_0 - \frac{k}{R_0^2 H_0^2}, \qquad (3.26)$$

where the second line defines Ω_0 the dimensionless matter density today. In units of critical density $3H_0^2/(8\pi G)$, $\Omega_0 = 1$ corresponds to flat space, $k = 0$. The curvature term can be expressed as $-\Omega_k \equiv \Omega_0 - 1$. As a function of the expansion rate, it will behave as $\Omega_k a^{-2}$. Thus, while curvature corresponds to the (im)balance of energy density, formally, it behaves as an energy density with $w = -1/3$. Note that $\rho \simeq 1/t^2$ for any $w \neq -1$.

With the above definition, the Friedmann equation (3.11) has an intuitive form

$$\frac{H^2(a)}{H_0^2} = \Omega_m a^{-3} + \Omega_r a^{-4} + \Omega_w a^{-3(1+w)} + \cdots + \Omega_k a^{-2}, \qquad (3.27)$$

where m, r, and w stands for matter, radiation, and DE, and the $\cdots$ represent any additional energy density for any more extended model. The curvature term is conveniently expressed as if it were an energy density with $\Omega_k = -(\Omega_0 - 1)$, and total energy density today $\Omega_0 = \Omega_m + \Omega_r + \Omega_w + \cdot$ is the sum of all individual energy densities. For most of this book, we assume $\Omega_k = 0$ consistently with most measurements. The most important special case for DE is the cosmological constant $w = -1$, yielding $\Omega_\Lambda = \text{const.}$ independent of the expansion rate a.

The above explicit expression for $H(a)$ is also a differential equation in $\dot{a}$, since $H(a) = \dot{a}/a$. Once we specify the ingredients of the universe, our cosmological model, we can solve Eq. (3.27) using the initial (or final) condition $a_0 = a(t_0) = 1$.

The standard ΛCDM concordance model contains matter (including DM, and baryons, or ordinary matter both behaving the same

Table 3.1.   Cosmological parameters from Planck measurements.

Parameter	Without curvature	With curvature
Hubble constant (H_0)	67.66 ± 0.42 km/s/Mpc	$67.3^{+2.1}_{-2.3}$ km/s/Mpc
Baryon density ($\Omega_b h^2$)	0.02237 ± 0.00015	0.02249 ± 0.00016
CDM density ($\Omega_c h^2$)	0.1200 ± 0.0012	0.1185 ± 0.0015
Curvature density (Ω_k)	Not applicable	0.0007 ± 0.0037

way), radiation, and cosmological constant. At the time of preparation of this manuscript, all measurements are statistically consistent with the concordance model, although some favour a more complex DE.

Table 3.1 illustrates the evidence for a flat universe using Planck 2018 data: a fit for the curvature parameter produces a result consistent with zero. Therefore, in what follows, we will focus on the flat concordance model, but keep in mind that the model space might open up in the future with more precise measurements. The radiation density $\Omega_r h^2 = 4.15 \times 10^{-5}$. As we will see later, this can be derived from the cosmic microwave temperature.

3.1.7   *Flatness problem*

From the Friedmann equation (3.27), if $\Omega_k = 0$, i.e. the universe is flat today, it will always remain flat within the framework of the Friedmann model.

On the contrary, the universe has positive or negative curvature if $\Omega_k \neq 0$ today. According to Table 3.1, measurements constrain such a curvature to be tiny today, contributing at most a few times 10^{-4}. At the time of equality, when radiation and matter had equal contributions, the universe was about $a \simeq 1/3000$ times smaller than today. Since the curvature contribution scales with $1/a^2$, while matter scales with $1/a^3$, the curvature contribution can't be more than 10^{-7}. Before equality, the universe's energy is dominated by radiation, with its energy density scaling as $1/a^4$. In the past, when the universe was $a \simeq 10^{-30}$ smaller around or right after the Big Bang, further decreasing the curvature contribution to $\simeq 10^{-67}$. This number is so tiny that it is difficult to explain if it is not zero. This is the *flatness problem*, and we will see that inflation provides a reasonable solution.[6] For now, it should be clear that curvature is either zero or close enough to zero that we do not need to deal with it. Thus, as interesting and beautiful as they are, we omit the general metrics for closed (compact like a sphere) and open (infinite but curved like a saddle) universes.

3.1.8   *Distances*

Distances are tricky in an expanding universe even if we use the flat metric of Eq. (3.3).

As a light ray propagates to us from an object, the universe keeps expanding while the light ray passes through. As noted earlier in Eq. (3.4), for light rays $ds = 0$, therefore

$$dr = \frac{dt}{a} = \frac{dt}{da}\frac{da}{a} = \frac{da}{a^2 H(a)} = -\frac{dz}{H(z)}. \tag{3.28}$$

Once the ingredients of the universe are specified, we can use Eq. (3.27) to specify $H(a) \equiv \dot{a}/a$ and integrate over a to obtain the *current proper* distance associated with an observed light ray starting at a particular expansion factor a_e and observed today at $a_o = 1$. We can associate a time with the current proper distance integrating

$$dt = \frac{dt}{da}da = \frac{da}{aH(a)} = -\frac{dz}{(1+z)H(z)} \tag{3.29}$$

over the same interval. These equations can be equivalently written in terms of the redshift using $a = 1/(a + z)$ and $-dz = da/a^2$, where the second identity follows from the first. The last equations above follow from $H(z) = H(a)|_{a=1/(1+z)}$. We included expressions as a function of z for convenience. The numerical calculations can be performed using either form.

The *angular diameter distance*, D_A, is the effective radius to multiply the angular extent of, say, a galaxy to obtain its physical size. In a flat universe with metric, Eq. (3.3), we can use Euclidean geometry (Euclidean is essentially a synonym for flat). Thus, the current proper distance to an object multiplied by its size in radians would

---

[6]It will be helpful to reread this section after we better understand the Big Bang, radiation–matter equality, and inflation.

give us the object's present (comoving) extent. However, at the time of emission, the universe was smaller by $a = 1/(1 + z)$, therefore

$$D_A(z) = \frac{r}{1 + z}, \tag{3.30}$$

giving the physical extent of the galaxy at the time of emission.

The *luminosity distance*, D_L, is defined such that standard formula between flux f and luminosity L would hold:

$$f = \frac{L}{4\pi D_L^2} = \frac{L}{4\pi r^2(1 + z)^2}. \tag{3.31}$$

The second equation follows from the change of apparent flux due to the expansion of the universe: one factor of $1 + z$ comes from the redshifted energy of individual photons and another from the time dilation of the average emission time between photons. If we want to know the original luminosity, we must consider these factors, which effectively means $D_L = r(1 + z)$.

Analytical integration of the above distances to a particular redshift or time is possible for specific pedagogical universes, but our universe is complex enough, cf. Table 3.1, that distances are always calculated numerically in practice. Fortunately, many packages exist for this purpose, such as `astropy` in `python`.

3.1.9 *Local Universe*

The time dependence of $a(t)$ determines the expansion history for the Local Universe in terms of phenomenological parameters. Let us expand $a(t)$ into a second-order Taylor series around the present time t_0:

$$a(t) = 1 + H_0(t - t_0) - \frac{1}{2}q_0 H_0^2(t - t_0)^2, \tag{3.32}$$

where we used $a(t_0) = 1$, $H_0 = (\dot{a}/a)|_0 = \dot{a}|_0$, and the deceleration parameter defined as

$$q_0 = -\left.\frac{\ddot{a}a}{\dot{a}^2}\right|_0. \tag{3.33}$$

Indeed, plugging the definition of q_0 into Eq. (3.13) and summing up the components of the universe assuming a simple equation of state $\rho_i = w_i p$ for each,

$$q_0 = \frac{1}{2} \sum \Omega_i (1 + 3w_i) \simeq \frac{1}{2} \Omega_m - \Omega_\Lambda. \qquad (3.34)$$

The second (approximate) equality is true since radiation has a negligible contribution in the Local Universe, and we used $w_m = 0$ and $w_\Lambda = -1$. The above is the reason for the strange negative sign in the definition of the deceleration parameter. In the absence of DE, measuring q_0 would determine (half of) the total matter content of the universe. The negative sign of the measured deceleration parameter means that the expansion of our universe accelerates. Imagine if Newton found that the apple pitched up would accelerate away from Earth. This is precisely what cosmologists found. The surprise was a smoking gun of DE worth a Nobel.

Given the above expansion, we can express cosmological distances using the two Taylor expansion parameters, H_0, and q_0. For instance, the present-day proper distance

$$r = \int_t^{t_0} \frac{dt}{a(t)} = \int_t^{t_0} dt(1 + z)$$

$$= \int_t^{t_0} (1 - H_0(t - t_0)) + (1 + \frac{1}{2} q_0) H_0^2 (t - t_0)^2 + \cdots$$

$$= t_0 - t + \frac{1}{2} H_0 (t - t_0)^2 + \mathcal{O}(\Delta t)^3. \qquad (3.35)$$

This is the distance to an object sending a light signal to the observer at time t. This time is not a measurable quantity; it would be more interesting to express the distance using the measurable redshift z. For that, we need to express the time difference $\Delta t = t_0 - t$ as a function of z corresponding to the inversion of a Taylor series. To second order

$$\Delta t = b_1 z + b_2 z^2 + \cdots$$

$$= b_1 \left(-H_0 \Delta t + \left(1 + \frac{1}{2} q_0 \right) H_0^2 \Delta^2 t + \cdots \right) + b_2 H_0^2 \Delta^2 t + \cdots$$

$$\qquad (3.36)$$

On the second line, both sides are a Taylor in Δt (with foresight, we assumed $b_0 = 0$). Collecting terms on the RHS, $b_1 = -1/H_0$ and

$b_2 = (1+\frac{1}{2}q_0)/H_0$, thus we inverted the Taylor series. Plugging $\Delta t(z)$ into Eq. (3.36), we obtain an expression of the present-day proper distance in terms of the observable z to second order:

$$r = \frac{z}{H_0}\left(1 - \frac{1+q_0}{2}z\right) + \mathcal{O}(z^3). \tag{3.37}$$

The linear term is precisely what we expect. If $q_0 > -1$, the correction is negative, i.e. distances are closer than the linear expression without the effect of q_0.

The angular diameter distance is $D_A = r/(1+z) \simeq r(1-z)$, yielding

$$D_A = \frac{z}{H_0}\left(1 - \frac{3+q_0}{2}z\right). \tag{3.38}$$

Finally, similar arguments for the luminosity distance $D_L = r(1+z)$ yield

$$D_L = \frac{z}{H_0}\left(1 + \frac{1-q_0}{2}z\right). \tag{3.39}$$

The above two equations can be used for distance measurements using objects of known size or luminosity, a.k.a *standard rods* or *standard candles*.

Astronomers define the magnitude of an object as proportional to the logarithm of its luminosity $m = -2.5\log_{10} f/f_0$. The constants f_0 2.5 and the minus sign (!) are arbitrary and determined by (somewhat annoying) tradition. The absolute magnitude M is the magnitude of an object at 10 pc[7] From these definitions, the distance modulus of a standard candle

$$m - M = 5\log_{10}(D_L/M\text{pc}) + 25$$

$$\simeq 5\left(\log_{10} 3000 - \log_{10} h^{-1} + \log_{10} z\right.$$

$$\left. + \log_{10}\left(1 + \frac{1-q_0}{2}z\right)\right) + 25$$

$$= 43.22 - 5\log_{10}(h/0.68) + 5\log_{10} z + 1.086(1 - q_0)z. \tag{3.40}$$

---

[7] 1 pc is where one astronomical unit, the Earth–Sun distance, subtends an angle of 1 arcsecond. It is approximately 3.26 lightyears.

The second equation follows from substituting Eq. (3.38) for the luminosity distance, and the final line collects constants and expands the logarithm for small redshifts. The most ubiquitous standard candles are type Ia supernovae. The last term in the last equation allowed the Nobel-prize-winning discovery of the universe's acceleration due to dark energy from a collection of supernovae observations.

3.2   Inflation and Big Bang

If the universe expanded, it must have been denser and, therefore, hotter in the past. Since $\rho \simeq 1/t^2$ (we can safely neglect the tiny contribution of DE in the early universe), the density will become infinite at the beginning of time. While infinity is an abstraction not likely to be realised by nature, it is clear that the universe started from an extremely hot and dense initial state and expanded from there. This is not unlike an explosion, colloquially known as the "Big Bang." The apocryphal story is that Fred Hoyle, the inventor of the steady-state model that had since fallen out of favour, wanted to discredit the now standard model with children's language. Astronomers liked it, and the hot Big Bang model became famous instead. It is still worth noting that the steady-state model corresponds to a continuous creation, where particles would be constantly created out of nothing. In contrast, at least formally, the Big Bang model corresponds to an instantaneous creation of infinite energy density. They both appear to be somewhat strange ideas until we get used to them (the story of cosmology). The hot Big Bang model has become standard among cosmologists, but the steady-state model-inspired cyclic models are still in contention.

Quantum gravity resolves the infinity at the Big Bang. We expect that known physics breaks down when the de Broglie wavelength becomes smaller than the Schwarzschild radius: this is the quantum gravity regime, where quantum fluctuations produce black holes, and, conversely, black holes behave in a quantum fashion. Thus, the transition occurs when $h/mc \simeq 2Gm/c^2$, which defines the Planck mass (omitting a π) as

$$m_P = \sqrt{\frac{\hbar c}{G}}, \tag{3.41}$$

where G is Newton's gravitational constant, c is the speed of light ($c = 1$ in our natural units), and $\hbar = h/2\pi$, the reduced Planck's constant. We will use $\hbar = 1$ units to avoid confusion with the unfortunate notation of $h = H_0/100$ for Hubble's constant. The Planck mass is an energy scale of $m_P \simeq 10^{19}$ GeV, and Newton's constant in these units is $G = 1/m_P^2$, length and time are inverse energy $1/m_P$. In more conventional units, we can define Planck length and time $l_P = 10^{-35}$ m and $t_P = 10^{-43}$ s. Thus, the Big Bang starts around $t \simeq 10^{-43}$ s or energy/temperature of 10^{19} GeV.[8]

At the time of writing this book, we do not understand quantum gravity. We know that conventional physics (quantum field theory in a background metric) will break down simply because we cannot define a smooth (coordinate) background where we can write our usual equations. Conversely, we expect known physics to work down to that time extremely close to the formal infinity that defines the Big Bang. Since space–time makes no sense at that point, time is probably not a useful variable "before" Planck time. In other words, time is not fundamental but an *emergent* variable, not unlike temperature. Perhaps quantum mechanics should be understood without time and space (De Bianchi & Szapudi, 2025), which would negate the possibility of the classical notion of quantum gravity.

3.2.1 *Inflation*

Inflation is the dominant theory of the early universe simply because it is the only one that explains several observations. We have seen the flatness problem: the universe appears to be critical ($\Omega \simeq 1$) today; therefore, it should have been even more critical earlier.

A related issue is the generation of initial conditions that later give rise to the large-scale structure of the universe: what generated the tiny fluctuations that grew by gravitational amplification? In addition, the homogeneity of these initial fluctuations generates the horizon problem: we will see that the opposite lines of sights of

---

[8]With a slight abuse of notation, $1eV \simeq 1.1 \times 10^4$ K, and we use energy scales to express temperature and time in the early universe, remembering that $1\,\mathrm{MeV}$ is at $1\,\mathrm{s}$.

the cosmic microwave background have the same temperature within one part in 10^5. Typically, we would think there had to be a thermalisation process that equalised the temperatures. Yet, the opposite lines of sight are not in causal contact. Therefore, naively, no causal process could equalise their temperatures.

Finally, a lesser-known problem arising from Big Bang cosmology is the monopole problem: phase transitions in the early universe should produce at least one monopole per horizon volume. Since the horizon in the early universe was tiny, we should see many of these exotic particles. Yet, we see none.

Inflation kills three of these birds with one stone: it explains the flatness, horizon, and monopole problems with an extremely fast (faster than the speed of light) expansion of the early universe. If you blow up a ball (the early universe) to a vast size, the part we observe will appear flat. The original small ball could have been in causal contact and thermalised before the expansion, and if it were small enough, it would contain one or no monopoles. In addition, we will see that inflationary theories provide initial conditions in the form of small quantum fluctuations consistent with observations. No other known theory can resolve these four problems so elegantly.

The fastest way any interaction can proceed is the speed of light. At any time t after the Big Bang corresponding to an expansion factor $a(t)$, there is a causal radius around a point it can influence causally. This is the particle horizon, and according to Eq. (3.28), we can calculate the horizon r as

$$r = \int_0^t \frac{dt}{a} = \int_0^a \frac{da}{a^2 H(a)} = \int_0^a \frac{d\log a}{aH(a)}. \tag{3.42}$$

In this equation, $1/aH(a)$ is the *comoving Hubble horizon*. It always grows for ordinary matter; thus, r will also increase. Conversely, the horizon r will become smaller in the past, and at any point in the universe's history, two points outside the horizon could not have been communicated. This is why the perfect homogeneity of the cosmic microwave background is a puzzle: the opposite lines of sights are outside the horizon now, so they must have been outside the horizon ever; how could they have the same temperature?

The answer is that there must have been a time when the universe was dominated by something other than ordinary matter. We have seen that DE causes exponential expansion. If something like DE dominated the early universe, the exponential expansion could cause the comoving Hubble horizon to decrease, and an area that was once in causal contact could come out of causal contact. This is the future of our universe; the comoving radius we can explore will close in, and if inflation happened, it has occurred in the early universe.

For the Hubble horizon to decrease, we need the inverse of it to grow $d/dt(aH(a)) = \ddot{a} > 0$, which corresponds to negative deceleration (acceleration). According to Eq. (3.13),

$$\frac{\ddot{a}}{a} = -\frac{4\pi G}{3}(\rho + 3p) = -\frac{4\pi G}{3}\rho(1 + 3w) > 0. \qquad (3.43)$$

An exotic material with $w < -1/3$, often called the *inflaton field*, would be needed. The standard solution is assuming a scalar field. Particle physics has many examples of scalar fields, such as the Higgs field, and scalar fields can be set up to provide the required negative pressure. Nevertheless, we don't know precisely the inflaton field, and learning about its properties to the extent that we can identify it with a known or well-motivated scalar field that would be part of the particle physics lore is a principal goal of observational cosmology.

The energy density and pressure of a scalar field in curved space–time correspond to the diagonal elements of its energy–momentum tensor. Without repeating the calculation here, we can easily understand the physical motivation for the energy density of a spatially homogeneous scalar field ϕ^0:

$$\rho = \frac{1}{2}\left(\frac{d\phi^0}{dt}\right)^2 + V(\phi^0). \qquad (3.44)$$

The equation is analogous to the sum of kinetic and potential energy in classical mechanics. The potential $V(\phi)$ usually contains a *mass term* $m^2\phi^2/2$ and possibly other terms; the simplest textbook example would be b $V \propto \phi^4$. The pressure in the same theory would be

$$p = \frac{1}{2}\left(\frac{d\phi^0}{dt}\right)^2 - V(\phi^0). \qquad (3.45)$$

The opposite sign of V moves the field towards potential minima. These equations predict that if $d\phi^0/dt \simeq 0$, $p \simeq -\rho$, or $w = -1$. This class of models, where the field initially sits high above in the potential with no kinetic energy, are called *slow-roll* models. Generically, for a wide range of assumptions on the potential, they will slowly roll down towards the minimum. Initially, by analogy with our earlier calculations for DE, they will produce an exponential expansion, but eventually, $w \simeq -1$ will no longer be true, and the universe exits inflation. While the details of such *reheating* are complex and hotly debated, the consensus is that slow-roll models are the simplest contenders in the vast landscape of possible inflationary models. Reviewing all the possibilities would be beyond the scope of our treatise. Therefore, we will restrict ourselves to slow-roll models.

In our model $\rho \simeq V(\phi^0)$, i.e. our field starts in a false vacuum. This constant energy density causes exponential expansion $a(t) = a_e \exp H(t - t_e)$, where the subscript e refers to the end of inflation. The horizon reads

$$r_{\text{prim}} = \int \frac{dt}{a} = \int_{t_i}^{t_e} \frac{dt}{a_e} e^{-H(t-t_e)} = \frac{1}{Ha_e}(e^{H(t_e-t_b)} - 1) \qquad (3.46)$$

After inflation, we can calculate the horizon integrating $r = \int_{t_e}^{t} dt/a$, but that does not contain the primordial horizon added by the exponential expansion. The word *inflation* comes from an analogy with the exponential decrease in the value of money during monetary inflation. The inflation potential inflates away the horizon, curvature, and monopoles. How much inflation is enough to solve the horizon problem?

The common assumption is that inflation ended around $T_e \simeq 10^{15}$ GeV. The present temperature (as measured by the cosmic microwave background) is $T_0 = 3$ K $\simeq 0.24$ MeV, therefore $a_e \simeq T_0/T_e \simeq 10^{-28}$. If we crudely assume that the universe has always been dominated by radiation, i.e. $H \propto a^{-2}$, then the ratio of Hubble horizons is

$$\frac{a_0 H_0}{a_e H_e} = a_e =\simeq 10^{-28}. \qquad (3.47)$$

We need to increase the comoving radius by this amount during inflation, so the opposite ends of our horizon would be in causal contact in the past. We usually express this as e^{64} or 64 e-foldings. This calculation changes slightly because the late universe is dominated by dark matter instead of radiation; we need at least $H(t_e - t_b) > 60$ or 60 e-foldings. This is a constraint for any specific inflationary model. Interestingly, at the end of inflation, our observable was about $\simeq 1$ cm (or less than half an inch).

Assuming homogeneity, we can calculate the time evolution of the inflation field by inserting ρ from Eq. (3.44) into Eq. (3.11) and differentiating

$$\ddot{\phi} + 3H\dot{\phi} + V'(\phi) = 0. \tag{3.48}$$

This equation is equivalent to the usual acceleration equals gradient of the potential, except the universe's expansion causes a "Hubble drag" represented by the middle term. A specific inflationary model corresponds to picking a $V(\phi)$ function. We can characterise the most popular slow-roll models generically with dimensionless parameters

$$\epsilon \equiv \frac{d}{dt}\left(\frac{1}{H}\right) = \frac{M_P^2}{2}\left(\frac{V'}{V}\right)^2, \tag{3.49}$$

$$\eta \equiv \frac{\ddot{\phi}}{H\dot{\phi}} = M_P^2 \frac{V''}{V}. \tag{3.50}$$

The first parameter characterises the slow change in H, and the second one is small when the second derivative of the inflaton field is smaller than its (already small) first derivative. One definition of the slow-roll models is $\epsilon < 1$. Neglecting the second derivative in Eq. (3.48) and using reduced Planck mass $M_P^2 = 3/8\pi G$, the second equation translates these parameters into requirements for the potential.

As foreshadowed, inflation solves a third problem beyond the horizon and monopole problems: it provides initial conditions from which the large-scale structure of the cosmological density field grows. These fluctuations are quantum fluctuations, and for sufficient rigour, they need quantum field theory calculations in a curved background space–time. Nevertheless, we can write the equivalent of Eq. (3.48) for the small fluctuations $\delta\phi$:

$$\delta\ddot{\phi} + 3H\dot{\phi} + k^2\phi, \tag{3.51}$$

where the last term is the Fourier transform of $\nabla^2\delta\phi$ which was neglected from Eq. (3.48) since ϕ is assumed to be homogenous. This is an equation of harmonic oscillator[9] with the Hubble drag that can be quantised. After some non-trivial calculations, for large scales or small k, the quantum fluctuations around the zero point behave as

$$P(k) \simeq \frac{1}{k^3}, \tag{3.52}$$

with small deviations due to the slow-roll parameters. Using this general behaviour, the traditional parametrisation for the fluctuations of the gravitational potential Φ (scalar modes) and gravitational waves T (a.k.a. tensor modes) is

$$P_\Phi(k) = \frac{50\pi}{9k^3} \left(\frac{k}{H_0} \right)^{n-1} \delta_H^2 \left(\frac{\Omega_m}{D_1(a=1)} \right)^2, \tag{3.53}$$

$$P_T(k) = A_T k^{n_T - 3}, \tag{3.54}$$

where δ_H and A_T characterise the amplitudes of initial fluctuations, $\frac{\Omega_m}{D_1(a=1)}$ describes the growth of fluctuations, and n and n_T are the slopes of scalar and tensor fluctuations, with defaults defined as $n = 1, n_t = 0$. More detailed calculations show that the ratio of these power spectra also should be ϵ, and the slow-roll parameters will control slight deviations of the slope as

$$n = 1 - 4\epsilon - 2\eta, \tag{3.55}$$

$$n_T = 2\epsilon. \tag{3.56}$$

The ratio of the amplitudes and the slopes provide consistency conditions for the class of slow-roll models that can be contrasted with future measurements. We have a good idea of the scalar power, and future CMB polarisation experiments will constrain the much smaller tensor power spectrum. If the above conditions are violated, the class of simple slow-roll models will be rejected.

The above ideas just scratch the surface of inflationary science. The correct calculation within the framework of general relativity incorporates the ambiguity of a general coordinate system: the coordinate system, or *gauge*, needs to be fixed appropriately. The quoted equation is valid in a "spatially flat" gauge, where the clever choice of coordinates eliminates small curvature fluctuations. Since we know

---

[9]Written in spatially flat gauge.

quantum theory in flat space, we can safely calculate quantum fluctuations, but the results must be interpreted in another coordinate
system. The transition between gauges is a bit technical for the
present. Still, the results are pretty intuitive: the fluctuation level
in each mode is constant since the number of modes increases as k^3.
This behaviour was foreseen by Zeldovich, motivated by naturalness,
before the invention of inflation. Therefore, such initial fluctuations
are named a Zeldovich scale-invariant spectrum.

While inflation explains a lot and, at the moment, has no serious
competitor, it has a "graceful exit" problem. It is hard to imagine inflation ending in concert at all causally disconnected places. If
inflation stops in a causal region, the surroundings will inflate further, rendering the non-inflating region negligible. Inflation got its
name from the exponential rise of bill denominations. The end of
inflation is analogous to exponential growth in capitalism, where the
one with the highest exponent still grows exponentially compared
to the second best, thus demolishing everybody else. This illustrates
why a graceful exit is so hard. If these ideas piqued your interest, you
can read a more in-depth discussion on inflation by Liddle & Lyth
(2000) and Dodelson & Schmidt (2020).

3.3 Thermal history

After the hot Big Bang, the radiation dominated the energy density
due to its steepest a^{-4} dependence. The energy and number density
of thermal (black body) photons is

$$\rho_\gamma = \frac{\pi^2 g_\gamma}{30} T^4, \tag{3.57}$$

$$n_\gamma = \frac{\zeta(3) g_\gamma}{\pi^2} T^3, \tag{3.58}$$

where $g_\gamma = 2$ is the number of spin states (photons are vectors
with spin 1, but the $m = 0$ state is not realised due to its massless nature); $\zeta(3) \simeq 1.202$, the Riemann zeta function. Comparing
the formula with the energy density of radiation, $T \simeq 1/a = 1 + z$,
photon temperature will redshift due to the redshift of each photon
in the ensemble.

These formulae apply to any relativistic (i.e., high temperature) bosons, and they need a trivial modification for fermions:

a multiplication with 7/8 and 3/4, respectively. These equations come from an integration over the Bose and Fermi functions $f = 1/(e^{(E-\mu)/T} \pm 1)$, and the modification comes from the sign difference between the two distributions. Therefore, it is customary to treat the energy density of all species during radiation domination with an effective number of spin states

$$g_* = \sum_{\text{bosons}} g_b + \frac{7}{8} \sum_{\text{fermions}} g_f. \tag{3.59}$$

Equations (3.57) and (3.58) are applicable with g_* replacing g_γ when all the species are in thermal equilibrium with the same temperature. While remaining relativistic, neutrinos drop out of equilibrium with photons at later times. In general, if some species have a different temperature than photons, g_* is modified when applied to entropy as

$$g_{*S} = \sum_{\text{bosons}} g_b \left(\frac{T_b}{T}\right)^3 + \frac{7}{8} \sum_{\text{fermions}} g_f \left(\frac{T_f}{T}\right)^3. \tag{3.60}$$

The entropy density $s = (\rho + p)/T$ is written as

$$s = \frac{4}{3}\frac{\rho}{T} = \frac{2\pi^2 g_*}{45}T^3 \simeq 1.8 g_* n_\gamma. \tag{3.61}$$

The second equation shows that the photon number counts are interchangeable with entropy density as long as g_* is constant. Since the entropy in a comoving volume conserves according to Eq. (3.12), the temperature of the universe scales as

$$T \propto (1+z)g_*^{1/3}. \tag{3.62}$$

When g_* is constant, the temperature scales with the expansion. But the effective number of relativistic species, hence g_*, changes (Table 3.2). This will happen when relativistic species drop out, turning non-relativistic as the universe cools. The universe will not cool according to the expansion rate for a short while since species that drop out dump their entropy into the plasma. In addition, g_{*S} can change if a species stays relativistic but decouples from the rest of the radiation and its temperature changes. The signature case is the 2 K neutrino background that decoupled from photons.

Table 3.2. Effective number of spin states.

Energy	Species	g_*
$\ll$ MeV	γ, ν	$2 + 7/8 \times 6 \times (4/11)^{4/3} \simeq 3.36$
1–100 MeV	above +e, e^+	$2 + 7/8 \times (6 + 2 + 2) \simeq 10.75$
$\gg$ 300 GeV	standard model	106.75

Notes: The calculation for g_* comes entirely from particle physics. The table contains the three most essential ranges in the thermal history of the universe. In the very early universe, above 300 GeV, all standard model particles contribute to the effective number of spin states, such as 8 gluons, $W^\pm$, and Z^0 particles $3\times$ generations of quarks, leptons, Higgs. In the MeV range photons (2), three generations of neutrinos (6), and electrons and positrons $(2 + 2)$ are relativistic. Later, only photons and neutrinos contribute below MeV scales, but neutrinos have a different temperature, hence the $(4/11)^{4/3}$ factor. The 7/8 factors account for the effect of the Fermi distribution for fermions.

Decoupling is ubiquitous in the early universe. For a particular species to be in thermal equilibrium, its reaction rate must be $\Gamma \geq H$. Once the Γ drops below the (inverse) Hubble time, it will drop out from equilibrium and thus decouple. While the detailed calculation of $g_*(T)$ and $g_{*S}(T)$ for the standard model of particle physics and its extensions is beyond our scope, we can consider two generic cases: interactions mediated by massless (e.g., electromagnetic force) or massive (e.g., weak force) gauge bosons. For massless mediation, the cross-section scales as $\sigma \simeq \alpha/T^2$, where α is the gauge coupling strength:

$$\frac{\Gamma}{H} = n\sigma|v| \times m_{pl}T^{-2} \propto \frac{\alpha m_P}{T}, \qquad (3.63)$$

where $v \simeq c = 1$ in our units and $n \simeq T^3$ generically. Once a species is kept in equilibrium through a massless gauge boson, it will not decouple. Early on, around 10^{16} GeV $\simeq 10^{-38}$ s, such an interaction could not keep a thermal equilibrium. Therefore, inflation must end, and reheating must occur after this time. For a massive gauge boson m_X, where $X = W^{\pm,Z}$ for the weak interaction, $\sigma \simeq G_X^2 T^2$, where $G_x \simeq \alpha/m_X^2$. Therefore,

$$\frac{\Gamma}{H} = n\sigma|v| \times m_{pl}T^{-2} \propto \frac{\alpha^2 m_P}{m_X^4}T^3 \simeq \left(\frac{T}{\text{MeV}}\right)^3, \qquad (3.64)$$

where the last equation is for weak interactions. At around 1 MeV, neutrinos will decouple.

When baryon non-conserving processes are slow, n_b/s is conserved. According to the above, $\eta_b = n_b/n_\gamma = 1.8g_{*S}n_b/s$. After electron–positron annihilation, $\eta_b \simeq 7n_b/s$, thus the baryon to photon ratio is interchangeable with entropy. At around MeV temperatures, neutrinos decouple and do not participate in interactions, but they still have the same temperature as other species. After that, electrons and positrons annihilate and dump their entropy into photons. As anticipated, $g_{*S}T^3$ remains constant (it happens fast enough that we can neglect the change in $(1+z)$. Before annihilation, $g_{*S}^b = 2 + 7/8(2+2)$ for photons, electrons and positrons. After annihilation, $g_{*S}^a = 2$, therefore

$$\frac{T_a}{T_b} = \left(\frac{g_{*S}^b}{g_{*S}^a}\right)^{1/3} = \left(\frac{11}{4}\right)^{1/3}, \qquad (3.65)$$

as noted in Table 3.2.

The thermodynamics of the early universe are best described in detail in many books (e.g., the classic book by Kolb & Turner (1990)). We summarised the top-level facts in this section.

3.3.1 *Concise timeline*

After inflation, the universe goes through eras classified through the forces dominating. The changes between these eras are often marked by *spontaneous symmetry breaking* that are supposed to leave a few relics, such as cosmic strings or textures that should be observable today:

- Grand Unified Theory (GUT) era $\simeq 10^{14}$–10^{16} GeV;
- Electroweak era $\simeq 300$ GeV electroweak force decouples from the strong force; mass acquired by the Higgs mechanism;
- quark-hadron phase transition 100–300 MeV–1 GeV, where baryons might have been produced;
- 0.1–1 MeV nucleosynthesis;
- matter–radiation equality $5.5\Omega^2$ eV;
- recombination/decoupling 0.26 eV;

We only discuss the last three stages in more detail since our understanding of the earlier events is still incomplete.

3.3.2 *Cosmological baryogenesis*

Regardless of the details of baryogenesis, Sakharov famously established three conditions for baryogenesis:

- *Baryon number violation*: If B were conserved, the relic abundance of baryons would only amount to $\eta_b = 7 n_b / s \simeq 10^{-20}$, about 9–10 orders below observations.
- *C,CP violation*: If charge conjugation and charge-conjugation parity were conserved, baryons and anti-baryons would be produced by equal numbers. In our observed universe, baryons dominate.
- *Non-equilibrium conditions*: In equilibrium, both baryons and anti-baryons will have the same Fermi distribution, resulting in the same number of baryons and anti-baryons.

Any theory attempting to explain baryogenesis will abide by the above conditions.

3.3.3 *Nucleosynthesis*

Nucleosynthesis, the origin of basic ingredients for ordinary matter that was later processed by stars to become most things we know, including people, famously happened in the first 3 minutes of the universe's history. During that time, isotopes of hydrogen, helium, and some lithium and beryllium were produced. A detailed understanding of this process entails virtually all of nuclear physics and a calculation keeping track of all the elements that can be produced in the hot plasma. When nucleosynthesis is finished because the universe is no longer hot enough for nuclear processes that happen on the MeV scale, only H and He are produced in appreciable levels; everything else is a small fraction of a per cent. Therefore, we can say that if we understand helium, we understand more than 99% (even though the details are nonlinear and complicated). As a starting point, we can orient ourselves using a form of Eq. (3.25) that is useful for the early universe:

$$t = \left(\frac{\text{MeV}}{T} \right)^2 \text{ s.} \tag{3.66}$$

When the universe was about 1 second old, and its temperature was around $\simeq 1\,\mathrm{MeV}$, baryogenesis had already happened, and the universe had a few relativistic and non-relativistic ingredients.

- γ, e^+, e^-: Photons, electrons, and positrons are in a relativistic equilibrium through the process $e^+e^- \leftrightarrow \gamma\gamma$.
- ν: Neutrinos are relativistic and decoupled since at 1 Mev the reaction rate $\Gamma \ll H$ for weak interactions like $\nu e \leftrightarrow \nu e$. At this time, the neutrino temperature is the same as that of photons (but photons will heat up after electron–positron annihilation).
- Baryons are non-relativistic due to their higher mass. As we have seen, the baryon to photon ratio is extremely small $\eta_b = n_b/n_\gamma \simeq 10^{-10}$ and conserved.

The situation is complex because it involves all isotopes, such as neutrons, protons, deuterium, tritium, and helium. Numerical studies confirm that we can understand this nuclear soup by taking some simplifications:

- We disregard elements beyond He. This means that we only have to worry about H and He isotopes. ^{4}He is a local maximum in binding energy; only a few light elements beyond it are produced at minuscule levels that can be neglected.
- Until $T \simeq 0.1\,\mathrm{MeV}$, the universe has only neutrons and protons.

Neutrons and protons are kept in equilibrium via weak interactions, such as

$$p + e^- \leftrightarrow n + \nu, \tag{3.67}$$

$$p + \bar{\nu} \leftrightarrow n + e^+, \tag{3.68}$$

$$p + \bar{\nu} + e^- \leftrightarrow n. \tag{3.69}$$

The appearance of (anti)neutrinos signals that this is a weak interaction. To proceed beyond neutrons and protons, the first crucial step is deuterium production

$$n + p \to D + \gamma. \tag{3.70}$$

The appearance of photons is the hallmark of an electromagnetic reaction. If this reaction turns on, helium production can proceed through $D+D \to n+{}^3He$ and $^3He+D \to p+{}^4He$, which is the result we are targeting. We can use equilibrium statistics to estimate the reaction rate at 1 MeV since the electromagnetic interaction can keep the thermal equilibrium (cf. Eq. (3.63)). We can recall from thermodynamics that $dE = TdS - pdV + \mu dN$. In equilibrium, the intensive parameters should equalise. If the temperature and the pressure are already the same, the chemical potential μ has to be equal on both sides of a reaction that changes species. The resulting equation has multiple names in different fields: the equation of chemical equilibrium, nuclear statistical equilibrium, or Saha equation. Regardless of the name, the physical meaning is the same. Applying it to the above reaction

$$\frac{n_p n_n}{n_p^0 n_n^0} = \frac{n_D}{n_D^0}, \tag{3.71}$$

where the number density of a non-relativistic species is

$$n_i = g_i e^{(\mu_i - m_i)/T} \left(\frac{m_i T}{2\pi}\right)^{3/2}, \tag{3.72}$$

with $n_i^0 = n_i(\mu_i = 0)$. Note that we can neglect the chemical potential of photons due to their thermal distribution, thus $n_\gamma = n_\gamma^0$. Since $g_D = 3$ (spin 1) and $g_n = g_p = 2$ (spin 1/2), and $n_p \simeq n_p \simeq n_b$, we can express the ratio of deuterium and baryon density as

$$\frac{n_D}{n_b} = n_b \left(\frac{1}{m_b T}\right)^{3/2} e^{B_D/T} \tag{3.73}$$

$$= \eta_b \left(\frac{T}{m_b}\right)^{3/2} e^{B_D/T} \tag{3.74}$$

where we used the approximation of $m_b \equiv m_p \simeq m_n \simeq m_D/2$, except when calculating the deuterium binding energy $B_D = m_n + m_p - m_D = 2.2\,\text{MeV}$. The second equation follows from $n_b = \eta_b n_\gamma$ and Eq. (3.57). The small value of the prefactor $\eta_b \simeq 10^{-10}$ inhibits deuterium production when $B_d \simeq T$. It is easy to show that $n_D/n_b \simeq 1$ is satisfied only around $T \simeq 0.07\,\text{MeV}$. At that point, deuterium production can start and proceed to helium. The reaction chain freezes there because there is no element with the atomic number $A = 5$, thus no reaction like $^4He + p \to X$ exists.[10]

---

[10]Luckily for us, the triple α reaction $^4He + ^4He + ^4He \to ^{12}C$ in stars continues the chain towards higher elements needed for biological life.

Our next step is to estimate the neutron–proton fraction, where we have to consider the reactions $n + \bar{\nu} \leftrightarrow n + e^+$ and $p + e^- \leftrightarrow n + \nu$, i.e. the reaction is a weak interaction mediated by massive boson. While in equilibrium, the proton to neutron ratio is extremely simple

$$\frac{n_p}{n_n} = \frac{n_p^0}{n_n^0} = e^{Q/T}, \tag{3.75}$$

where $Q = m_n - m_p = 1.293\,\mathrm{MeV}$, the mass difference between the two species. As noted earlier, the weak interaction cannot keep the equilibrium: detailed non-equilibrium calculations from the Boltzmann equation show that around $0.5\,\mathrm{MeV}$, the neutron fraction freezes out at around 0.15. At that point, neutron decay further reduces the neutron fraction through $n \to p + e^- + \bar{\nu}$. The decay rate is $\exp(-t/\tau_n)$, where according to the latest measurements of the neutron lifetime is $\tau_n = 877\,\mathrm{s}$ (other measurements claim about 10 s higher; therefore, we adopt 880 s). This means that by the time corresponding to $T \simeq 0.07$, the fraction is only $0.15 \times \exp\left(-((1/0.07)^2 - (1/0.5)^2)/880\right) \simeq 0.12$ (cf. Eq. (3.66)). This is the neutron fraction present when nucleosynthesis starts in earnest with deuterium. Since the binding energy of helium is larger than deuterium, all deuterium quickly turns into $^3\mathrm{He}$ and finally into $^4\mathrm{He}$. Since we need two neutrons for each $^4\mathrm{He}$, their abundance is half of what we calculated. The traditional measure is the mass fraction, i.e. it multiplies the (half) neutron abundance fraction by 4 (for two neutrons and two protons in the helium atom). The final result is thus 0.24, fairly close to observations, given the above crude estimates!

The observed value of helium mass fraction from more precise calculations is (Olive *et al.* (2000) and Kolb & Turner (1990))

$$Y_p = 0.2262 + 0.0135 \ln\left(\eta_b/10^{-10}\right). \tag{3.76}$$

Observations corroborate this prediction, which is a triumph of standard cosmology. Note that a small fraction of deuterium is not burned up into helium corresponding to a residual fraction of $10^{-5} \ldots 10^{-4}$. Since stars destroy deuterium, the highest (undepleted) level of observed deuterium abundance in pristine systems can be used to constrain Ω_b along with the light elements additionally produced.

3.4 The cosmic microwave background

3.4.1 *The surface of last scattering*

The cosmic microwave background is a relic black body radiation from the time of recombination around $z = 1100$. The universe was opaque for photons until then, and suddenly it became transparent.

The temperature of the cosmic microwave background is $T = 2.725\,\text{K} \pm 0.002$ (Mather *et al.*, 1999). Equation (3.57) with $g_{*S} = 2 + 3 \times 7/8(4/11)^{4/3}$ using Table 3.2 gives the radiation density

$$\Omega_r h^2 = 4.15 \times 10^{-5}, \qquad (3.77)$$

assuming massless neutrinos. As a side note, if $m_\nu > 0$, its contribution to the energy density is

$$\Omega_\nu h^2 = \frac{m_\nu}{94\,eV}. \qquad (3.78)$$

At least one of the neutrino masses is likely larger than $0.05\,\text{eV}$ from neutrino oscillations.

Nevertheless, at high redshift, we can neglect the mass of neutrinos and calculate the epoch of equality when radiation and matter contribute about the same to the universe's energy density. We can define the matter–radiation equality when their contributions are equal to the Friedmann equations $\Omega_r a_{\text{eq}}^{-4} = \Omega_m a_{\text{eq}}^{-3}$, therefore

$$a_{\text{eq}}^{-1} = 1 + z_{\text{eq}} = 2.4 \times 10^4 \Omega_m h^2, \qquad (3.79)$$

where $\Omega_m = \Omega_c + \Omega_b$, the total matter content, including DM and baryons.

Soon after equality, we have recombination: somewhat of a misnomer designating the process of electrons and protons of the hot plasma combining into hydrogen atoms *for the first time*. For the process $e^- + p \leftrightarrow H + \gamma$, we can write the Saha equation

$$\frac{n_e n_p}{n_e^0 n_p^0} = \frac{n_H}{n_H^0}, \qquad (3.80)$$

where n_i is the number density of species $i = e, p, H$ (cf. Eq. (3.72)), and $n_i^0 = n_i(\mu_i = 0)$ is the corresponding number density with chemical potential $\mu_i = 0$. As before, photons are not in the equation since the photon number is not fixed, and their chemical potential in this reaction is negligible. Introducing $X_e = n_e/(n_e + n_H) = n_p/(n_p + n_H)$, with the second equation following from the neutrality of the universe $n_e = n_p$, we get

$$\frac{X_e^2}{(1 - X_e)} = \frac{1}{n_e + n_H} \left(\frac{m_i T}{2\pi} \right)^{3/2} e^{-\epsilon_0/T}, \qquad (3.81)$$

where $\epsilon_0 = m_e + m_p - m_H = 13.6 \, \text{eV}$ the hydrogen binding energy. Since the denominator $n_p + n_H \simeq \eta_b n_\gamma \simeq 10^{-9} T^3$, even at $T \simeq \epsilon_0$, the right-hand side will be huge, implying that $X_e \simeq 1$, the neutral hydrogen fraction is negligible. The temperature has to drop an order of magnitude below ϵ_0 such that the exponential starts compensating for the large value of η_b. This is when X_e can become appreciably smaller than 1. The hydrogen fraction increases, *recombination* happens around $z_* \simeq 1000$ soon after equality.

As X_e drops, the scattering rate of photons, $X_e n_b \sigma_T$, drops as well, where $\sigma_t \simeq 0.665 \times 10^{-24}$ is the Thompson cross-section. Expressing n_b with Ω_b and dividing with the Hubble parameter lead to

$$\frac{n_e \sigma_T}{H} \simeq 100 X_e \qquad (3.82)$$

for concordance model parameters. The left-hand side compares the reaction rate with the Hubble time. Photons decouple from baryons when less than one reaction happens in a Hubble time, i.e. $X_e \lesssim 0.01$. Once X_e drops exponentially, decoupling occurs extremely fast after recombination.

In summary, equality is at around $z \simeq 3000$, and soon, recombination and decoupling proceed in a rapid sequence at around $z_* \simeq 1000$ or $T \simeq 1/4 \, \text{eV}$. Before decoupling, photons and baryons are strongly coupled in the hot plasma. Photons have a small mean free path. This means that the universe is entirely opaque. After decoupling, the universe suddenly becomes transparent due to the drop in the electron density and the plasma temperature. The cosmic microwave background (CMB) radiation is the image of this *last scattering surface* at around z_*. In all the above-simplified discussions, we ignored the Helium fraction corresponding to an order of 10% correction to our argument capturing the fundamentals.

To the first approximation, the cosmic microwave background is an isotropic black body radiation of $T_0 = 2.725\,\mathrm{K}$. This corresponds to an energy density of about 400 photons per cubic centimetre. The movement of our Sun compared to the cosmic rest frame with about $v \simeq 370\,\mathrm{km/s}$ causes a dipole anisotropy pattern through the Doppler effect. The amplitude of the Doppler effect is approximately $v/cT_0 = 3.36\,\mathrm{mK}$. Once this is subtracted, the last scattering surface has tiny anisotropies of order 10^{-5}, originating from inflationary initial conditions. Thus, the CMB anisotropies encode precise information about the early universe.

3.4.2 *CMB anisotropies*

To understand CMB anisotropies, recall that photons and baryons are tightly coupled before recombination. They form a hot *photon–baryon fluid*. After recombination and decoupling, photons are *free streaming* to us, possibly modified slightly with foregrounds (secondary anisotropies discussed later).

The horizon size at recombination was around $\simeq 100\,h^{-1}$ Mpc. Above the horizon size, no causal physics can shape the CMB fluctuations. Thus, the observed anisotropy on the largest scales directly corresponds to the gravitational potential fluctuations Ψ generated during inflation. Denoting the temperature fluctuations $\theta = \delta T/T$, we will observe $\theta_0 + \Psi$ since the photons have to climb out of the potential wells before reaching us: so a negative Ψ will lower the fluctuations we see. With some algebra (and GR), we can show t at $\theta_0 + \Psi = -\delta/6$. Thus, perhaps surprisingly, hotspots on the CMB correspond to negative fluctuations in the original inflationary initial conditions. In other words, climbing out takes more energy than the excess photons have within potential wells. The Sachs–Wolfe effect is the CMB fluctuations on superhorizon scales imprinting initial conditions.

Below the horizon size, gravity and pressure oppose each other. Density increases as we compress the photon–baryon plasma; thus, gravity attracts even more matter. However, compression also raises temperature and pressure. Eventually, pressure will overcome gravity, pushing the plasma outward in the opposite direction. The interplay of the two forces leads to oscillation, termed baryonic acoustic oscillations (BAO). Acoustic refers to sound waves in the plasma that have a characteristic equation of a forced, damped oscillation

$$\ddot{\theta}_0 + H\frac{R}{R+1}\dot{\theta}_0 + k^2 c_s^2 \theta_0 = F, \qquad (3.83)$$

where F represents gravity and the sound speed $c_s^2 = 1/3(1 + R)$ depends on the photon baryon ratio $R = 3\rho_b/4\rho_\gamma$. The damping from the expansion of the universe is the usual Hubble drag. With a negligible amount of baryons, the speed of sound is $1/\sqrt{3}$ fraction of the speed of light. Adding baryons, often termed *baryon loading*, adds mass and slows down the propagation of waves in the medium. Note that BAO is also present in the statistics of galaxy distributions but to a lesser extent than for the CMB. BAO only modulates the observed power spectrum dominated by dark matter, which has 5.45 times more density than baryons.

The characteristic scale of these waves is the *sound horizon* $r_s(t)$, the distance a sound wave could propagate during time t. This is approximately $1/\sqrt{(3)}$ times the particle horizon or less. Without solving the above equation, it is not surprising that solutions behaving like $\cos kr_s$ and $\sin kr_s$ exist. The first one is relevant for the angle averaged, or monopole, temperature fluctuations θ_0 as above, while the second corresponds to the dipole or Doppler part. The Doppler nomenclature emphasises that a dipole moment implies velocity fluctuation. The higher moments of temperature fluctuations are negligible well before decoupling.

The simple story of BAO gets subtle as we approach decoupling: photons undergo a diffusion process as their mean free path is getting longer $\lambda_{MFP} = 1/n_e\sigma_T$. In a Hubble time, the mean number of scattering is $N = n_e\sigma_T/H$. When a photon undergoes a random walk, it gets $\sqrt{N}$ times the mean free path from its original position. This means that the diffusion length scale is

$$\lambda_D = \lambda_{MFP}\sqrt{N} = \frac{1}{\sqrt{n_e\sigma_T H}}. \qquad (3.84)$$

Since this diffusion length scale acts as a smearing length, below this scale, we cannot see any detail of the last scattering surface.

After decoupling (neglecting any foregrounds), the photons will free-stream to us. We will see the Gaussian fluctuations imprinted by inflation: on superhorizon scales unchanged, on subhorizon scales modulated by the BAO oscillations, and on small scales smoothed by the diffusion scale. The small-scale smearing of the CMB fluctuations is called Silk damping, which refers to Joe Silk, who first understood this process. The hot and cold spots on the CMB sky with characteristic scale α correspond to

$$\alpha \simeq \frac{1}{\ell} \simeq \frac{1}{kr_*}, \tag{3.85}$$

where r_* is (angular diameter) distance to the CMB and ℓ is the multipole of spherical harmonics.

In addition, radiation near recombination causes the gravitational potential to decay. If a photon leaves a small potential well, decay means it does not have to climb out as much; therefore, it will be hotter than it would be without decay. This is called the integrated Sachs–Wolfe effect because its amplitude is proportional to an integral over the derivatives of the gravitational potential.

Usually, we expand the fluctuations of the CMB into spherical harmonics

$$\frac{\delta T}{T}(\vec{n}) = \sum a_{\ell m} Y_{\ell m}(\vec{n}). \tag{3.86}$$

Since the ensemble average $\langle \delta T \rangle = 0$ by definition, the harmonic coefficients $\langle a_{\ell m} \rangle = 0$. For a Gaussian field, and the CMB fluctuations inherit the Guassianity of inflationary quantum fluctuations, each a_{lm} is uncorrelated. Therefore, the variance for each mode ℓ

$$C_\ell = \langle a_{\ell m} a_{\ell m}^* \rangle = |a_{\ell m}|^2 \tag{3.87}$$

defines the angular power spectrum. Since the formula is true for each m, the simplest estimator (for full sky and no noise) is

$$C_\ell = \frac{1}{2\ell + 1} \sum_m |a_{\ell m}|^2, \tag{3.88}$$

where the prefactor compensates for the number of modes from $-m \ldots$ to m. It is customary to plot $\ell(\ell + 1)/2\pi C_l$, the contribution of temperature fluctuations per logarithmic interval. Were it not for the BAO oscillations and Silk damping, we'd expect this to be close to constant for Harrison–Zeldovich-type inflationary initial conditions. Indeed, the first few ℓ's in the Sachs–Wolfe regime are roughly constant.

The CMB power spectrum, C_l, constrains cosmological parameters by responding to parameter changes. The angular spectrum's response to each parameter, all else being held constant, makes a qualitative understanding possible.

The overall amplitude of the CMB power spectrum depends on the amplitude of the initial power. Traditionally, C_{10}, the amplitude of the 10th mode characterises this normalisation. A rise in initial amplitude will carry a corresponding increase in C_{10}. While $\ell = 10$ is an arbitrary pick, it is far from the first CMB peak at $\ell \simeq 200$ and much better measured than the quadrupole ($\ell = 2$).

The slope of the initial power spectrum, n, or the *tilt* of the initial power spectrum will similarly tilt the angular power spectrum as a function of ℓ.

The opacity parameter, τ, quantifies the scattering of light after reionisation. Scattering is a causal process corresponding to scales larger than the horizon at reionisation. On smaller scales, the power spectrum will be suppressed.

The tensor-to-scalar ratio parametrises the power ratio in tensors (or gravitational waves) versus scalars, e.g., gravitational potential. It is defined as the ratio of the quadrupole moment of the power spectra $r = C_2^T / C_2^S$. Gravitational waves increase the power on superhorizon scales since they decay after entering the horizon. Since the power in gravitational waves adds to the scalar perturbations, the effect for the CMB power spectrum is a boost for $\ell < 100$ corresponding to the horizon at recombination. The future of CMB science is the precise measurement of the B-polarisation power spectra to separate the contribution from gravitational waves. This will yield an independent constraint on r.

The baryon energy density $\Omega_b h^2$ influences the power spectrum in two distinct ways. Increasing the baryon content will slow the sound speed, thus lowering the sound horizon and, with it, the wavelength of the oscillations, causing a lateral shift to the right. In the wave equations, baryons act as a mass (baryon loading), boosting the odd peaks of the power spectrum.

If the matter density, $\Omega_m h^2$, decreases, it will bring equality closer to recombination, increasing the effect of radiation. The decaying of potentials will increase the driving force; therefore, the amplitude of the oscillations will increase. In addition, the ISW effect will also increase due to the increased decay of the potentials.

The cosmological constant, the simplest form of DE, is negligible around recombination. Therefore, it only has an effect by lengthening the photons' free streaming path, ultimately pushing the power spectrum towards smaller scales or large ℓ's compared to a matter-dominated flat model with the same sound horizon.

Photons will travel on curved paths if the universe has a curved geometry parametrised with $\Omega_k = 1-\Omega_m-\Omega_\Lambda \neq 0$. For instance, photon trajectories will diverge in an open universe, demagnifying hot and cold spots. Ultimately, that scales the power spectrum towards higher ℓ's. The CMB angular power spectrum is one of the most robust evidence for a flat universe.

So far, we assumed that photons will reach us undisturbed by the foregrounds after the last scattering. We mentioned that if the universe re-ionised, for instance, due to supernovae and AGN at late times, the free electrons would scatter the CMB photons, as described by the parameter τ.

In addition, DE causes gravitational potentials to decay, which is analogous to the early ISW effect, called the late-time ISW effect. If a photon crosses a potential well, corresponding to a supercluster in the galaxy distribution and a positive density fluctuation in the inflationary initial conditions, it will gain energy descending. As the photon climbs out, normally, it cools back to the same temperature as before entering, but if the potential decays, it does not have to climb as much, and it will gain energy. Thus, superclusters will correspond to a hot spot. Photons crossing a void will undergo an opposite change and correspond to a cold spot. In this way, the super-large-scale structure of the universe is imprinted on the CMB map and revealed by coanalysing with galaxy surveys that track the structure in the foreground. Cross-correlation of the two density fields and stacking the CMB map in the direction of superstructures, super-voids and superclusters in the galaxy survey are some of the leading methods used to study the late ISW effect.

The gravity of foreground structures will also bend the path of CMB photons slightly. This *gravitational lensing* effect introduces slight non-Gaussianities in the measured CMB maps. The lensing effect must be removed for high-precision constraining primordial non-Gaussian statistics that arise, for instance, from inflationary theories with more than one inflaton field.

The Sunyaev and Zeldovich (SZ) effect is an inverse Compton scattering of CMB photons crossing galaxy clusters. The low-energy CMB photons will be scattered by the high-energy electrons heated up in the deep potential well of a cluster. The result is a spectral distortion, heating up beyond the peak of the CMB and cooling in the low-frequency Rayleigh–Jeans tail. The Compton y-parameter characterises the effect

$$y = \int \sigma_T n_e \frac{T}{m} dl, \qquad (3.89)$$

a line integral over the photon path. While the distortion is slightly complicated, at low frequencies, $\delta T/T = -y$. This way, galaxy clusters imprint on the CMB map as cold (low frequency) or hot (high frequency) spots without affecting the 218Ghz Planck maps.

So far, we have only considered the CMB temperature fluctuations. The *last scattering* produces linear polarisation at the 10% level. Measuring it is thus an order of magnitude more challenging, and it is the next frontier of CMB research. The E and B mode polarisation do not refer to electric and magnetic fields but rather to the polarisation characteristics of the radiation. For light propagating in the z-direction, E-modes refer to polarisation in the x- or y-direction, while B-modes are polarised in the $\pm 45°$ direction. E-modes have been detected for 20 years, with DASI, WMAP, and Planck, among others. Scalar perturbations generate E-modes but no B-modes, while tensor perturbations, expected to be lower by r according to inflation, generate B-modes. A confirmed discovery of B-modes from tensors by a CMB mission is one of the most predictable Noble prizes of the near future.

3.5   Gravitational instability

Due to the neutrality of the universe and the short-range nature of weak and strong forces, gravity is the dominating force on large scales. Newtonian gravity is a good approximation below the Hubble scale $1/H_0 \simeq 3000h^{-1}$ Mpc and above the Schwartzschild radius of any black hole. Since the universe expands, it is convenient to express the motion of particles on comoving coordinates x, where the physical (or proper) separation between particles is $r = ax$. For a completely homogeneous universe, x is a constant; the smooth large-scale expansion of the universe entirely determines the motion of test particles. The comoving coordinate will change in response to local perturbations of the gravitational potential due to (initially small) inhomogeneities. To describe dynamics in comoving coordinates, let's consider the Lagrangian of a test particle in gravitational potential Φ:

$$L = \frac{1}{2}m(a\dot{x} + \dot{a}x)^2 - m\Phi = \frac{1}{2}ma^2\dot{x}^2 + \frac{d}{dt}\left(\frac{1}{2}ma\dot{a}x^2\right)$$

$$- m\left(\Phi + \frac{1}{2}a\ddot{a}x^2\right). \tag{3.90}$$

We can omit the derivative from the Lagrangian to obtain

$$L = \frac{1}{2}ma^2\dot{x}^2 - m\phi, \tag{3.91}$$

with $\phi = \Phi + \frac{1}{2}a\ddot{a}x^2$. The Poisson equation for Φ in physical coordinates transforms for the modified gravitational potential ϕ in terms of comoving coordinates as

$$\nabla^2\phi = 4\pi Ga^2(\rho(x,t) - \rho_b(t)) = 4\pi Ga^2\rho_b\delta, \tag{3.92}$$

with $\delta = \rho(x,t)/\rho_b(t) - 1$ representing local density fluctuations (cf., Eq. (3.13)). The standard equations of motion are

$$p = ma^2\dot{x}, \quad \dot{p} = -m\nabla\phi. \tag{3.93}$$

The hydrodynamical continuity and Euler equations approximate the motion of an ensemble of test particles as long as single streams dominate. It is straightforward to transform the equations into a comoving frame:

$$\frac{\partial\delta}{\partial t} + \frac{1}{a}\nabla_i(1+\delta)v_i = 0, \tag{3.94}$$

$$\frac{\partial v_i}{\partial t} + \frac{1}{a}(v_j\nabla_j)v_i + \frac{\dot{a}}{a}v_i = -\frac{1}{\rho a}\nabla p - \frac{1}{a}\nabla_i\phi, \tag{3.95}$$

where we used index notation for the Euler equation due to the vector nature of v; Latin indices run over the values $1, 2, 3$, and repeated (lower) indices carry Einstein summation. We assumed a single stream fluid. For DM, the pressure is zero. When fluctuations are small, $\delta, v \ll 1$, we can neglect all non-linear terms and solve the equations in linear theory. Substituting $\nabla v = a\dot{\delta}$ in the divergence of the second equation and using (3.92) yield

$$\frac{\partial^2 \delta}{\partial t^2} + 2H(a)\frac{\partial \delta}{\partial t} = 4\pi G \rho_b \delta, \qquad (3.96)$$

with $H(a)$ as the Hubble parameter. The Friedmann equation (3.11) encodes a cosmological model through $H(a)$. It is easy to show that there are two solutions to this equation. A decaying mode, $\delta \propto H(a)$ and a growing mode (usually more important),

$$\delta \propto H(a) \int^a \frac{db}{(H(b)b)^3}. \qquad (3.97)$$

For an Einstein–deSitter universe, or early times ($z \simeq 10$) for any model, $\delta \propto a \propto t^{2/3}$. In the concordance ΛCDM model, growth will stop at late times when DE starts dominating. We often express these results in terms of the *growth function* $\delta = D(a)\delta_0$, isolating the time dependence of the initial fluctuations. Using $H = H_0\sqrt{\Omega_m a^{-3}}$, the proper normalisation is

$$D(a) = \frac{5\Omega_m}{2}\frac{H(a)}{H_0}\int_0^a \frac{db}{(H(b)b/H_0)^3}. \qquad (3.98)$$

The linear growth factor D describes the growth of fluctuations when Newtonian physics is adequate: when the fluctuation is well within the horizon, i.e., late times.

The solution determines the velocity field in linear theory. Assuming rotation-free velocity, its divergence $\theta = \nabla v$ suffices for characterisation. From Eq. (3.95), considering growing modes,

$$\theta = -a\frac{d\delta}{dt} = -a\frac{\delta}{D}\frac{dD}{dt} = -aH(a)\delta\frac{d\log D}{d\log a} = -aHf\delta, \qquad (3.99)$$

where δ/D does not depend on time. A useful approximation for $f = d\log(D)/d\log(a) \simeq \Omega^{0.6}$. The above equation shows that linear velocity fields can tightly constrain Ω through growth.

In early times, the horizon was much smaller than today. It is crucial to consider when a fluctuation (we can imagine it as a wave) *enters the horizon.*[11] In particular, the most interesting question is whether a fluctuation enters the horizon before or after radiation equality. Only a full general relativistic treatment, through numerical integration or analytic solutions with approximations in different regimes, will give the full picture (e.g., Dodelson & Schmidt, 2020). Large fluctuations entering the horizon late in the matter-dominated regime always grow with the growth function above (3.98).

Small-scale fluctuations enter the horizon earlier during radiation domination. We can understand their growth around equality by solving Eq. (3.96). The Friedmann equation at that time $H^2 = 8\pi G(\rho_r + \rho_m)/3$ (neglecting DE). Introducing the variable $y = \rho_m/\rho_r = a/a_{eq}$ (since the two terms are equal at a_{eq} by definition), we can express the equation in terms of y as

$$\frac{d^2\delta}{dt^2} + \frac{2+3y}{2y(1+y)}\frac{d\delta}{dt} - \frac{3\delta}{2y(1+y)} = 0. \tag{3.100}$$

This is the Meszáros equation. One solution is the growing mode (with $d^2\delta/dt^2 = 0$) is

$$\delta \propto D_1 = \frac{2}{3} + y. \tag{3.101}$$

The decaying mode is complicated (easily found with computer algebra) but is logarithmic when $y \ll 1$.

$$\delta \propto D_2 \simeq \frac{2}{3}\ln\frac{4}{y} - 2 \quad y \ll 1. \tag{3.102}$$

With these solutions, we can understand the behaviour as fluctuations pass equality and enter the horizon. Mathematically, we can define the transfer function as

$$T(k) = \frac{\delta(z=0)}{\delta(z)D(z)}, \tag{3.103}$$

factoring out the growth function of Eq. (3.98). Large-scale fluctuations stay above the horizon during most of the universe's history.

---

[11] A bit of a misnomer, it means the moment when the horizon outgrows the scale of the fluctuation.

While above the horizon, gravitational potentials are frozen. According to the Poisson equation, Eq. (3.92), $\delta \propto a \propto y \propto D(a)$ at early times. This stays even after the fluctuations enter the horizon late, deep in the matter-dominated regime. Therefore, the transfer function is $T(k) \simeq 1$ for them (general relativistic treatment gives a factor $9/10$, which is traditionally factored out from the transfer function such that it tends to unity on large scales $k \ll 1$).

According to the above Meszáros solutions, small scales enter the horizon during radiation domination and undergo only a slow logarithmic growth until matter domination. Since $y \propto a \propto t^2$ (cf., Eq. (3.23)), the smaller the wavelength $1/k$ is compared to t (which is also the horizon), the more growth is missing. The combination of the missed growth and a small logarithmic growth in the radiation-dominated regime gives a transfer function as

$$T(k) \simeq 12 \left(\frac{k}{k_{\rm eq}} \right) \ln \frac{k}{8 k_{\rm eq}}, \tag{3.104}$$

where $k_{\rm eq} = a_{\rm eq} H(a_{\rm eq}) = H_0 a_{\rm eq}^{-1/2}/\sqrt{2}$ (Dodelson & Schmidt, 2020). More accurate approximations exist, including the famous BBKS approximation, but for today's high-precision cosmology, only numerical calculations suffice. Many *Boltzman codes* calculate various power spectra for CMB and small DM fluctuations (CAMB is still the most popular one). Here, we only attempt to recount the basic physics behind these codes. Equation (3.92) in Fourier space yields

$$\delta_k = \frac{k^2 \phi}{4\pi G a^2 \rho} = \frac{2k^2 \phi a}{3\Omega H_0^2} = \frac{2k^2}{3\Omega H_0^2} \frac{9}{10} T(k) D(a) \phi_P, \tag{3.105}$$

where we added the $9/10$ factor to the transfer function and included late growth in terms of the growth function. Using the power spectrum of the primordial inflationary potential ϕ_P from Eq. (3.54), the linear power spectrum $P(k) = |\delta_k|^2$ grows as

$$P(k) = 2\pi^2 \frac{k^n}{H_0^{n+3}} T^2(k) \left(\frac{D(a)}{D(a=1)} \right)^2 \delta_H^2. \tag{3.106}$$

The normalisation of this traditional form comes from defining the dimensionless power spectrum as $\Delta^2 \equiv k^3 P(k)/2\pi^2$. Then the normalisation is $\delta_H^2 = \Delta^2(H_0)$ for a horizon size fluctuation today.

3.6 The power spectrum as a statistical probe

The power spectrum $P_g(k)$ of the galaxy density field is the principal statistical probe of cosmology. When we measure it, we distinguish it from that of the DM density field. Since galaxies are biased tracers of DM, in the simplest case of linear bias $\delta_g = b\delta$, the galaxies present b-times larger fluctuations than the underlying DM field. Thus, the measured power spectrum is

$$P_g(k) = b^2 P(k). \tag{3.107}$$

We can interpret such a measurement according to Eq. (3.106). On the largest scales, the slope $n \simeq 1$ of the power spectrum comes from inflation. Any deviation from 1 might tell us about slow-roll parameters. δ_H is the normalisation that also comes from inflation. The *break* in the spectrum from the $1/k^2$ cut-off comes from the time of equality $k_{eq} = a_{eq}H(a_{eq})$. Since from the temperature of the CMB, we know the radiation density, a_{eq} determines the matter density Ω_m. Measuring the power at different times constrains $D(a)$, although uncertainties in the bias b from a combination of selection effects and galaxy formation can severely limit how much we can use the amplitude of $P_g(k)$.

Similar to the CMB angular power spectrum, the galaxy power spectrum preserves the imprint of BAO. The physics is the same as in the CMB: pressure and gravity create oscillations in the photon–baryon plasma. The baryon fraction determines the scale of the oscillations through the sound horizon. After the decoupling from photons, baryons will move under the influence of gravity, behaving like DM but with slightly different initial conditions. The oscillation imprinted on the distribution at decoupling maintains a constant comoving scale during the universe's expansion history. The contrast of the BAO imprint on the power spectrum depends on the ratio of baryons to DM, which is about 1:5.45. Under the influence of gravity, baryons fall into DM potential wells. As a result, the initial contrast of BAO waves in the power spectrum diminishes with time and becomes a small effect at lower redshift. The BAO scale, about $100h^{-1}$ Mpc, is a *standard ruler* and constrains cosmological parameters through the angular diameter distance.

The power spectrum contains all the information for a Gaussian distribution. Each Fourier mode δ_k has a zero mean; its variance is $P(k)$. Thus, each mode contains an equal amount of information. The number of modes will grow as k^3, the volume of Fourier space. Each layer between $k, k + dk$ represents the independent directions of k vectors within our resolution $dk \simeq 2\pi/L$ determined by the real space size of our survey. Therefore, it seems that there is an infinite amount of information available as we reach larger k's or smaller scales $2\pi/k$ for a given volume L^3. Unfortunately, there are several reasons why this is not the case.

In any survey, there is a finite number of N galaxies. If the cell size or smallest scale we use for our analysis is l, the number of cells is $N_{\text{cell}} = (L/l)^3$. Increasing the resolution, the average number of galaxies per cell becomes $n = N/N_{\text{cell}} \simeq 1$. At this point, shot noise (Poisson) fluctuations dominate the statistics, and we cannot extract more information from the survey.

In many instances, the information content of a survey peters out before reaching the Poisson limit. The *information plateau* (Rimes & Hamilton, 2005; Wolk *et al.*, 2015) of the power spectrum happens because of non-linearities. The covariance matrix of modes for a Gaussian distribution is diagonal: each mode carries the same amount of information about the power spectrum independent of other modes. Emergence of non-linearities is a scattering process, where two modes k_1 and k_2, interact. Thus, the k-vectors' sum and difference appear in the distribution, introducing correlations between different k-modes. The final result is that the covariance matrix becomes non-diagonal due to these correlations. In particular, large-scale modes (or small k's) will be correlated with small-scale modes. Since the number of modes in a sphere of radius k grows as k^3, there are many small-scale-large k modes. Naively, their variance would be small, but unfortunately, the large variance of small k modes will transfer to the large k modes due to the covariance between them. The same phenomenon is called *beat coupling* if it happens with the large scale modes within a survey or *super-survey effect* if modes larger than the survey influence the modes within the survey.

The final result is that we can extract much less information from the power spectrum after we reach non-linear scales. Where will that information go? The conventional wisdom is that it goes into higher-order statistics. Since $P(k)$ corresponds to the variance of the modes, one can consider the joint moment of three modes, confusingly called a bispectrum, since it depends on two k-vectors. The theory and measurement of higher-order statistics are exponentially complicated. From N pixels, it typically takes $N \log N$ operations to measure the power spectrum, but $N^2 \log N$ for the bispectrum, and each higher order takes N-times more operations. A relatively modest 1 Gpc volume survey analysed at 10 Mpc resolution has $N \sim 10^6$. State-of-the-art surveys can be two orders of magnitude larger and, due to excellent sampling, of higher resolution. Therefore, calculating higher-order statistics can quickly result in diminishing returns versus effort.

Fortunately, new analysis techniques use a local transformation to diagonalise the covariance matrix of power spectra at least approximately to higher k's. One idea is to take $A = \log(1 + \delta)$ and calculate the power spectrum of A. The resulting *log-power spectra* contain significantly more information due to the suppression of the most nonlinear high-density regions. An even more efficient technique that includes log spectra as a special case calculates power spectra of different levels in a density field (Fig. 3.1).

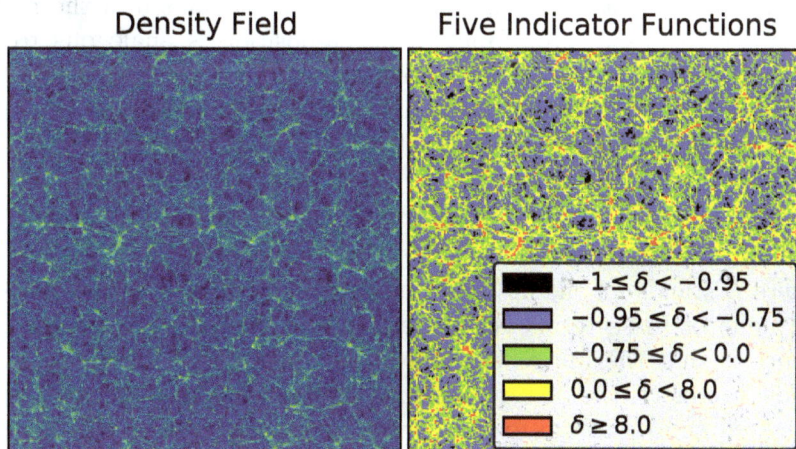

Density Field   **Five Indicator Functions**

■	$-1 \leq \delta < -0.95$
■	$-0.95 \leq \delta < -0.75$
■	$-0.75 \leq \delta < 0.0$
■	$0.0 \leq \delta < 8.0$
■	$\delta \geq 8.0$

Fig. 3.1. Left: The density field of a slice (one pixel thick) through the millennium simulation, **with density computed in cubical pixels of side length** $\mathbf{1.95}h^{-1}$ **Mpc**. Right: Indicator functions (on the same slice) for five density bins, with each colour marking the locations where the corresponding indicator function is non-zero.

3.7 Indicator power spectra

We have seen that the non-linear information plateau of the traditional power spectrum severely limits the available cosmological information in the power spectrum. Non-linearity is characterised by $\nu = \delta/\sigma$, the density ratio versus the linear variance. The bigger the density fluctuation $|\nu|$ is, the more non-linear the statistics are and the more they contribute to the covariance of large- and small-scale modes that erase cosmological information. Therefore, dispersing the density field into levels of different densities is advantageous. Mathematically, an indicator function $\mathcal{I}_B$ takes a value of 1 when the density fluctuation is in a particular bin B and 0 elsewhere. Thus,

$$\mathcal{I}_B(\delta) = \begin{cases} 1 & \delta \in B, \\ 0 & \text{otherwise,} \end{cases} \tag{3.108}$$

where the bin $B = (B_-, B_+)$ denotes a particular level of the density field $B_- < \delta \leq B_+$. In simple terms, we turn the pixelised density field δ into several binary fields containing only 0, 1's, the 1's marking pixels that belong to a particular bin B. Each pixel appears in each binary field. It has 1 in only one of them; it contains 0 in the rest. Therefore, the indicator function decomposition is analogous to a Haar transform in the vertical direction.

From these binary indicator fields, we can trivially reconstruct the original density field (at a particular *vertical* resolution):

$$\delta(x) = \sum_B B\mathcal{I}_B(\delta(x)), \tag{3.109}$$

where B stands for the midpoint of the bin. According to the above, only one of these terms is non-zero in this formal sum. Its significance is that the full δ-field is a linear combination of the binary fields; therefore, the power spectrum $|\delta_k|^2$ can be reconstructed as a weighted sum of all (auto- and cross-) power spectra of the binary indicator fields.

The fraction of 1's in an indicator field for bin B is identical to the probability distribution function (PDF). The PDF and its (higher) moments are the primary statistics to extend the standard δ power spectrum (e.g., Szapudi, 2009). Since indicator functions (Repp & Szapudi, 2022) encode most leading statistical measures, such as the power spectrum, log-power spectrum (Neyrinck *et al.*, 2009), inverse-mapped power spectrum (Wang *et al.*, 2011), box-cox power spectrum (Joachimi & Taylor, 2011), the PDF, and even more, it is sometimes called the grand unified spatial statistics (GUSS) for cosmology. Since it encodes many previous probes, it contains at least as much information as jointly measuring all of those statistics and perhaps more.

The simplicity of using indicator fields is that their power spectra, measured by standard codes, are proportional to the linear theory power spectrum, at least up to a certain k. More precisely, the cross-power spectra of two indicator fields belonging ν_1, ν_2 is

$$P_{\mathcal{I}_1, \mathcal{I}_2} = \frac{\nu_1 \nu_2}{\sigma_{\text{lin}}^2} P_{\text{lin}}(k), \qquad (3.110)$$

where the logarithmic density contrast $\nu = (A - \bar{A})/\sigma_A$, with $A = \log(1 + \delta)$. We can think of the indicator fields having a bias $b = \nu/\sigma$. This is analogous to the theory of galaxy bias by the late Nick Kaiser, who proposed that galaxies form in regions above a certain threshold (Kaiser, 1984) in a Gaussian field. In the Gaussian limit where the logarithmic density contrast well approximates the linear contrast, an indicator field above a certain threshold will result in the mathematical equation (3.110). This is identical to the famous Kaiser bias.

The above result means that we can recover the linear power spectrum from only the indicator fields, as long as we know the variance σ_{lin} on the pixel scale, and k is below the scale of non-linearity; otherwise, the power spectrum is no longer $P_{\text{lin}}(k)$ of Eq. (3.106). The key observation is that the higher the fluctuation $|\delta|$, the more non-linear the field. Thus, each indicator field's maximum k for non-linearity is different, cf. Fig. 3.2. It turns out that moderately high values of δ will stay linear up to twice as large k's. Since information grows with k^3, this means about eight times the information extracted compared to using δ. Normally, (a fraction of) this information would be accessible through higher-order statistics, but standard and fast power spectrum codes can recover it simpler from indicator fields.

Indicator functions from galaxy fields inherit their bias. On linear scales, such a bias is likely to be linear. Similarly, indicator levels adhering to the linear theory of Eq. (3.110) are likely to be well described with linear bias in contrast with higher-order statistics, where higher-order non-linear bias becomes important. Since $P_{\text{lin}}(k)/\sigma_{\text{lin}}^2$ is independent of linear bias, indicator functions depend

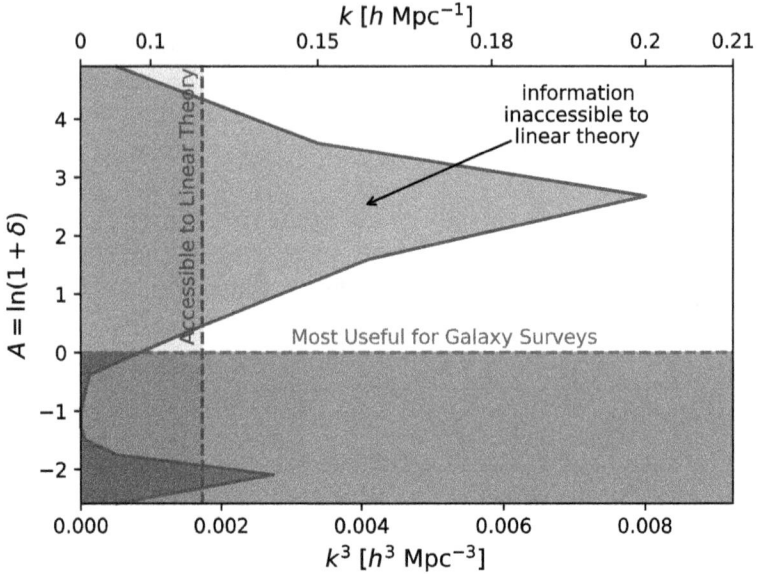

Fig. 3.2. The double triangular region: estimated applicability of Eq. (3.110) +const. (for autospectra), which incorporates only linear analysis; Left of the vertical line: applicability of standard linear theory. Since Fisher information in the linear regime scales roughly as k^3, we use this scale on the x-axis for a visual impression of the information gain from linear indicator-function analysis.

on the bias non-linearly through ν in the non-Gaussian limit and are independent of the bias in the Gaussian limit.

In Section 6.1, we introduce higher-order moments, such as skewness and kurtosis. Perturbation theory will predict the value of these moments, and we can use them to supplement the information extracted by the power spectrum and thus improve constraints on cosmological parameters. Nevertheless, Fig. 3.2 unveils the limits of traditional perturbation theory: each level of δ has a different k-range in which perturbation theory is valid (double triangular region). High (and, to a lesser extent, very low) density regions will go non-linear at larger scales (or smaller k's). Since perturbation theory mixes all density levels, including extremely non-linear regions, its applicability is severely limited by them.

In Section 5.2.4, we discuss the differences between the 2D and 3D density field. It is worthwhile to note here that when we project the original 3D field onto a plane, the result is dominated by features with the highest density contrast. At the same time, moderate fluctuations of the cosmic web might get buried in the noise. According to Fig. 3.2, those are some of the most information-rich regions. This effect is partly the reason for the graver-than-expected information loss through projection.

ARIEL: I have trouble understanding the... gravity of the flatness problem. Unless we talk about my bicycle tyre.

CALLISTUS: I remember I used to feel the same way, but I got used to it.

RHEA: It's more like I got used to some things, but they made other things feel weird.

CALLISTUS: Right. Things that before you got used to those other things had not been weird.

ARIEL: Ugh. You could not confuse me more.

RHEA: The problem is with this awfully small number. 10^{-67}. It's just hard to imagine where an initial curvature this small could come from.

ARIEL: But why?

CALLISTUS: We have no theories providing such a small number.

ARIEL: Wait, you can imagine a big number like the size of the universe. But you cannot imagine a small number?

CALLISTUS: I am talking about the initial or primordial curvature. We want simple initial conditions.

RHEA: It's intuitively clear that physics uses fundamental constants of nature –

CALLISTUS: And maybe combine them with some mathematical constants, like π, or e multiplied with integers, so you typically don't get very small numbers. [*Arial calculating*]

ARIEL: $e^{-\pi^5/2}$ is close to 10^{-67}, it satisfies all your conditions. I just don't buy why this is a problem.

CALLISTUS: You're right. The flatness problem is ultimately a prejudice based on our physical intuition.

RHEA: As a last resort, we use Occam's razor. Zero is simpler than an *ad hoc* formula of π's powers.

CALLISTUS: Unless it comes from theory.

ARIEL: I don't fully get it, but maybe I'll get used to it...

CALLISTUS: Or don't.

RHEA: Yes. Getting too comfortable with weirdness can stop you from discovering new stuff. We support you.

ARIEL: Too late, now I'm totally used to it...

[*On desperate looks from Rhea and Callistus*]

ARIEL: Just kidding.

Chapter 4

Formation and Evolution of the Cosmic Web

According to the presently accepted cosmological paradigm, the Universe started from a quantum fluctuation, followed by rapid exponential expansion — inflation. Thereafter, quantum fluctuations were magnified into macroscopic density fluctuations. In some regions, the Universe contained somewhat less matter than their surroundings, while in other regions somewhat more matter. Most fluctuations had been blown up to scales larger than the horizon. During the regular Friedmann expansion, the fluctuations gradually entered the horizon and started to grow. The growth can be described by the gravitational instability theory discussed in Chapter 3.

Equations of the gravitational instability are complicated and cannot be solved analytically. The only way to follow the formation and evolution of the Universe is to use numerical simulations. In the first numerical simulations of the evolution, the authors applied the direct integration of equations of motion under the influence of mutual gravitational interaction of particles. The number of particles was in the interval from ∼300 to ∼1000. As for initial conditions, particles were mostly put at random locations with either zero or random initial velocities. Presently, a hydrodynamical method is applied, which allows us to follow a large number of test particles.

In this chapter, we discuss first the numerical simulations of the formation and evolution of the cosmic web. To describe the pattern of particle distribution, respective density fields are used. Further, we discuss the methods on how to find simulated galaxies, which can be compared with real observed galaxies.

4.1 Numerical simulations

ARIEL: Is the universe infinite?

RHEA: We will only ever know a finite part of it.

CALLISTUS: We don't know, really... even in principle, as long as our universe adheres to Einstein's General Relativity, we can only see until the horizon.

ARIEL: The distance light has travelled since the birth of the universe.

RHEA: The horizon is larger. Imagine a point where a light ray initiated near the horizon. That point is now further from us since the universe expanded while the photon travelled towards us.

ARIEL: Sad. We cannot ever know what's behind the horizon.

CALLISTUS: The simplest educated guess would be the same thing that's inside the horizon.

RHEA: But that will always remain an assumption. Just as we cannot peak into a black hole, we cannot peak behind the curtain of the horizon.

ARIEL: So if the universe is infinite and computers are finite, how can we ever simulate the universe?

CALLISTUS: Clever question and, indeed, we must be clever to fit the universe into our computer memory.

RHEA: Cosmological N-body simulations follow Dark Matter particles interacting through gravity with periodic boundary conditions.

CALLISTUS: Equivalent to a three-dimensional torus topology. Imagine your car tyre with one more dimension.

ARIEL: I don't have a car. I use a bicycle to avoid climate-destroying carbon emissions.

RHEA: A way to think about a periodic boundary is that a particle leaving the cubic simulation on the left side returns on the right. Ditto for top-bottom and front-back.

CALLISTUS: Or, we can think of the same cube repeating infinitely many times all over the universe, which is the definition of a 3-D torus. Like tyres on an emission-friendly 3-D bicycle.

ARIEL: or tricycle? I feel uneasy about this, and not just because of the electricity usage. In the real universe, every cube is a random realisation of some cosmological statistics, but we repeat the same pattern. This means extreme correlations on very large scales.

RHEA: You're right. When we sum gravitational forces for one par-
            ticle, we usually add the contribution from several images of
            each particle. This is called Ewald summation, and it is stan-
            dard in cosmological simulations.

ARIEL: We are back to the problem of infinities: we surely cannot add
            an infinite number of images of the same particle.

CALLISTUS: Luckily, gravitational force decreases with $1/r^2$, and we
            only have to add up a few images of a particle before the
            effect of these extreme correlations becomes negligible.

ARIEL: What about the horizon?

RHEA: Most cosmological simulations use Newtonian gravity, which is
            a great approximation above black hole scales and well below
            the horizon. The only GR input is the Friedmann equation for
            controlling the expansion of our simulated universe.

ARIEL: Up to numerical errors, such a simulation is a perfect rendition
            of the evolution of fluctuations in our universe.

CALLISTUS: Not quite. A cube is not spherically symmetric.

RHEA: Periodic simulations break the assumed isotropy of the uni-
            verse, although that's a tiny effect.

CALLISTUS: There are measurable consequences. Instead of being sta-
            tistically invariant over the $SO(3)$ group of rotations, a cubical
            simulation is subject to the O_h octahedral group.[1] Thus, grav-
            ity in periodic simulations is anisotropic, inserting anisotropy
            into gravitational dynamics and collapse.

RHEA: Fortunately, for a simulation with a size larger than $L \geq$
            $300\ h^{-1}$ Mpc the effect is subpercent. We can safely use our
            cubical simulations for most cosmological purposes.

CALLISTUS: Or there is another solution: instead of imagining the uni-
            verse repeats itself, we can borrow a page from string theory:
            we can compactify the infinite universe.[2] The Stereographic
            Projected simulations (StePS) use a stereographic projection
            to project the infinite universe onto a finite Riemann sphere.
            Such a simulation has isotropic gravitational forces.

RHEA: You can imagine stereographic projection as a translucent ball
            with a light bulb at the North Pole sitting on an infinite table.
            Light rays will intersect the ball and the infinite plane. We
            identify points that belong to the same ray. As a result, the
            image of the plane on the ball is distorted.

CALLISTUS: Stereographic projection was invented by cartographers.
            Near the South Pole, the distortion is negligible, but further
            away, it is sizeable, especially on the Northern hemisphere of
            the ball.

---

[1] Rácz *et al.* (2021).
[2] Rácz *et al.* (2018).

RHEA: As a result, StEPS simulations are multiresolution.

CALLISTUS: Like zoom-in simulations, where you run smaller and smaller size nested cubes with higher and higher resolutions to achieve a high dynamic range. Compactified simulations smoothly do the same.

ARIEL: My head is spinning.

RHEA: And we just talked about the boundary conditions of cosmological simulations.

CALLISTUS: There are other possibilities: mirror image simulations, compact space-filling topologies other than the cube, like the Platonian solids.

RHEA: Not to mention multi-component simulations with baryon, neutrino, etc. species in addition to Dark Matter.

CALLISTUS: And my favourite: GR simulations!

RHEA: For most purposes, a periodic Newtonian Dark Matter simulation is good enough. And since they simulate the infinite universe with finite energy, their emission per simulated volume tends to zero.

ARIEL: That's cheating. According to the internet, the Millennium MXXL used 12,000 cores for an equivalent of 300 years. If I'm generous and assume 5W cores and 0.8 pounds of carbon per kWh, I get 4.8 metric tons, approximately the same as the yearly carbon emission of a typical car at 4.6 metric tons.[3]

CALLISTUS: A great deal if you ask me. Especially compared to the 23 megatons/y of carbon emission of BitCoin.

ARIEL: Hmmm. I'll think about it on the way home *riding my bicycle*.

4.1.1 *Initial conditions and simulation methods*

In numerical simulation of the evolution of structure, two issues are of crucial importance: (1) the method to evolve the ensemble of particles and (2) initial conditions for calculations.

The primordial matter is a continuous medium. For this reason, the evolution of the medium must be treated as a hydrodynamical problem. Since the early Universe was almost uniform, it is convenient to use the perturbations of coordinates (displacements from the uniform state) and the perturbations of velocities (departures from the Hubble velocities) with respect to the uniform (unperturbed) state. Initial perturbations were small and the evolution was in the

---

[3]https://www.epa.gov/greenvehicles/greenhouse-gas-emissions-typical-passenger-vehicle.

linear regime. In the linear regime, the gravitational instability amplifies a particular combination of coordinate and velocity perturbations. This combination is called the growing mode. In the growing mode, the initial displacements in the medium and its initial velocities are proportional to each other. Therefore, it requires only one function to describe both displacements. The other combination of the displacements and velocities comprises the decreasing mode that decays in the course of the evolution and can be neglected.

To evolve the ensemble of particles, the "cloud-in-cell" (CIC) method can be applied, which makes it possible to study the collective effects of a large number of particles, while suppressing two-body effects (Hockney & Eastwood, 1981). In this case, the medium is considered as a continuous one, i.e. a fluid, and particles are used only as markers or test objects to show the evolution of the medium. Simulation particle masses are distributed over a finite volume (cloud), their movement is followed in a mesh. On this mesh, Poisson equations are solved using the fast Fourier transform (FFT) with periodic boundary conditions. Since the FFT works very fast, it is possible to use much more particles than in the direct integration method to simulate the evolution of the Universe.

Early evolution of the Universe can be described by the linear theory. When density fluctuations approach unity, the linear theory ceases to be valid. In the non-linear clustering process, three processes are apparent:

- hierarchical clustering;
- anisotropic collapse;
- the formation of the cellular morphology of the cosmic web.

4.1.1.1 *Hierarchical clustering*

CALLISTUS: Top-down, or bottom-up, that is the question. Or that was for a while, anyway...

RHEA: Ah, the legendary debate by Peebles and Zeldovich. A profound disagreement of great minds.

ARIEL: Science is about facts. How can great scientific minds disagree?

CALLISTUS: Only temporarily. That was a different time: less data but also more imagination. Worried today about a Hubble constant being 68 or 72km/s/Mpc? At the time, de Vaucouleurs measured 100km/s/Mpc, while Sandage-Tammann insisted on 50km/s/Mpc.

Ultimately, it turned out to be a problem of methods and systematics. If history says anything about the present Hubble puzzle...

RHEA: I'm glad to be living in the era of high-precision cosmology. Must have been a stressful period.

CALLISTUS: In some ways, it was. But also more fun. Zeldovich imagined first that Dark Matter to be neutrinos. As light particles, they tend to stream out of potential wells, erasing small-scale fluctuations. Structure formation thus starts on large scales, with pancakes forming and later breaking up. *Top-down*. A contrario, Peebles thought Cold Dark Matter was more likely. With enough small-scale power, small objects collapse first, proceeding towards large scales. *Bottom-up*.

RHEA: I guess Peebles won. Data today support Cold Dark Matter and hierarchical clustering.

CALLISTUS: The quest for knowledge is not a competition. Zeldovich and Peebles met last time in 1987, at IAU Symposium in Balatonfüred, Hungary. By then, it was clear neutrino masses were not large enough for the pancake model, but that did not make any difference. They were equally friendly and happy to see each other. This was Zeldo's last conference before his passing.

RHEA: Jim Peebles got the Nobel prize 32 years later!

CALLISTUS: Finally. Palmam qui meruit ferat.[4]

ARIEL: I thought people of this calibre were never wrong.

RHEA: Data rule in science. The final word in any argument.

CALLISTUS: And yet, Zeldovich wasn't wrong either. Today, we know neutrinos have a small mass in the fraction of an eV range and probably have a measurable effect on the galaxy power spectrum. They are not the whole story, but they certainly contribute at an interesting level.

RHEA: Our picture of hierarchical structure formation has also gotten more nuanced. The cosmic web of clusters, filaments, walls and voids contains pancake-like structures.

ARIEL: I see. People of this calibre are never *completely* wrong.

Perturbations correspond to density waves of different scale and phase, and have a certain power spectrum. Zeldovich assumed that perturbations consist of common motion of photons and baryons. These perturbations conserve entropy and are therefore called

---

[4]Let him who has earned it bear the reward.

"adiabatic". The amplitude of fluctuations on different length scales is described by the power spectrum. The primordial power spectrum is usually assumed to have a power law dependence on scale: $P(k) \sim k^n$, where k is the wavenumber. The index n determines the balance between small- and large-scale power. A popular choice is the scale-invariant spectrum with spectral index $n = 1$ proposed by Zeldovich and Harrison. In this case, fluctuations on different length scales correspond to the same amplitude of fluctuation in the gravitational potential. Small-scale fluctuations collapse first to form bound objects before larger-scale structures do, resulting in a gradual building-up of successively larger structures by the clumping and merging of smaller-scale structures.

A model of the hierarchical structure formation is the Press–Schechter theory (Press & Schechter, 1974), which describes the sample average characteristics of the population of non-linearly collapsed objects.

4.1.1.2 *Anisotropic collapse: The Zeldovich approximation*

Zeldovich (1970) investigated how far the linear theory can be applied. He found that the linear theory can also be used for further stages of the evolution, and this approach is called the "Zeldovich approximation". Using this approximation, the density and velocity are calculated for a continuous medium, not for a finite number of discrete point masses, as in the CIC method. For practical purposes, it is convenient to place test particles in a regular grid and calculate perturbations for each particle. Initial perturbations are random, thus at any point, the perturbation tensor is three-axial, i.e. in one direction, the growth of perturbations is faster. The essential aspect of the Zeldovich approximation is the understanding that in this direction, the growth goes to the non-linear regime faster and leads to the formation of flat structures — pancakes. This result is independent of the scale of perturbations: both on large and on small scales, pancakes do form.

If the matter is baryonic, as was believed in the early 1970s, then during the radiation-dominated era of the evolution, small-scale fluctuations are damped due to photon viscosity (Silk damping). For this reason, the power spectrum contains only waves larger than a critical length. First, large-scale density enhancements grow, and thereafter, they fragment into smaller units. This scenario is thus called top-down.

The Zeldovich formalism is a first-order Lagrangian approximation to the formation and evolution of cosmic structure. In the Zeldovich approximation, the motion of a fluid element is determined by the primordial density fluctuations, following a ballistic displacement approach. At some time t, the Eulerian position $\mathbf{x}(t)$ of the fluid element is given by

$$\mathbf{x}(t) = \mathbf{q} + D(t)\,\nabla\psi(\mathbf{q}), \tag{4.1}$$

where $\mathbf{q}$ is the initial or Lagrangian position of the element. The quantity $D(t)$ denotes the linear growth factor and ψ is the Lagrangian displacement potential (Peebles, 1980). The second term is the primordial linearly extrapolated gravitational potential up to a constant multiplication factor. Using this prescription, we can describe how an initial mass element $\bar{\rho}d^3\mathbf{q}$ gets mapped at a later time t to $\rho(\mathbf{x})d^3\mathbf{x}$. The mass within the mapped volume is conserved, i.e. $\bar{\rho}d^3\mathbf{q} = \rho(\mathbf{x})d^3\mathbf{x}$, which leads to

$$\rho(\mathbf{x}) = \frac{\bar{\rho}}{[1 - D\,\lambda_1(\mathbf{q})]\,[1 - D\,\lambda_2(\mathbf{q})]\,[1 - D\,\lambda_3(\mathbf{q})]}, \tag{4.2}$$

where $\rho(\mathbf{x})$ denotes the density at Eulerian position $\mathbf{x}$ and $\bar{\rho}$ is the mean cosmic density. The three $\lambda_1 \geq \lambda_2 \geq \lambda_3$ quantities denote the eigenvalues of the deformation tensor

$$\psi_{ij}(\mathbf{q}) = \frac{\partial^2\psi(\mathbf{q})}{\partial q_i \partial q_j}. \tag{4.3}$$

The Zeldovich formalism can be used to identify the cosmic web components. This can be done using Eq. (4.2), which describes the evolution of the density at a later time in terms of the primordial matter distribution. The formation of pancakes, filaments and clusters is given by the eigenvalues of the deformation tensor. As time evolves, the overdensity contracts along all directions, but first along the direction corresponding to the largest eigenvalue λ_1. The collapse along this axis takes place when $1 - D(t)\,\lambda_1 \to 0$, resulting in a sheet. Thereafter, the contraction follows along the second axis and forms a filamentary configuration and later with the collapse along the third direction to form a 3D virialised object.

First *N*-body simulations, using Zeldovich approximation in the early phase of the evolution and spectra cut at large scales, were carried out by Doroshkevich *et al.* (1980) in two dimensions (2D) and by Klypin & Shandarin (1983) in three dimensions (3D). Both simulations correspond to the hot dark matter (HDM) scenario, as assumed in the early 1980s. As discussed in Chapter 2, this scenario is in conflict with several observational data (the age of clusters of galaxies). These difficulties are avoided in the CDM model of the universe. First simulations with CDM universe were achieved by Centrella & Melott (1983), Melott *et al.* (1983) and White *et al.* (1983). Davis *et al.* (1985) described the evolution of large-scale structure in a CDM-dominated Universe, and White *et al.* (1987) analysed the distribution of clusters, filaments, and voids in a CDM-dominated Universe. An even better description of the evolution is achieved in the ΛCDM universe.

4.1.1.3 *Cellular morphology*

Bond *et al.* (1996) analysed the formation of the cosmic web and found that the tidal field plays an essential role in its formation. They showed that the cosmic web is largely defined by the position and primordial tidal fields of rare events in the medium, with the strongest filaments between nearby clusters whose tidal tensors are nearly aligned. They showed that the order in which significant structures arise is basically the inverse of that in the classical pancake picture: first high-density peaks, then filaments between them, and afterwards the walls. Voids mark the transition scale at which density perturbations have decoupled from the Hubble flow. As voids expand, matter is squeezed in between them, and sheets and filaments form the void boundaries.

The Zeldovich approximation breaks down when different matter streams cross paths, since then the motion is dominated by the gravitational field of these non-linear structures. This limitation is overcome in the adhesion model by Kofman & Shandarin (1988) via an artificial viscosity term. The adhesion model describes approximately the non-linear stage of gravitational instability and results in a better description of the later stages of anisotropic collapse. It is based on the Burgers equation — adhesion or sticking model. As shown by Kofman *et al.* (1990, 1992), the adhesion approximation yields for the present epoch structures that are very similar to the structures calculated with N-body numerical simulations with the same initial fluctuations.

Hidding (2010) performed the calculations of the evolution of the cosmic web using three models: the usual ΛCDM model, a model calculated with Zeldovich approximation, and the adhesion model. All models had the box size 100 h^{-1} Mpc and identical set of initial fluctuations. The cosmic web as found for the present epoch by all three simulations is presented in Fig. 4.1. We see that major elements are almost identical in all three models. The Zeldovich approximation model lacks the fine structure seen in the actual N-body simulation and adhesion models.

4.1.2   *Simulations with cosmological constant*

A flat cosmological model with $\Omega_{tot} = 1$ is theoretically preferred. On the other hand, direct observational data suggested that the density of matter, including dark matter in galaxies and clusters, yields a value $\Omega_m \approx 0.2$ (Einasto *et al.*, 1974a; Ostriker *et al.*, 1974). Thus, the reminder must lie in some smoothly distributed background. The most suitable candidate for such a background is the cosmological term, $\Omega_\Lambda = \Omega_{tot} - \Omega_m \approx 0.8$. Additional arguments in favour of a flat cosmological model are data on the Hubble constant and on the age of oldest stellar populations in Galaxy. Using these arguments, a flat cosmological model with Λ-term was suggested by Gunn and Tinsley (1975), strong evidence for this model was presented by Peebles (1984), Rees (1984b), Turner *et al.* (1984), Kofman & Starobinskii (1985) and Kofman *et al.* (1986).

In order to understand the evolution of the supercluster-void phenomenon, it is needed that numerical simulations are performed in a box which contains both small and large waves. The most common systems of galaxies are groups and clusters of galaxies with a characteristic scale of ~ 1 h^{-1} Mpc, thus the simulation must have at least a resolution of this scale. On the other hand, the largest non-percolating systems are superclusters of galaxies, which have a characteristic scale up to ~ 100 h^{-1} Mpc. Superclusters have rather different richness from small systems like the Local supercluster to

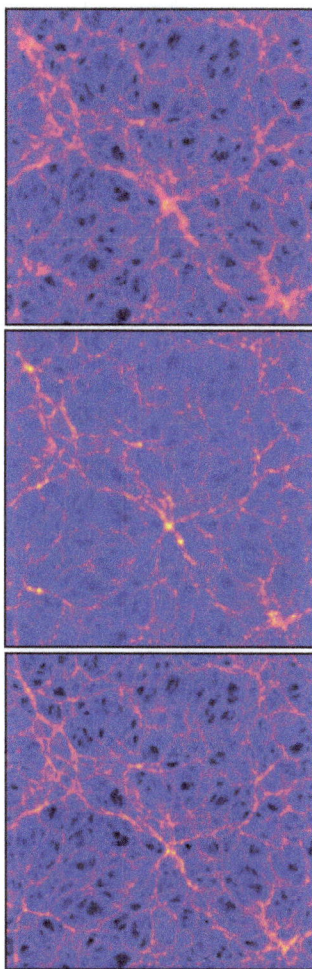

Fig. 4.1. The large-scale structure of the universe as predicted by the Zeldovich approximation (top panel), an N-body simulation (centre panel) and the adhesion model (bottom panel). For each case, the initial conditions are the same, which leads to the formation of the same large-scale pattern (Hidding, 2010).

very rich systems like the Shapley supercluster. It is clear that this variety has its origin in density perturbations of still larger scales. Thus, in order to understand the supercluster-void phenomenon correctly, the influence of very large density perturbations should also be studied.

To have both high spatial resolution and the presence of density perturbations in a large-scale interval, we use in the present analysis a series of simulations of the ΛCDM model. We use the GADGET code (Springel *et al.*, 2005b) with three different box sizes $L_0 = 256$, 512, 1024 h^{-1} Mpc with $N_{grid} = 512$, and number of particles $N_{part} = 512^3$. We call these simulations as L256, L512, and L1024. Cosmological parameters for all simulations are $(\Omega_m, \Omega_\Lambda, \Omega_b, h, \sigma_8, n_s) = (0.28, 0.72, 0.044, 0.693, 0.84, 1.00)$. Initial conditions were generated using the COSMICS code by Bertschinger (1995), assuming Gaussian fluctuations. Simulations started at redshift $z = 30$ using the Zeldovich approximation. We extracted density fields and particle coordinates for eight epochs, corresponding to redshifts $z = 30$, 10, 5, 3, 2, 1, 0.5, 0.

The ΛCDM model of structure formation and evolution combines all essential aspects of the original structure formation models, the pancake and the hierarchical clustering scenario. First structures form at very early epochs soon after recombination in places where the primordial matter has the highest density. This occurs in the central regions of future superclusters. The first objects to form are small dwarf galaxies, which grow by infall of primordial matter and other small galaxies. Thus, soon after the formation of the central galaxy, other galaxies fall into the gravitational potential well of the supercluster. These clusters have had many merger events and have "eaten" all its nearby companions. During each merger event, the cluster suffers a slight shift of its position. As in central regions of superclusters, merger galaxies come from all directions, the cluster settles more and more accurately into the centre of the gravitational well of the supercluster. This explains the fact that very rich clusters have almost no residual motion with respect to the smooth Hubble flow. According to the old paradigm galaxies and clusters form by random hierarchical clustering and could have slow motions only in a very low-density Universe.

One difficulty with the original pancake scenario by Zeldovich was the shape of objects formed during the collapse. It was assumed that forming systems are flat pancake-like objects, whereas the dominant features of the cosmic web are filaments and knots (Einasto *et al.*, 1980a). This discrepancy was explained by Arnold *et al.* (1982) and Shandarin & Zeldovich (1989), showing the importance of phase space in understanding cosmic structures. The most profound descriptions of the cosmic mass distribution, involving the full phase-space structures, are presented by Shandarin (2011), Shandarin *et al.* (2012), Abel *et al.* (2012) and Falck *et al.* (2012). They involved tessellation-based techniques to produce razor-sharp images of the cosmic web density field. Falck *et al.* (2012) presented the ORIGAMI structure finding algorithm, which finds haloes by testing whether particles have undergone shell-crossing. In recent years, progress has been made towards the development of a fully nonlinear analytical description of the evolution of the cosmic web in terms of the caustic skeleton, see Hidding *et al.* (2014) and Feldbrugge *et al.* (2018, 2019, 2023).

4.1.3 *Simulating galaxy samples*

During the simulation, we calculated for all particles and all simulation epochs the local density values at particle locations, ρ, using the positions of the 27 nearest particles, including the particle itself. Densities were expressed in units of the mean density of the whole simulation. Local densities were used to select particles for biased samples. The full ΛCDM simulation includes all particles. Following Einasto & Saar (1987), we formed biased samples that contained particles above a certain limit, $\rho \geq \rho_0$, in units of the mean density of the simulation.

This model to select particles for simulated galaxies is often called the censoring model (Mann *et al.*, 1998). It is similar to the Ising model discussed by Repp & Szapudi (2019a, 2019b). Actually, galaxy formation is a stochastic process, thus the matter density limit which divides unclustered and clustered matter is fuzzy (Dekel (1998), Tegmark & Bromley (1999)). However, a fuzzy density limit has little influence on properties of correlation functions of biased and non-biased samples, as shown by Einasto *et al.* (2019a). Biased model samples include particles with density labels, $\rho \geq \rho_0$. These samples are found from the full DM sample by excluding particles of density labels less than the limit ρ_0. In this way, biased model samples mimic observed samples of galaxies, where there are no galaxies fainter than a certain luminosity limit. We use a series of particle-density limits ρ_0 to imitate samples of galaxies with various luminosity limits. As shown in hydrodynamical simulation by Cen & Ostriker (1992) and Springel *et al.* (2005b, 2018), galaxies do not form in low-density regions of the cosmic web. However, this phenomenological method to select simulated galaxy samples from pure DM simulations is not identical to the physical method of forming galaxies in hydrodynamical simulations. Some of these differences shall be discussed in Chapter 6.

The first large modern simulation was the *Millennium Simulation*, conducted in the Max–Planck Institute for Astrophysics in Garching by Volker Springel and collaborators (Gao *et al.*, 2005a; Springel *et al.* 2005b, 2006). The simulation assumes the ΛCDM initial power spectrum. A cube of the comoving size of 500 h^{-1} Mpc was simulated using about 10 billion DM particles that allowed the evolution of small-scale features in galaxies to be followed. Using a semi-analytic model, the formation and evolution of galaxies was also analysed (Croton *et al.*, 2006; Di Matteo *et al.*, 2005; Gao *et al.*, 2005b). For simulated galaxies, photometric properties, masses, and luminosities were found.

More recent hydrodynamical simulations of galaxy formation included essential physical processes to simulate and follow galaxy formation and evolution in more detail. These simulations have in addition to DM particles baryonic particles influenced by processes described in detail in Chapter 7. Examples of such simulations are Illustris (Vogelsberger *et al.*, 2014), EAGLE (Schaye *et al.*, 2015), HorizonAGN (Dubois *et al.*, 2016), and Illustris The Next Generation (TNG) series of hydrodynamical simulations by Springel *et al.* (2018). The first analysis of matter and galaxy clustering using TNG100 and TNG300 simulations was made by Springel *et al.* (2018).

4.2 Smoothing density fields

4.2.1 *Smoothing density fields with constant smoothing length*

To compare morphological properties of simulations of different box sizes L_0, we used density fields with various smoothing lengths. The simulation procedure includes calculation of density and gravitational potential fields. For selected epochs z, we also extracted density fields along with particle coordinates and local density values. These density fields correspond to the filter of size, equal to the size of the simulation cell, $R_0 = L_0/N_{\text{grid}}$, where L_0 is the size of the simulation box and N_{grid} is the number of grid elements in one coordinate. The density fields were normalised to the average matter density, providing us with the relative density $D(\mathbf{x})$:

$$D(\mathbf{x}) = \frac{\rho(\mathbf{x})}{\overline{\rho}}, \tag{4.4}$$

where $D(\mathbf{x})$ is the density at location $\mathbf{x}$ and $\overline{\rho}$ is the mean density. The second moment of the density contrast $\delta(\mathbf{x}) = D(\mathbf{x}) - 1$ is the variance of the density field, σ^2. In the following, we call σ as the dispersion of the density field.

For smoothing, we used the B_3 spline (see Martínez & Saar, 2002):

$$B_3(x) = \frac{1}{12} \left[|x - 2|^3 - 4|x - 1|^3 + 6|x|^3 - 4|x + 1|^3 + |x + 2|^3 \right]. \tag{4.5}$$

The spline function is different from zero only in the interval $x \in [-2, 2]$. The smoothing with index i has a smoothing radius $R_i = L_0/N_{\text{grid}} \times 2^i$. The effective scale of smoothing is equal to $2 \times R_i$. We calculated density fields of simulations up to smoothing scale $R_i = 8 \ h^{-1}$ Mpc. The B_3 kernel of radius $R_B = 1 \ h^{-1}$ Mpc corresponds to a Gaussian kernel with dispersion $R_G = 0.6 \ h^{-1}$ Mpc. We note that the B_3 spline smoothing allows us to restore the continuous density field of matter in galaxies better than the Gaussian filter since it has no extended low-density stellar windows.

Figure 4.2 shows the density fields of the model L512 at the present epoch for three smoothing levels. The left panel is for the original density field without smoothing, and it corresponds to smoothing scale $1 \ h^{-1}$ Mpc. Following panels show for the same simulation density fields smoothed with 4 and 8 h^{-1} Mpc scale.

4.2.2 Defining cosmic web populations

The growth of structures on various scales during the evolution of the cosmic web is shown in Fig. 4.3. We plot here the density field of simulation L256 at epochs $z = 30$, $z = 10$, $z = 3$, and $z = 0$. This simulation has the highest resolution and allows showing the evolution on galaxy up to supercluster scales better. The upper panels show the

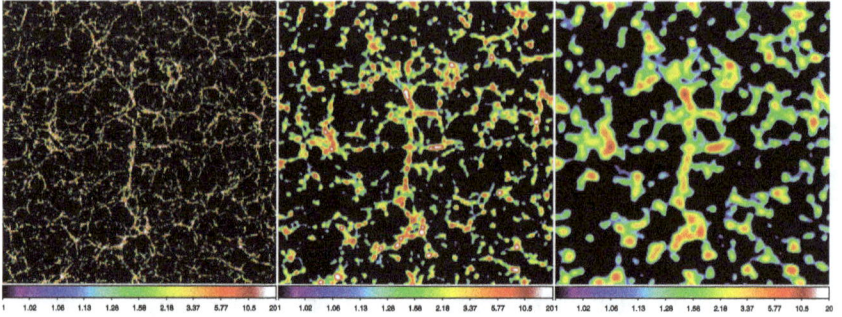

Fig. 4.2. Density fields of simulation L512 at the present epoch. Left panel shows the original field without smoothing, and central and right panel fields smoothed with kernels of size 4 and 8 h^{-1} Mpc, respectively. Density fields are shown in logarithmic scale; only overdensity regions are shown.

original L256 density fields with a cell size $L_0/N_{\mathrm{grid}} = 0.5\,h^{-1}$ Mpc, and the lower panels show the density fields smoothed with a length $R_t = 8\,h^{-1}$ Mpc, using the B_3 spline with a resolution 512^3. This smoothing method and scale are often used to determine galaxy superclusters, see Liivamägi *et al.* (2012). We show only overdensities where $D \equiv \delta + 1 \geq 1$.

In the upper panels, we show the evolution of small-scale elements of the cosmic web, galaxies and clusters of galaxies. During the evolution, they merge to form a sharp filamentary structure at the present epoch. This evolution is predicted by the theoretical models of Arnold *et al.* (1982), Bond *et al.* (1996), and others. In the lower panels, the evolution of supercluster-scale elements of the cosmic web is shown. Superclusters alter their pattern very little during the evolution, only the amplitude of density fluctuations increases.

The density field method was used to select and study various components of the cosmic web: clusters, filaments, superclusters, voids, etc. These elements are individual objects located in different areas of the universe. Their volumes do not overlap, but in sum, these components fill the whole Universe.

The moments of the density field are integrated quantities that characterise properties of the whole web. Objects of various compactness of the cosmic web can be highlighted using a smoothing of the density field with different scales. Examples of various smoothing scales were shown in Fig. 4.3. Smoothing with small lengths, $R_t \leq 1\,h^{-1}$ Mpc, highlights the whole cosmic web in the volume

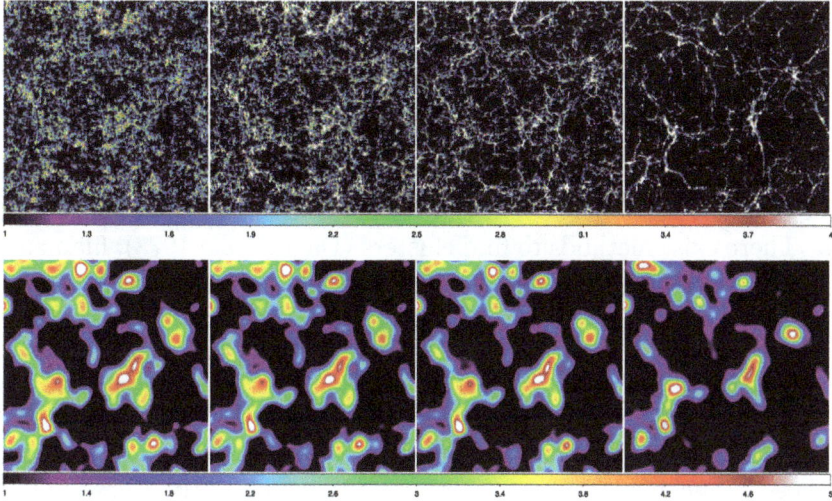

Fig. 4.3. Density fields of simulation L256 without additional smoothing and with a smoothing length $8\, h^{-1}$ Mpc are shown in the upper and bottom panels, respectively. The panels from left to right show fields for epochs $z = 30$, $z = 10$, $z = 3$, and $z = 0$ presented in slices of size $200 \times 200 \times 0.5\ h^{-1}$ Mpc. Only overdensity regions are shown with colour scales from left to right $1 - 1.4$, $1 - 2$, $1 - 4$, and $1 - 8$ in the upper panels and $1 - 1.08$, $1 - 1.25$, $1 - 1.8$, and $1 - 5$ in the lower panels.

under study on scales of haloes and subhaloes of ordinary galaxies and poor clusters. Medium smoothing lengths, $R_t = 2,\ 4\,h^{-1}$ Mpc, are suited to highlighting the cosmic web on scales of rich clusters of galaxies and the central regions of superclusters. A large smoothing with $R_t = 8-16\,h^{-1}$ Mpc highlights the cosmic web on the supercluster scale. Populations cover the whole cosmic web, they characterise the web on the selected smoothing scale. The smoothing scale is a physical parameter that allows highlighting the cosmic web at the scale of interest. Using various smoothing lengths, we can study the hierarchy of structures in the cosmic web.

4.3 Tracing cosmic web populations

Structure in the Universe emerged as a result of the gravitational growth of small amplitude primordial density and velocity perturbations. The most prominent property is its hierarchical nature.

The gravitational clustering proceeds such that small structures form first and subsequently merge into larger entities. As a result, cosmic structure consists of various levels of substructure. Hence, we need to take into account a range of scales. The second prominent aspect of the structure of the Universe is its web-like geometry, marked by elongated filaments, flattened planar structures and under-density regions — voids.

There exist methods to find clusters (Davis *et al.*, 1978), filaments (Tempel *et al.*, 2014a), and voids (Platen *et al.*, 2007), for overviews, see Martínez & Saar (2002) and Libeskind *et al.* (2018). In this section, we describe methods to find once all main populations: blobs, filaments and walls. Such methods are widely used in other fields such medicine to detect the web of blood vessels in medical images. The vascular system is a complex pattern of elongated tenuous features whose branching closely resembles a fractal network. These methods use the density field and its derivatives, and require continuous first and second derivatives of the field. Cosmological data include galaxies and simulation particles. These discrete data must be first transformed to continuous density fields using adaptive smoothing. Most methods discussed in the following section have been applied for DM particles in DM only simulations.

4.3.1 *Adaptive smoothing methods*

The main problem in the quantitative description of the cosmic web is the transition from points (particles or galaxies) to the continuous density field. The most straightforward way involves the partition of space into bins centred on the sampling points. The field is then assumed to have the constant value equal to the one at the sampling point. Evidently, this yields a field with unphysical discontinuities at the boundaries of the bins. A first-order improvement is the linear interpolation between the sampling points, leading to a fully continuous field. But we note that adaptive smoothing methods described in the following can be easily applied only for DM simulations.

In the calculation of the continuous density field, an important step is interpolation. There exist several interpolation schemes: grid-based Cloud in Cell (CIC), spherically symmetrical SPH schemes, and Triangular-Shaped Clouds (TSC) schemes, for details see Hockney & Eastwood (1981) and Schaap & van de Weygaert (2000).

Grid-based density field reconstruction depends on the grid resolution. At low resolution, high-density regions are poorly resolved, but regions of low density are recovered more accurately. At high resolution, high-density regions are better resolved, but regions of low density become hampered by shot noise.

These problems are avoided in adaptive tessellation schemes. A Voronoi tessellation of a set of spatially distributed nuclei is a space-filling network of polyhedral cells, each of which delimits that part of space that is closer to its nucleus than to any of the other nuclei (van de Weygaert, 2002). Each set of nuclei, corresponding to a Voronoi vertex, defines a unique tetrahedron, which is known as Delaunay tetrahedron. In two dimensions, the Delaunay tessellation consists of a volume-covering tiling of space into triangles; in three dimensions, these are tetrahedra, whose vertices are formed by three specific points in the dataset. These three points are selected such that their circumscribing circle does not contain any of the other data points. The Voronoi and Delaunay tessellations are intimately related, being each other's dual in that the centre of each Delaunay triangle's circumcircle is a vertex of the Voronoi cells of each of the three defining points, and conversely each Voronoi cell nucleus is a Delaunay vertex. The relation between Voronoi and Delaunay tessellations is illustrated in Fig. 4.4.

The Voronoi tessellation field estimator (VTFE) results in a reconstructed field which is constant inside the Voronoi cells and discontinuous at the boundaries of these cells. A continuous density field may be obtained with the Delaunay Tessellation Field Estimator (DTFE). The Delaunay tessellation field varies linearly inside each of the Delaunay tetrahedra. It is volume-covering and continuous, but whose derivative is discontinuous at the boundaries of the Delaunay triangles.

The construction of DTFE fields consists of three main steps: (i) construct the Delaunay tessellation corresponding to the point distribution; (ii) from tessellation to set of density estimates; (iii) from set of density estimates to volume-covering density field. These steps are illustrated in Figs. 4.5 and 4.6. As seen from Fig. 4.5, the DTFE is adaptive — at high-density regions, Delaunay tetrahedra are small, and at low-density regions large. Figure 4.6 shows that the reconstructed field is continuous. Given a point distribution (top left), one

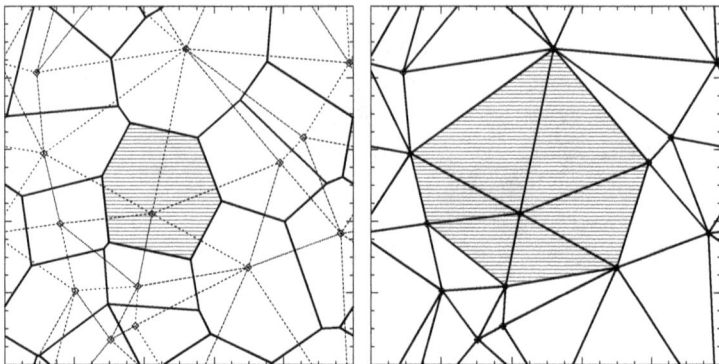

Fig. 4.4. A set of 20 points with their Voronoi (left frame: solid lines) and Delaunay (right frame: solid lines) tessellations. Left: the shaded region indicates the Voronoi cell corresponding to the point located just below the centre. Right: the shaded region is the "contiguous Voronoi cell" of the same point as in the lefthand frame (Schaap & van de Weygaert, 2000).

has to construct its corresponding Delaunay tessellation (top right), estimate the density at the position of the sampling points by taking the inverse of the area of their corresponding contiguous Voronoi cells (bottom right) and finally assume that the density varies linearly within each Delaunay triangle, resulting in a volume-covering continuous density field (bottom left).

Schaap (2007) calculated density and velocity profiles along one-dimensional sections through density fields of an N-body simulation of box size 141 h^{-1} Mpc and parameters: $\Omega_m = 0.3$, $\Omega_\Lambda = 0.7$, $H_0 = 70$ km/s/Mpc. In Fig. 4.7, a typical void-like region is shown, together with the DTFE density and velocity field reconstructions. The top left-hand frame shows the particle distribution in a thin slice through the simulation box. Top right-hand frame gives the two-dimensional slice through the three-dimensional DTFE density field reconstruction. Bottom left-hand frame gives two-dimensional slice through the three-dimensional DTFE velocity field reconstruction. Bottom right-hand frame shows the density and velocity reconstructions along the thick line shown in the other frames. We see that in the void region, velocities are directed out of voids.

Figure 4.8 shows density and velocity fields through a typical filament region. In the top left-hand frame, we see particle distribution in

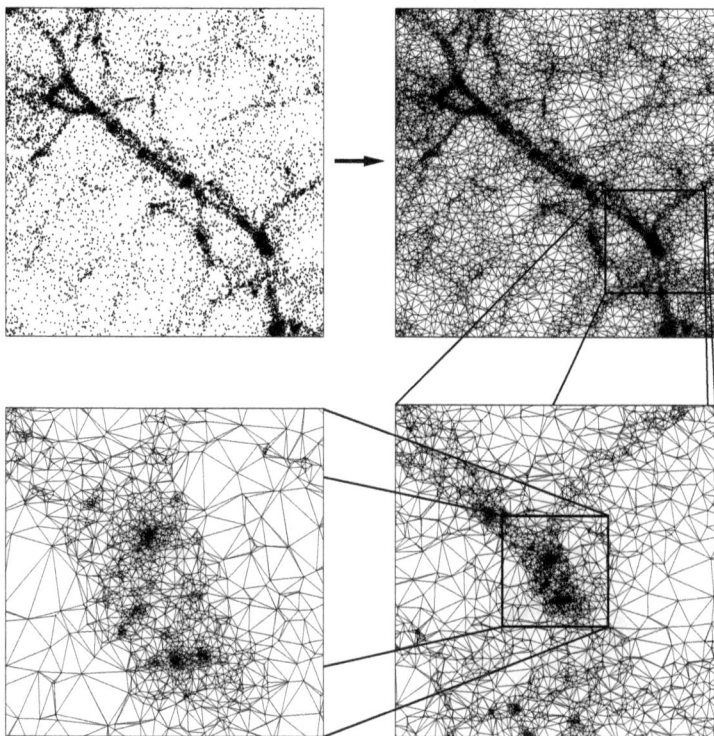

Fig. 4.5. Step 1 of the DTFE reconstruction procedure. Given a point distribution (top left), one constructs the corresponding Delaunay tessellation (top right). The bottom frames illustrate the adaptive properties of the Delaunay tessellation. These frames zoom in on the regions indicated by the squares (Schaap, 2007).

a thin slice through the simulation box. Top right-hand frame shows a two-dimensional slice through the three-dimensional DTFE density field reconstruction. Bottom left-hand frame shows two-dimensional slice through the three-dimensional DTFE velocity field reconstruction. Bottom right-hand frame shows density and velocity reconstructions along the thick line shown in the other frames. Here, velocities are directed towards the filament axis away from surrounding voids.

4.3.2 *Cosmic web populations*

Aragón-Calvo *et al.* (2007) devised the multiscale morphology filter (MMF), which allows us to identify and extract spatial patterns in

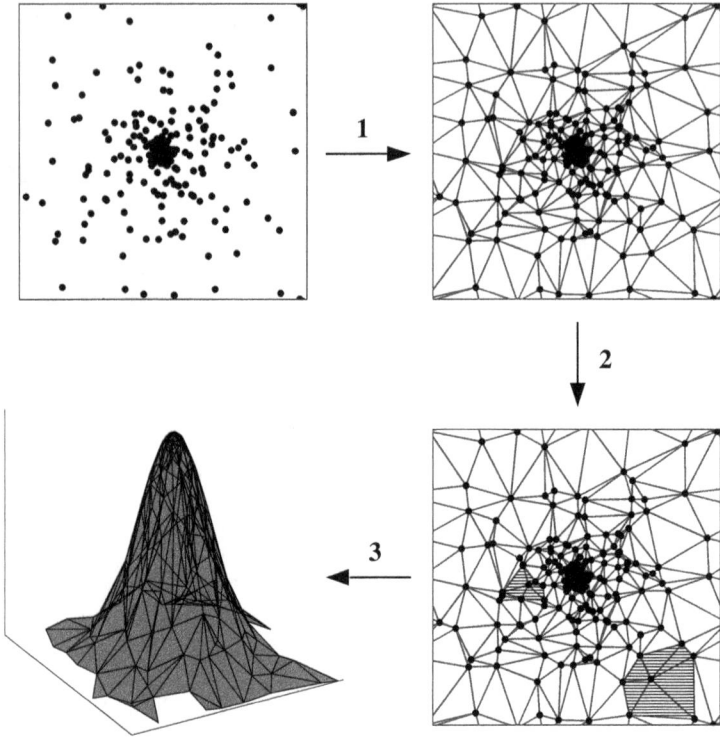

Fig. 4.6. Overview of the first-order DTFE reconstruction procedure (Schaap, 2007).

the galaxy distribution. The method uses local Hessian matrix eigen-values on various scales. The data are filtered to produce a hierarchy of maps having different resolution. At each point, the dominant parameter value is selected. Authors use the principal components of the local curvature of the density field at each point as a morphology type indicator. This requires that the density is defined at all points of a grid. To calculate the density field, the adaptive DTFE method was used.

The MMF analysis consists of several cycles: the analysis cycle, scale-space filtering, feature detection and extraction, assigning point to features. The analysis cycle requires three passes through the data, each time eliminating the features found in the previous pass. In the first pass, the blobs in the dataset are identified along with their enclosed datapoints. The points that are in blobs are eliminated and

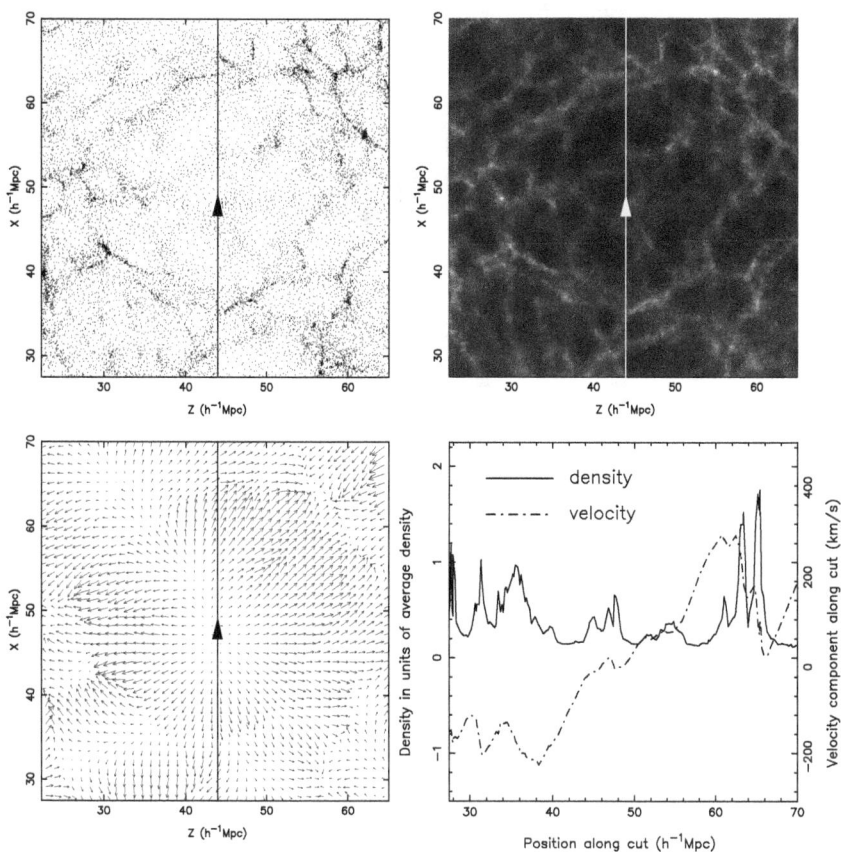

Fig. 4.7. A typical void-like region in the simulation shown (Schaap, 2007).

then the filaments are identified with their constituent points. After eliminating the filament points, the walls and their constituent points can be identified.

Aragón-Calvo *et al.* (2007) describes the analysis cycle as follows. Each pass involves 13 components and procedures: further, Aragón-Calvo *et al.* (2007) discussed in detail morphology filters. In the first pass, blob particles are found. These particles are removed for the second pass, which selected filament particles, and removed for the final third pass, which selected walls. When blob, filament and wall particles are removed, the rest are field or void particles.

Cautun *et al.* (2014) analysed the evolution of the cosmic web using the Millennium simulation (MS) by Springel *et al.* (2005b) and

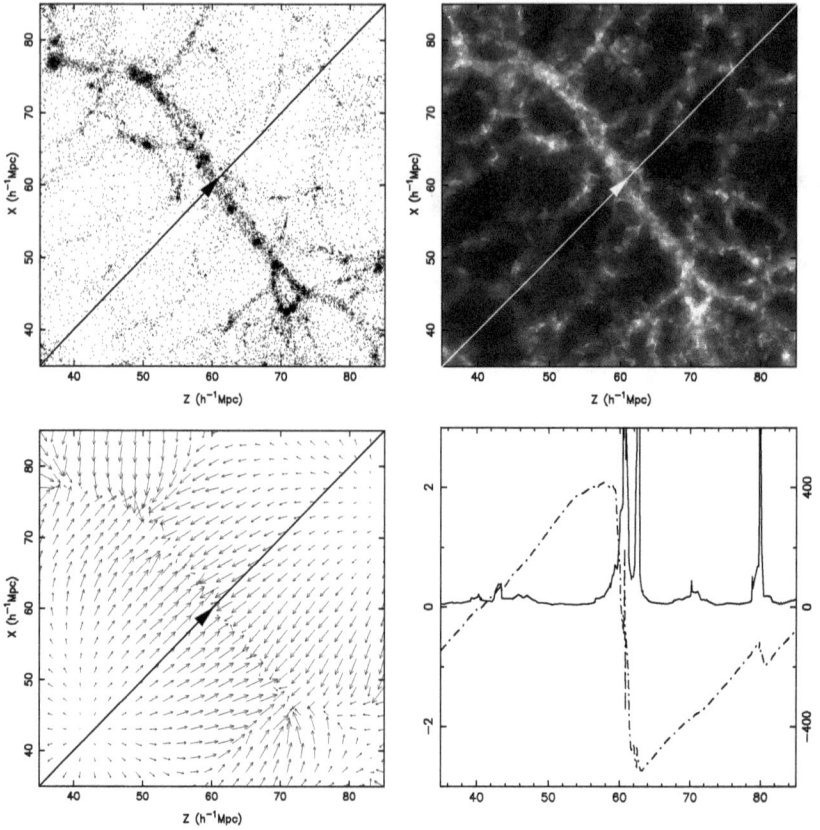

Fig. 4.8. A typical filamentary region in the simulation (Schaap, 2007).

Millennium II (MS-II) simulation by Boylan-Kolchin *et al.* (2009). Both simulations are DM only and used cosmological parameters: $\Omega_m = 0.23$, $\Omega_\Lambda = 0.75$, $h = 0.72$, $n_s = 1$ and $\sigma_8 = 0.9$. The MS simulation was made in a box of size 500 h^{-1} Mpc and the MS-II simulation in a smaller box of size 100 h^{-1} Mpc. To quantify the structure of the cosmic web, authors applied the NEXUS multiscale morphology filter technique, and its NEXUS+ version. The NEXUS and NEXUS+ methods are inspired by scale-space analysis techniques used in the medical imaging field for the detections of nodules and blood vessels. NEXUS extends the MMF formalism by Aragón-Calvo *et al.* (2007) to incorporate not only the density field but also tidal, velocity divergence and velocity shear fields. This offers a consistent and physically motivated framework for the detection of the cosmic web components using the full 6D phase-space information. The NEXUS+ method uses the log-Gaussian type filter to density field.

The NEXUS and NEXUS+ methods consist of six steps (Cautun *et al.* 2014). The population finding algorithm performs the environment detection by applying the steps first to clusters, then to filaments and finally to walls. This sequence must be followed to make sure that each volume element is assigned only a single environment characteristic. The remaining regions that are not identified as nodes, filaments or sheets, are classified as cosmic voids. Both algorithms are implemented on a grid using the density and velocity divergence fields found by the DTFE method. This means that each grid cell is classified as being part of a node, filament, wall or void. The presence of a grid implies a finite scale given by the grid spacing Δx below which we cannot study the cosmic web.

The easiest way of studying large-scale environments involves evaluating global quantities, like the mass and volume content of these structures. Such a determination is shown in Fig. 4.9, where we show the mass and volume fractions occupied by the cosmic web components identified by NEXUS+. Here, the nodes have a significant fraction of the total mass content of the universe, but occupy a negligible volume. The filamentary network contains a relatively small volume, but half of the total cosmic matter distribution. The walls have an approximately equal share of both the mass and volume fractions, with a mean density $1 + \delta \approx 1$. Voids take the largest volume fraction in the universe, but they only have $\sim15\%$ of the total mass content. The average density in voids is 0.2 in mean density units. But we note that distributions presented in this and following figures apply to DM only simulations.

Cautun *et al.* (2014) analysed the density distribution of the matter content in cosmic web populations. Figure 4.10 shows the probability distribution function (PDF) of the density field separately for each main population. The DTFE density is extrapolated to a regular grid with spacing $\Delta x = 0.4\ h^{-1}$ Mpc, with no additional smoothing. The figure shows that various environments are characterised by different density values. The node regions have typically the highest density, with values ≥ 100. The filaments also represent

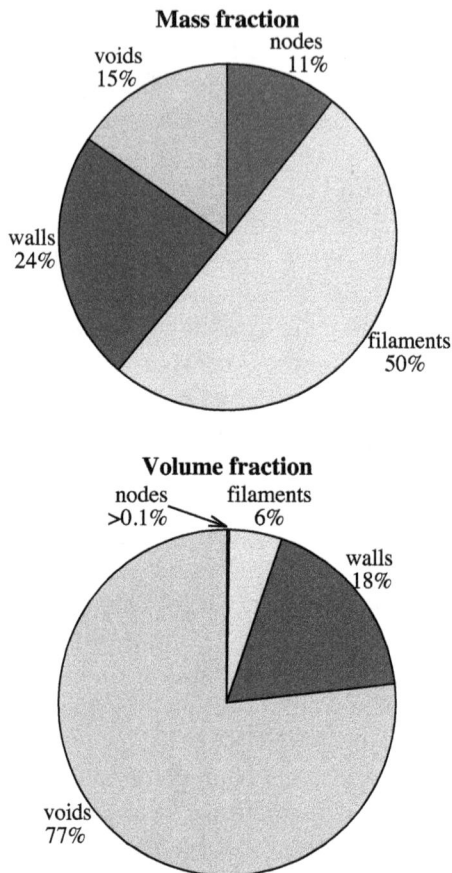

Fig. 4.9. The mass and volume fractions occupied by cosmic web environments detected by the NEXUS+ method (Cautun *et al.*, 2014).

overdense environments, but to a lesser degree than clusters. Next, we have the walls for which the density PDF peaks just below an overdensity of 1. The voids have significantly lower density values, with the distribution from $1+\delta = 10^{-2}$ to $1+\delta = 3$ with a maximum at $1+\delta \approx 0.1$ There are significant overlaps between the density PDF of different environments. This means that a simple density threshold is not sufficient in identifying the cosmic web components.

Haloes play a dominant role in the theories of structure formation, since they are the sites of galaxy formation (White & Rees, 1978). The differences in the halo population across the cosmic web

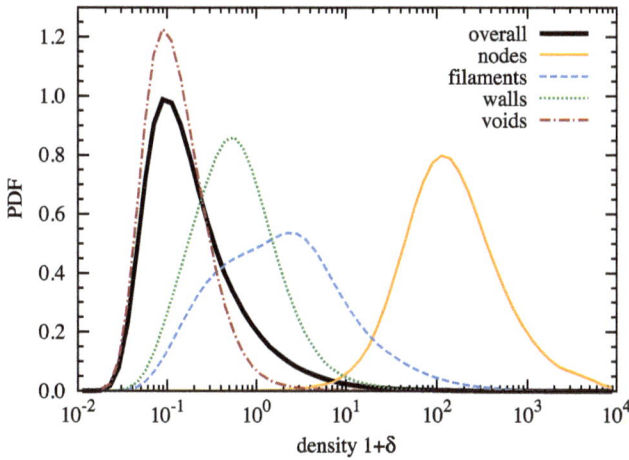

Fig. 4.10. The density PDF in each environment of the cosmic web as detected by NEXUS+. The histogram was obtained using the DTFE density field on a regular grid with spacing $\Delta x = 0.4 \, h^{-1}$ Mpc. No additional smoothing was used (Cautun *et al.*, 2014).

components are indicative of variations within large-scale environment in the population of galaxies. The halo mass function is segregated according to the morphological component in which the haloes reside. At the higher mass end, the most massive $M \geq 5 \times 10^{13} \, h^{-1} \mathrm{M}_\odot$ haloes are exclusively located in cluster (node) regions. Less massive haloes are found in filaments. Less than 10% of haloes more massive than $10^{12} \, h^{-1} \mathrm{M}_\odot$ are found in sheets. This implies that very few luminous galaxies are found in walls. This suggests that most of the galaxies in typical galaxy redshift surveys are found in filament and cluster regions, with only a small fraction of them in walls. For voids, only 5% of $10^{11} \, h^{-1} \mathrm{M}_\odot$ and higher mass haloes are located in this environment. This explains why redshift surveys find large regions of the Universe almost devoid of galaxies. Actually, all galaxies form in haloes, but very faint haloes cannot be detected in wall and void environment.

The simplest way to characterise the cosmic web evolution is to follow the mass and volume content of each of its components. This evolution is presented in Fig. 4.11. It shows that cluster (node) environment starts to contain a significant fraction of matter only at late times, and at the present time, they contain 10% of the mass.

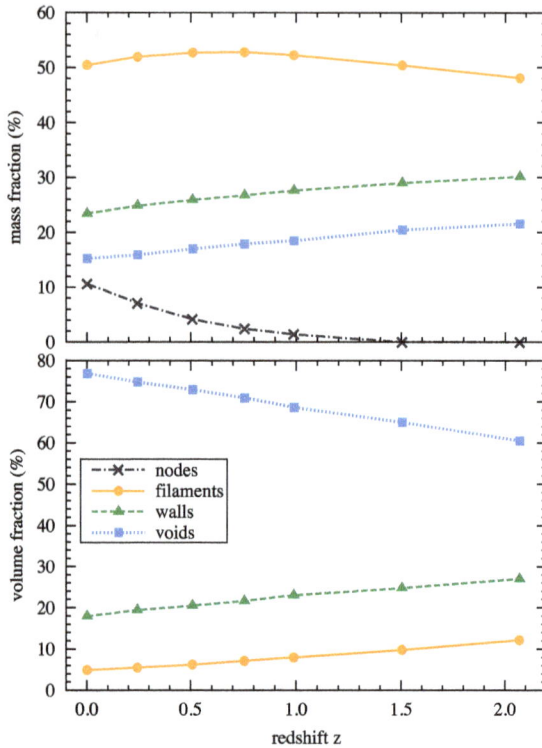

Fig. 4.11. The time evolution of the mass (top panel) and volume (bottom panel) filling fractions for nodes (crosses), filaments (solid with circles), walls (dashed with triangles) and voids (dotted with squares). The effect of cosmic variance on the mass and volume fraction is smaller than the size of the symbols and it is not shown (Cautun *et al.*, 2014).

The filaments have a more complex evolution, with an initial increase in mass until around $z \sim 0.5$, after which we find a slight decrease. The reduction in mass is due to the formation of the cosmic web nodes that accumulate a considerable share of mass, predominantly from filaments. Filaments show a factor of 2 decrease from $z = 2$ to present in volume fraction. This means that the same mass fraction gets accumulated into fewer, but more massive filaments. The cosmic sheets (walls) have a decreasing mass and volume content, and contain at present time about 20% less mass and volume than at $z = 2$. The void environments show a similar decrease in mass fraction, but show an opposite trend in volume fraction. This suggests that voids

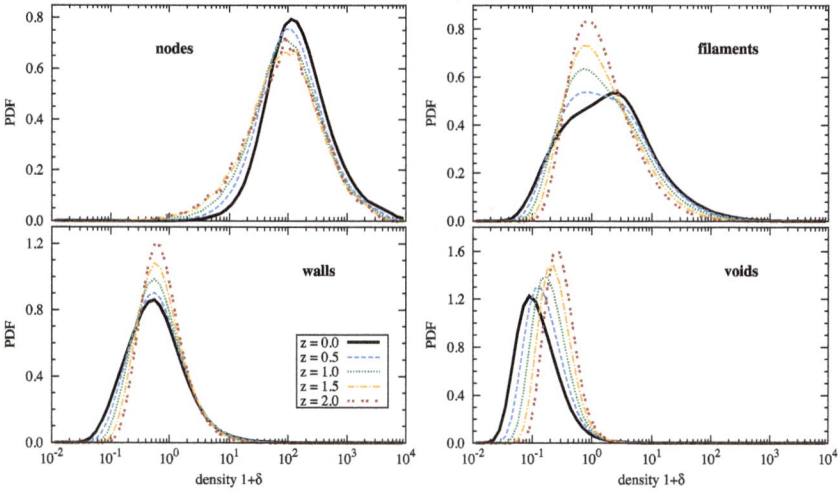

Fig. 4.12. The evolution with redshift of the density PDF for each cosmic web environment: nodes (top left), filaments (top right), walls (bottom left) and voids (bottom right). The density $1 + \delta$ is expressed in units of the mean background density at each redshift. The histogram was obtained using the DTFE density field on a regular grid with spacing $\Delta x = 0.4\ h^{-1}$ Mpc. No additional smoothing was used (Cautun *et al.*, 2014).

do not only increase by merging with other voids but also by taking over regions that were previously identified as walls and possibly filaments. The main conclusion from this study emphasises the dominant role played by filaments, which contain the largest share of mass at all redshifts.

The evolution of density PDF of each cosmic web component is shown in Fig. 4.12. Here, the area under each curve is constant and equal to unity. A clear evolution is seen in walls. Here, the high density tail does not change significantly, but there are more underdense regions. Since the mean wall density is almost constant with time, it suggests that most of the mass content of sheets is located in a few very massive regions, but small in volume. Many of the wall regions have low to very low densities, with no massive haloes in them. This explains why cosmic sheets are so difficult to identify in both simulations and observations. The voids show a clear shift of the density PDF towards lower $1 + \delta$ values at present time, which suggests a significant emptying of void regions.

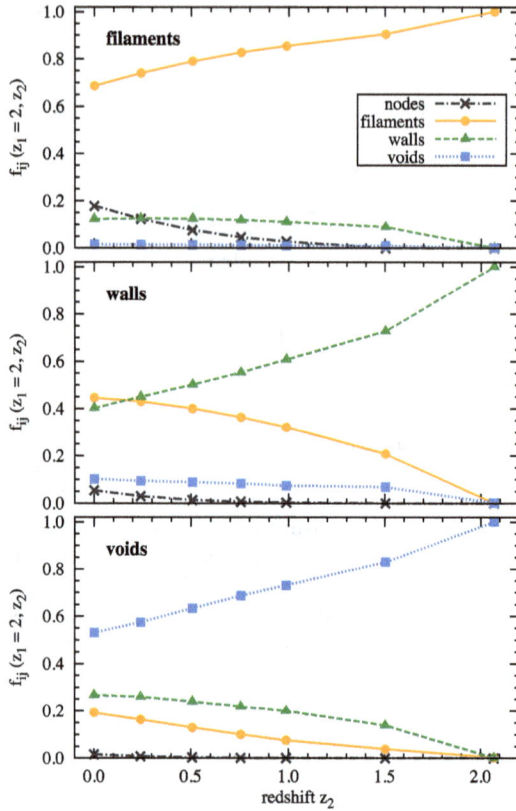

Fig. 4.13. Tracking the final destination of the $z = 2$ mass found in cosmic web filaments (top), walls (centre) and voids (bottom). The y-axis gives how the mass in a given environment at $z_l = 2$ was split among the cosmic web components at lower redshift z_2. Note that at $z = 2$, we do not find any cosmic nodes Cautun *et al.* (2014).

Figure 4.13 shows another way of looking at the evolution of matter in the cosmic web. It shows the successive destinations of the matter that is initially, at $z = 2$, identified as being part of filaments, walls and voids. The main effect is the outflow of mass from walls and voids. Only about 40% of the walls redshift $z = 2$ mass is still part of present-day's sheets, while most of the mass was flowing into filaments. For voids, around half of their high redshift mass has streamed out into sheets and filaments, showing the outflow of mass from underdense regions.

One possibility to characterise the nature of the distribution of matter in the cosmic web components is to measure their fractal dimension. Cautun *et al.* (2014) made an analysis of the structure using the box-counting method by Mandelbrot (1986). It involves overlying the simulation box by a regular grid with spacing l and counting how many of the grid cells intersect the pattern that we measure, i.e. the filamentary and wall networks. The number of intersecting grid cells gives the box count N at scale l. The method works by measuring the box counts at different scales and then investigating the dependence of N on l. In the case of a fractal, there is a well-defined relation $N \propto l^{-d}$, where d is the fractal dimension of the pattern. For example, the fractal dimension of an infinitely thin line is $d = 1$, of a zero thickness plane is $d = 2$ and that of a filled box is $d = 3$. In general, a fractal pattern has a non-integer fractal dimension, showing an intermediate behaviour between the ideal cases.

To obtain the fractal dimension, Cautun *et al.* (2014) measured the box count N for the largest possible box, which is the simulation box. After this, in each successive step, authors reduced the box length l by a factor of 2 and measured N again. This process was stopped when l was equal to the grid spacing used to obtain the filamentary and wall networks. This procedure was applied to the MS-II data since it allowed us to obtain a larger dynamical range at small l, which shows the most interesting behaviour. When comparing between MS-II and MS results, authors could not find any important differences, suggesting that the MS-II findings are not significantly affected by cosmic variance.

Figure 4.14 shows the time evolution of the fractal dimension d and that of the breaking length l_0 for both filaments and walls. Filaments are characterised by $d \sim 2.2$ which shows that they have a fractal dimension higher than that of a thin plane. This seems puzzling at first sight, since the filamentary network is made of many line-like objects and therefore we would expect $d < 2$. Actually, filaments are not infinitely thin lines, but have an intrinsic width, and hence, we can have $d > 2$. Figure 4.14 shows a strong time evolution of the fractal dimension, with larger values at high redshift. It shows the decrease in complexity of filaments, with lower values suggesting a simpler network with fewer branches, in agreement with the visual impression of the cosmic web. In the case of walls, we find only a

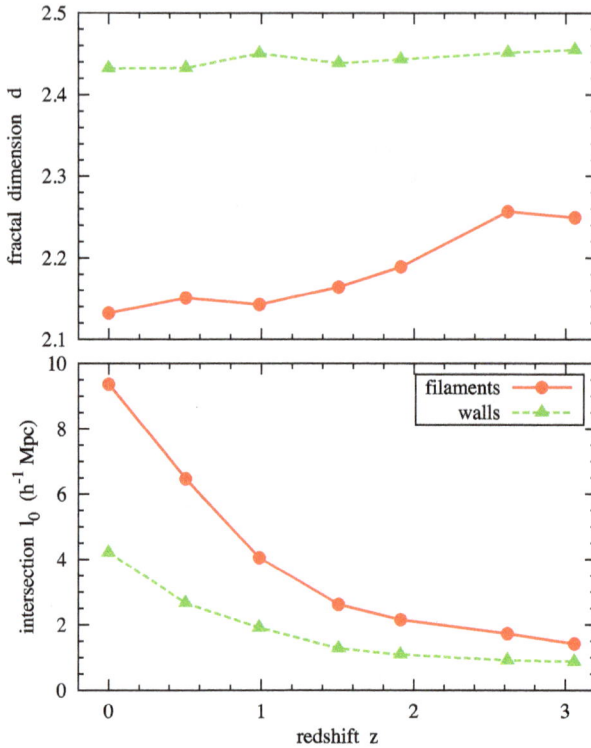

Fig. 4.14. Top panel: the variation with redshift z of the fractal dimension d for filaments (solid curve) and walls (dashed curve). Bottom panel: the variation with redshift of the breaking scale l_0 for both filaments and walls (Cautun *et al.*, 2014).

very weak time evolution of the fractal dimension, with slightly lower values at present day.

But note that here we discuss the fractal structure of various cosmic web populations, not of the whole web, as done later in Chapter 5.

4.4 Density fields of galaxies and matter

4.4.1 *The luminosity density field*

To understand better differences between real and simulated density fields, we have to study in detail the real luminosity density field. As an example of the real density field, we can chose the equatorial slice of the Sloan Digital Sky Survey (SDSS). The density field is usually described by the amplitudes of Fourier components of the field expressed by the power spectrum of density perturbations. However, the spatial structure of the field depends both on the amplitudes and on the phases of the density field. Thus, both aspects of the field must be investigated.

The importance of the phase information in the formation of the cosmic web has been understood long ago. To demonstrate the role of phase information, we extracted from the density field of the SDSS Northern equatorial slice a rectangular region with box size 512 Mpc calculated for Hubble constant $h = 0.8$. We Fourier-transformed the density field and randomised phases of all Fourier components and thereafter Fourier-transformed it back to see the resulting density field. The modified field has on all wavenumbers k the same amplitudes as the original field, only phases of waves are different.

Results are shown in Fig. 4.15. Upper left panel shows the rectangular regions extracted from the high-resolution luminosity density field of the Northern equatorial slice of the SDSS. The density was calculated using Gaussian smoothing with an rms scale 0.8 h^{-1} Mpc. The observer is located at the lower left corner. Colour-coded density levels used in plotting were in the interval from 0 to 10 in mean density units. The upper right panel shows the same density field as in the previous panel, but phases are randomly shifted. Densities are here expressed in a linear scale. We see that the whole structure of superclusters, filaments and voids has gone, the field is fully covered by tiny randomly spaced density enhancements. There are no clusters of galaxies in this picture, comparable in the luminosity to real clusters of galaxies. This simple test shows the importance of phase information in the understanding of the structure of the cosmic web.

It is well known that the Fourier space is not sensitive to the location of particular high-density features in real space, such as filaments, clusters, and superclusters. To have a better understanding of the texture of the cosmic web, the web must be studied in real space. For this purpose, wavelet analysis can be applied, which analyses properties of waves of various scales in real space. In the wavelet analysis, the density field is decomposed into several frequency bands as follows. The high-resolution (zero level) density field is calculated with the kernel of width equal to the size of one cell of the field.

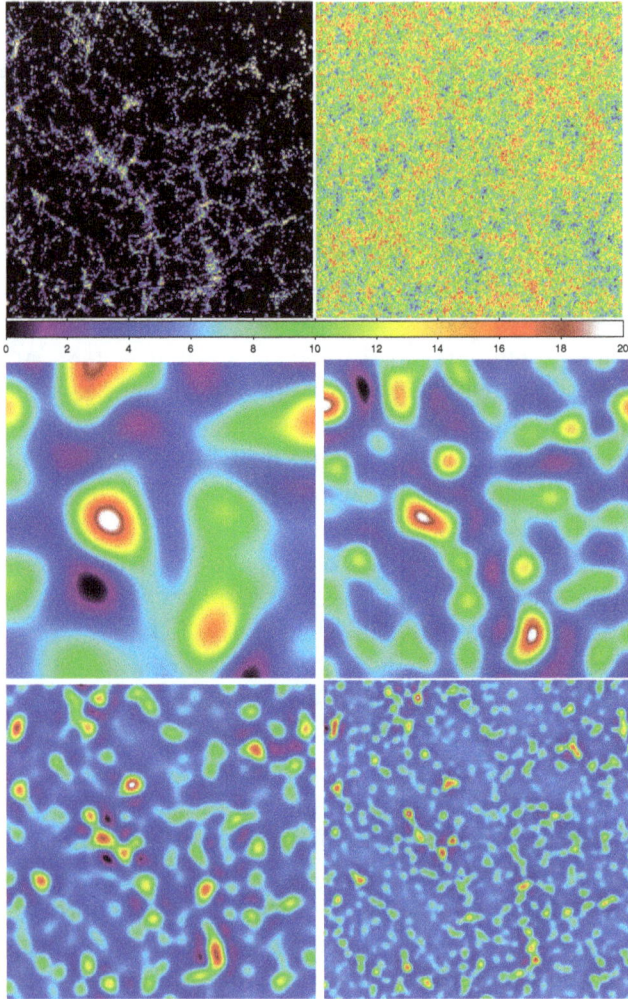

Fig. 4.15. Role of phase information. For explanation, see text (Einasto, 2024).

Every next field is calculated with the twice larger kernel. Wavelets are found by subtracting higher level density fields from the previous level fields. In such a way, each wavelet band contains waves twice longer the previous band in the range $\pm\sqrt{2}$ times the mean (central) wave (Martínez & Saar, 2002). The sum of all wavelets is by construction the original high-resolution density field.

Wavelets w7, w6, w5 and w4 of the slice are shown in middle and lower panels of Fig. 4.15, colour coding is linear. Supercluster numbers are according to the catalogue by Einasto *et al.* (2001). The middle left panel of Fig. 4.15 shows the waves of length about 256 Mpc. Here, highest density regions are three very rich superclusters: N20 from the list by Einasto *et al.* (2003), located in the upper left part of the figure, supercluster N13 (SCL126 from the list by Einasto *et al.*, 2001) near the centre (a member of the Sloan Great Wall), and supercluster N02 (SCL82) in the lower right part of the panel. The middle right panel shows waves of scale about 128 Mpc. Here, the most prominent features are superclusters N13 (SCL126) and N02 (SCL82) and also the supercluster N23 (SCL155) in the upper left part of the panel is fairly strong (seen as weak density peak already in the previous panel). In addition, we see the supercluster N15 just above N13 near the minimum of the wave of 256 Mpc scale and a number of poorer superclusters located mostly in voids defined by waves of larger size.

The lower panels plot waves of scale about 64 and 32 Mpc. Here, all superclusters seen on larger scales are also visible. A large fraction of density enhancements are either situated just in the middle of low-density regions of the previous panel, or they divide massive superclusters into smaller subunits. This property is repeated in the lower right panel. Here, the highest peaks are substructures of rich superclusters, and there are numerous smaller density enhancements between the peaks of the previous panel.

For comparison, we present in Fig. 4.16 the high-resolution luminosity density field for a spherical layer at a distance 240 h^{-1} Mpc from us. This picture shows the Sloan Great Wall in the context of the whole supercluster-void network. The thickness of the shell is 10 h^{-1} Mpc. In the plotting, we used SDSS coordinates η, λ. Since we use a spherical shell at a fixed distance, distance-dependent selection effects are excluded. We see that in the lower part of the figure, a huge complex of several superclusters is located — this complex is called the Sloan Great Wall; actually, it consists of three very rich superclusters, see Liivamägi *et al.* (2012).

When we compare density waves of all scales, then we come to the conclusion that superclusters form in regions where large density waves of various scales combine in *similar over-density phases*: the

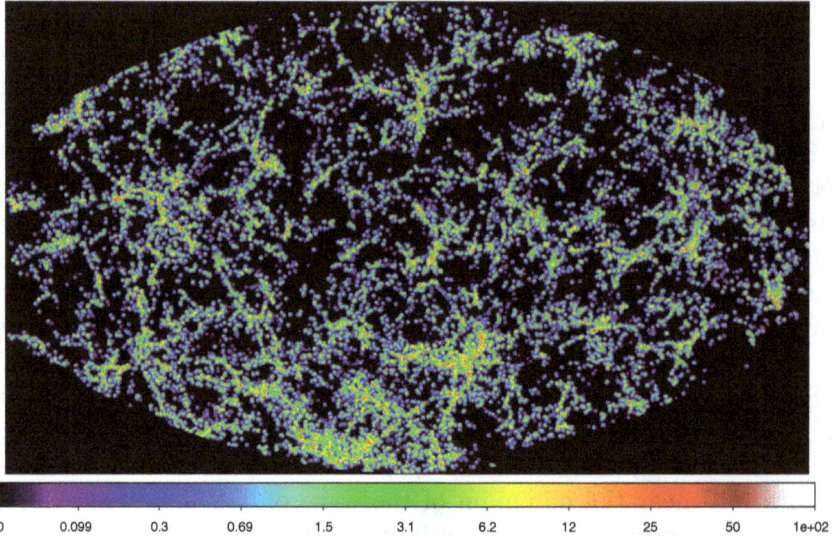

Fig. 4.16. The luminosity density field of the SDSS in a spherical shell of $10\,h^{-1}$ Mpc thickness at a distance of $240\,h^{-1}$ Mpc. To enhance the faint filaments in voids between the superclusters, the density scale is logarithmic in units of the mean luminosity density for the whole DR7. The rich complex in the lower area of the picture is part of the Sloan Great Wall; it consists of three very rich superclusters (Einasto, 2024).

larger is the scale of the wave where this coincidence takes place, the richer the superclusters are. Similarly, voids form in regions where large density waves of various scale combine in *similar under-density phases*. In large voids, medium-scale perturbations also generate a web of filamentary structures with knots. But in voids the overall density level is low and decreases with the time, thus the inflow of primordial DM-particles to filaments and knots inside large voids is prevented.

This simple analysis demonstrates very clearly the role of phase synchronisation of density waves in the formation of the supercluster-void network.

ARIEL: The cosmic web looks amazing. Like an abstract expressionist painting.

RHEA: Probably due to the range of scales that exhibit a fractal nature.[5]

---

[5]Taylor *et al.* (1999).

CALLISTUS: I'm glad to see you warming up to the f-word.

ARIEL: But it also looks so much like a slice of the brain. Ever wondered why?

RHEA: Whoa. That's just a superficial similarity; I'd make nothing of it.

ARIEL: You're upset!

RHEA: It isn't hate, it's just indifference.

CALLISTUS: There is a paper[6] claiming a similarity between the birth and evolution of biological and cosmological systems. It is called a Darwinian cosmology.

RHEA: I call bull. In what sense is it Darwinian? How does the fittest galaxy proliferate its DNA? And what is DNA in this analogy?

CALLISTUS: They claim DNA corresponds to matter. But colour me sceptical. The parallels are unconvincing, if not superficial.

ARIEL: Is birth the Big Bang?

CALLISTUS: The inflationary expansion of space with constant uniform density is only vaguely similar to the exponential growth of an organism from a single cell.

RHEA: A cell uses compressed information to take energy and matter from their environment for growth. The universe is self-contained and does not contain compressed code, at least to the best of our knowledge.

CALLISTUS: I'm open to ideas and can detect vague similarities, but they lose me at cosmic fertilisation and male and female galaxies.

RHEA: I'm pretty sure galaxies are non-binary. Let's turn the tables on the biologists: we should use Einstein's general relativity for biological systems.

ARIEL: Yeah. Everything is relative. I see a Nature paper followed by a quick media blitz. Are you in?

CALLISTUS: Right. Let's end sarcasm here. It does reflect the open-mindedness of our community that this paper was published. I found another paper[7] that corroborates your visual impression statistically: the power spectrum of brain nodes is eerily similar to the power spectrum of dark matter from high-resolution simulations, but it is dissimilar from other natural processes like sky clouds, trees, even MHD and water turbulence, which display a more straightforward power law correlation function usually associated with fractals.

RHEA: And Pollock paintings.

ARIEL: But can we conclude anything definite from such statistical similarities?

---

[6]Kleinmann (2016).

[7]Vazza & Feletti (2020).

CALLISTUS: Probably not. Fractals and multi-fractals are ubiquitous in complex systems. Unless you unearth credible parallels between the dynamics of gravitational evolution and network growth in a brain, the morphological similarities count as extremely feeble evidence.

RHEA: Moreover, visual impression and resemblance depend on how you display and analyse the data.

ARIEL: I guess our sense of beauty must be rooted in nature due to evolutionary psychology. We will find maps of complex natural processes beautiful, whether it's the universe or the brain, clouds, or trees. Given the plethora of complex biological systems, it could as well be pure chance that one of them is similar to the universe.

4.4.2   The evolution of density perturbations of various scales

Our next task was to look at how the phase coupling or synchronisation may arise. To see this process, we use the wavelet decomposition of the density field. Figure 4.17 shows the high-resolution density fields of the model L256 at four redshifts: $z = 10$, 5, 1, 0. Wavelet decompositions at levels w6, w5, and w4 for the same redshifts are shown in the same figure. Colour-coding of wavelets at different redshifts is chosen so that a certain colour corresponds approximately to the density level corrected by the linear growth factor for that redshift. Blue wavelet colours correspond to under-dense regions of density waves, green colours to slightly over-dense regions, and red and white colours to highly over-dense regions.

We see that the pattern of the cosmic web on the scale of the largest wavelet w6 is almost identical at all redshifts; only the amplitude of the density waves is increasing approximately in proportion to the linear growth factor. This linear growth is expected for density waves of large scales, which are in the linear stage of growth. The pattern of the web of the wavelet w5 changes little, but the growth of the amplitude of density waves is more rapid. The pattern of the wavelet w4 changes much more during the evolution, and the amplitude of density waves increases more rapidly, but essential features remain unchanged, i.e. the locations of high-density peaks and low-density depressions are almost independent of the epoch. In other words, *density perturbations of medium and large scales have a tendency of phase coupling or synchronisation at peak positions.* Figure 4.17 shows that the synchronisation of medium and large scales also applies to underdense regions.

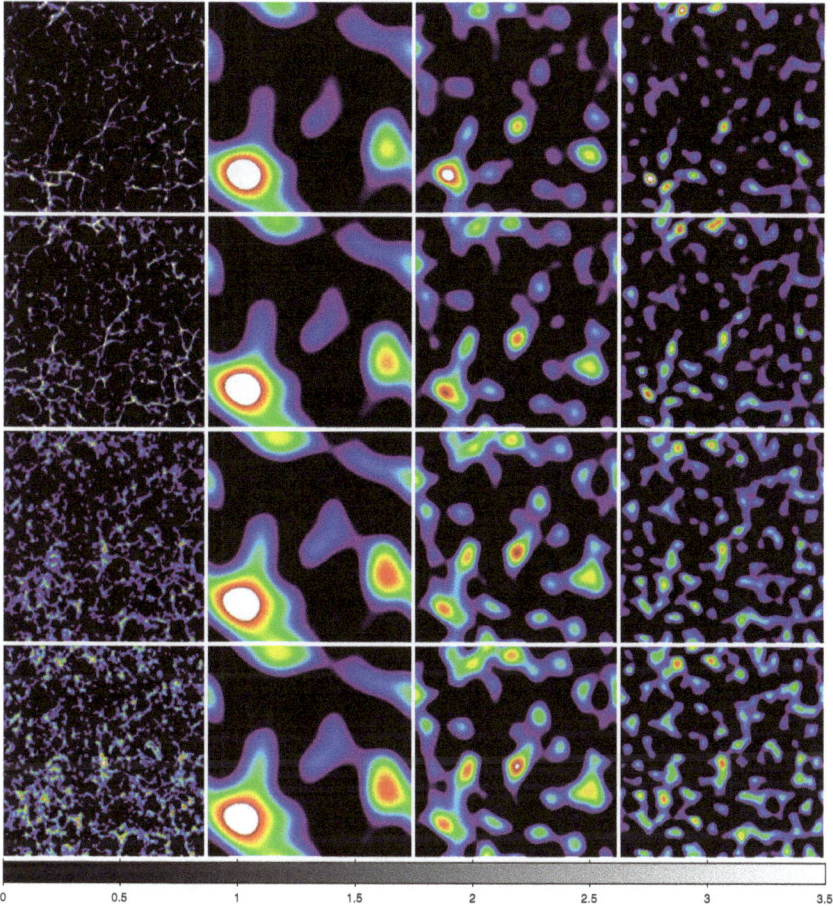

Fig. 4.17. The high-resolution density field of the full model L256 (where density perturbations of all scales are present) is shown in the left column at $k = 153$ coordinate. The second, third, and fourth columns show the wavelet w6, w5, and w4 decompositions at the same k, respectively. The upper row gives data for present epoch, $z = 0$, the second row for redshift $z = 1$, the third row for redshift $z = 5$, and the last row for redshift $z = 10$. Densities are expressed in a linear scale (Einasto, 2024).

This analysis suggests that the synchronisation of peak positions of wavelets of *various scales* represents a general property of the evolution of the density field of the Universe.

Chapter 5

Integrated Properties of the Cosmic Web

In this chapter, we discuss the basic statistical descriptors of the cosmic web, such as the percolation function and the correlation function and its derivatives. These functions characterise the connectivity and clustering properties of the web. Before the recent emergence of wide-field redshift surveys, correlation functions were calculated from the two-dimensional (2D) distribution, and three-dimensional (3D) functions were estimated from inverting the 2D function. To estimate the accuracy of this procedure, we study the relationship between 2D and 3D correlation functions.

5.1 Percolation properties of galaxies and matter

ARIEL: Percolation is... interesting. Especially when I make coffee. But it's a surprise cameo in cosmology.

RHEA: Sometimes, technologies proliferate for no good reason. Other times, more is more.

ARIEL: We studied the power spectrum, then correlation functions. They carry almost the same information. And now percolation. Where does it end?

CALLISTUS: It's sometimes hard to tell the difference between a want and a need. Like for coffee or percolation. But I do see a need for a wide array of techniques because the universe is highly non-Gaussian. The power spectrum perfectly describes a Gaussian distribution: it's the variance in each mode with zero mean.

RHEA: The modes are uncorrelated, even independent, in a homogeneous spatial Gaussian distribution.

ARIEL: Like in the early universe.

CALLISTUS: At least to a very good approximation. CMB measurements constrain primordial non-Gaussianity to a degree that it's negligible for most purposes.

RHEA: But then gravity comes into play. It amplifies fluctuations first linearly and then non-linearly.

ARIEL: Like a guitar amp that starts to distort!
             [*Ariel imitates a rock guitar player.*]

RHEA: Exactly. The same guitar I always played. Distortion creates non-Gaussianity. To paraphrase Tolstoy,[1] all Gaussian distributions are alike, but non-Gaussian distributions are non-Gaussian in their own ways.

CALLISTUS: Hence, a priori, we don't know which statistical tools are the most efficient to characterise them. Different methods reveal different properties of the cosmic web. Percolation analysis is one of them, a brilliant technology transfer from solid-state physics to cosmology.

RHEA: After all this talk about percolation, I need a coffee. Does anybody want to join me?

CALLISTUS: I'm all for turning caffeine into scientific theories.

ARIEL: The mathematician Erdős was also famous for turning coffee into mathematical theorems. Later in his life, he switched to amphetamines and produced even more papers. It's true!

RHEA: Ha, don't believe everything you read on the Internet. Let's have coffee.

5.1.1 *Extended percolation analysis*

One essential geometrical property of the cosmic web is the connectivity of its components. Clusters are connected by filaments to superclusters and the whole web. Similarly, voids form a complex system connected by tunnels. The connectivity property is analysed in percolation theory and applied in physics, geophysics, medicine, etc.; for an introduction to percolation analysis, see Stauffer (1979). Zeldovich *et al.* (1982) and Melott *et al.* (1983) applied the percolation method to particles and galaxies.

A natural extension of the method is to use the density field instead of particles, permitting the study of the connectivity of

---

[1] *All happy families are alike; each unhappy family is unhappy in its own way.* This is also known as the *Anna Karenina principle.*

overdense and underdense regions above and below an appropriate threshold density. Connected high-density regions are defined as clusters and connected low-density regions as voids. If the threshold is high, then clusters are small and are isolated from each other. As the threshold density decreases, clusters start to merge, and at a certain threshold, the largest cluster spans the whole volume under study. This threshold is the percolating density threshold. Analogously, the connectivity of voids is characterised by low-density regions below the threshold density.

Initially, percolation studies have been concentrated on clusters and voids near the percolating threshold density (Shandarin & Zeldovich, 1983, 1984). An extended version of the percolation analysis, suggested by Gott *et al.* (1986) and Einasto & Saar (1987), differs from previous analyses as follows: (i) we use a wide threshold density interval to find cluster and void lengths and filling factors; (ii) we use a large range of smoothing lengths to describe the density field of DM and galaxies in a complex way. We use only positional data that are available for DM particles and galaxies, and ignore velocities that are available only for particles. We consider DM as a physical fluid with a continuous density distribution. Thus, simulated DM particles are only markers of the field. Similarly, we consider observed galaxies sampling a smooth luminosity density field. The extended percolation analysis is one of the few methods that allows for comparing the dark matter and real galaxy density fields.

We divide the cosmic web under study at each threshold density into high-density and low-density systems. For each threshold, we find catalogues of clusters and voids, and select the largest clusters and the largest voids. Lengths and volumes of the largest clusters and voids, and numbers of clusters and voids at their respective threshold density levels, as functions of the threshold density, are used as percolation functions. Percolation functions allow an easy, compact, and intuitive representation of the integral geometry of the whole web, i.e. ensembles of all clusters and voids for a particular parameter set. Another outcome of the method is a catalogue of clusters and voids with individual information on each object.

As is traditional in percolation analyses, we identify high-density regions as clusters and low-density regions as voids. In our context, these terms mean geometric clusters and voids. Geometric clusters can contain sub-clusters as physical clusters (the standard definition of clusters outside of percolation theory) and superclusters connected by filaments and sheets. Similarly, geometric voids may consist of physical voids that are connected by intermediate-density tunnels. The extended percolation analysis is based on the clustering Friends-of-Friends (FoF) algorithm, which also identifies physical clusters and voids, but not filaments and walls.

In the first application of the extended percolation method, Gott *et al.* (1986) and Einasto & Saar (1987) analysed the spatial distribution of simulated and observed density fields, as given by numerical simulation of the cold dark matter (CDM) model by Melott *et al.* (1983) and the observational CfA sample of galaxies. Both studies demonstrated that the topological properties of simulated and real samples depend on the density threshold level to select over- and under-density regions. At low threshold densities, voids form isolated bubbles in a single high-density region. This topology is similar to a *Swiss cheese* model. At medium threshold levels, the topology is sponge-like: both clusters and voids form intertwined continuous regions throughout the space. At high threshold densities, there are numerous isolated clusters in continuous voids, often called *meatball* topology.

Gott *et al.* (1986, 1987, 1989) applied the topological analysis to find the genus of density fields. A more detailed analysis of the genus statistics is given by Martínez & Saar (2002). The genus of the density field is defined as the number of holes minus the number of isolated regions plus one.

5.1.2 *Evolution of percolation functions*

To understand the geometric properties of the cosmic web at various stages of its evolution, Einasto *et al.* (2018) calculated percolation functions for five epochs of the evolution of the cosmic web corresponding to redshifts $z = 30$, 10, 3, 1 and the present epoch, $z = 0$. Functions were calculated for the model L512, using original density fields without additional smoothing, as found from simulations.

Figure 5.1 shows how percolation functions change during the evolution of the cosmic web and obtain the form at the present epoch. Left panels show the lengths of the largest clusters and voids, $\mathcal{L}(D_t) = L_{max}/L_0$; central panels display the filling factors of the largest clusters and voids, $\mathcal{F}(D_t) = V_{max}/V_0$; and right panels show the numbers of clusters and voids, $\mathcal{N}(D_t)$; all as functions

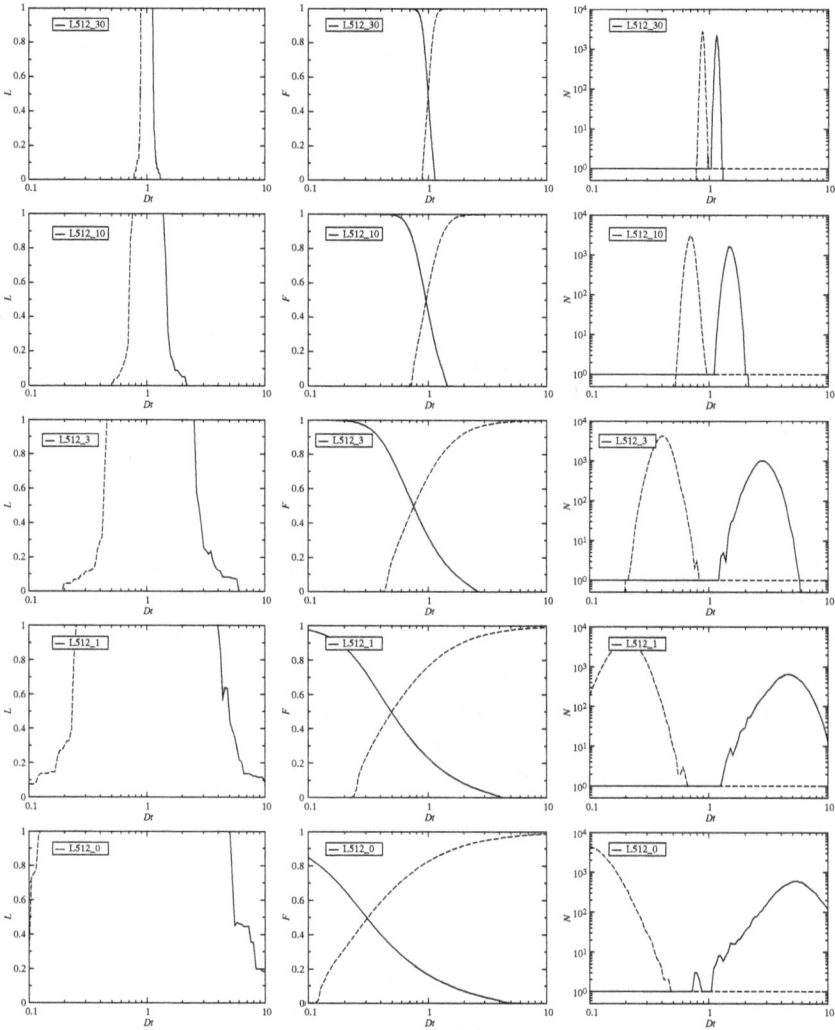

Fig. 5.1. Change of percolation functions of the model L512 with simulation epoch. Left, central and right panels indicate the lengths of largest clusters and voids, $\mathcal{L}(D_t) = L_{\max}/L_0$, filling factors of largest clusters and voids, $\mathcal{F}(D_t) = V_{\max}/V_0$, and numbers of clusters and voids, $\mathcal{N}(D_t)$, respectively, as functions of the threshold density, D_t. Panels from top to bottom denote the initial epoch, $z = 30$, and epochs $z = 10$, $z = 3$, $z = 1$, and $z = 0$ (Einasto *et al.*, 2018).

of the threshold density, D_t. Functions are calculated for the original non-smoothed density fields, which correspond to a resolution of $1\,h^{-1}$ Mpc. Functions for clusters are plotted with solid lines and for voids with dashed lines.

At the early epoch $z = 30$, there are no voids at threshold densities $D_t \leq 0.8$ and no clusters at $D_t \geq 1.2$. As the evolution proceeds, the interval of threshold densities, where clusters and voids exist, increases. At early epochs, the log-percolation functions of clusters and voids are rather symmetrical. This symmetry is gradually lost during evolution. At the early epoch $z = 30$, the distribution of densities is almost Gaussian and symmetric around the mean density, $D_t = 1.0$. The percolation functions at early epochs are very close to those of purely random samples. Thus, the change of percolation threshold density with epoch characterises the transition of the initial Gaussian density field to its present non-Gaussian form.

Gott *et al.* (1987) analysed how isodensity surfaces evolve. The authors used numerical simulations of the standard $\Omega = 1$ CDM model with identical initial fluctuations and calculated isodensity surfaces for several fractions of matter in low- and high-density regions. Results of the analysis are shown in Fig. 5.2. The density contours at both epochs are very similar. As expected, we see the isolated bubbles (Swiss cheese) topology at the low, the sponge topology at the medium, and the meatball topology at the high threshold.

5.1.3 Percolation functions of DM and SDSS clusters and voids

Percolation functions of DM and SDSS clusters and voids at the present epoch are shown in Fig. 5.3. Panels from left to right present lengths of largest clusters and voids, $\mathcal{L}(D_t) = L_{\max}/L_0$, filling factors of largest clusters and voids, $\mathcal{F}(D_t) = V_{\max}/V_0$, and numbers of clusters and voids, $\mathcal{N}(D_t)$, as functions of the threshold density, D_t. Lengths and filling factors are expressed in units of the total lengths and volumes of samples. The top panels represent the DM model L512; the bottom panels represent the SDSS samples. Percolation functions were found using smoothing kernels of radii 1, 2, 4, and $8\,h^{-1}$ Mpc, plotted with green, blue, black and red lines, respectively. Solid lines show data on clusters, and dashed lines show

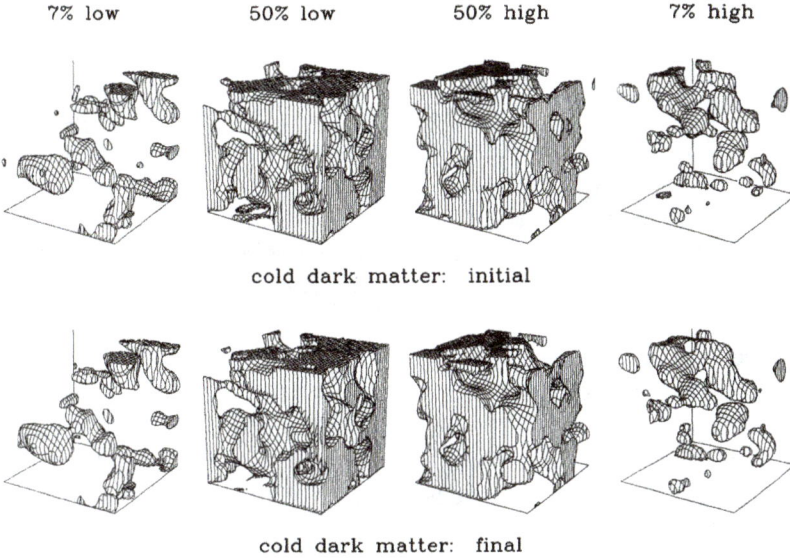

cold dark matter: initial

cold dark matter: final

Fig. 5.2. Density contours at $z = 25$ (top row) and at $z = 0$ (bottom row) of a standard CDM model with $H_0 = 50\,\mathrm{km\,s^{-1}}$ $\mathrm{Mpc^{-1}}$. The cube has a comoving side length of $128\,\mathrm{Mpc}$ (Gott *et al.*, 1987).

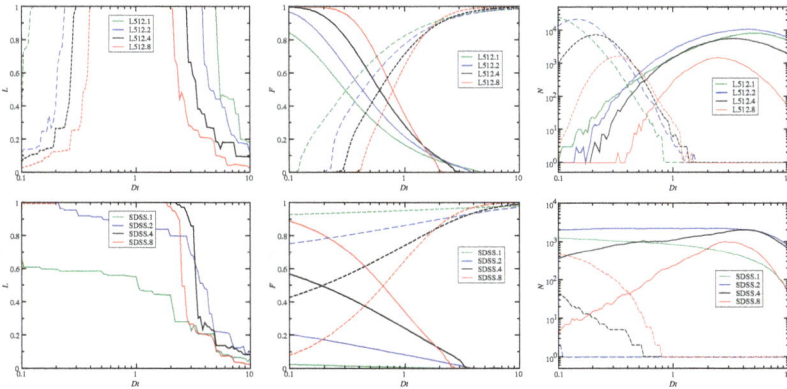

Fig. 5.3. Percolation functions. Left: lengths of the largest clusters and voids, $\mathcal{L}(D_t) = L_{\max}/L_0$; Central: filling factors of the largest clusters and voids, $\mathcal{F}(D_t) = V_{\max}/V_0$; Right: numbers of clusters and voids, $\mathcal{N}(D_t)$. The top panels show L512 model function, the bottom panels show the SDSS sample functions; solid lines denote clusters, dashed lines denote voids, colours indicate smoothing length (Einasto *et al.*, 2018).

data on voids. Indices and colours show the smoothing kernel length in h^{-1} Mpc.

We consider first the cluster percolation functions, i.e. high-density regions above the threshold density D_t in our simulation L512 and in the SDSS cluster samples; then we repeat the analysis for voids. At a very high threshold density, there exist only a few high-density regions — the peaks of the density field represent ordinary clusters of galaxies. These isolated peaks cover a small filling factor in space. When we lower the threshold density, the number of clusters increases, as does the filling factor of the largest cluster. At a certain threshold density, $D_t \approx 5$, depending on the smoothing scale, the number of clusters reaches a maximum. At this threshold density, large clusters still cover a low filling factor and have lengths less than the sample size. Most large clusters are conventional superclusters, consisting of high-density knots woven by filaments into a single system.

When the threshold density decreases, the length of the largest cluster increases very rapidly, supercluster-like systems merge, and at threshold $D_t = P_C$, the largest cluster reaches the opposite walls of the simulations. In periodic boxes, this means that the cluster is infinite. Note that the percolation threshold depends on the smoothing length.

Let us now consider the percolation functions of voids. At a very small threshold density, $D_t \ll 0.1$, there are no voids at all. Voids appear at a certain threshold density level, depending on the smoothing kernel, and their number rapidly increases with increasing D_t. At a very low threshold density, void sizes are small; they form isolated bubbles inside the large over-density cluster, and the filling factor of the largest void is very small. Voids in the density field have at the smallest threshold density the length of the largest void, $\mathcal{L}_V(D_t) \leq 0.5$, depending on the smoothing scale. Void bubbles are separated from each other by DM sheets. Some sheets have tunnels that permit the formation of larger connected voids. With increasing threshold density, tunnels grow rapidly, joining neighbouring voids. At a certain threshold density ($D_t \approx 0.2$ for small smoothing lengths), the largest void percolates, but still does not fill a large fraction of the volume. With larger smoothing lengths, expanded high-density regions block tunnels between voids at a small threshold density, and percolation occurs at higher D_t. The number of voids has a maximum at threshold density $D_t \approx 0.1$ for the density field of smoothing scale $1\ h^{-1}$ Mpc. With increasing smoothing scale, the maximum shifts to larger D_t. For $D_t \geq 1$, there is only one large percolating void; its filling factor increases with D_t.

The bottom panels of Fig. 5.3 show percolation functions of SDSS clusters and voids. The essential difference between models and observations is the absence of fine structure of voids in the SDSS samples. At all smoothing lengths, the SDSS voids percolate; thus, the percolation threshold is not defined. For small smoothing lengths, the percolating SDSS void is the only void. As the smoothing scale increases, additional small SDSS voids appear at low threshold densities. The total number of voids, $\mathcal{N}_V$, increases with increasing smoothing length. These isolated small voids are artificial and are created by blocking tunnels between sub-voids with increasing sizes of clusters by smoothing. At low D_t, the largest SDSS void forms a filling factor $\mathcal{F}_V \approx 0.1$ for small smoothing kernel. With increasing D_t, the filling factor of voids rapidly increases with D_t and reaches a value $\mathcal{F}_V \approx 0.98$ at the highest D_t. There are also differences between the models and observations of clusters. For smoothing length, $R_B = 1\ h^{-1}$ Mpc, the SDSS cluster samples do not percolate at all, and the length of the largest cluster is smaller than the sample size, $\mathcal{L}_C(0.1) = 0.6$. At this smoothing length, clusters form isolated systems due to the absence or weakness of galaxy filaments between high-density knots.

It is remarkable that the number of L512 and SDSS clusters at smoothing length $R_B = 8\ h^{-1}$ Mpc reaches maximal values at identical $D_t \approx 5$ and that the number of clusters is also approximately the same. The smoothing length $R_B = 8\ h^{-1}$ Mpc and the threshold density $D_t \approx 5$ are often used to find superclusters of galaxies (Liivamägi *et al.*, 2012). Samples L512 and SDSS have approximately the same total volumes, thus the comparable number of superclusters in both samples suggests that the L512 model represents the real cosmic web on supercluster scales well.

Our analysis shows that DM models have all three types of topology: cellular at small percolating threshold densities P, sponge-type at medium P, and meatball-type at large P. Limits for the sponge topology are broadest for the smallest smoothing length, $1\ h^{-1}$ Mpc. The voids of SDSS samples always percolate, thus at small and medium threshold densities until the percolation of clusters, $P \approx 2$, the topology is of sponge-type. At a larger threshold density, it becomes a meatball-type. In classical topology studies by Gott *et al.* (1986, 1987, 1989) and Vogeley *et al.* (1994), relatively large smoothing scales were applied. The authors found that the topology of both model and SDSS samples is in good agreement with the random Gaussian fields hypothesis. In a more recent study, Park *et al.* (2005) used a large ΛCDM simulation to investigate topology beyond simple Gaussian models. They found that the topology of simulated galaxies and SDSS samples depart from simple Gaussian models, as expected for biased models. These results are in good agreement with our percolation and bias analyses.

5.2   Correlation function

Early data on the distribution of galaxies came from the counts of galaxies on the sky. The available data suggested that the distribution of galaxies on the sky is dominated by an almost random field population of galaxies with randomly spaced clusters. This distribution can be described by the two-point correlation function of galaxies (Peebles, 1980). It is a function of galaxy separations and describes the excess probability of finding two galaxies separated by this distance. It is not sensitive to the pattern of the cosmic web because it measures only the amplitude of the density fluctuations of the density field of galaxies, not the phase information, which is essential to describe the cosmic web.

RHEA: In a galaxy survey, the probability of finding a galaxy in a small volume is proportional to the galaxy density. If we sit at a distance r from a galaxy, the probability of finding another one is enhanced (or decreased) by a factor defined as the correlation function.

ARIEL: Like this couple in our class, they do everything together. If you see one, you're likely to see the other. They are correlated.

RHEA: And do you like them?

ARIEL: I'm anti-correlating with them.

RHEA: Anti-correlating feels a bit unkind.

ARIEL: Kindness is a finite state machine. TBH, I'm a bit annoyed they're always together and ignore anybody else. I guess I was kinder in my ground state, but after some interactions with them, I formed some memories. I'd rather avoid them outside class.

RHEA: Fair enough... Yet, sometimes you are all together in your class, regardless of your two-point interactions. Such complexity comes with non-Gaussianty. For instance, the whole class is clumped together, not unlike galaxy clusters. That external potential well overwrites any likes or dislikes between the pairs and generates non-linear N-point interactions.

ARIEL: I hear you loud and clear. Correlation functions are intuitive, but I would have a hard time telling the same story with power spectra. Are they the same as correlation functions?

RHEA: The Fourier modes are a linear transformation of the data, and the power spectrum is the correlation function of the Fourier modes. Therefore, they are the same.

CALLISTUS: But if we looked at them in more detail, they are different. They package the information differently. The power spectrum is the Fourier transform[2] of the correlation function. But think about a Dirac delta, a high peak in real space. In Fourier space, it corresponds to a plane wave, i.e. the information becomes delocalised. Sometimes that's not what we want.

ARIEL: I see. This is why it would be difficult to think about correlations in my class in Fourier space: we are localised in the classroom, and that localisation would be much less obvious in Fourier space.

CALLISTUS: Since in astronomy we do our measurements in real space –

RHEA: As a recovering radio astronomer, I beg to disagree.

CALLISTUS: I stand corrected. Still, most astronomical measurements with optical telescopes map objects, such as stars, galaxies, and quasars, in real space. Therefore, the information is localised in real space, and correlation functions are more convenient. Moreover, systematics, such as cut-out holes around bright stars, ragged edges of the survey, and large-scale backgrounds related to our Galaxy appear in real space. Thus, it is easier to correct these effects for the correlation function.

RHEA: Yes, that's true. Observational systematics are often easier to contemplate and correct in real space. On the other hand, theory lives in Fourier space, which is also conveniently uncorrelated. There is a way to combine the advantages of the two methods, obtaining an estimate of the power spectrum from correlation function measurements,[3] e.g., using the `SpICE` software. To me, this means that they are ultimately the same for all intents and purposes.

---

[2] Or more precisely a Hankel transform in flat space and Legendre polynomial transform on the sky.

[3] Szapudi *et al.* (2001).

CALLISTUS: In practice and straightforwardness of interpretation, they are different.

ARIEL: You guys disagree too politely. This is not TikTok-worthy. I'd like to see some more passionate anti-correlation flaring up.

CALLISTUS: Nope, it's getting late; it's time to correlate with our colleagues at the pub.

RHEA: I could not agree more.

ARIEL: Can I correlate with you?

CALLISTUS: Are you above 21? Of course, you are. Let's go!

5.2.1 *Calculation of correlation function and its derivatives*

The natural estimator to determine the two-point spatial correlation function is

$$\xi_N(r) = \frac{DD(r)}{RR(r)} - 1, \qquad (5.1)$$

where r is the galaxy pair separation (distance), and $DD(r)$ and $RR(r)$ are normalised counts of galaxy–galaxy and random–random pairs at a distance r of the pair members.

In the following analysis, we use the spatial correlation function, $\xi(r)$. To investigate fractal properties of the cosmic web in more detail, we can use a relative of the galaxy correlation function — the structure function,

$$g(r) = 1 + \xi(r), \qquad (5.2)$$

and its log–log gradient, the gradient function,

$$\gamma(r) = \frac{d \log g(r)}{d \log r}, \qquad (5.3)$$

which we call simply as $\gamma(r)$ function. It is related to the effective fractal dimension function $D(r)$ of samples at mean separation of galaxies at r (Martinez & Jones, 1990; Pietronero, 1987):

$$D(r) = 3 + \gamma(r). \qquad (5.4)$$

Martínez & Saar (2002) defined the correlation dimension $D_2 = 3 + d\log \hat{g}(r)/d\log r$, where $\hat{g}(r)$ is the average of the structure function, $\hat{g}(r) = 1/V \int_0^r g(r')dV$. For our study, we prefer to use the Local value of the structure function to define its gradient. The effective fractal dimension of a random distribution of galaxies is $D = 3$ and $\gamma = 0$; in sheets: $D = 2$ and $\gamma = -1$; in a filamentary distribution: $D = 1$ and $\gamma = -2$; and within clusters: $D = 0$ and $\gamma = -3$.

5.2.2   *Correlation functions of galaxies and matter*

It is clear that filaments and voids are important ingredients of the cosmic web, and it is interesting to look at how they influence the correlation function. Einasto *et al.* (2020a) used the ΛCDM model in a box of size 512 h^{-1} Mpc as described above. In the simulations, we selected particles from the same set for biased and unbiased samples. Thus, observational selection effects could be ignored. We took advantage of the fact that the positions and local density values of all particles are known for this model. This allowed the differential comparison of DM and simulated galaxy (local density selected DM particles) correlation functions. For the comparison, we analysed absolute magnitude (volume) limited SDSS samples, as found by Liivamägi *et al.* (2012).

To get a sample of DM particles that imitate the distribution of galaxies, we only used particles in high-density regions. We applied a sharp particle density limit, ρ_0, to select biased samples of particles as discussed in Chapter 4. We denoted biased samples as LCDM.i, where i denotes the particle-density limit ρ_0. The full DM model includes all particles and corresponds to the particle-density limit $\rho_0 = 0$, and therefore, it is denoted as LCDM.00.

LCDM model samples are based on all particles of the simulation and contain detailed information on the distribution of matter in regions of different densities. Figure 5.4 shows the correlation functions, $\xi(r)$, Fig. 5.5 shows the structure functions, $g(r) = 1 + \xi(r)$, and Fig. 5.7 shows the fractal dimension functions, $D(r) = 3 + \gamma(r)$ for the basic samples of the model. Correlation and structure functions of the SDSS galaxies are also shown in these figures for five luminosity thresholds.

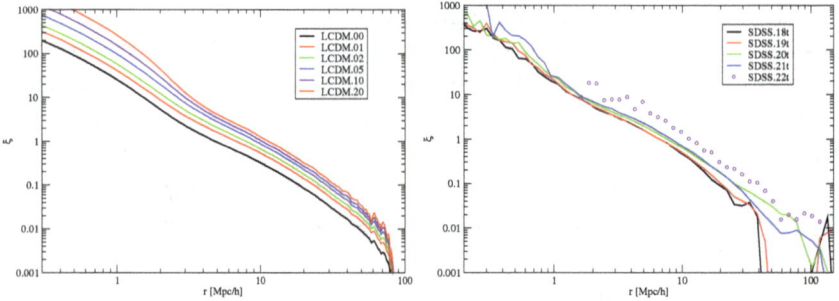

Fig. 5.4. Correlation functions of galaxies, $\xi(r)$. Left panel shows LCDM models for different particle selection limits; right panel shows SDSS galaxies using five luminosity thresholds (Einasto *et al.*, 2020a).

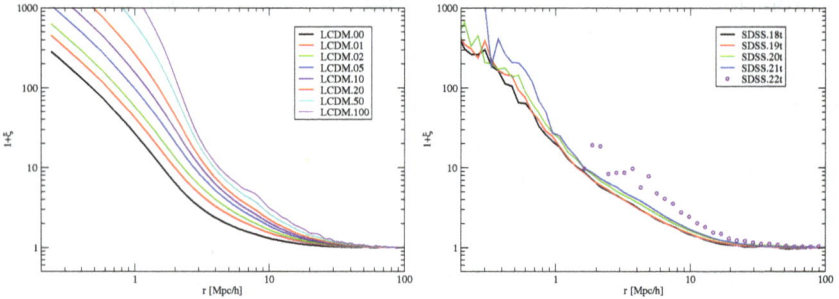

Fig. 5.5. Structure functions, $g(r) = 1 + \xi(r)$. The location of the panels is the same as in Fig. 5.4 (Einasto *et al.*, 2020a).

Figure 5.4 shows that for galaxy samples, the amplitudes of the correlation functions are almost constant for low-luminosity samples and rise for samples brighter than approximately $M_r = -20$. The behaviour of ΛCDM model particle density selected correlation functions is different — with increasing particle density threshold ρ_0, the amplitudes rise continuously. The luminosity dependence of the correlation functions is the principal factor of the biasing phenomenon, as discussed by Kaiser (1984). Here, we compare the luminosity dependence of the correlation functions of real galaxies and particle density-limited simulated galaxies, as found in the Einasto *et al.* (2020a) analysis.

In Fig. 5.6, we show the correlation lengths, r_0, of the SDSS, EAGLE, and Millennium samples as functions of magnitude M_r.

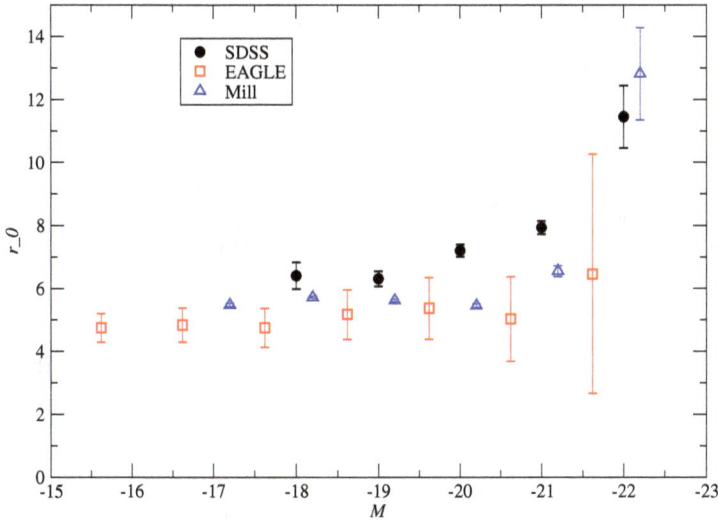

Fig. 5.6. Correlation length r_0 of SDSS galaxies. For comparison, we also show the correlation lengths of the EAGLE and Millennium simulations for the luminosity bins (Einasto *et al.*, 2020a).

The correlation length r_0 has a rather similar luminosity dependence for all samples. The correlation lengths and amplitudes of the SDSS samples are slightly higher than in the simulated galaxy samples by a factor of about 1.2. Amplitudes of correlation functions and their derivatives are defined in Sections 5.2.4 and 6.2.7.

An important detail in the luminosity dependence is the fact that the correlation lengths and amplitudes are almost constant at low luminosities, $M \geq -20.0$. It is seen in observational samples SDSS.19 and SDSS.18, but cannot extend to fainter luminosities because very faint galaxies are absent from SDSS luminosity-limited samples. In the EAGLE and Millennium model galaxy samples, this very slow decrease of r_0 with decreasing luminosity can be followed until very faint galaxies, $M \approx -15.6$ for the EAGLE sample and $M \approx -17.2$ for the Millennium sample. This tendency indicates that very faint galaxies follow the distribution of brighter ones, i.e. faint galaxies are satellites of bright galaxies. The tendency of the clustering dwarf galaxies to form satellites of brighter galaxies has been known for a long time (Einasto *et al.*, 1974b).

5.2.3 *Fractal properties of the cosmic web*

ARIEL: Okay. I went to this cosmology conference over the weekend. Everything went smoothly until somebody mentioned fractals. Oh-ohh.

RHEA: Yes, the f-word.

CALLISTUS: The resistance in the community is a mystery to me.

ARIEL: I did not realise it was so bad.

RHEA: The universe is not a mathematical fractal.

CALLISTUS: And nobody claims that... any more.

ARIEL: We've seen that the brain and galaxies have a power spectrum that is more complex than a simple power-law.

CALLISTUS: A mathematical fractal is simply something that looks the same no matter what scales you look at.

ARIEL: The universe on solar system scales certainly looks different from galactic or intergalactic scales.

CALLISTUS: We've seen that fractals are everywhere in nature. Coastlines, clouds, trees, and even some human tissue, like lungs. But mathematical fractals cannot exist in nature. It's always an approximation between the lower and upper cut-off scales. The behaviour is scale-invariant only between those scales.

RHEA: Volume scales as a power smaller than the two or three dimensions of the sky or space.

CALLISTUS: Fractal geometry yields a great insight into the nature of the universe. It's a consequence of the scale invariance of the initial conditions and gravity. However, some caveats are important to keep in mind. Above the "scale of homogeneity", a few hundred megaparsecs, a homogeneous assumption is a better approximation than an ever-evolving fractal. And if fractal geometry is due to gravity, it would certainly break down or change on small scales, where the infamous gastrophysics[4] kicks in.

RHEA: It's fractal within some reasonable scale range, where a power-law well approximates the correlation function. I can live with that.

ARIEL: So why is the f-word so scary?

CALLISTUS: Historically, some scientists pushed the fractal theory to extremes, predicting that we can never measure the average density of a galaxy catalogue and would always find a straight power-law correlation function. Such a theory is incompatible with general relativity and the Friedmann equation. As soon as galaxy surveys became larger than a gigaparsec, a pure fractal geometry was clearly rejected.

---

[4] A word coined by Dick Bond to describe the complex physics of interacting gas and dark matter essential to galaxy formation.

RHEA: Somehow, the word "fractal" has become synonymous with the extreme version of the theory. A pure mathematical fractal that cannot exist in nature.

ARIEL: You're saying cosmology spearheaded cancel culture before it was a thing.

CALLISTUS: It's ancient. Cultura deleticia.

ARIEL: Seriously?

RHEA: You just translated it to Latin.

CALLISTUS: Quidquid latine dictum sit, altum videtur.[5]

The study of fractal properties of the cosmic web has a long history. The fractal description of galaxy distribution was suggested by Mandelbrot (1982). Soneira & Peebles (1978) made the first essential step studying fractal properties of the cosmic web. They constructed a fractal model that allowed reproducing the angular correlation function of the Lick catalogue of galaxies, see Fig. 2.2. Seldner *et al.* (1977) found that the spatial correlation function can be well represented by a power law:

$$\xi(r) = (r/r_0)^\gamma, \tag{5.5}$$

over a wide range of mutual separations (distances), $0.01 \leq r \leq 10\ h^{-1}$ Mpc. Here, $r_0 = 4.5 \pm 0.5\ h^{-1}$ Mpc is the correlation length, $\gamma = -1.77$ is a characteristic power index, and $D = 3 + \gamma = 1.23 \pm 0.04$ is the fractal dimension of the sample (Peebles, 1989, 1998).

A simple fractal Universe, as suggested by Pietronero (1987), is not a likely option. If the structure is described by a simple fractal, then the mean density of the Universe is zero, which contradicts all other independent data. Thus, the fractal description can be valid only in a limited range of scales because on very large scales the distribution of galaxies is rather uniform — there are no extremely large systems of galaxies. The cosmic web itself is the largest system, and on large scales, it is statistically homogeneous.

For ΛCDM particle density-limited samples, and SDSS luminosity-limited galaxy samples, the structure functions $g(r)$ are shown in Fig. 5.5, and fractal dimension functions are plotted in Fig. 5.7. The last figure shows clearly that the fractal dimension function r has two well-separated regions with transition around

---

[5]Anything said in Latin sounds profound.

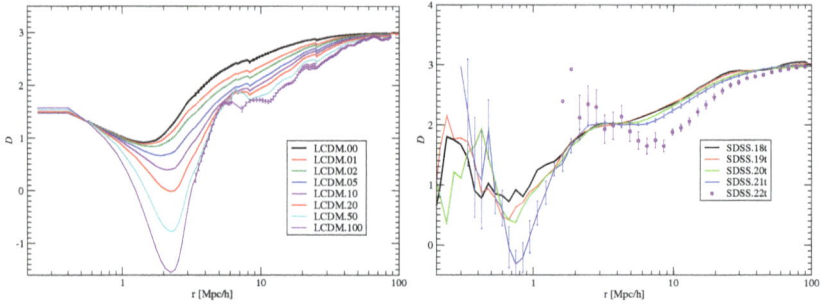

Fig. 5.7. Fractal dimension functions, $D(r) = 3 + \gamma(r)$. The location of the panels is the same as in Fig. 5.4. Errors are given for several representative samples (Einasto *et al.*, 2020a).

separation $r \approx 3\ h^{-1}$ Mpc. This feature was discussed already by Zeldovich *et al.* (1982): at small mutual separations r, the correlation function characterises the distribution of matter within DM haloes, and on larger separations, the distribution of haloes.

The left panel of Fig. 5.7 shows that the fractal dimension function of the LCDM samples strongly depends on the particle density limit ρ_0 used in selecting particles for the sample. All ΛCDM samples have an identical value of the gradient function, $\gamma(0.5) = -1.5$, at a distance $r = 0.5\ h^{-1}$ Mpc, and the respective local fractal dimension, $D(0.5) = 1.5$. At a distance of about $2\ h^{-1}$ Mpc, the gradients have a minimum that depends on the particle density limit ρ_0 of the samples. After the minimum, the gradient function increases and reaches the expected value $\gamma(100) \approx 0\ (D(100) \approx 3.0)$ for all samples at the largest distance.

The identical values of the gradient functions at $r = 0.5\ h^{-1}$ Mpc, followed by a minimum at $r \approx 2\ h^{-1}$ Mpc, can be explained by the internal structure of the DM haloes. DM haloes have almost identical density profiles, which can be described by the NFW (Navarro *et al.*, 1997) and by the Einasto (1965) profile. As shown by Wang *et al.* (2020), the density of DM haloes is better fitted by the Einasto profile. Wang *et al.* (2020) showed that density profiles of haloes of very different masses are almost identical over a very wide range of halo masses and have the shape parameter value $\alpha = 1/N = 0.16$. The authors calculated the logarithmic density slope $d\log\rho/d\log r$ of the density profile over a distance range $0.05 \leq r/r_{-2} \leq 30$. Near the halo centre, the logarithmic slope is $d\log\rho/d\log r = -1.5$, and it decreases to $d\log\rho/d\log r = -3.0$ near the outer boundary of the halo.

At small distances, the correlation functions are the mean values of the sums of the particle mutual distances within the DM haloes averaged over the whole volume of the samples. At the very centre, all DM haloes have identical density profiles, which explains the constant value, $\gamma(0.5) = -1.5$ ($D(0.5) = 1.5$). At the halo outer boundary, the gradient changes to $d\log\rho/d\log r = -3.0$ ($D(r) = 0$). At various particle density limits, the fraction of haloes of different masses dominates. In the full sample ΛCDM.00, small DM haloes dominate, and the minimum of the gradient function $\gamma(r)$ has a moderate depth. With increasing particle density limit ρ_0, low-density haloes are excluded from the sample, and more massive haloes dominate. This leads to an increase in the depth of the minimum of $\gamma(r)$ and $D(r)$ functions. More massive haloes have larger radii; thus the location of the minimum of the $\gamma(r)$ function shifts to higher distance r values. More massive haloes also have a higher density, thus the minimum of the $\gamma(r)$ function is deeper.

After the minimum at higher separation r values, the distribution of DM particles in filaments outside the haloes dominates. This leads to an increase in the $\gamma(r)$ function. In these regions, the distribution of the DM particles of the whole cosmic web determines the correlation function and its derivatives. Figure 5.7 and Fig. 2 of Zehavi *et al.* (2004) show that the transition from one DM halo to the general cosmic web occurs at $r \approx 2\ h^{-1}$ Mpc, which agrees well with the characteristic sizes of DM haloes. We conclude that correlation functions of ΛCDM models describe in a specific way the internal structure of DM haloes and also the fractal dimension properties of the whole cosmic web.

Figure 5.7 shows that the shapes of fractal dimension functions of SDSS galaxy samples are rather similar to the shapes of the ΛCDM sample, except that the scatter at small separations is much larger. Small differences between shapes show that the internal structure of the DM haloes of the ΛCDM models differs from the internal structure of real and simulated galaxy clusters. In our ΛCDM model samples, all DM particles with density labels $\rho \geq \rho_0$ are present. Thus, we see the whole density profile of the haloes up to the outer boundary of the halo. In the real and simulated galaxy samples, only galaxies brighter than the selection limit are present. This means that in most luminous galaxy samples, only one or a few brightest galaxies are located within the visibility window, and the true internal structure of clusters up to their outer boundary is invisible.

5.2.4   Relation between 2D and 3D correlation functions

Early data allowed the measurement of the two-dimensional (2D) CFs from observational data. In the 1980s, redshift data became available and it was possible to measure the spatial three-dimensional (3D) CFs. It is known that redshifts are distorted by local movements of galaxies in clusters, the Finger-of-God effect, and that galaxies and clusters flow towards attractors, the Kaiser (1987) effect. To avoid the Kaiser effect, Peebles (1976), Davis *et al.* (1978), and Davis & Peebles (1983) suggested using the galaxy position and velocity information separately. In this case, pair separations can be calculated parallel to the line of sight, π, and perpendicular to the line of sight, r_p. The angular correlation function, $w_p(r_p)$, can be found by integrating over the measured $\xi(r_p, \pi)$, using the equation

$$w_p(r_p) = 2 \int_{r_{\min}}^{r_{\max}} \xi(r_p, \pi) d\pi, \tag{5.6}$$

where $r_{\min}$ and $r_{\max}$ are the minimum and maximum distances of the galaxies in the sample. This equation has the form of the Abel integral equation and can be inverted to recover the spatial correlation function (Davis & Peebles, 1983):

$$\xi(r) = -\frac{1}{\pi} \int_{r}^{r_{\max}} \frac{w_p(r_p)}{\sqrt{r_p^2 - r^2}} dr_p. \tag{5.7}$$

Davis & Peebles (1983) found that the spatial correlation function can be well represented by a power-law equation (5.5) with parameters $r_0 = 5.4 \pm 0.3\ h^{-1}$ Mpc and $\gamma = -1.77$.

The inversion equation (5.7) assumes that spatial 3D and projected 2D density fields are statistically similar. This similarity has been accepted by the astronomy community, and the use of Eqs. (5.6) and (5.7) to calculate 3D CFs was a standard procedure. However, a visual inspection of 2D and 3D density fields shows the presence of essential differences between these fields. The 3D field is dominated by the filamentary cosmic web, whereas the 2D field is much more random. Thus, it is unclear how accurate the standard procedure for calculating CFs is in this way. To find the relation between 2D and 3D CFs, these functions must be compared using identical data. The results of this comparison were published by Einasto *et al.* (2021a) and are presented here.

To compare 2D and 3D CFs, Einasto *et al.* (2021a) used ΛCDM simulations and Millennium simulations. Three-dimensional correlation functions for both models are shown in Fig. 5.4. The ΛCDM model was calculated in a box of size $512\ h^{-1}$ Mpc. We took advantage of the fact that for this model, the positions of all particles are known, and we can use these particles as objects in the cosmic web to study the properties of the web. The Millennium simulation has a box of size $500\ h^{-1}$ Mpc. For our study, we use the galaxy catalogue based on a semi-analytical model of galaxy formation. For the correlation analysis, we use simulated galaxies as test particles. To calculate CFs, we used the method developed by Szapudi *et al.* (2005). This method uses density fields on 3D or 2D grids as input data and applies fast Fourier transforms (FFT) to calculate CFs.

To compare 2D and 3D correlation functions, we first constructed the 2D density fields on a 2048^2 grid by integrating the 3D field:

$$\delta_2(x, y) = \int_{z_1}^{z_2} \delta(x, y, z) \mathrm{d}z. \tag{5.8}$$

This *distant observer approximation* neglects the conical shape of a real galaxy sample. Then we divided the cubic sample into n sequentially located 2D sheets of size $L_0 \times L_0 \times L\ h^{-1}$ Mpc, where $L = L_0/n$ is the thickness of the sheets and $n = 1, 2, 4, \ldots, 2048$ is the number of sheets. For each n, we calculated 2D CFs for all n sheets and then found the mean CF and its error for a given n. It is clear that $n = 1$ corresponds to the whole sample in the z-direction of thickness, $L = L_0 = 512\ h^{-1}$ Mpc, $n = 2$ corresponds to thickness $512/2 = 256\ h^{-1}$ Mpc, and $n = 2048$ corresponds to thickness $L = 512/2048 = 0.25\ h^{-1}$ Mpc. In this way, we calculated 2D correlation functions for a range of particle density limits ρ_0 for our ΛCDM model and for a number of magnitude limits for the Millennium sample. Two-dimensional CFs depend on two parameters, the thickness of the sheets, $L = 512/n\ h^{-1}$ Mpc, and the particle density limit of LCDM samples, ρ_0, or the magnitude limit, M_r, of Millennium samples.

Our calculations showed that the luminosity dependences of the 2D CFs of the LCDM model and the Millennium model are similar to the luminosity dependence of 3D CFs. With increasing luminosity, the amplitude of CFs increases; this is the well-known biasing effect, quantified by Kaiser (1984). This means that 2D density fields and respective CFs preserve the information on the luminosity dependence contained in 3D density fields and 3D CFs.

For our study, the important factor is the dependence of 2D correlation functions on the thickness of the samples. In Fig. 5.8, we present 2D correlation functions for a series of sample thicknesses $L = L_0/n \ h^{-1}$ Mpc, using number of sheets $n = 1, 2, 4, 8, 16, 32, 64,$

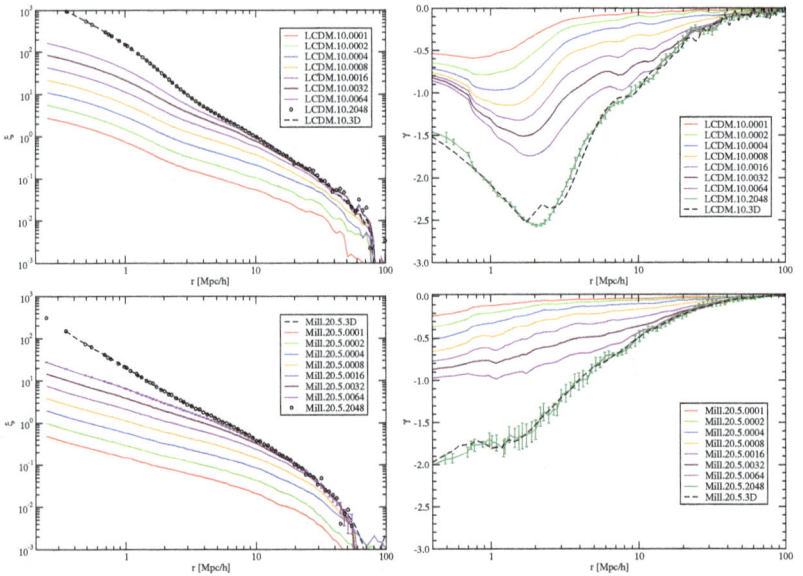

Fig. 5.8. Two-dimensional CFs and gradient functions for models for different thicknesses of 2D samples in left and right panels. Top row: LCDM model with particle density limit $\rho_0 = 10$. Bottom row: Millennium samples with magnitude limit $M_r = -20.5$ in real space. As arguments, we use pair separations perpendicular to the line of sight, $r_p = \sqrt{(\Delta x)^2 + (\Delta y)^2}$. As parameters, we use the thickness of samples, L. Lines of various colours mark 2D samples of different thicknesses. For comparison, we show by dotted lines 3D functions for samples with $\rho_0 = 10$ and $M_r = -20.5$. Error bars are shown for 2D samples with $n = 2048$ (Einasto *et al.*, 2021a).

and 2048. The first one with $n = 1$ corresponds to the whole sample thickness $L = L_0$. Thus, it has the lowest amplitude. The last one corresponds to the mean 2D CF of the thinnest sheets, each of a thickness of $L = 0.25\ h^{-1}$ Mpc. This is the curve with the highest amplitude identical to the 3D CF. Correlation functions are shown for the fixed particle density limit $\rho_0 = 10$ of LCDM.10 samples, and for Millennium samples Mill.20.5 with luminosity limit $M_r = -20.5$. Both limits correspond approximately to L^* galaxies (Einasto *et al.*, 2019a).

The essential parameters of correlation functions are the amplitude and the slope. We calculated amplitudes of the 3D and 2D CFs at separation $r = 6\ h^{-1}$ Mpc, $A = \xi(6)$. Figure 5.9 shows in the left panel the amplitudes of 2D CFs for three LCDM simulations and one Millennium simulation as a function of the thickness L of the samples. Amplitudes of LCDM samples are shown for three particle density limits, $\rho_0 = 0,\ 2,\ 5$; amplitudes from the Mill.17.4 sample are shown for real and redshift space. The analysis shows that for large thickness, the amplitudes are approximately inversely proportional to the thickness; the relation $A = 50/L$ is shown in a black dashed line. Short horizontal lines on the left axis show amplitudes of 3D CFs of the same samples. At small thickness values, amplitude curves of 2D model samples smoothly approach the limits given by 3D samples.

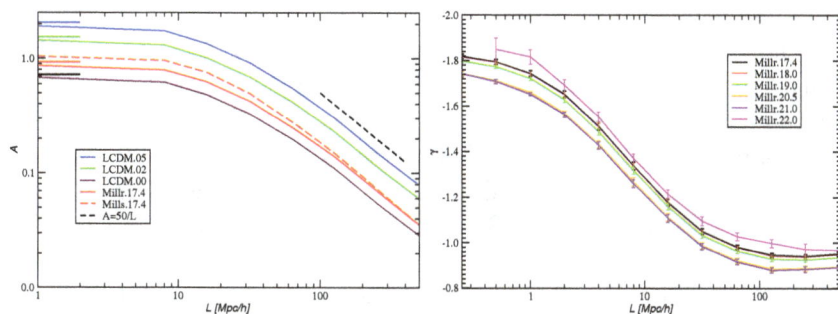

Fig. 5.9. Left: Amplitudes $A = \xi(6)$ of several LCDM and Millennium simulation 2D CFs as functions of the thickness of samples, L. For the Mill.17.4 sample, amplitudes are shown for real space and redshift space CFs. Short lines show 3D CF amplitudes of the same samples. The black dashed line shows the approximation $A = 50/L$. Right: Dependence of the slope γ of Millennium 2D CFs of various magnitude limits on the thickness of samples L (Einasto *et al.*, 2021a).

The figure shows that the decrease of amplitudes with increasing thickness is very regular and almost identical in all samples relative to the amplitude of the 3D CF. The most essential aspect of Fig. 5.9 is the suggestion that the amplitudes of 2D CFs depend significantly on sample thickness and are *always* lower than the amplitudes of the 3D CFs of the same sample. In this way, 2D CFs of samples of various thicknesses imitate 3D CFs for various luminosity limits. The behaviour of the 2D CFs of samples of different thicknesses of LCDM and Millennium model samples is very similar: the greater the thickness of the 2D sample, the greater the difference between the amplitudes of 2D and 3D CFs. This behaviour is almost identical in all our samples, in LCDM and Millennium samples in real space and in Millennium samples in redshift space. This similarity is remarkable because test particles in samples are defined differently: DM particles with local density labels versus simulated galaxies with luminosity labels.

Now, let us turn to the shape of 2D CFs. We see that 2D correlation functions of thinnest samples with $n = 2048$ are practically identical to 3D correlation functions, as shown by dotted lines in Fig. 5.8. This means that the information on the structure of the DM haloes and the cosmic web in general is fully preserved in thin 2D correlation functions. The characteristic minimum of gradient functions near $r \approx 2 \ h^{-1}$ Mpc is well visible. The fine structure of DM haloes is gradually erased with increasing thickness of 2D slices. At large separations, gradient functions approach zero values, as expected for random samples.

The slope of 2D CFs depends on the thickness L of the samples. This dependence is shown in the right panel of Fig. 5.9 for Millennium 2D samples in real space for various luminosity limits. For thick samples with $L \geq 32 \ h^{-1}$ Mpc, the slope is $\gamma \approx -0.9$, close to the characteristic slope of 2D CFs, which is well known from early studies of CFs: Peebles (1973), Peebles & Groth (1975), and Peebles (2001). With decreasing thickness of 2D samples, the negative slope increases and for very thin sheets reaches a value of $\gamma \approx -1.8$, which is characteristic of 3D CFs. There is only a weak dependence of the slope on the magnitude limit, M_r.

This analysis shows that in Millennium samples, the information on the internal structure of the clusters is lost in 2D CFs of simulated galaxies of the sample Mill.20.5 (and of samples with a brighter luminosity limit). At this luminosity threshold, clusters contain only a few galaxies. The amplitudes of the 2D CFs of the Millennium samples are rather low, and in the calculation of the gradient function, the first constant term of the function $g(r) = 1 + \xi(r)$ dominates. As seen in the bottom panels of Fig. 5.8, for thick samples, the slope of the 2D CF is almost constant over a large separation range, $r_p > 1$ h^{-1} Mpc. This explains the well-known observation that the CF can be approximated by a simple power-law function equation (5.5).

The CF characterises the geometrical properties of the cosmic web in a general and global way. To understand the influence of the pattern of the cosmic web on properties of CFs, let us look at the geometry of the density field of the cosmic web, as given by 2D and 3D data. Examples of 2D density fields for various thicknesses L are given in Fig. 5.10. The top panels present 2D density fields of thin

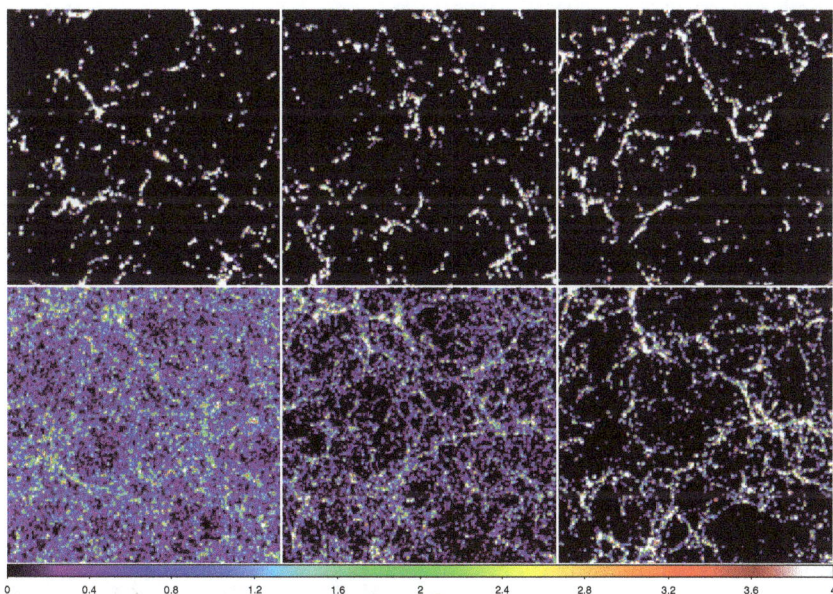

Fig. 5.10. Two-dimensional density fields of LCDM.10 models with particle density threshold $\rho_0 = 10$. Top panels: 2D density fields of thickness $L = 8$ h^{-1} Mpc at z-coordinates 100, 200, 300. Bottom panels (from left to right): Thickness $L = 512$, 128, 32 h^{-1} Mpc. We show the central sections of 2D density fields of size 256×256 h^{-1} Mpc. The colour scale is linear, and the code is identical in all panels (Einasto *et al.*, 2021a).

slices of thickness $L = 8 \, h^{-1}$ Mpc in x, y coordinates at three z locations. These density fields are so thin that their morphological properties are close to those of the 3D density field of the same model, namely the fraction of high- and low-density regions and mutual distances between high-density regions. Here, knots and filaments are surrounded by zero-density voids, occupying most of the volume of the density field. In the bottom panels, we show much thicker 2D density fields. In the bottom left panel, the thickness of the 2D density field is $L = 512 \, h^{-1}$ Mpc, i.e. the whole cube of the LCDM.10 model. In the bottom middle and bottom right panels, the thicknesses are $L = 128 \, h^{-1}$ Mpc and $L = 32 \, h^{-1}$ Mpc, respectively. With increasing thickness L of 2D density fields, the fraction of cells with zero density rapidly decreases. This is clearly seen in Fig. 5.10 visually and in Fig. 5.9 graphically. In the 2D field, clusters are superposed by galaxies from different vertical locations in projection; see Fig. 5.10, and fill in the voids between clusters and filaments.

It is well known that the amplitudes of correlation functions depend on several factors: (i) cosmological parameters: matter–energy densities Ω_b, Ω_m, Ω_Λ and the *rms* matter fluctuation amplitude averaged over a sphere of radius $8 \, h^{-1}$ Mpc, σ_8; (ii) luminosities of galaxies; (iii) systematic motions of galaxies in clusters, the FOG effect, and (iv) the flow of galaxies toward attractors, the Kaiser effect. Systematic motions decrease the amplitudes of 3D CFs on small scales and increase them on large scales. The increase in CF amplitudes on large scales is well described by the dynamical model by Kaiser (1987).

The present analysis shows that amplitudes of 2D CFs are influenced by an additional factor: the thickness of samples, L. The dependence of the amplitudes of 2D CFs on sample thickness follows from the spatial structure of the cosmic web. The cosmic web consists of galaxies located in an intertwined filamentary pattern, leaving most of the space void of galaxies. In projection, clusters and filaments fill in the voids, depending on the thickness of the sample. For this reason, the cosmic web patterns in 2D and 3D are qualitatively different. The thicker the 2D sheets, the greater the difference.

The difference between 2D and 3D density fields is the same as the difference between trees and forests. In trees, we see gaps between branches that occupy most of the tree's volume. When we look at trees from a distance, we see the forest, where branches from different trees overlap. The information on gaps between branches and the detailed structure of the forest is lost.

5.2.5 *Cosmological implications*

In this chapter, we analysed the properties of the cosmic web as characterised by percolation and correlation functions and their derivatives. Both functions describe *global* properties of the cosmic web, not its components such as clusters, filaments or voids. Here, we summarise essential results of this analysis.

The percolation analysis characterises the connectivity of the cosmic web. Einasto & Saar (1987) and Einasto *et al.* (2018) extended the percolation analysis to a wide range of smoothing lengths and threshold densities to find connected clusters and voids. Extended percolation functions are sensitive to faint DM filaments of the cosmic web present in DM models, but absent in SDSS samples. At low and medium threshold densities, and smoothing length $\sim 1 \ h^{-1}$ Mpc, the percolation functions of the SDSS sample are different from the percolation functions of the DM model, both for clusters and voids. Most notably, the SDSS sample has only one large percolating void filling almost the whole volume due to a lack of small-scale resolution.

The correlation function is probably the most commonly used statistic in cosmology. It characterises the clustering properties of galaxies. The correlation function was applied first to the 2D projected distribution of galaxies and clusters, available in 1970s. Peebles (1973) and Peebles & Groth (1975) found that the correlation function can be expressed by a simple formula, Eq. (5.5), where the index $\gamma \approx -1.8$ of the spatial CF is the same for all samples studied.

Three-dimensional data on the distribution of galaxies were available by the 1980s from positions of galaxies on the sky and their redshifts used to find distances. Distances from redshifts are distorted by the flow of galaxies towards attractors due to the Doppler shift. To avoid this effect, Peebles (1976) and Davis & Peebles (1983) suggested using the galaxy position and velocity information separately. In this case, the pair separation of galaxies was found using 2D positional data, and the spatial correlation function was calculated by inversion of the measured angular correlation function, see

Eqs. (5.6) and (5.7). This procedure assumes that 2D and 3D distributions of galaxies are statistically identical.

The analysis by Einasto *et al.* (2021a) shows that the 2D correlation functions preserve the information on the luminosity dependence contained in the 3D correlation functions. However, the amplitudes of 2D and 3D correlation functions are different. Amplitudes of 2D correlation functions depend on the sample thickness and are always lower than the amplitudes of the respective 3D correlation functions. In this way, 2D CFs for a fixed magnitude limit imitate 3D CFs of fainter galaxies. This insight is essential in interpreting bias functions, as discussed in the following chapter.

The slope of the 3D correlation function is the gradient function $\gamma(r)$, defined as the logarithmic derivative of the structure function $g(r) = 1 + \xi(r)$, see Eqs. (5.2) and (5.3). The shape of the gradient function shows clearly that the 3D correlation function and its derivatives describe at small separations r the distribution of galaxies within clusters, and at large separations, the distribution of galaxies and clusters themselves. This changes the gradient and the respective fractal dimension function, $D(r) = 3 + \gamma(r)$, as shown in Fig. 5.7. Both functions have two well-separated regions with a transition around separation $r \approx 3\ h^{-1}$ Mpc.

The information content of the 2D correlation functions varies. Our study shows that integration according to Eq. (5.7) yields 2D CFs with decreasing amplitudes, depending on the thickness of the sheets. This means that, because of projection effects, information on zero-density regions in the 3D density field is partly lost, as demonstrated in Figs. 5.8–5.10. More importantly, the information on the internal structure of clusters of galaxies is lost, as shown by the behaviour of gradient functions of simulated galaxies in Fig. 5.8. This explains the observational result by Peebles (1973) and Davis & Peebles (1983) that the gradient function is featureless with almost constant value $\gamma \approx -0.7$ for 2D CF, which corresponds to $\gamma \approx -1.7$ for 3D CF. This property was confirmed by Maddox *et al.* (1990), who found that the angular CF of the Automatic Plate Measuring (APM) survey of galaxies has a constant slope, $\gamma \approx -0.7$, over a wide range of angular scales, $0.01 \leq \theta \leq 2$ degrees, and galaxy apparent magnitudes, $17 < b_j < 20$. Similar results were obtained by Connolly *et al.* (2002) and Wang *et al.* (2013) for the SDSS galaxy angular correlation function.

As the information on the internal structure of clusters and the presence of voids is lost, analyses by Davis & Peebles (1983), Maddox *et al.* (1990) and others were explained as the evidence for a mean constant fractal dimension of the spatial distribution of galaxies of the survey, independent of galaxy separations. The almost constant slope of the 2D CF over a large range of angular scales was interpreted by Peebles (2001) as an evidence that the spatial CF is well represented by the law equation (5.5) over the range of separations 10 kpc$\leq hr \leq$ 10 Mpc. The featureless CF and its derivative, γ, were considered as a common property of the distribution of galaxies, as seen from the discussion of these results by Longair (2023) in Section 2.2.1. This interpretation is actually a reminder from the past when only the angular positions of galaxies were known.

Three-dimensional density fields contain more information than 2D density fields, and therefore, the inversion to get 3D CFs from 2D CFs (Eq. (5.6)) adds information. The relation between projected and spatial CFs is similar to the relation between the projected and spatial densities of galaxies, which are used to calculate their dynamical models, see Kuzmin (1952). The information gain comes in this case from the assumption that galaxies are spatial ellipsoids of rotation, as noted by Kuzmin (1952). The apparent information gain in the calculation of 3D CFs from 2D CFs comes from the tacit assumption that the spatial and projected structures of the cosmic web are statistically identical, which is not the case.

Analysis shows that 3D correlation functions of galaxies of various luminosities have an important property — their correlation lengths and amplitudes are almost constant at low luminosities, as shown in Fig. 5.6. This tendency indicates that the very faint galaxies follow the distribution of brighter ones, i.e. faint galaxies are satellites of bright galaxies.

Chapter 6

Evolution of the Cosmic Web

In this chapter, we discuss first the evolution of the shape of the cosmic web using dark matter (DM) population as its main constituent. We apply simple parameters to describe the asymmetry of the web as customary in mathematical statistics — the skewness and the kurtosis. The definition of these parameters in cosmological studies differs from definitions applied in mathematical statistics, and we show how they are related. The second problem discussed in this chapter is the relation between the distribution of DM and galaxies, which is the basis to understand the biasing phenomenon. We quantify differences in the distribution of galaxies and matter using the correlation descriptor of the cosmic web.

6.1 Evolution of the shape of the density field

According to the currently accepted cosmological paradigm, the evolution of the structure in the Universe began from small perturbations that were created during the epoch of inflation. The structure evolved by gravitational amplification to form the cosmic web that is observed now. It is also accepted that initial density fluctuations were random (but correlated) and had a Gaussian distribution. The Gaussian random field is symmetrical around the mean density, i.e. positive and negative deviations from the mean density are equally probable. On the other hand, it is well known that the current density field of the cosmic web is highly asymmetric: positive density departures from the mean density can be very strong, while the negative deviations are restricted by the condition that the density cannot be negative. The asymmetry of the density field can be studied with a one-point probability distribution function (PDF) of the density field and its moments.

The asymmetry and flatness of the PDF are measured by the third (skewness S) and fourth (kurtosis K) moments of the distribution functions. The moments are the most simple forms of the three-point and four-point correlation functions (CFs). In mathematical statistics, skewness and kurtosis of a random variable are defined as dimensionless parameters and can be called mathematical skewness S and mathematical kurtosis K. They change during the evolution and can be used to characterise the evolution. In cosmology, there is a tradition to define skewness and kurtosis in a different way.

CALLISTUS: Ariel, the mean of a distribution is ...

RHEA: He means to test you.

ARIEL: Means testing? That would be mean.

CALLISTUS: That means something completely different.

RHEA: They mean well.

CALLISTUS: I mean to give you a good grade, but you have to answer my question. Let's step back. What is a distribution?

RHEA: Let's consider a galaxy density field.

ARIEL: If I pixellise my field, count the number of galaxies in each pixel and then average the numbers I get: that's the mean. On the other hand, I could create a histogram of all those counts. That's the distribution. The first moment of that distribution is the same mean as before. The physical meaning is the average galaxy count. I could call it the typical count.

CALLISTUS: When studying the galaxy density field, we typically look at fluctuations, so the mean is zero.

RHEA: I'm at variance with your idea of it being typical. It's only true if the distribution is strongly peaked around the mean.

ARIEL: Indeed, if the variance of the distribution is big, we can have different values than the mean. We estimate the variance sigma as the second moment of the distribution. It corresponds to the width of the histogram. For a Gaussian distribution, we would find values within one sigma about 68% of the time.

[*Rhea skews her head.*]

RHEA: The cosmic web is really non-Gaussian. Therefore, it's −

ARIEL: skewed.

CALLISTUS: What is the meaning of the skewness?

ARIEL: It is calculated as the third moment of the distribution.

CALLISTUS: Omne trium perfectum.[1] My favourite moment.

ARIEL: For a symmetric distribution, the third moment would change sign for positive and negative values above and below the mean. Therefore, its presence signals asymmetry.

CALLISTUS: Why is the cosmological distribution asymmetric?

ARIEL: Gravity.

RHEA: That's a curt reply.

ARIEL: The rich get richer, the poor get poorer. Like capitalism. To anticipate your next question, the kurtosis is the fourth moment. I have no idea what it means because different sources say different things.

RHEA: Kurtosis comes from Greek, meaning bulging or convex.

CALLISTUS: There is a common misconception that kurtosis means flatness. But what happens if we add an infinitely large and infinitely narrow peak at the middle of the distribution?

ARIEL: It would add a zero contribution to any of the moments, including the kurtosis. If not flatness, what is it?

CALLISTUS: It tells us about tail leverage: it is really sensitive to the distribution's tails. We use it because the cosmic web is asymmetric and has a long tail.

RHEA: It's close to a lognormal.

ARIEL: Like the income distribution of society. Are we going to go even higher than kurtosis?

RHEA: Sometimes, we go to fifth, sixth, or even higher moments.

CALLISTUS: Some people call these superskewness and superkurtosis, but it's an uncommon terminology.

RHEA: Once we need higher than the fourth moment, we might as well use a local non-linear transformation, like a logarithmic mapping,[2] to approximately Gaussianise the distribution.

ARIEL: Therefore, skewness and kurtosis make sense in the weakly non-linear perturbative regime, while non-perturbative techniques are more useful in the highly non-linear regime.
[*Callistus looks at his watch.*]

CALLISTUS: I don't mean to be at variance with you, or being skewed and curt, but I have a class to teach.
[*Callistus exits.*]

ARIEL: Did I pass?

RHEA: With high probability. If you tell me about the difference between the mathematical and cosmological definitions of skewness and kurtosis?

---

[1] Everything that comes in threes is perfect.
[2] Neyrinck *et al.* (2009).

ARIEL: That's easy, in math, we divide the nth moment with the nth power of the variance. In cosmology, we follow a scaling motivated by perturbation theory, and we divide it by $(2n-2)$th power of the variance.

RHEA: If you explain to me why, I guarantee you an A+ for the midterm.

ARIEL: In leading-order perturbation theory, we expand the non-linear fluctuation δ as a function of the initial linear Gaussian δ_L and count the powers of it. The skewness is a third moment, which is zero to linear order due to the Gaussian symmetry. The leading order follows from expanding one of the three final δ-s to the second order, which results in the fourth power δ_L^4 in total. Thus, it makes sense to divide with the fourth power by sigma to scale out the variance. Kurtosis has one more power of σ^2 and so forth for higher-order moments.

         [*Rhea smiles.*]

RHEA: Callistus is famous for never giving A+ to anybody. But I think this time, he will.

ARIEL: Maybe if you ask him in Latin.

6.1.1   *Mathematical and cosmological estimates of the shape of the density field*

In the following, we discuss the asymmetry of the cosmic density field on the basis of a DM-only numerical simulation of the evolution of the density field by Einasto *et al.* (2021c). Each N-body simulation provides us with a population of DM particles in a box of size L_0 at redshift z. The density field is estimated using a filter of size R_t with a total number of independent elements N. The density field is normalised to the average matter density, providing us with the density contrast δ:

$$\delta = \frac{\rho_{DM}}{\Omega_m \rho_{cr}} - 1, \qquad (6.1)$$

where Ω_m is the density parameter for the cosmological model and ρ_{cr} is the critical density of the Universe. The density distribution function $P(\delta)$ is defined as a normalised number of elements of the density field with a density contrast in the range $[\delta, \delta + d\delta]$:

$$P(\delta) \equiv \frac{\Delta N}{N \Delta \delta}.$$

(6.2)

The second moment of $P(\delta)$ is the dispersion of the density field:

$$\sigma^2 = \frac{1}{N} \sum_{j=1}^{N} \delta_j^2 = \langle \delta^2 \rangle.$$

(6.3)

The third and fourth moments of the PDF are defined as the skewness and kurtosis parameters S and K:

$$S = \frac{1}{N} \sum_{j=1}^{N} \left(\frac{\delta_j}{\sigma} \right)^3 = \frac{\langle \delta^3 \rangle}{\sigma^3}, \quad K = \frac{1}{N} \sum_{j=1}^{N} \left(\frac{\delta_j}{\sigma} \right)^4 - 3 = \frac{\langle \delta^4 \rangle}{\sigma^4} - 3.$$

(6.4)

The additional term -3 in Eq. (6.4) causes the value $K = 0$ for the Gaussian distribution. In statistics, this is called excess kurtosis. These definitions are used in mathematical statistics and many scientific fields. The skewness S characterises the degree of asymmetry of the distribution, while the kurtosis K measures the presence of heavy tails and peaks in the distribution.

In the cosmological literature, another definition of the PDF moments is used, see Peebles (1980), Bernardeau *et al.* (2002) and Szapudi (2009):

$$S_p = \langle \delta^p \rangle / \sigma^{2(p-1)},$$

(6.5)

where

$$\langle \delta^p \rangle = \int_0^{\infty} d\delta \, P(\delta) \, \delta^p.$$

(6.6)

These moments determine the S_p parameters (Bernardeau & Kofman, 1995). Specifically, the third moment defines the skewness:

$$S_3 = \langle \delta^3 \rangle / \langle \delta^2 \rangle^2,$$

(6.7)

and the fourth moment defines the kurtosis:

$$S_4 = \left(\langle \delta^4 \rangle - 3 \langle \delta^2 \rangle^2 \right) / \langle \delta^2 \rangle^3.$$

(6.8)

The second term in the last equation has the goal to obtain $S_4 = 0$ for the Gaussian distribution.

By comparing mathematical and cosmological definitions, it is easy to see that

$$S(\sigma) = S_3 \times \sigma \tag{6.9}$$

and

$$K(\sigma) = S_4 \times \sigma^2. \tag{6.10}$$

Equations (6.9) and (6.10) show that mathematical skewness and kurtosis can be considered as power-law functions of the standard deviation σ, where cosmological skewness and kurtosis, S_3 and S_4, play the role of amplitude parameters of mathematical skewness and kurtosis, $S(\sigma)$ and $K(\sigma)$. By definition, both mathematical and cosmological skewness and kurtosis parameters are dimensionless quantities, since the density contrast δ is a dimensionless quantity (e.g., Bernardeau & Kofman, 1995; Kofman *et al.*, 1994a).

The perturbation theory provides the following approximations (see Juszkiewicz *et al.* (1993), Bernardeau (1994) and Kofman *et al.* (1994a)):

$$S_3 = \frac{34}{7} + \gamma, \quad S_4 = \frac{60712}{1323} + \frac{62\gamma}{3} + \frac{7\gamma^2}{3}, \quad \gamma_1 = \frac{d\log \sigma^2(R)}{d\log R}. \tag{6.11}$$

Here, γ is the logarithmic slope of the dispersion function $\sigma^2(R)$ with the filtering radius R. The parameter γ is related to the effective slope of the power spectrum of perturbations at radius R as $\gamma = -(n_{\text{eff}} + 3)$. The approximations in Eqs. (6.11) are expected to work only for small amplitudes of perturbations, $\sigma \lesssim 0.1$.

6.1.2 *The growth of structures of various scales*

ARIEL: Why are so many distributions in nature close to a lognormal? Even the US income distribution. Even though it's way outside of nature.

RHEA: The reason is Gibrat's law, a multiplicative version of the central limit theorem.

CALLISTUS: The logarithm of a lognormal variable is Gaussian. Therefore, central limit theorem applies to the log.

ARIEL: The sum of logs corresponds to the multiples of the original numbers. Just like the sum of independent distributions[3]
tends to a Gaussian, the product of independent processes
tends to a lognormal.

RHEA: That's Gibrat's law. People argue that factors such as education, talent, family, existing wealth, and geographic location
modify your income multiplicatively, and this would explain
the lognormal distribution of incomes.

ARIEL: But these factors are not independent. If you come from a
richer family, you will most likely get a better education at a
more famous university and be in a more desirable location.

CALLISTUS: The actual income distribution is not quite lognormal. The
tail fits very well, but, sadly, there are even more poor people
than lognormal would suggest. They often model the distribution with a relative of the lognormal, the Pareto distribution.

ARIEL: I conjecture that the correlations are more important at low
income: it is harder to break out since being poor correlates
highly with other income limiting factors... However, what are
the independent factors of the cosmological density field that
drive it towards a lognormal distribution?

CALLISTUS: It's easy to understand the large tail qualitatively. The
density can be very large, but it cannot go below zero.

ARIEL: Like income: you can be a billionaire, but you cannot make
less than zero.

RHEA: You can have a loan with larger repayment than your income,
but that's rare.

ARIEL: Be like a cosmic void and declare bankruptcy.

CALLISTUS: In units of the zero mean cosmological fluctuation field,
the lowest possible value is minus one, while the high densities can go as high as 200. Virialisation in spherical collapse
sets the high limit approximately. There is also a hard limit
corresponding to the Schwarzschield radius of black holes.

ARIEL: Ugh. 200 is extremely high compared to minus one, and I
can see that a logarithmic mapping would somewhat equalise
these differences. I still don't see the relevance of Gibrat's law.

RHEA: The physics of the cosmological density field is somewhat lognormal for an entirely different reason than a bunch of independent factors: the continuity equation. We can put it in a
form where the total time derivative of the log-density equals
the (negative) divergence of the velocity field.[4] Since the initial peculiar velocities are Gaussian, the log-density field is
also Gaussian.

---

[3]Under some mathematical conditions.
[4]Coles & Jones (1991).

ARIEL: A lognormal distribution for the densities follows. Very pretty.

CALLISTUS: Of course, lognormal is only an approximation. It is often quite accurate for the projected 2D distribution, but the 3D distribution has an even fatter tail. It is well approximated by the generalised extreme value[5] distribution.

RHEA: Naively, one would suspect a process where the most extreme fluctuations within a pixel would dominate its final value. However, if the densities within pixels were uncorrelated, they would still produce a normal distribution according to the central limit theorem.

ARIEL: But intrapixel densities are highly correlated!

CALLISTUS: It has been shown that sums of correlated variables often produce extreme value distributions even with non-extremal underlying processes.[6]

ARIEL: Another example where entirely different physical processes can forge similar distributions.

Due to its integrated nature, the PDF does not allow selecting individual components of the cosmic web, such as clusters, filaments or superclusters. The PDF of the density field and its moments are integrated quantities that characterise properties of the whole web. As discussed in Chapter 4, objects of various compactness of the cosmic web can be highlighted using a smoothing of the density field with different scales. Instead of "cosmic web at smoothing scale R_t", we can use the term "populations of the cosmic web" according to the smoothing length that is applied to calculate the density field. Populations in this meaning cover the whole cosmic web, they characterise the web on the selected smoothing scale.

Einasto *et al.* (2021c) calculated density fields for DM ΛCDM simulations for cubic samples of box size $L_0 = 256$, 512 and $1024\ h^{-1}$ Mpc with resolution $N_{\mathrm{grid}} = 512$, named L256, L512 and L1024, and described in Chapter 4. Figure 6.1 shows the evolution of the PDF of the model L1024. We use as argument the reduced density $\nu = \delta/\sigma$. This presentation is useful to show how the density distributions of our simulations can be represented by a Gaussian distribution. We show the density fields of DM simulation L1024 using two smoothing lengths $R_t = 4$, $8\ h^{-1}$ Mpc on the left and

---

[5]Repp & Szapudi (2018).
[6]Bertin & Clusel (2006).

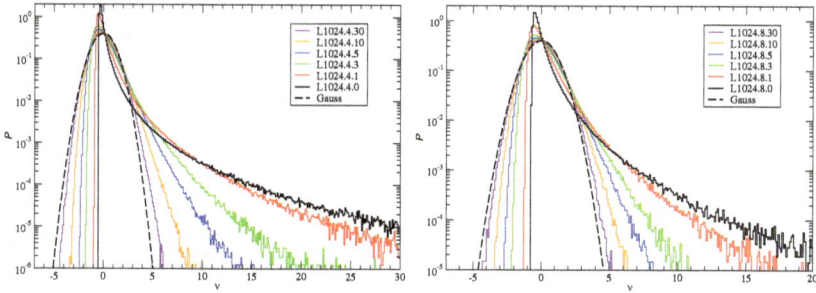

Fig. 6.1. Density distribution functions $P(\delta)$ as functions of the DM density contrast δ normalised to the rms of density fluctuations $\nu = \delta/\sigma$ for DM simulation L1024. Left and right panels are for smoothing lengths $R_t = 4$ and $8h^{-1}$ Mpc, respectively, indicated as the first index in the simulation name. Redshift is the second index. Dashed curves show the Gaussian distribution (Einasto *et al.*, 2021c).

right panels, respectively. The figure shows that (i) PDFs are asymmetric in the sense that high-density regions extend much farther than low-density regions and (ii) smoothing has a dominant role in determining the width of the PDF distribution. Both the asymmetry and the importance of smoothing role of PDFs have been known for a long time; for early works, see Bernardeau (1994), Kofman *et al.* (1994a), and Bernardeau & Kofman (1995). The dependence of PDFs on smoothing length was recently studied by Klypin *et al.* (2018). In our study, we see the growth of the asymmetry in a very broad redshift interval from $z = 30$ to $z = 0$, see also Fig. 5.1.

6.1.3 *Evolution of the skewness and kurtosis of DM density fields*

Next, we show the evolution of skewness and kurtosis parameters with cosmic epoch. We call the graphs in which lines join simulations with a given smoothing length R_t at various epochs z as "evolutionary tracks" of cosmic web populations, and the graphs in which lines join simulations with various smoothing lengths R_t at a given epoch z as "evolutionary diagrams". These terms follow the analogy with stellar evolution tracks and Hertzsprung–Russel diagrams. Evolutionary tracks show evolutionary trajectories of populations of the cosmic web at various characteristic scales in the $S(z)$, $K(z)$, $S(\sigma)$, $K(\sigma)$, and $S_3(z)$, $S_4(z)$, $S_3(\sigma)$, $S_4(\sigma)$ plots. Evolutionary diagrams

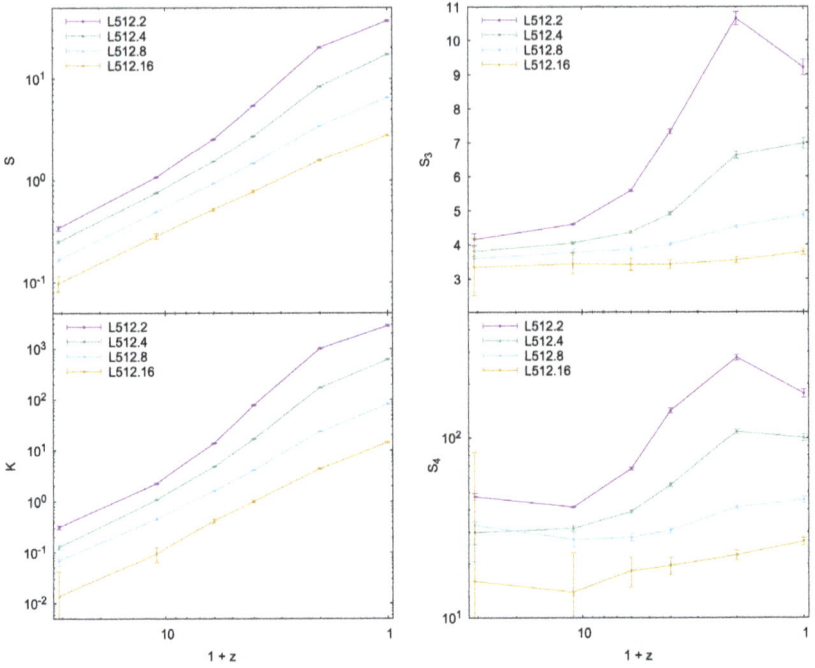

Fig. 6.2. Evolutionary tracks of the simulation L512. The left top and bottom panels show the evolution of the mathematical skewness $S(z)$ and the mathematical kurtosis $K(z)$, respectively, and the right panels show the evolution of their cosmological equivalents $S_3(z)$ and $S_4(z)$. The index in the simulation name is the smoothing length in h^{-1} Mpc (Einasto *et al.*, 2021c).

show where characteristic populations of the cosmic web are situated in these diagrams at various epochs. Evolutionary tracks and diagrams are based on identical data, only the data points are joined differently by the lines to show different aspects of the evolution.

Figure 6.2 presents the evolutionary tracks of the PDF moments of simulation L512. In the left panels the dependence is shown the skewness $S(z)$ and the kurtosis $K(z)$, and in the right panels the respective cosmological functions $S_3(z)$ and $S_4(z)$. Coloured lines joining symbols are the evolutionary tracks of populations of various richness identified by the smoothing lengths. The horizontal axes are inverted to show the evolution from left to right as in the following figures. If we join points at given epochs z, we obtain evolutionary diagrams. In this representation, they are vertical lines.

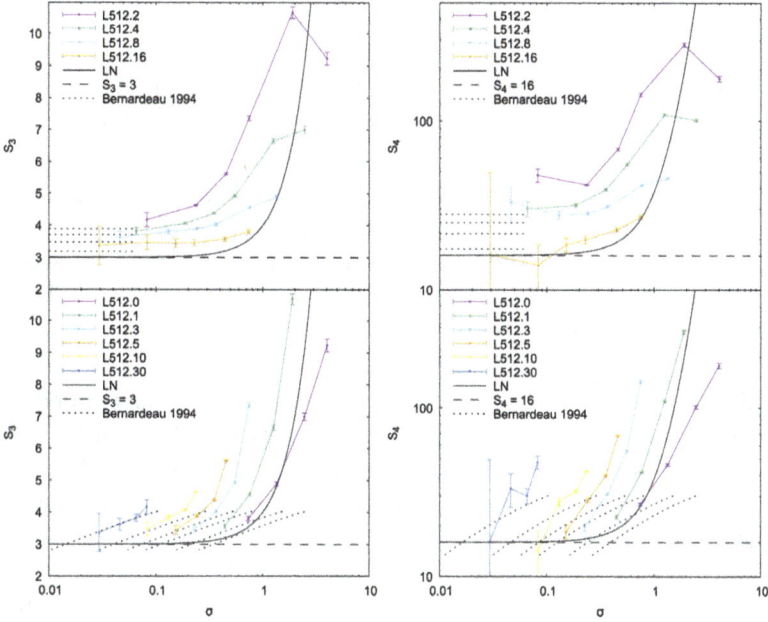

Fig. 6.3. Top panels: evolutionary tracks; bottom panels: evolutionary diagrams. Left panels: cosmological skewness S_3; right panels: cosmological kurtosis S_4. In the evolutionary tracks, the symbols along the tracks from the left are for redshifts $z = 30$, 10, 5, 3, 1, and 0. In the evolutionary diagrams, the symbols from top to bottom correspond to smoothing lengths 2, 4, 8, and 16 h^{-1} Mpc. The solid black curves show the cosmological skewness S_3 and kurtosis S_4 according to the lognormal distribution. The dotted curves show the predictions of the perturbation theory Eqs. (6.11) (Einasto *et al.*, 2021c).

In Fig. 6.2, we use as argument the evolutionary epoch z. The evolution of the skewness S and kurtosis K is dominated by changes in the rms of the density field σ. A much more compact presentation of the evolution is possible when we use cosmological parameters according to the definitions in Eqs. (6.9) and (6.10): $S_3(\sigma) = S(\sigma)/\sigma$ and $S_4(\sigma) = K(\sigma)/\sigma^2$. This version is presented in the right panel of Fig. 6.2 and in Fig. 6.3. But note that in Fig. 6.2, we use as argument the epoch of simulation $1 + z$, and in Fig. 6.3, the rms of the density field σ. σ itself can be approximated by a power-law function of $1 + z$, thus both quantities can be used as arguments.

In *evolutionary tracks* (top panels of Fig. 6.3), the coloured curves join the $S_3(\sigma)$ and $S_4(\sigma)$ values for various smoothing lengths R_t.

Moving along the tracks from left to right, the asymmetry and flatness parameters of the cosmic web at given smoothing lengths change with redshift. The populations, selected with small smoothing length $R_t \leq 2\ h^{-1}$ Mpc, are shown by cyan curves with highest amplitudes. They have maxima $S_3 \approx 11$ at $z \approx 2$ and decrease for later epochs, $z \leq 2$. The populations selected with a smoothing length $R_t = 4\ h^{-1}$ Mpc and shown by green curves reach the maximum amplitude $S_3 \approx 7$ at the present epoch. The populations selected with a smoothing length $R_t = 8\ h^{-1}$ Mpc and shown by light blue curves have a moderate increase in the cosmological skewness $S_3(\sigma)$.

The bottom panels in Fig. 6.3 show *evolutionary diagrams* of the populations for simulation L512: curves joining symbols connect simulations of identical redshift z. For each epoch, the symbols from top to bottom correspond to smoothing lengths $R_t = 2$, 4, 8, and $16\ h^{-1}$ Mpc. The evolutionary diagrams for different ages are well separated from each other and are located at approximately similar mutual distances along the σ-axis, the curve for the present epoch has the highest amplitude σ. This conclusion is valid for both $S_3(\sigma)$ and $S_4(\sigma)$.

The dotted curves in Fig. 6.3 also present evolutionary diagrams as predicted by the perturbation theory. The figure shows that the perturbation theory represents actual evolutionary diagrams only for early epochs and large smoothing lengths. The lognormal distribution is only a very crude approximation of evolutionary tracks. Horizontal dashed lines show limits, $S_3 = 3$ and $S_4 = 15$, found from perturbation theory.

6.1.4   *Cosmological implications*

Asymmetry parameters shown in Figs. 6.2 and 6.3 were obtained using one L512 simulation and are influenced by shot noise. To check these results, Einasto *et al.* (2021c) made a series of GLAM simulations. Figure 6.4 shows the evolutionary tracks and diagrams of skewness $S_3(\sigma)$ and kurtosis $S_4(\sigma)$, as found from GLAM simulations. Here, we combine data from six GLAM simulations, all with many realisations, thus the shot noise is much smaller. Data for different boxes agree quite well within 2–3%. In top panels, coloured curves join simulations with identical smoothing lengths to show evolutionary tracks, in bottom panels they join simulations of identical

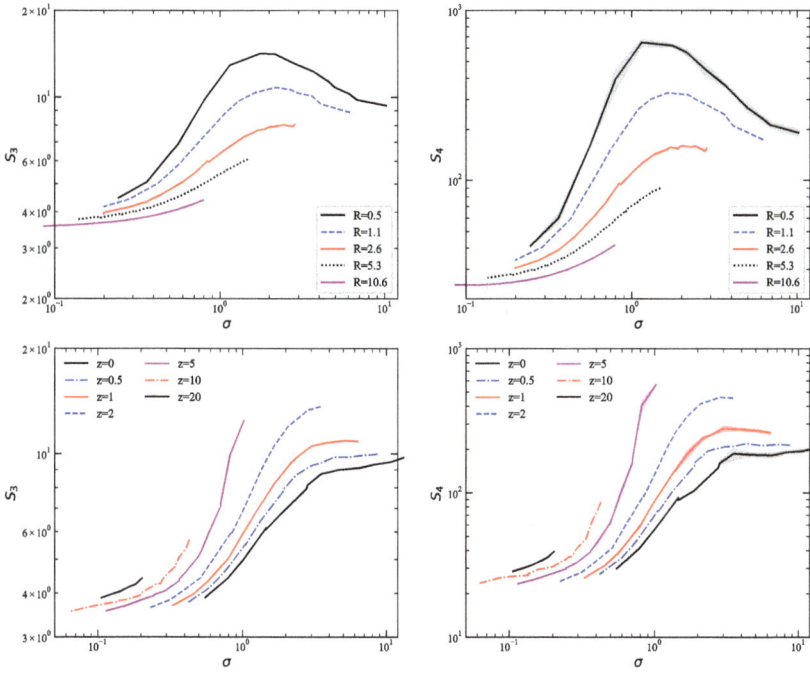

Fig. 6.4. Top panels: evolutionary tracks; bottom panels: evolutionary diagrams. Left panels: skewness S_3; right panels: kurtosis S_4. Shaded areas correspond to 1σ statistical uncertainties (Einasto *et al.*, 2021c).

redshift to show evolutionary diagrams. New simulations confirm earlier results.

Cosmological skewness and kurtosis, S_3 and S_4, are amplitude parameters of mathematical S and K. The asymmetry of the cosmic web is measured by mathematical skewness and kurtosis, shown in left panels of Fig. 6.2. The dispersion parameter σ is a power-law function of $1+z$ with negative slope. When we approach the inflation period while perturbations were created, $z \Rightarrow \infty$, then $\sigma \Rightarrow 0$ and mathematical skewness and kurtosis approach zero values since perturbations had a Gaussian distribution. Cosmological skewness and kurtosis parameters approach finite values $S_3 \Rightarrow 3$ and $S_4 \Rightarrow 15$, as expected in the perturbation theory. This means that the asymmetric evolution started immediately after the creation of initial Gaussian perturbations. The increase in the asymmetry is almost linear in log–log diagram for large smoothing scales. For small smoothing scales, the increase in the asymmetry of the cosmic web speeds up until the epoch $z \leq 2$, thereafter the increase continues with lower speed. This transition is due to the influence of the dark energy, which dominates the matter–energy balance after this epoch.

One of the findings of our study was the contrast between the evolution of the cosmic web on small and large scales, as defined by the smoothing length R_t. The cosmic web populations defined by a small smoothing length $R_t \leq 4\ h^{-1}$ Mpc have at all cosmic epochs $S_3 \geq 4$ and $S_4 \geq 35$. Populations with large smoothing length converge at earliest epochs to values $S_3 \approx 3$ and $S_4 \approx 15$, as predicted.

To understand the reason for this difference, we recall that the cosmological moments S_3 and S_4 are actually amplitude parameters of the mathematical skewness $S = S_3 \times \sigma$ and kurtosis $K = S_4 \times \sigma^2$, as defined by Eqs. (6.9) and (6.10). The data presented in Fig. 6.3 can be expressed as functions of redshift of the mathematical skewness $S(z)$ and kurtosis $K(z)$, as shown in Figure 6.2. The speed of the evolution of the small-scale elements, characterised by a small smoothing length R_t, is much faster: for $R_t = 1\ h^{-1}$ Mpc, the gradient of S and K is $\langle \gamma_S \rangle \approx -1.5$ and $\langle \gamma_K \rangle \approx -3$. The increase of the gradient in the interval $z \geq 3$ is due to the non-linear growth of density perturbations, which is important for small-scale perturbations. The relative decrease of the gradient for $z \leq 1$ is due to the effect of the Λ term.

We calculated asymmetry parameters from the perturbation theory, following Bernardeau (1994). We show in all panels of Fig. 6.3 via dotted curves cosmological skewness and kurtosis parameter functions $S_3(\sigma)$ and $S_4(\sigma)$, as predicted by the perturbation theory, for the same set of smoothing lengths and redshifts. The theory suggests the dependence of S_3 and S_4 on the effective slope of the power spectrum of the perturbations at radius R as $\gamma = -(n_{\text{eff}} + 3)$. According to the perturbation theory, during the evolution-only the height of values S_3 and S_4 changes, see the ladder of dotted lines in the top panels of Fig. 6.3 for different R_t, and the lines in bottom panels of Fig. 6.3 for different z. No strong increase in S_3 and S_4 for later evolutionary epochs, $z \geq 3$, and smaller smoothing scales, $R_t \leq 4$, is predicted. This comparison shows that for early epochs with redshift $z > 10$ and large smoothing lengths, the perturbation theory is in fairly good agreement with the results of numerical simulations. For later epochs with lower redshifts and smaller smoothing lengths, the perturbation theory is not able to reproduce the $S_3(\sigma)$ and $S_4(\sigma)$ functions found in simulations. Differences increase with the decrease in smoothing length, as expected, see Fig. 3.2 of Chapter 3.

The almost identical pattern of the cosmic web on the supercluster scale at epochs with $z \geq 3$ suggests that the same pattern existed at earlier epochs, even soon after the creation of density fluctuations, see Fig. 4.17 of Chapter 4. The development of the density field in the early phase is well described by the Zeldovich (1970) approximation and its extension, the adhesion model by Kofman & Shandarin (1988). As shown by Kofman *et al.* (1990, 1992), the adhesion approximation for the present epoch yields structures that are very similar to the structures calculated with N-body numerical simulations of the evolution of the cosmic web with the same initial fluctuations, see Fig. 4.1 of Chapter 4. *Thus, the combination of theoretical models and numerical simulations suggests that the asymmetry of the PDF started to form soon after the creation of fluctuations in the early period of the evolution of the Universe.*

6.2   Biasing phenomenon

RHEA: Oh, bias, bias... such a loaded word.

ARIEL: To be honest, I'm completely confused by it. It can mean so many different things.

CALLISTUS: It depends on the context. Bias is commonly used when somebody has a prejudice for or against one person or a group. Like I'm favourably biased towards you two.

RHEA: I'd hope so. In statistics, bias means distortion. We might have an estimator for the mean of a variable. The statistical error might decrease as $1/\sqrt{N}$, but no matter how many measurements we perform, we still might be off a little bit. That residual distortion is a statistical bias.

CALLISTUS: However, in spatial statistics of cosmology, we use bias in yet another sense. Our inflationary concordance models constrain matter fluctuations, primarily dark matter with some baryon content. However, typically, we observe starlight from galaxies. In the simplest theories, the number of galaxies would be proportional to the amount of dark matter.

RHEA: A fun fact is that this sense of bias comes from early studies of noise in telephone networks. With amplification b, the noise power increases as b squared. Similarly, galaxy density fluctuations could be an amplified version of the dark matter density fluctuations. If the galaxy over-density is b times the dark matter over-density, its power spectrum will be amplified by b squared. In an ideal world, b would be 1, and we would say that galaxies trace dark matter.

ARIEL: This appears to be a simplistic theory; we would be extremely lucky if the galaxy overdensity were simply equal to that of the dark matter. I have a bad feeling that it will be worse.

CALLISTUS: Do you want the good or the bad news first?

ARIEL: I always want to hear the good news!

CALLISTUS: Well, if we assume linear bias, that is, the galaxy overdensity is proportional to the dark matter overdensity, gravitational interaction will evolve the bias value asymptotically toward $b = 1$ as time goes on.[7]

ARIEL: Oh-ohh. You said linear. That means trouble –

RHEA: The bad news is non-linear local bias. It means that the galaxy overdensity field is a non-linear function of the galaxy density.

CALLISTUS: We can model it with a specific non-linear function, such as an exponential or expand it in Taylor series and define bias coefficients, $b_1, b_2, \ldots$

ARIEL: And I deduce that non-local bias would mean that the galaxy density at a particular location depends on the density at another location. Any more good news?

CALLISTUS: First, I'm afraid there's more bad news. The number of galaxies in a cell is not a deterministic value based on the local or non-local dark matter density. Galaxy formation is a stochastic process. The simplest of such processes is the Poisson process, where we assume that the dark matter density determines the mean of a Poisson distribution.

RHEA: In this case, you have a Poisson bias in the variance proportional to one over the average galaxy count $1/ < N_g >$. The next simplest assumption is that all large haloes have a central galaxy, while outskirts produce galaxies with a Poisson process. This simple idea can explain sub-Poisson scatter found in galaxy catalogues.

CALLISTUS: At a slightly higher level are semianalytic models, or SAMs, which take into account merger history when generating galaxies in haloes.

ARIEL: I guess analysing hydrodynamical simulations, which track galaxy formation with sub-grid physics, such as radiation transfer, feedback from supernovae and black holes, would be the next, and probably the highest, level of approximation.

RHEA: For completeness, there is, of course, Lagrangian bias, which captures the effect of the initial neighbourhood of a galaxy on

---

[7]Tegmark & Peebles (1998).

its emergence, and Laplacian bias, which encodes the effect of tidal fields.

CALLISTUS: Sometimes, simple physics provides powerful insights. We can picture the matter distribution as the sum of a clustered fraction, F_c, of the galaxy distribution, and the rest is unclustered. Then the bias is the inverse of the clustered fraction $1/F_c$.

ARIEL: Which theory should I use?

CALLISTUS: It depends on your goals and the accuracy of your data.

RHEA: Remember, one person's bias is another person's galaxy formation theory. Nature is complex and non-linear on galaxy scales.

CALLISTUS: Another simple idea is the Ising bias,[8] we can dig into it next time. As a preview, consider your sub-haloes: they either host or do not host a galaxy with a fixed probability.

ARIEL: Will this be on my comp?

CALLISTUS: Cosmologists came up with all these ideas paying scant attention to the issue of collateral damage to graduate students studying for comprehensive exams.

RHEA: It's nature that's doing it to you, not cosmologists.

ARIEL: Roger that. Loud and clear.

6.2.1 *Statistical biasing*

An important element of the classical cosmology paradigm is the distribution of galaxies. Early data on the distribution of galaxies on the sky suggested that this distribution is essentially a random one (field galaxies) to which some clusters and perhaps even super-clusters were added; see the angular distribution of the numbers of galaxies brighter than $B \approx 19$ by Soneira & Peebles (1978) in Fig. 2.1. The angular distribution of galaxies can be considered as a random Gaussian process described by the angular CF of galaxies, as done by Peebles (1973, 1974).

The term "biasing" was suggested by Kaiser (1984) to denote the difference between distributions of galaxies and clusters of galaxies and can be characterised by the difference between CFs of galaxies and clusters of galaxies. Subsequently, this term was used in a more general sense to measure differences in the distribution of galaxies of various luminosity and matter. An early use of bias was in CDM simulations of the evolution of the Universe by Davis *et al.* (1985) and

---

[8]Repp & Szapudi (2019b).

White *et al.* (1987), where a fairly high bias parameter was applied to bring CFs of model galaxies into agreement with observed CFs.

A very detailed review of the galaxy biasing was given by Desjacques *et al.* (2018), who considers the bias as the statistical relation between the distribution of galaxies and matter. Usually, the bias parameter b is defined using the ratio of the density contrast of galaxies of various luminosity and matter at location $\mathbf{x}$:

$$b = \frac{\delta_{gal}(\mathbf{x})}{\delta_m(\mathbf{x})}. \tag{6.12}$$

Desjacques *et al.* (2018) focused on large scales where cosmic density fields are quasilinear. On these scales, the clustering of galaxies can be described by a perturbative bias expansion, and the physics of galaxy formation is absorbed by a set of coefficients of the expansion called bias parameters.

The definition in Eq. (6.12) is based on the tacit assumption that galaxies are randomly placed. In other words, voids are just regions of lower density of galaxies. But there is a problem with this interpretation. Observations show that there are no galaxies in voids, except faint galaxy filaments crossing voids, determined by clusters and superclusters. To apply the formula in Eq. (6.12), and to find the mean value of the bias parameter, smoothing of the density field using a rather large smoothing length is applied, so that the density of galaxies is everywhere non-zero. But this procedure smooths galaxies into regions which are actually empty. When we apply smoothing on scales, comparable to sizes of galactic haloes, we expect $b = 0$ outside haloes. If the distribution of galaxies follows the distribution of matter in high-density regions, then in these regions $b \sim 1$.

6.2.2 *Physical biasing*

We consider the biasing phenomenon is a physical problem concerning the relation between distributions of matter and galaxies. The bias function b is a fundamental cosmological function, which quantitatively relates differences between distributions of matter and galaxies. The distributions of matter and galaxies can be modelled by numerical simulations.

The expansion of the Universe in its early phase is an adiabatic process (Peebles, 1982b; Zeldovich, 1970). The growth of adiabatic perturbations proceeds at a low temperature of the primordial "gas" and the flow of particles is very smooth (Zeldovich *et al.*, 1982). Small initial perturbations combined with the smooth flow of particles develop into the non-linear stage and dense regions will be built up by the concentration of matter into caustics by intersection of particle trajectories (Arnold *et al.*, 1982; Zeldovich *et al.*, 1982).

Galaxy formation is a two-stage process — gravitating material in the Universe condenses first into DM haloes (White & Rees, 1978). To form a galaxy, the density of matter must exceed a critical value, the Press & Schechter (1974) limit, $\delta_{cr} = 1.686$. This result is confirmed by hydrodynamical models of galaxy formation (for early model, see Cen & Ostriker, 1992). These considerations are supported by direct observational evidence — all galaxies are DM-dominated, especially dwarf galaxies. The luminous content of galaxies results from the combined action of gravitational and hydrodynamical processes within potential wells provided by DM haloes. Arguments by Zeldovich, White, and Rees suggested that galaxy formation is a threshold phenomenon: *in low-density regions, no galaxy formation occurs at all, here the matter is still in pre-galactic unclustered form.* Thus, the matter can be divided into two components: clustered in high-density regions and non-clustered in low-density regions. As described in Chapter 2, differences in the distribution of galaxies and matter were noticed already by Jõeveer *et al.* (1977) and discussed in more detail by Zeldovich *et al.* (1982) and Einasto & Saar (1987).

To avoid problems with the definition of bias parameter according to Eq. (6.12), it is better to define the bias parameter b using power spectra or CFs of matter and galaxies. Following these considerations, we define the bias function $b(r, M_r)$ through the ratio of CFs of the galaxy (particle) samples with luminosity limits M_r (particle density limits ρ_0) to correlations functions of all DM, both at identical separations r:

$$b^2(r, M_r) = \xi(r, M_r)/\xi_{DM}(r), \qquad (6.13)$$

and a similar formula for $b(r, \rho_0)$, where the limiting luminosity M_r is replaced by the particle density limit ρ_0. Bias depends on the luminosity M_r of galaxies (particle density limit ρ_0) used in the calculation of CFs.

Einasto *et al.* (1994a, 1999, 2023b) found the following relation between the power spectra of the matter, $P_m(k)$, and that of the clustered population (galaxies or clusters of galaxies), $P_c(k)$:

$$P_m(k) = \langle|\delta_m(k)|^2\rangle = F_c^2 P_c(k), \qquad (6.14)$$

where k is the wave number and F_c is the fraction of matter in the clustered population. A similar relation is also valid between CFs of galaxies and matter. The derivation of this equation is given in the following section. The last equation gives, for the bias parameter,

$$b_c = 1/F_c. \qquad (6.15)$$

In other words, the bias parameter depends on the fraction of matter in the clustered population. Numerical simulations showed that this relation is more accurate for galaxy samples with low-luminosity limits, see the following.

Einasto *et al.* (2023a, 2023b) investigated the time evolution of bias functions using The Next Generation (TNG) series of hydrodynamical simulations by Springel *et al.* (2018), see Fig. 6.8, high-resolution versions of simulations, TNG100-1 and TNG300-1, have particle numbers 1820^3 and 2500^3, respectively, both for DM and gas particles, and the same number of cells. Low resolutions versions, TNG100-3 and TNG300-3, have 455^3 and 625^3 cells and numbers of DM and gas particles. TNG100 and TNG300 simulations are done in boxes of sizes $L_0 = 75 \ h^{-1}$ Mpc and $L_0 = 205 \ h^{-1}$ Mpc, respectively.

Density fields for DM and galaxies are shown in Fig. 6.5 for the TNG100-3 simulation in x-, y-coordinates in a sheet of a thickness of 11 simulation cells, $1.8 \ h^{-1}$ Mpc, across a massive cluster for epochs $z = 0$, $z = 2$ and $z = 5$. The top panels are for full DM samples, bottom panels for DM particles with densities $\log \rho \geq \log \rho_0 = -7.8$, i.e. all particles of the clustered population. Densities are given in solar masses per cubic comoving kpc. Colour codes are for grid cells of different spatial densities.

The figure demonstrates the presence of the cosmic web from the early epoch $z = 5$ to the present epoch $z = 0$. All principal elements of the cosmic web are present already at the epoch $z = 5$. The contraction of superclusters (low-mass systems flow towards central clusters of superclusters) is also visible in these plots. The top panels of Fig. 6.5 show the presence of faint DM filaments, which are absent in the bottom panels in the density fields of the clustered matter. Here, only galaxy filaments exist, and most of the

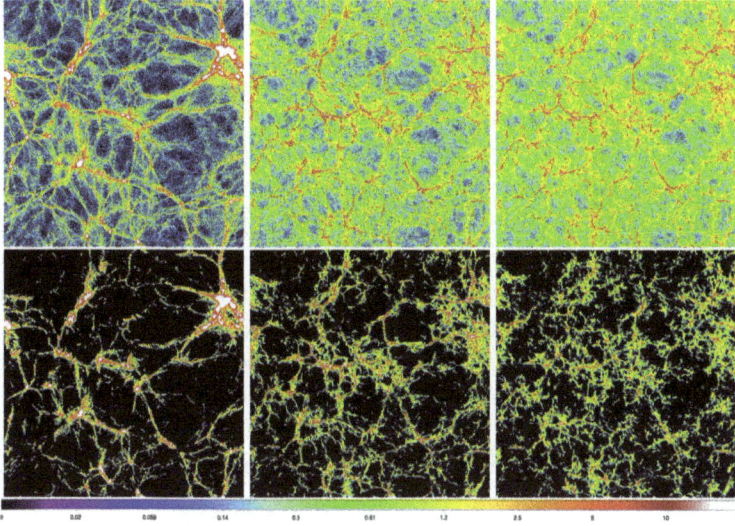

Fig. 6.5. Density fields of DM of the TNG100-3 simulation of size $L_0 = 75 \, h^{-1}$ Mpc and thickness $1.8 \, h^{-1}$ Mpc. The top panels show density fields of the full DM, and the bottom panels show density fields of the clustered population of DM using DM particles with densities $\log \rho \geq \log \rho_0 = -7.8$, i.e. all particles of the clustered population. The left, central and right panels show density fields at epochs $z = 0$, $z = 2$ and $z = 5$. The densities are given in the logarithmic scale to see better the distribution of faint filaments (Einasto *et al.*, 2023a).

volume has zero density. Already the visual inspection of the bottom panels shows the increase of the fraction of the volume of zero-density cells with time (decreasing z).

6.2.3 *The influence of a homogeneous population in voids*

The dependence of the bias parameter on the fraction of particles in the clustered population was studied by Einasto *et al.* (1986), Einasto & Saar (1987), Gramann & Einasto (1992), Einasto *et al.* (1994a) and Einasto *et al.* (1999). This factor is crucial to understand our results, thus we give here two simple toy models to explain the idea.

The natural estimator to determine the two-point spatial CF is

$$1 + \xi(r) = \frac{DD(r)}{RR(r)}, \tag{6.16}$$

where $DD(r)$ and $RR(r)$ are normalised counts of galaxy–galaxy and random–random pairs at separation r. Consider a volume of size V_0, containing galaxies and systems of galaxies like supercluster central regions. Denote counts of galaxy–galaxy and random–random pairs as $DD_0(r)$ and $RR_0(r)$. Now, surround this volume with empty space with no galaxies, and denote the total volume of this sample as V_1. Galaxy–galaxy counts in the new volume are identical to counts in the original volume, $DD_1(r) = DD_0(r)$. Random–random counts at separation r, $RR_1(r)$ are, however, lower since the random sample is diluted over a larger volume V_1. This rises the amplitude of CF $1 + \xi(r)$, the rise is proportional to the ratio of volumes, V_0/V_1. To illustrate this effect, we calculated CFs for two samples: one for a sample with galaxies of size 100 h^{-1} Mpc and the other where this sample is located in a cube of size 200 h^{-1} Mpc, containing outside the inner cube with no galaxies. The bias function is given by the ratio of correlation functions of both samples: $1 + \xi(r) = RR_0(r)/RR_1(r)$. Over most of the r range, it is proportional to the square root of the ratio of volumes, V_1/V_0, i.e. the presence of voids increases the amplitude of the bias function.

The second model was suggested by Einasto *et al.* (1999), and we give here a summary of the discussion. Consider an idealised density field, which consists of a fluctuating clustered component and a background of constant density, so that

$$D_m(\mathbf{x}) = D_g(\mathbf{x}) + D_s(\mathbf{x}), \qquad (6.17)$$

where subscript m stands for all matter, g stands for galaxies (the clustered component), and s stands for the smooth component. The density contrast of the matter is

$$\delta_m = \frac{D_m - \overline{D}_m}{\overline{D}_m}, \qquad (6.18)$$

or, applying (6.17),

$$\delta_m = \frac{D_g + D_s - (\overline{D}_g + \overline{D}_s)}{\overline{D}_g + \overline{D}_s}. \qquad (6.19)$$

Since $D_s = \overline{D}_s$, we get

$$\delta_m = \frac{D_g - \overline{D}_g}{\overline{D}_m} = \delta_g \frac{\overline{D}_g}{\overline{D}_m}. \qquad (6.20)$$

In the last equation, $\overline{D}_g/\overline{D}_m$ is the fraction of matter in the clustered population, F_c, and we get

$$\delta_m = \delta_g F_c. \tag{6.21}$$

According to traditional definition, $\delta_g/\delta_m = b$, and we get for the bias parameter

$$b = \frac{1}{F_c}. \tag{6.22}$$

Equations (6.21) and (6.22) show that the subtraction of a homogeneous population from the whole matter population increases the amplitude of the CF (power spectrum) of the remaining clustered population. In this approximation, biasing is linear and does not depend on scale. These equations have a simple interpretation. If we subtract from the density field a constant density background but otherwise preserve density fluctuations, then amplitudes of *absolute* density fluctuations remain the same, but amplitudes of *relative* fluctuations with respect to the mean density increase by a factor which is determined by the ratio of mean densities.

To check how accurately the relation (6.22) holds, Einasto *et al.* (2019a) calculated ratios $1/F_c$ for the present epoch $z = 0$ for a range of ρ_0 values. These calculations showed that bias parameter values for particle density threshold $\rho_0 \leq 3$ are rather close to values found from the relation (6.22). This is expected since bias functions $b(r)$ are almost constant for $\rho_0 \leq 3$, see Fig. 6.8, i.e. correlation functions of galaxies and matter have a similar shape and differ only by the amplitude, which defines the bias parameter. For higher ρ_0 values, differences in shapes of CFs influence amplitudes of bias functions, and the relation (6.22) is not valid.

In this way, we see that the amplitude of the CF, which defines the bias parameter, measures the emptiness of samples of galaxies, quantified by the fraction of particles in voids and in the clustered population.

6.2.4 *Evolution of the luminosity function of galaxies*

Differential luminosity functions of galaxies of the TNG300-1 simulation are shown on the left panel of Fig. 6.6. The integral over

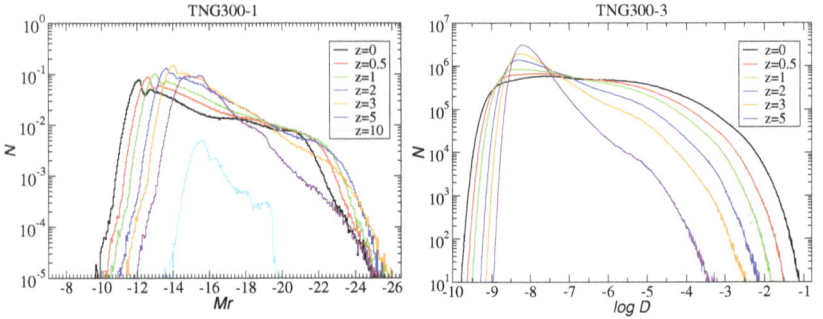

Fig. 6.6. Left: differential galaxy luminosity distribution in photometric system r for the TNG300-1 simulation. Colour codes show distributions at various z. Right: differential distribution of DM particles of the TNG300-3 simulation as a function of the total density at the location of particles, $\log \rho_0$ (Einasto *et al.*, 2023a).

all luminosities defines the number of sub-haloes/galaxies, which did not change considerably during the period $0 \leq z \leq 5$. The number of resolved galaxies, including dwarf satellite galaxies, reaches a maximum at $z = 3$, and decreases slightly thereafter. Authors also calculated luminosity functions for the TNG100-1 simulation, which has ~ 20 times smaller volume, but contains only ~ 4 times less galaxies. The increase in the number of galaxies per unit magnitude interval and cubic megaparsec in the TNG100-1 simulation comes essentially from dwarf galaxies with luminosities $M_r \geq -18.0$, where the number of dwarf galaxies per cubic megaparsec is up to 10 times larger than in the simulation TNG300-1. The faint end tail of luminosity distribution of the TNG100-1 simulation is two magnitudes fainter than that of the simulation TNG300-1. For both simulations, there is a rapid increase of the luminosity of the brightest galaxies between redshifts $z = 10$ and $z = 5$. This increase is due to the merging of galaxies in the centres of massive haloes. Note also that the luminosity of most luminous galaxies with $M_r < -22.0$ increases between redshifts $z = 10$ and $z = 3$ and thereafter decreases slightly. The most essential difference between simulations at epochs $z = 10$ and $z \leq 5$ is in the number of simulated galaxies: at redshift $z = 10$, it is much lower than in smaller redshifts, as seen from Fig. 6.6. This result may be a property of the TNG simulation, since direct observational data show that $L^\star$ type galaxies were present in the very young Universe at $z \approx 12$, see Chapter 8. In the further analyses, we shall use only simulated galaxies of TNG300-1 simulation with redshifts $z \leq 5$.

The right panel of Fig. 6.6 shows the distribution of DM particle densities at various redshifts $z \leq 5$. We see that DM density in most dense regions increases from redshift $z = 5$ to $z = 0$ about 100 times, much more than luminosities of most luminous galaxies M_r from redshift $z = 5$ to $z = 0$.

6.2.5 *Evolution of CFs of simulation galaxies and particle density-limited DM samples*

In Chapter 5, we analysed CFs of DM simulations and SDSS samples using DM-only simulations with the number of particles $N_{\text{part}} = 512^3$. In this chapter, we analyse CFs of simulated galaxies and particle density-limited DM samples using most recent hydrodynamical simulations by Springel *et al.* (2018), as described above. Einasto *et al.* (2023a, 2023b) used the definition of the bias function through correlation functions of clustered and non-clustered particles and calculated correlation and bias functions for the TNG simulations. As the reference sample, authors used DM particles from simulation TNG300-3.

In the top row of Fig. 6.7, we present CFs of *particle density-limited DM samples* of TNG300-3 simulations for epochs $z = 0,\ 5$. Here, we used particle density limits ρ_0 in $\log \rho_0$ units, starting from $\log \rho_0 = -8.3$. The limit $\log \rho_0 = -8.3$ corresponds to DM particles, which are located in regions too low to form galaxies at the epoch $z = 0$; the corresponding CF is marked with dashed orange lines. DM samples, corresponding to the faintest simulated galaxies at $z = 0$, have the limit $\log \rho_0 = -7.8$. DM samples corresponding to most luminous galaxies have at present epoch $\log \rho_0 \approx -3.3$.

In the bottom row of Fig. 6.7, we show CFs of *galaxies* of the TNG300-1 simulation for evolutionary epochs $z = 0$ and $z = 5$. For all samples, a large range of limiting luminosities M_r was applied from $M_r = -11.0$ to $M_r = -23.0$. Bold black lines show DM CFs, and coloured lines present galaxy CFs for various M_r limits shown as labels.

Figure 6.7 shows CFs of DM samples for various density limits and of galaxies for various M_r limits form sequences of increasing amplitude with increasing particle density and luminosity limits.

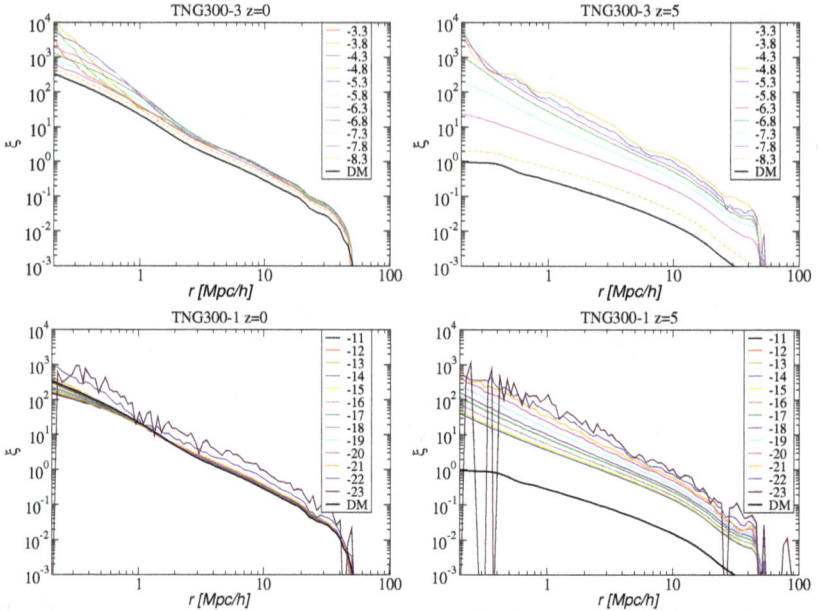

Fig. 6.7. CFs $\xi(r)$ for epochs $z = 0$, 5 are shown in the left and right panels. The top row shows the TNG300-3 DM simulations. Coloured lines show functions for various DM particle density limit $\log \rho_0$. The stellar mass of galaxies is in units 10^9 $M_\odot$. The bottom row is for the TNG300-1 galaxy simulation. Magnitude limits in r photometric system are shown as symbol labels. Separations r are in comoving units. Black bold lines show CFs of DM for respective epochs (Einasto *et al.*, 2023a).

This is the conventional biasing effect (Kaiser, 1984). The number of galaxies in most samples is large, thus random errors of CFs are very small. The figure also shows that for the present epoch $z = 0$, CFs of the faintest galaxies almost coincide with CFs of DM. For earlier epochs, there exists a gap in the amplitudes of CFs of galaxies relative to DM, which increases with the simulation epoch z.

6.2.6 *Evolution of bias functions of simulated galaxies and DM particles*

CFs of galaxies divided by CFs of DM define bias functions, see Eq. (6.13). For TNG samples, bias functions are shown in Fig. 6.8. In top panels, we show the bias functions for particle density-limited

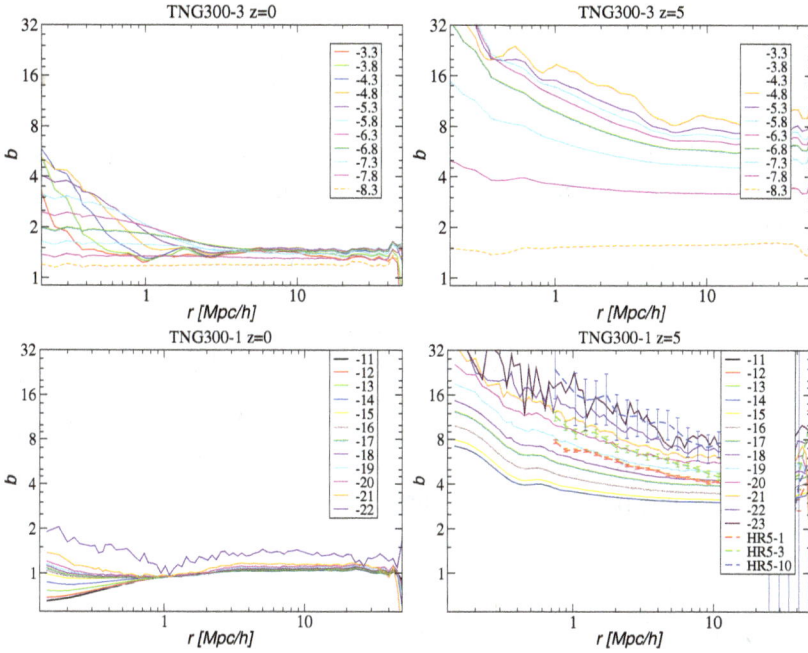

Fig. 6.8. Bias functions of galaxies, $b(r)$, are shown for epochs $z = 0$, 5 in the left and right panels, respectively. Top panels show the TNG300-3 DM simulation. Coloured lines show functions for various particle density limits $\log \rho_0$. The limit $\log \rho_0 = -8.3$ is marked with dashed orange lines, and it corresponds to DM particles, not associated with galaxies. Star masses of galaxies are in units $10^9 \, M_\odot$. Bottom panels stand for the TNG300-3 galaxy simulation. Magnitude M_r limits are shown as symbol labels. Level $b = 1$ is the bias function of DM (Einasto *et al.*, 2023a).

DM samples of TNG300-3 simulation for two epochs, $z = 0$, 5, using particle density limits in $\log \rho_0$ units, starting from $\log \rho_0 = -8.3$. In the bottom row of Fig. 6.8, we present the bias functions of TNG300-1 galaxy simulation. Various colours stand for bias functions of galaxies of different luminosity limits. In bottom right panel, we also show with dashed lines bias functions of HR5 simulations by Park *et al.* (2022) for simulation epoch $z = 5$.

Bias functions, presented in Fig. 6.8, have three important properties. The first property is as follows: bias function curves for galaxies of low luminosity, $M_r \geq -18.0$, are almost identical. We discuss this effect in more detail in the following.

The second important feature is the shape of bias functions for separations, $r \leq 5 \ h^{-1}$ Mpc. In this separation range, bias functions have larger values than on medium and larger separations. This is due to the effect of haloes, which have characteristic diameters up to $r \approx 5 \ h^{-1}$ Mpc (in comoving coordinates), both in particle density-limited DM samples as well as luminosity-limited galaxy samples.

The third feature is the amplitude at very small separations, $r \leq 2 \ h^{-1}$ Mpc, for galaxy samples of the lowest luminosities. Here, bias functions are lower than at higher separations. This effect is observed for recent epochs, $z \leq 3$. For the present epoch $z = 0$, this means anti-biasing since $b < 1$. Bias functions of particle density-limited DM samples of simulation TNG300-3 do not have this feature, and in this separation interval and particle density limits, $\log \rho_0 \leq -6.8$, their bias functions are almost parallel lines with amplitudes, increasing with increasing particle density limit $\log \rho_0$.

6.2.7 *Evolution of bias parameters of simulated galaxies and DM particles*

Bias functions have a plateau at $6 \leq r \leq 20 \ h^{-1}$ Mpc, see Fig. 6.8. We use this plateau to measure the relative amplitude of the bias function as the bias parameter,

$$b(M_r) = b(r_0, M_r), \qquad (6.23)$$

and a similar formula, where M_r is replaced by particle density limit ρ_0 and r_0 is the value of the separation r to measure the amplitude of the bias function. We calculated for all samples bias parameters for the comoving separation, $r_0 = r_{10} = 10 \ h^{-1}$ Mpc, as functions of the galaxy absolute magnitude in r colour, M_r, or particle density limit ρ_0 for DM simulations. In earlier analysis of bias functions for the present epoch, we used $r_0 = 6 \ h^{-1}$ Mpc to define bias parameters, but for epochs $z > 0$, the limit $r_0 = 10 \ h^{-1}$ Mpc is better, since then DM haloes were larger in comoving coordinates. At smaller distances, bias functions are influenced by the distribution of particles and galaxies in haloes, and at larger distances, the bias functions have some wiggles, which makes the comparison of samples with various galaxy luminosity limits difficult.

In Fig. 6.9, we present the evolution of bias parameters with cosmic epoch z. In the left panel, we show the evolution of bias

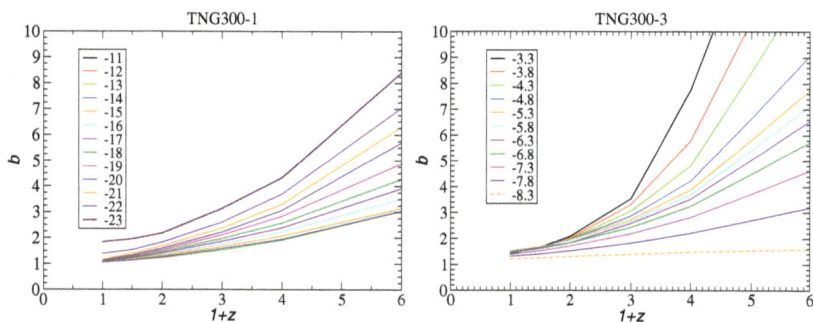

Fig. 6.9. Evolution of bias parameter values of TNG simulations with the epoch z. The left panel is for galaxy simulation TNG300-1, magnitude limits in M_r photometric system are shown as symbol labels. The right panel shows the evolution of the bias parameter of DM samples of simulation TNG300-3 with various limits of substructure density $\log \rho_0$. The limit $\log \rho_0 = -8.3$ is marked with dashed orange lines, it corresponds to DM particles, not associated with galaxies. Level $b = 1$ is the bias of DM (Einasto *et al.*, 2023a).

parameters of *galaxy simulation* TNG300-1, and different colours show bias parameters for galaxies of various luminosity, M_r. As we see, bias parameters of galaxies form smooth curves, $b(z, |M_r|)$, with amplitudes, increasing with luminosity M_r. With decreasing luminosity, bias parameters $b(z, |M_r|)$ approach asymptotic low limits $b(z, |M_r|) \to b_0$. A similar effect was found for correlation lengths of SDSS galaxies, see Fig. 5.6.

The right panel of Fig. 6.9 shows the evolution of bias parameters of *particle density-limited DM samples* of simulation TNG300-3 for various limits $\log \rho_0$. Samples with DM particle density limit $\log \rho_0 = -7.8$ correspond to the faintest clustered population similar to the faintest galaxies of simulation TNG300-1. The dashed orange line corresponds to DM samples with a limit $\log \rho_0 = -8.3$, i.e. to DM particles below the density limit, needed to form stars and galaxies.

Figure 6.10 presents the bias parameter as a function of the luminosity $b(z, M_r)$ for various epochs from $z = 5$ to $z = 0$. In the left panel, the evolution of galaxy bias parameter is shown as a function of luminosity M_r. It demonstrates the essential property of bias parameters of galaxies — at low and medium luminosities, the bias parameter $b(z, M_r)$ approaches asymptotically to a low-luminosity limit, b_0, and rises for higher luminosities from $M_r \leq -15$

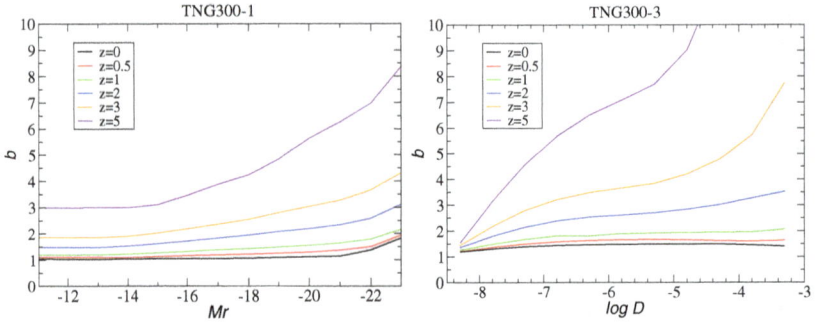

Fig. 6.10. Left: the evolution of the bias parameter b of simulation TNG300-1 as a function of the magnitude limits M_r. Right: the evolution of bias parameter for particle density-limited DM samples of simulation TNG300-3 as a function of the logarithm of particle density limit ρ_0. Data are given for various simulation epochs z (Einasto *et al.*, 2023a).

to $M_r \leq -18$; corresponding stellar masses are $1.7 \times 10^8\,\mathrm{M_\odot}$ and $2.7 \times 10^9\,\mathrm{M_\odot}$; details depend on simulation epoch z.

The right panel of Fig. 6.10 shows the dependence of evolution of the bias parameter $b(\log \rho_0)$ of clustered DM samples of simulation TNG300-3. The comparison of panels shows that there are essential differences between bias parameters of clustered DM samples (TNG300-3) and bias parameter of galaxy samples (TNG300-1). In the clustered DM samples, there is no flat region of the bias function $b(\log \rho_0)$ at low particle density limits, ρ_0. Rather, the amplitudes of $b(\log \rho_0)$ curves rise continuously with increasing $\log \rho_0$ ($\log D$ in Fig. 6.10). At very low particle density limit, $\log \rho_0 \approx -10$, essentially all particles are included, see Fig. 6.6. Thus, by definition the bias parameter should be $b = 1$ at this particle density limit. The $b(\log D)$ curves of TNG300-3 DM samples really converge to 1 for low-density limit, but not for TNG300-1 galaxy samples.

At the particle density limit $\log \rho_0 = -7.8$, the bias parameter values of DM simulation TNG300-3 are almost equal to the bias values of the faintest galaxies of the TNG300-1 simulation, see the left panel of Fig. 6.10. At higher particle density limits, the $b(\log \rho_0)$ curves of the DM-selected TNG300-3 samples are rather similar to the $b(M_r)$ curves of the galaxy TNG300-1 simulations. The basic difference lies in the bias values for earlier epochs — here, $b(\log \rho_0)$ curves lie higher than $b(M_r)$ curves. This difference is due to the fact that at earlier epochs particle density limits, $\log \rho_0$ correspond to more luminous galaxies, see Fig. 6.6.

6.2.8 *Bias parameter of faintest galaxies and the fraction of matter in the clustered population*

Our analysis shows that the bias parameter of faintest galaxies, b_0, is a well-defined quantity, almost independent of the luminosity of galaxies, but dependent on the evolutionary epoch z. The bias parameter of faintest galaxies, b_0, approaches at the present epoch to an asymptotic value $b_0 = 1.045$ for TNG300-1 galaxy simulation, see the left panel of Fig. 6.11.

The right panel of Fig. 6.11 shows the evolution of the bias parameter b_0 of DM simulation TNG300-3. The particle density ρ, extracted from the TNG website, includes all matter: DM plus baryonic gas and stellar matter. The ρ value for stellar matter, is not given on the TNG website. To determine the lower ρ_0 limit for stellar matter, Einasto *et al.* (2023a) used a two-step procedure. First, authors found the distribution of the fraction of matter in the clustered population, $F_c(\rho_0)$, of the DM-only simulation TNG300-3, using the cumulative distribution of particle densities, see in Fig. 6.6 differential distributions of particle densities, $N(\log \rho)$, for redshifts

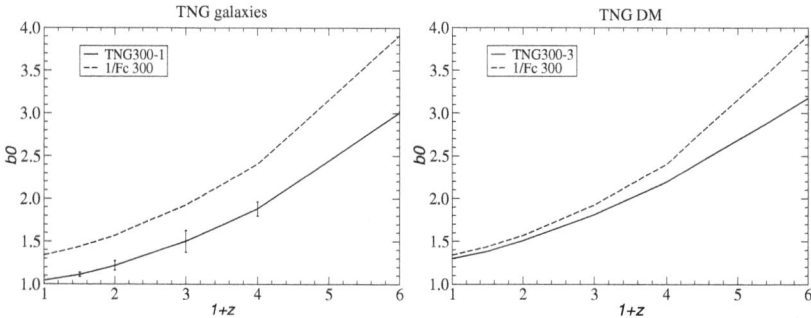

Fig. 6.11. Left: Evolution of the bias parameter b_0 of the faintest galaxies from simulations TNG300-1. Dashed curves show the evolution of the inverse of the fraction of matter in the clustered population $b_0 = 1/F_c$ for simulation TNG300-3. Right: Evolution of the bias parameter b_0 of the faintest galaxies from simulations TNG300-3 at a particle density limit $\log \rho_0 = -7.8$. The dashed curves show the evolution of the inverse of the fraction of matter in the clustered population $b_0 = 1/F_c$ for simulations TNG300-3 (Einasto *et al.*, 2023a).

$z \leq 5$. In the next step, authors compared cumulative distributions of particle densities $F_c(\rho_0)$ with bias parameter distributions $b(\log \rho_0)$ for various particle density limits ρ_0, as shown in Fig. 6.10. This comparison showed that for particle density limit $\log \rho_0 = -7.8$, functions $b(z)$ and $b_c(z) = 1/F_c(z)$ are very close and also close to the b_0 value for TNG300-1 galaxy simulations. This particle density limit $\log \rho_0 = -7.8$ was used as the limit for faintest galaxies in TNG300-1 simulation. Dashed curves in Fig. 6.11 show the inverse of the fraction of the clustered population, $b_c(z) = 1/F_c(z)$ for the TNG300-3 simulation. The figure shows that at the present epoch $z = 0$, DM simulations yield for b_0 a value $b_0 = 1.30$, very close to the expected value from the fraction of the clustered population, $b_0 = 1/F_c = 1.34$.

We use the same $b_c(z) = 1/F_c(z)$ curves, found for the TNG300-3 DM particle simulations, also for TNG300-1 galaxy simulations, as shown by dashed curves in the left panel of Fig. 6.11. We see that $b_c(z) = 1/F_c(z)$ curves lie higher than the actual $b_0(z)$ curves of TNG300-1 galaxy simulations. We conclude that the bias analysis of simulated galaxies in the TNG simulation, based on the correlation function of simulated galaxies and DM, is in conflict with the $b_c(z) = 1/F_c(z)$ criterion.

6.2.9 *Cosmological implications*

Correlation functions of galaxies of the 2dF Galaxy Redshift Survey (2dFGRS) were calculated by Norberg *et al.* (2001) using the conventional procedure, as described in Eqs. (5.6) and (5.7) of Chapter 5. Lahav *et al.* (2002) compared the amplitudes of fluctuations of the 2dFGRS and the CMB and found for the bias parameter of $L^\star$ galaxies at the present epoch a value $b(L^\star) \approx 0.96$. A similar result was found by Verde *et al.* (2002), $b = 1.04 \pm 0.11$. To study biasing properties of galaxy samples authors applied angular CFs of the form (5.5) in the separation interval $0.1 \leq r \leq 10 \ h^{-1}$ Mpc. A similar result was found by Tegmark *et al.* (2004) for SDSS galaxy sample and by Springel *et al.* (2018) comparing TNG simulations with observations. Benson *et al.* (2000) found bias parameter values of $L^\star$ galaxies $b(z) = 1.07,\ 1.24,\ 1.47,\ 2.05,\ 2.91,\ 5.21$ for redshifts $z = 0.0,\ 0.5,\ 1.0,\ 2.0,\ 3.0,\ 5.0$, respectively. Verde *et al.* (2002) summarised these results as follows: *optically selected galaxies do indeed trace the underlying mass distribution.* This conclusion was accepted by the astronomical community.

As noted above, Einasto and his collaborators studied the biasing phenomenon and found that the biasing parameter depends on the fraction of matter in voids. To understand the possible reason for the difference, Einasto *et al.* (2023a, 2023b) studied the distribution of galaxies and matter using density fields and applying CF analyses for TNG simulations, as described in the previous section. This analysis is based on two assumptions: (i) the ΛCDM model is a good approximation of the real Universe on scales where gravitational forces are dominating and the local density is the dominant factor in galaxy formation, and (ii) hydrodynamical processes of galaxy formation and evolution are simulated correctly in TNG simulations. As a result of the gravitational clustering, there exists at all cosmological epochs a low-density population, consisting of a mixture of dark and baryonic matter. Phase synchronisation, which leads to the formation of small, filamentary high-density regions and large contiguous regions with very low spatial densities, plays a crucial role in the evolution of the Universe, as discussed in Chapter 4. This allows us to divide DM and baryonic particles to clustered and unclustered populations. The clustered population forms haloes where the galaxy formation is possible. The unclustered population forms weak filamentary web and populates cosmic voids with no galaxies.

Using TNG simulations, it is possible to study the clustering of simulated galaxies as well as the clustering of DM in high-density regions. In the second case, authors applied a sharp particle density limit, $\rho \geq \rho_0$, to select particles for biased DM model samples, see Chapter 4 for details. To characterise biasing properties of simulated galaxies and clustered DM particles, authors applied CFs of galaxies and DM, as presented in Fig. 6.7. In simulated galaxy samples, redshift distortions by Kaiser effect are absent, thus the correlation analysis can be made using 3D data. The ratio of correlation functions of galaxies (high-density DM regions) to correlation functions of DM defines bias functions, as presented in Fig. 6.8. The amplitude of bias functions defines bias parameters, which we use instead of correlation lengths to characterise the galaxy bias.

As discussed above, the bias parameter is related to the fraction of matter in the clustered population $F_c(z)$: $b_c(z) = 1/F_c(z)$.

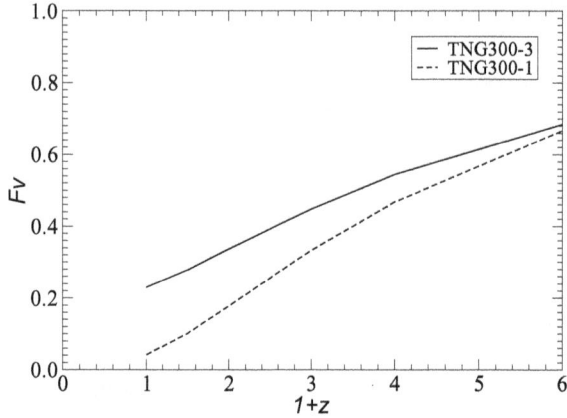

Fig. 6.12. The fraction of void particles F_v for DM TNG300-3 simulations and TNG300-1 galaxy simulations plotted with solid and dashed lines, respectively.

The evolution of the bias parameter $b(z)$ and the fraction of matter in the clustered population, $F_c(z)$, are shown in Fig. 6.10. Using data presented in this figure, we calculated the evolution of the fraction of matter in voids, $F_v(z) = 1 - F_c(z)$, using data on the bias parameter $b(z)$, separately for TNG300-3 DM simulations (solid line) and for TNG300-1 galaxy simulations (dashed line). Results are shown in Fig. 6.12. During the evolution, some non-clustered void matter flows to haloes and becomes clustered matter, thus the fraction of void particles decreases with time. We see that the bias function of DM particle samples leaves sufficient room for further evolution beyond the present epoch $z = 0$ and satisfies the $b = 1/F_c$ criterion. The evolution of the galaxy CF yields for the present epoch $F_v = 0.04$. This means that there is almost no room left for void particles. Since bias parameter decreases with time, the value of bias parameter at present epoch $b(z) \approx 1$ means that we live in an exceptional epoch with no void particles. For further evolution $1 + z < 1$, the extrapolated bias parameter is less than unity, and the $F_v(z)$ curve goes to negative values, which is impossible.

To understand the reason for such behaviour, let us look the shape of CFs of galaxies and DM, as presented in Fig. 6.7. The figure shows that at small separations $r < 5 \ h^{-1}$ Mpc, DM CFs have higher amplitudes than CFs of galaxies. A similar behaviour of the CF was detected by Benson *et al.* (2000) by comparison of a ΛCDM model with APM observation, by Zehavi *et al.* (2004) in the CF of SDSS galaxies, and by Springel *et al.* (2018) in the ΛCDM TNG model, see their Fig. 19. This means that DM is more concentrated to haloes than galaxies, i.e. relative to filaments in haloes, there are more faint DM minihalos than galaxies. An excess clustering of small DM minihalos without corresponding dwarf galaxies was also found by Sawala *et al.* (2017). This lowers bias functions of galaxies in halo regions relative to bias functions of DM, as shown in Fig. 6.7.

We come to the conclusion that the CFs of galaxies have properties not expected when they were suggested to characterise the clustering of galaxies and matter. Another possibility to describe differences in the distribution of galaxies and matter is to use percolation functions. As discussed in Chapter 5, percolation functions of galaxies and matter are very different, see Fig. 5.3. This difference is the largest when we use small smoothing length to calculate percolation functions. As discussed above, a smoothing length not exceeding the characteristic size of DM haloes allows us to see the actual differences between distributions of galaxies and matter, since galaxies are located in DM haloes. However, the extended percolation analysis finds only percolation functions, not parameters which could be used to characterise differences in the distributions of galaxies and DM.

If we want to use the bias parameter as an indication which takes into account the presence of non-clustered matter in voids, we should use CFs of DM and particle density-limited samples of clustered matter. The TNG300-3 DM model yields acceptable answer to this question, as demonstrated in Fig. 6.12. The fraction of matter in voids at the present epoch is according to our analysis, $F_v = 0.23$. This is higher than the fraction of matter in voids, as found by Cautun *et al.* (2014), $F_v = 0.15$, see Fig. 4.9. As mentioned above, our $\rho \geq \rho_0$ definition of clustered particles divides particles to two classes only. Thus, our low-density regions include particles in all low-density regions, including particles in low-density tails of walls in Cautun samples. In lower-density tails of walls, the density is too low for galaxy formation. Wall particles have mass fraction $F_v = 0.24$ (Cautun *et al.*, 2014). The sum of fractions of matter in voids and walls according to Cautun is $F_v = 0.39$. This is larger than our estimate $F_v = 0.23$. But only some Cautun wall particles belong to our non-clustered particles, and such difference is expected. Our estimate of the fraction of void particles, $F_v = 0.23$, is probably closer to reality.

All studies have shown that the bias parameter depends on the luminosity of galaxies. Recent study of the galaxy clustering by Norberg *et al.* (2001) found the dependence of the correlation length and the bias factor on the luminosity. The luminosity dependence of bias factor can be expressed according to Norberg *et al.* (2001) as follows: $b/b^\star = 0.85 + 0.15L/L^\star$, where $L^\star$ is the characteristic luminosity and $b^\star$ the bias factor of $L^\star$ galaxies. This relationship implies that $b^\star = b_0/0.85$, where b_0 is the limiting bias for faint galaxies. Tegmark *et al.* (2004) calculated the power spectrum of galaxies in the SDSS survey. For the relative bias, authors found $b(M)/b^\star = 0.895 + 0.150L/L^\star - 0.040(M - M^\star)$ rather close to the Norberg *et al.* (2001) result. Zehavi *et al.* (2011) found very similar luminosity dependencies. In these studies, CFs were calculated on the basis of 2D data. These determinations were influenced by two effects: (i) the absence of galaxy formation in voids and (ii) the distortion of CFs of galaxies by differences in clustering of DM and galaxies in halo regions. For this reason, only relative bias factors were found reliably.

If we accept for the bias factor of lowest luminosity galaxies the value found from DM analysis in agreement with the $b = 1/F_c$ criterion, $b_0 = 1.30$, and the relative bias factor as found by Norberg *et al.* (2001), we get for the bias factor of $L^\star$ galaxies at the present epoch, $b^\star = 1.53$. Since there exist non-luminous matter in voids, the bias parameter of $L^\star$-galaxies must be definitely higher than unity.

The analysis of TNG simulations brings us to the conclusion that the bias function of galaxies as defined by the comparison of CFs of galaxies and DM samples cannot be interpreted as functions which characterise differences in the distribution of galaxies and matter. A better characteristic is given by the comparison of CFs of particles density-limited DM samples with full DM samples. The main errors of this determination are the possible error in the particle density limit of faintest galaxies, ρ_0, and the error in the fraction of particles in voids.

Another difference between the behaviour of bias parameters of galaxy simulation TNG300-1 and DM simulation TNG300-3 is seen in Fig. 6.10 — at low and medium luminosities $b(M_r)$, the bias parameters of galaxy simulation TNG300-1 approaches asymptotically to a low-luminosity limit, b_0. A flat region of the bias function $b(M_r)$ at low luminosities suggests that faint dwarf galaxies are located in the same filamentary web as brighter galaxies, and there exist no population of faint galaxies in voids.

To conclude the discussion of the biasing problem, we can say that our results have three consequences:

(1) The bias parameter b is not a free parameter which can be chosen to bring models into agreement with observations, as done in early simulations of CDM models of critical cosmological density (Davis *et al.*, 1985; White *et al.*, 1987); actually, it depends on the fraction of matter in the clustered population and on the luminosity of galaxies.

(2) The absence of galaxy formation in low-density regions of the cosmic web and the flow of matter from low-density regions towards high-density regions are essential properties of the evolution of the Universe. The flow of matter from low-density regions is a slow process, and there is always some unclustered matter in voids. Thus, the bias parameter b of clustered matter (galaxies) is always larger than unity.

(3) There exists a difference between properties of correlation and bias functions of galaxies and DM particle samples — galaxy and DM particle samples represent different properties of the cosmic web.

Chapter 7

Physical Processes in Galaxies

RHEA: There was something fishy with this... fish and chips.

ARIEL: Ugghh. Thirteen used to be an amazing restaurant. It was always a dump, but their food –

RHEA: Thirteen was my lucky number. Until now. Something has changed.

CALLISTUS: Panta Rhei.[1] Your favourite restaurant has changed some of its internal parameters, and your observations have revealed that. Well, sic vita est.[2]

RHEA: Amicus meus,[3] your multilinguistic wisdom is profound as always. It's nearly Facebook-worthy, except nobody would understand it. But it doesn't help my disappointment that we had an absolutely horrifying, disgusting lunch.

CALLISTUS: Te video.[4] You will survive.

ARIEL: I will survive, and it's a movie, not a video. It's still sad that nothing is constant in this universe any more.

CALLISTUS: Hmm. The universe is not like a restaurant.

RHEA: Wait, are you discounting the possibility of, say, varying physical constants?

ARIEL: Ugh, changing of the physical laws in space and time? I like it. After all, Panta Rhei.

CALLISTUS: The simplest assumption approved by Occam would be that the physical laws provide a constant framework for our cosmology.

RHEA: Do I detect wishful thinking of a theorist here?

---

[1] Everything flows. Attributed to Heraclitus.
[2] Such is life.
[3] My friend.
[4] I see you.

CALLISTUS: Perhaps. We would be in good company thinking about changing physical constants. Dicke and later Dirac proposed that the gravitational constant might evolve with time. Their idea was prompted by earlier observations by Weyl, Eddington and others: the "large number coincidences". The ratio of an electron's mass and gravitational self-energy is the size of the universe in units of electron radii.

RHEA: We need approximately 10^{40} electrons to span the universe. That's big. But then again, that was before we learnt that the vacuum energy density is about 10^{120} times larger than its natural value from quantum field theory.

ARIEL: I don't see how these large numbers motivate changing physical laws.

CALLISTUS: The strength of gravity is about 10^{40} times smaller than the electromagnetic force. In units where the speed of light and the electron radius are unity, the Universe has an age of 10^{40}. Thus, Dirac assumed that the big G is proportional to the inverse age of the Universe, which perhaps explains why it's so weak.

ARIEL: A pretty idea.

CALLISTUS: Contrary to all observations to date.

ARIEL: Ohh. Too bad.

RHEA: Another parameter constrained through the years is the fine structure constant, aka $1/137$. It's related to the Rydberg constant, and if it changed sufficiently, it would alter the spectral lines of distant objects. No such change was detected, constraining it to a constant at the level of less than a part of a million.[5]

CALLISTUS: Corroborated by CMB and baryogenesis measurements, as well as the Okla natural nuclear reactor. Another interesting example is the variable speed of light theory.[6] Inflation can be mimicked by changing the speed of light during cosmic history.

RHEA: Many other physical constants have been considered to be changing.[7]

CALLISTUS: The large-scale cosmological surveys of the near future are an expensive bet that at least one constant is changing: the cosmological constant or vacuum energy. I doubt it, though.

RHEA: A bet that appears to be paying off.[8]

ARIEL: It's still only $\simeq 4\sigma$, I'd rather bet on Thirteen becoming a decent restaurant again.

RHEA: DESI will obtain more data, but I'll never sample Thirteen again.

---

[5]Chand *et al.* (2004).

[6]Albrecht & Magueijo (1999).

[7]Uzan (2003).

[8]DESI Collaboration *et al.* (2025b).

CALLISTUS: Right. Can we at least agree that Panta Rhei is the best progressive Hungarian rock band ever?

RHEA: Can I swiftly interject that I prefer –

ARIEL: We know. But does she have a renowned cosmologist[9] playing guitar in her band? Didn't think so.

CALLISTUS: Bene dictum.[10]

7.1 Formation and evolution of galaxies in dark matter haloes

7.1.1 *Physical conditions before and after the recombination*

By the time of recombination, the particle content of the universe was ready — the Universe consisted of positively charged nuclei, negatively charged electrons and photons. The Universe was more or less hot; its temperature was slightly over 3000–4000 K, and particles moved randomly in space. These nuclei were mostly just protons but a little of helium and rather small quantities of deuterium and lithium.

The photons scatter from the charged particles (the Thomson scattering), continuously changing direction and exchanging energies with the particles. As the Thomson scattering probability (the cross-section) is inversely proportional to the square of the mass m of the charged particles, the scattering occurs mainly from electrons. However, as the electron and proton motions are coupled through Coulombic forces, scattering also affects protons' distribution. Hence, matter and radiation are tightly bound before recombination due to scattering.

Let us look at the sound speed in the media. The sound speed determines how fast the information about the state of the media spreads. If a certain gas cloud starts to contract a little, gas pressure usually increases, and contraction stops. This is when the sound travelling time in the gas cloud is shorter than the contraction time. One can say that information about the contraction has time to spread all over the cloud, and the state of the gas reacts. But if the sound travelling time is longer, then the state of the gas has no time to

---

[9] Alex Szalay.
[10] Well said.

react, contraction continues, and the gas cloud collapses. By comparing these two timescales, it is possible to get a simple estimate of the Jeans mass.

The precise formula to calculate the Jeans length and mass and explanation of how it was derived is given by, e.g., Binney & Tremaine (2008, Section 5.2.3). So, to be precise, sound speed characterises how much the pressure changes when the density changes, i.e. $c_S = \sqrt{\partial p / \partial \rho}$. Partial derivatives should be taken with constant entropy, meaning that we look to adiabatic processes. The sound speed in this primordial environment before the recombination was determined mainly by the density and motions of the radiation (photons), but both components contribute. However, when comparing the contribution of baryons and photons to total pressure, one can find that radiation clearly dominates. Let us calculate. The pressure due to baryons is

$$p_b = \frac{\rho_b k_B T}{\mu m_H}$$

and the pressure due to photons is

$$p_r = \frac{\rho_r c^2}{3},$$

where k_B is the usual Boltzmann constant, μ is the mean atomic weight, and m_H is the hydrogen mass. Their ratio is very small, implying that baryons do not contribute in pressure.

We need to take both components to deal with density, i.e. $\rho = \rho_b + \rho_r$. Now, we have

$$\frac{\partial p}{\partial \rho} \simeq \frac{c^2}{3} \frac{\partial \rho_r}{\partial \rho_r + \partial \rho_b} = \frac{c^2}{3} \frac{1}{\frac{\partial \rho_r}{\partial \rho_r} + \frac{\partial \rho_b}{\partial \rho_r}}.$$

We may do this with partial derivatives as we have always assumed a constant entropy. In principle, we may replace even partial derivatives with ordinary derivatives.

Taking into account that

$$\frac{d\rho_b}{d\rho_r} = \frac{3}{4} \frac{\rho_b}{\rho_r},$$

we derived the answer we intend to have, namely

$$c_s^2 = \frac{c^2}{3} \left(1 + \frac{3}{4} \frac{\rho_b}{\rho_r} \right)^{-1}$$

or

$$c_s^2 = \frac{c^2}{3} \frac{4\rho_{\text{rad}}}{4\rho_{\text{rad}} + 3\rho_{\text{b}}}.$$

The ratio ρ_b/ρ_r at different redshifts can be written with the help of the redshift z_{eq} (the redshift when matter and radiation energy densities are equal) as

$$\frac{\rho_b}{\rho_r} = \frac{1 + z_{\text{eq}}}{1 + z},$$

giving us

$$c_s^2 = \frac{c^2}{3} \left[1 + \frac{3}{4} \frac{(1 + z_{\text{eq}})}{(1 + z)} \right]^{-1}.$$

In the case of radiation dominance in densities, this gives $c_s = c/\sqrt{3}$. Thus, the sound speed before the recombination is in the order of the speed of light.

This clearly affects the masses and dimensions of the contracting structures. Let us calculate the Jeans mass at, e.g., $z = 1500$. Using the Jeans mass formula from Binney & Tremaine (2008), we have

$$M_J = \frac{\pi^{5/2}}{6} \frac{c_s^3}{\sqrt{\rho} G^{3/2}}.$$

Sound speed at $z = 1500$ is (using $z_{\text{eq}} \simeq 3600$)

$$c_s = 1.04 \times 10^8 \, \text{m/s},$$

and the resulting Jeans mass is

$$M_J = \frac{\pi^{5/2}}{6} \frac{c_s^3}{\sqrt{\rho} \, G^{3/2}} = 2.57 \times 10^{48} \, \text{kg} \simeq 10^{18} \, M_\odot.$$

Let us continue. After the recombination, radiation and matter are no longer coupled, and the sound speed should drop rather steeply. Ordinary thermal sound speed is directly related to the temperature of the gas and, thus, should now be much less than the speed of light. If we take $z = 500$, the gas temperature is similar to the radiation temperature, i.e. $T = 2.73\,(1+z) = 1370\,\text{K}$. The sound speed is

$$c_s = \sqrt{\frac{5k_\mathrm{B}T}{3m_\mathrm{H}}} = 4350\,\text{m/s},$$

and corresponding Jeans mass is

$$M_J = \frac{\pi^{5/2}}{6}\frac{c_s^3}{\sqrt{\rho}\,G^{3/2}} = 1.9 \times 10^{36}\,\text{kg} \simeq 10^6\,\text{M}_\odot.$$

This is about the mass of globular clusters.

After about this redshift, gas simply cools adiabatically, i.e. $T_\mathrm{b} \sim (1+z)^2$. This means that $c_s \sim (1+z)$ and $M_\mathrm{J} \sim \frac{(1+z)^3}{(1+z)^{3/2}} = (1+z)^{3/2}$. We have that at $z = 50$ and Jeans mass is $M_\mathrm{J} \simeq 3 \times 10^4\,\text{M}_\odot$ and at $z = 20$, $M_\mathrm{J} \simeq 8000\,\text{M}_\odot$. The Jeans masses here are baryonic masses.

7.1.2 *First stars*

Let us have a DM subhalo with a typical mass of $10^7\,\text{M}_\odot$ and some gas within it. These subhaloes formed from relatively small density fluctuations, and there were many of them. It is hard to say how much is "some gas" but taking a typical ratio between dark and baryonic matter derived in MillenniumTNG simulations (Kannan *et al.*, 2023) for redshifts $z \sim 10$, one can say that for $M_{DM} \simeq 10^7\,\text{M}_\odot$, the bayonic mass within this DM subhalo is $M_b \simeq 10^4\text{M}_\odot$. This is more than the Jeans mass at that redshift.

As the gas cools, its pressure decreases, the density of the gas increases and the gas settles in the central region of the DM subhalo with the temperature corresponding to the virial temperature of the subhalo, $\sim(2-3) \times 10^3$ K. Probably that gas has some net rotation.[11] Contraction and rotation give us a rotating gas disk within the DM subhalo.

As we look at the formation of the first stars, we need to consider that the gas's chemical content is primordial, i.e. there are only H

---

[11]Rotation results due to interactions with other subhaloes or with some more extensive DM systems.

and He with a very small amount of D and Li. Heavier elements such as carbon, nitrogen, oxygen, etc., called "metals" in astronomy, are absent because the Universe expands and the temperature falls too fast to form heavier elements in nuclear synthesis. When using an ordinary notation, the chemical content of the primordial Universe is $X : Y : Z = 0.76 : 0.24 : 0.$[12] These are the dark ages, as there are no stars yet and no light.

Neutral gas cools because the atoms collide, and collision excites atoms to a higher energetic level. As a rule, very soon thereafter, this excited atom emits a photon and returns to the ground state. Emitted photon escapes the gas. Due to energy conservation, the thermal kinetic energy of the gas decreases, and the gas cools. Unfortunately, the gas temperature, a few thousands K, is too low for cooling via atomic hydrogen excitation. Due to the lack of metals, this primordial gas can cool only via the excitation of molecules, namely via transitions in H_2 molecules.[13] This molecule can cool gas until temperatures of about 400 K. The Jeans mass, corresponding to this temperature, is about $1400 \, M_\odot$.[14]

A small core forms within this gas, and gas starts to accrete to the core, forming an inside-out growing rotating disk around the core. Simulations show that formed in this way, the disk becomes unstable and fragments into a small number of protostars (Clark *et al.*, 2011a, 2011b, Prole *et al.*, 2022). In Fig. 7.1, it is seen how, in simulations around the first protostar, an accretion disk forms and within about a 100 years, three additional stars form.

It is difficult to say how massive these first stars can be — the formation stage is complicated; see Loeb & Furlanetto (2013) and a more recent review by Klessen & Glover (2023). The upper limits were derived from 30 to a few hundred solar masses.[15]

---

[12]Due to Li, the precise number of Z is not zero, but is about 10^{-12}–10^{-15}, i.e. some very small number. For comparison, the metallicity of the solar surface is 0.738:0.249:0.013.

[13]Formation of H_2 molecules in this primordial gas is not trivial, and we do not present it here and refer to the book by Loeb & Furlanetto (2013, Section 5.1).

[14]This is an approximate estimate, and we should keep in mind that during the subsequent evolution also, the Jeans mass evolves.

[15]There are studies where upper limits of 10^3–$10^4 \, M_\odot$ were derived (Chantavat *et al.*, 2023; Nandal *et al.*, 2024).

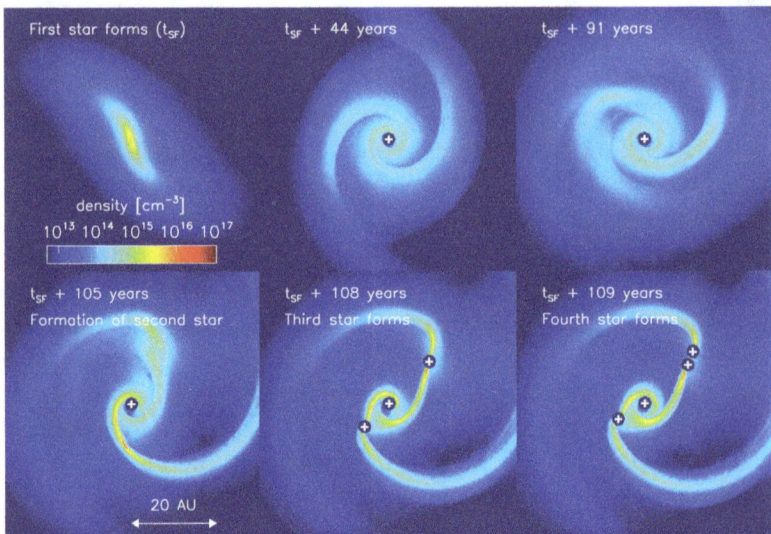

Fig. 7.1. The evolution of the density with $\dot{M}_* = 10^{-3}\,\mathrm{M_\odot yr^{-1}}$ in the simulation by Clark *et al.* (2011b), showing the build-up of the accretion disk around the central protostar and formation of three additional protostars (Clark *et al.*, 2011b). Reproduced from Clarke *et al.*, *Science*, DOI 10.1126/science.1198027, 2011, AAAS.

These first-generation stars are called population III stars (usually simply Pop III stars). Their formation and evolution era was at redshifts $z \sim 10-20$ or even up to 30. The initial mass function of these stars is different from that of today's stars. Details of physical conditions during their formation and the resulting initial mass function are uncertain. We do not know how large the turbulent velocities are in gas clouds; we do not know the strength of magnetic fields and radiation fields (see, e.g., Clark *et al.*, 2011b; Prole *et al.*, 2022; Stacy *et al.*, 2022; Toyouchi *et al.*, 2023).

After the first stars have formed, the situation in the primordial gas changes. The UV emission from the first stars ionises H. We remind you that there are still no metals. H_2 molecules both dissociate and form; thus, there exist still these molecules. In addition, a new molecule, the HD molecule, appears in the scene. The transition of this molecule can cool gas to lower temperatures than was possible before. Typical stellar masses are now smaller than before, $\sim 10-30\,\mathrm{M_\odot}$. These stars are called the second generation of Pop III stars (Clark *et al.*, 2011a).

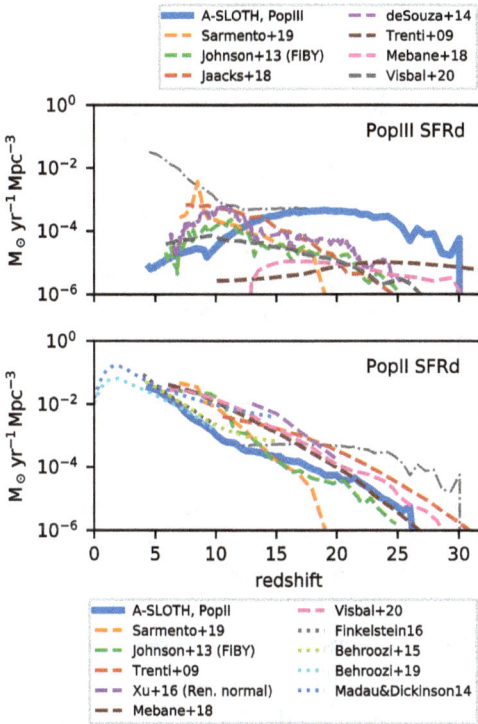

Fig. 7.2. Star formation rate densities (SFRd) as a function of redshift for the Pop III and Pop II stars according to semianalytic simulations by Hartwig *et al.* (2022). The bold line in the upper panel denotes the SFRd for Pop III stars and that in the lower panel denotes the SFRd for Pop II stars. For comparison, in both panels, the gray dashed-dotted line gives the total (Pop III + Pop II) SFRd. For other curves and references, see the original paper (Hartwig *et al.*, 2022).

After the gas is enriched with heavier elements (carbon, oxygen, nitrogen) by 10^{-3}–10^{-4} of the solar value, star formation changes. Cooling by ionised carbon or neutral oxygen allowed gas clouds to fragment much more, and lower-mass stars started to form. Oldest today's Pop II stars appeared (Bromm & Loeb, 2003). We see from Fig. 7.2 how the star formation rate density (SFRd) due to Pop III stars (upper panel) starts to increase at $z \sim 30$, then remains nearly constant between $z \sim 15 - 25$, and then starts to decrease. After the first Pop III stars produced some metals, the next generation of Pop II stars (lower panel) started to form at $z \sim 27$, and their SFRd nearly monotonically increased to $z \simeq 5$.

7.1.3　*First galaxies*

First, what are the first galaxies? Are they dark matter subhaloes with masses $\sim 10^6$–10^7 $M_\odot$ and some gas and primordial stars? Usually, it is agreed that a galaxy should have a mass to keep gas expelled by the stellar wind and explosion of the first stars within it. While estimating a dark halo binding energy $E_{\rm bind} \sim GM_{\rm vir}^2/r_{\rm vir}$ and demanding that it should clearly exceed a supernova energy release (although the energy releases of the first Pop III and Pop II supernovae are uncertain), it can be estimated that the first galaxies should have total virial masses $M_{\rm vir} \sim 10^9$ $M_\odot$. These masses can be achieved by merging many subhaloes and their gas.

7.1.3.1　*Dark matter haloes*

As an approximation, let us look first only at the dark matter. This is justified as the baryonic matter content is at least an order of magnitude smaller and does not contribute too much to gravitational potential. Dark matter is collisionless, which should be considered when studying the mergers of dark haloes. We already know that there are a lot of haloes with very different masses that have grown from an overdensity field. We know pretty well from simulations how these systems merge and grow. The formation of Milky Way-size dark haloes is nicely seen in the Millennium and Millennium-II simulations, as well as in the most recent MillenniumTNG (MTNG) simulation.

　　The main physical process in merging collisionless systems is violent relaxation (Lynden-Bell, 1967). Although the aim of Lynden-Bell in his paper was to explain the smooth, nearly universal luminosity distribution and Maxwellian velocity distribution of elliptical galaxies, the described process can be used in many other cases, including mergers of DM subhaloes. The essence of the process is that violently changing gravitational fields change particle energies,

$$\frac{d\epsilon}{dt} = \frac{\partial \psi}{\partial t},$$

and in this way, redistributes their orbits. Here, ϵ is the energy per unit mass of the particle, and ψ is the gravitational potential of the system. The relaxation timescale (the time when energy change is comparable to the energy) is approximately equal to the typical radial period of an orbit in the system or, e.g., free-fall time. This is very fast. Energies of particles may change in different ways, with some particles escaping the system. This is an important aspect, as together with the formation of merged systems, an overall background of DM particles forms.

In addition to the violent relaxation, phase mixing is also important in mergers (see Binney & Tremaine, 2008, Section 4.10.2). Let us look at two nearby, similar trajectories and a particle in both of these orbits. Initially, they are near to each other. Or it is said that they have similar phases. However, as their orbits are similar but still a little different (they are not the same but only nearby), in time, these two particles will separate, or, as it is said, their phases in all directions start to differ more and more. The same process takes place with all orbits and particles there, and it is called phase mixing.

In merging subhaloes, both processes, violent relaxation and phase mixing, occur. The theoretical considerations (rather complicated to the full extent) referred to above can be tested with numerical simulations. First, simulations carried out in the 1970s showed that when dealing with the merger of, for example, two systems, the resulting configuration was not very universal and still depended on the initial conditions.

In the case of continuous multiple mergers, systems grow in a more unified way, and the resulting density profile is nearly universal. In their highly cited flat CDM Universe simulation, Navarro *et al.* (1996) derived that resulting dark haloes were not spherical but had a triaxial shape and did not rotate significantly as a rule. This may not be related to overall relaxation processes but to a particular recent major merger. The main result of these simulations was that in rather large total mass ranges, the final density profile was quite universal. This is the NFW or Einasto profile.

In more recent Millennium (Springel *et al.*, 2005a) and Millennium-II (Boylan-Kolchin *et al.*, 2009) simulations, the evolution of DM systems in ΛCDM Universe was already followed. In the present context, we are more interested in Millennium-II simulation as the smaller systems were followed there. In both simulations, merging systems were followed starting from redshift $z = 127$, and merger trees were constructed. In addition to simple mergers of subhaloes, there is also a diffuse background of DM particles. Thus, seeds of galaxies grow both by mergers and by accretion from the background. Together with a protogalaxy growth, the DM background was also replenished with merger ejecta. According to simulations, the first systems with total masses $\sim 10^9$–10^{10} $M_\odot$ were reached by redshifts $z \sim 12$–15.

7.1.3.2 *Gas within the dark haloes*

Now, let us move to the most important aspect of early galaxy formation, the galaxy–halo connection and look at dark matter + baryonic gas. When dark matter subhaloes merge, their gas content also merges. The merging process of dark haloes disturbs the formation and physical properties of primordial gas clouds. The first paper studying the evolution of these two-component systems in detail was by White & Rees (1978). Although in their paper, DM was assumed to be low-mass or dead stars, it is important that this DM is collisionless (dissipationless), but gas is dissipational. This is also our current understanding. Violent collapse together with mergers heat gas by shock waves to the virial temperatures of millions of Kelvin being enough to ionise most of the gas. But ionised gas cools radiatively, contracts, fragments and forms stars. Simon White and Martin Rees demonstrated that this two-component model with reasonably selected parameters was able to describe the present-day luminosity function of galaxies and some properties of galaxy clusters.

This work was followed by studies of Fall & Efstathiou (1980), who concentrated their attention on the formation of gas disks in dark haloes. As the gas cools and settles at the centres of dark haloes, a cold rotating disk forms (initial rotation was acquired from tidal torques with neighbouring systems). In their detailed analytical model, Michael Fall and George Efstathiou derived that, indeed, the evolution of these collisionless DM and dissipational gas systems led to disk galaxies similar to the observed ones. In addition, it was necessary to assume that dark haloes formed first.

All this was, in essence, confirmed by later semianalytical and hydrodynamical simulations. Simulations range from volumes of cubes with sides of a few tens of megaparsecs to several hundreds of megaparsecs. Often, several different simulations are combined. For example, the most recent MTNG simulation combines the Millennium dark matter-only simulation of various box size resolutions with the hydrodynamical galaxy formation model of IllustrisTNG (Hernández-Aguayo *et al.*, 2023). An early generation of galaxies within these simulations was studied by Kannan *et al.* (2023). To improve the resolution of simulations for smaller masses, Kannan *et al.* (2023) combined MTNG with, e.g., the IllustrisTNG50 (Pillepich *et al.*, 2018) and THESAN (Kannan *et al.*, 2022) hydrodynamical simulations. The MTNG has a box size of 740 Mpc, DM particles mass of $m_{DM} = 1.6 \times 10^8 \, M_{\odot}$, and gas particles mass of $m_b = 3 \times 10^7 \, M_{\odot}$. This is excellent for studying the large-scale structure formation and galaxy clustering there. To study the formation, growth, and star formation processes in smaller galaxies in the early Universe, it was necessary to include also IllustrisTNG50 simulation with the smaller box size of 50 Mpc, but with much better mass resolutions: $m_{DM} = 4.6 \times 10^5 \, M_{\odot}$ and $m_b = 8 \times 10^4 \, M_{\odot}$. For THESAN, the mass resolutions are $m_{DM} = 3 \times 10^6 \, M_{\odot}$ and $m_b = 6 \times 10^5 \, M_{\odot}$.

An example of the output of this kind of combined simulation is the relation between the DM halo mass and the stellar mass given in Fig. 7.3 for the redshift $z = 10$ (Kannan *et al.*, 2023). Approximately linear relation in the log–log plot is seen up to $z = 15$.

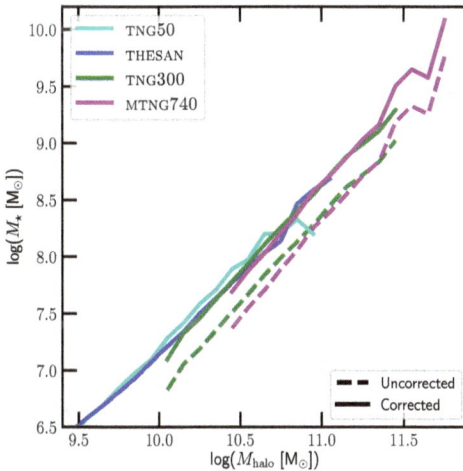

Fig. 7.3. Combined stellar mass M_* of galaxies as a function of their DM halo masses M_{halo} at $z = 10$ from four different simulations. Dashed curves show intrinsic predictions, while the solid lines are the results after correcting for the different resolutions of simulations (Kannan *et al.*, 2023).

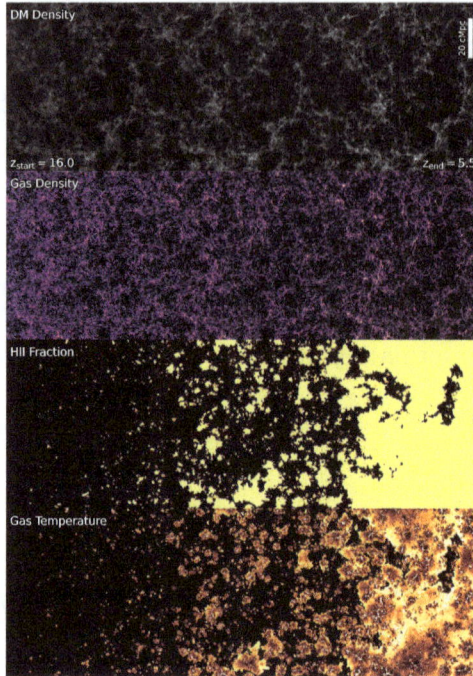

Fig. 7.4. A visualisation of the redshift evolution of the structure formation from $z = 16$ (left side) to $z = 5.5$ (right side) according to the THESAN simulation. Dark matter and gas clusterings are quite synchronous, but the gas is, due to its dissipational nature, more concentrated on small scales. HII fraction and gas temperature increase in time until eventually by $z = 5.5$, all gas is ionised (Kannan *et al.*, 2022).

In Fig. 7.4, a visualisation of the redshift evolution of the DM density (top row), gas density (second row), hydrogen ionisation fraction (third row) and gas temperature (fourth row) according to the THESAN simulation is given (Kannan *et al.*, 2022). Redshift runs from the left to the right (from $z = 16$ to $z = 5.5$) and it is seen how the DM structure builds up and how gas parameters change with redshift.

7.1.4 *Observations*

As the James Webb Space Telescope (JWST) has been operating for almost two years, it is justified to ask if we have succeeded in observing these first Pop III stars and first galaxies.

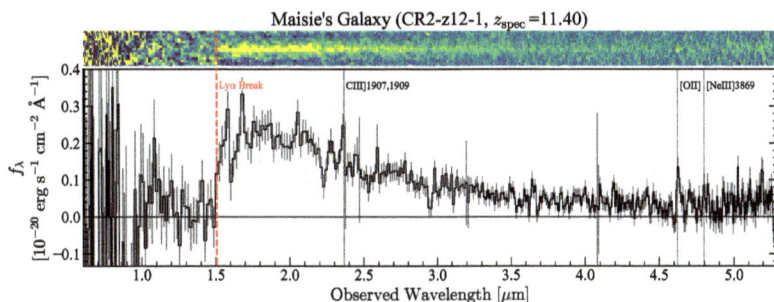

Fig. 7.5. NIRSpec spectra of Maisie's Galaxy (CR2-z16-1) at $z_{spec} = 11.40$. The top panel shows the two-dimensional spectrum, and the bottom panel shows the one-dimensional spectrum. Carbon, oxygen and neon lines are seen. The dashed line indicates the rest frame 1215.67 Å corresponding to the Lyα-break (Harikane *et al.*, 2024).

The fair answer to the Pop III stars is that they have not yet been found. As we remember, the distinction is that Pop III stars do not contain metals. Thus, we need to find a high redshift luminous object where we see, e.g., the He 1640 Å line, but no lines of carbon, oxygen, and nitrogen. JWST observers have not found them yet.

But, we have some indirect observational arguments for the existence of Pop III stars. First, by now, JWST has observed several galaxies at spectroscopically confirmed redshifts $z = 10$–15. A galaxy at $z = 11.4$ has carbon, oxygen, and neon lines in its spectrum (Harikane *et al.*, 2024) — by this redshift, gas is already enriched with metals (Fig. 7.5)! The spectrum of a young galaxy at $z \sim 11$ was observed to contain nitrogen lines. The enrichment with nitrogen is too slow to be done by stellar winds; thus, supermassive stars produced this element before.

JWST observers have found over a 100 galaxies at redshifts $z > 8$–15. Several have spectroscopic redshifts, most of them the photometric ones. Although one always likes to have larger statistics with confirmed redshifts, something can still be concluded by now.

First, what about the luminosity function? We talk here about the UV luminosity function as they are extremely distant galaxies. From the studies by Harikane *et al.* (2024), Finkelstein *et al.* (2024), Robertson *et al.* (2023, 2024), and references therein, we know that UV luminosity functions at redshifts $z > 8$–15 were surprising — there were many more luminous galaxies compared with

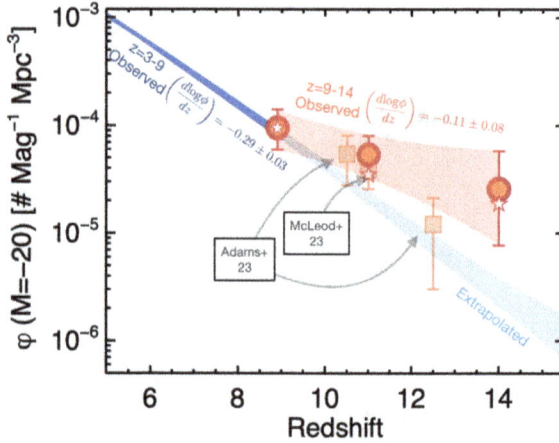

Fig. 7.6. The evolution of galaxy's observed number density at $M_{UV} = -20$ as a function of redshift from the study by Finkelstein *et al.* (2024), Fig. 11. The red circles show the observed number density at this absolute magnitude from their survey (connected by the light red shaded region; the small stars show the DPL fit values at this magnitude). The dark blue region shows the measured value from Finkelstein & Bagley (2022), and the lighter-shaded region shows the extrapolation of their results to higher redshift. See details from the original paper (Finkelstein *et al.*, 2024).

expectations. At redshifts $z < 8$, everything was consistent with expectations, but at higher redshifts, not so. This is seen in Fig. 7.6 from the paper by Finkelstein *et al.* (2024) where the UV luminosity function of bright galaxies is plotted as a function of redshift. The authors found that the number density of galaxies surprisingly flattens at $z > 9$, while JWST pre-launch expectations were that the number densities at $z > 9$ would either continue the observed trend at $z = 3$–9 or evolve more rapidly downwards with increasing redshift.

The next obvious question is how large their DM and baryonic masses are. Are they in accordance with what we expect? These galaxies are very small and faint, so the data are scanty. As expected, these galaxies have smaller radii than today's. Typical physical sizes are $0.2 - 1$ kpc, i.e. they are quite compact. Using chemical evolution models, Robertson *et al.* (2023, 2024) calculated their stellar masses to be $M_* \sim 10^7$–10^9 $M_\odot$. But keep in mind that stellar templates for those extreme cases are very uncertain, and one needs to be cautious. Comparison with simulations shows that M_{DM} is surprisingly constant for eight galaxies studied by Robertson *et al.* (2024) at $M_{DM} \sim (0.4\text{--}1.6) \cdot 10^{10} \, M_\odot$.

Looking at Fig. 7.6, a question arises, why does the UV luminosity function at very high redshifts not decrease as expected? Are the observations biased, our interpretation questionable, or are we missing in our theory something important?

We have no firm answer yet, and all the possibilities are open. Observations may be biased: we are able to observe only the most luminous galaxies, and the derived luminosity function does not take into account a possible myriad of fainter galaxies. Next, these UV luminosities may be temporary only due to a particular burst of a Pop III star (this also belongs to the interpretation fault). Interpretation can also be wrong as the correction of luminosities from dust attenuation may be wrong. By the way, this is, at present, quite a widespread explanation. Finally, the third explanation is that matter and gas cooling, gathering together and forming stars may be faster than calculated due to some unknown reason.

The method used by Robertson *et al.* (2024) should also reveal galaxies at redshifts $z = 15\text{--}20$, as mentioned by the authors. But these galaxies were not found. It may indicate that UV luminosity function of galaxies drops faster after $z \simeq 15$.

7.1.5 *Reionisation*

After the first stars and galaxies formed, the environment of subsequent star formation changes.

Although the evolution of primordial stars with masses $\sim 100\text{--}300 \, M_\odot$ is complicated and their parameters poorly known, we may estimate that their temperatures are $\sim 10^5 \, K$ and luminosities are $10^5\text{--}10^6 \, L_\odot$. High temperatures mean that they are excellent objects that can reionise all the gas in the universe (see a review by Gnedin & Madau (2022)).

Without free electrons, the photons can travel freely in space. However, Planck satellite CMB polarisation data show us that about 9% of CMB photons were scattered after the recombination.[16]

---

[16]Not all photons were scattered because the Universe has significantly expanded already.

This tells us there was a period with free electrons, or in another way, when gas was ionised again, and corresponding electrons were able to scatter photons again. This is called a reionisation period. Reionisation should be started as early as at least $z \sim 8$–9. Recent simulations indicate that reionisation may have already started at $z \sim 11$ or earlier, increased slowly and was finished by $z \simeq 6$. By that time, the interstellar gas was fully ionised.

Reionisation is a complicated process, where reionisation and recombination co-occur and, simultaneously, environmental conditions change.[17] Details of physical conditions at these times are not sufficiently well known unfortunately. For example, it may be that there were two periods of reionisation. According to the theory developed by Cen (2003), the first period was as early as at $z \sim 15$, the second at $z \sim 6$, see Fig. 7.7. The first ionisation results from Pop III stars and was much more efficient compared to the second

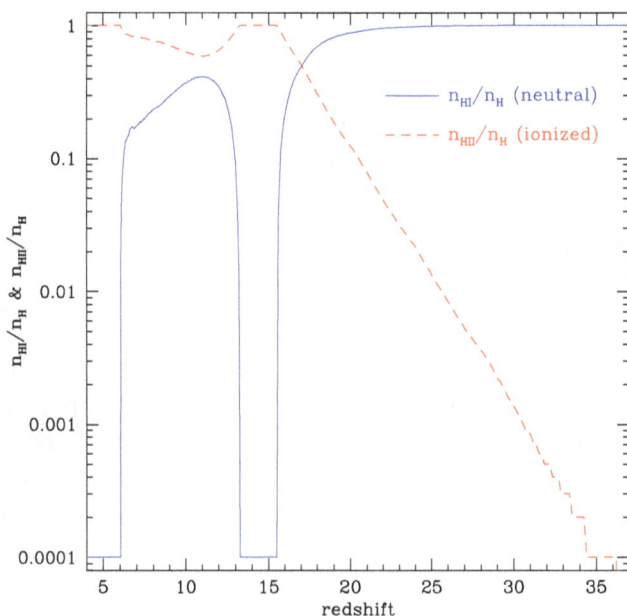

Fig. 7.7. Global mean fractions of neutral hydrogen (solid line) and ionised hydrogen (dashed line) as a function of redshift (Cen, 2003).

---

[17]Usually, the term "reionisation", is understood as the reionisation of the hydrogen only. To be precise, there is also helium reionisation.

one, resulting from the next generation, Pop II stars. We also see that the gas remained quite highly ionised during the intermediate times, the periods distinguished by the content of neutral hydrogen.

So what? A reader may ask, why should we worry about the reionisation? The answer is that the reionisation of the gas changes a lot. The temperature of the gas is higher; the gas does not concentrate so easily on the galaxies, and small galaxies do not grow as fast as larger ones (see, e.g., Wyithe & Loeb, 2006). Ignoring the reionisation, our picture of galaxy and star formation will be simply somewhat wrong.

7.2 Star formation

ARIEL: Have you seen Thirteen? Almost empty, no more long lines.

RHEA: Bye, bye, baby. Not gonna go back. I heard they put a ton of salt on everything. As if that helps.

CALLISTUS: Are you still going on about this restaurant? There are many restaurants, some will put more salt, some less. The ones that don't find the happy medium will go bankrupt. Kind of like the universes.

RHEA: In plural?

ARIEL: And why do I hear fine structure constant when you say salt?

RHEA: I heard that too. Although he might have meant cosmological constant. I worry he's trying to turn our gastronomical misfortune into a teaching moment about the anthropic principle.

ARIEL: Why the hate towards the... anthropic principle.

CALLISTUS: No respectable cosmologist likes it.

RHEA: Thank you.

CALLISTUS: But Rhea downright hates it. That's an overkill if you ask me.

RHEA: For the record, I did not. Ask you.

CALLISTUS: Cum grano salis.[18] Emotions about the Universe or restaurants will cloud your judgement.

ARIEL: Stop it, please. How is salt like the fine structure constant?

CALLISTUS: We established that no observations suggest changing physical constants. But what if there are many universes, each with its own salt? I mean, constants?

ARIEL: Uh. In my string theory class, we talked about all the possible compactifications of the theory, they call them Calabi–Yau manifolds, and there are mind-blowing 10^{500} possibilities, or

---

[18]With a grain of salt.

even exponentially more. Our task is to find the right theory in the string-theory landscape that describes our world.

CALLISTUS: Good luck with that.

RHEA: So far, it produced hopeless meandering.

CALLISTUS: For a second, let's assume all of these theories on the landscape have an equal chance of being realised in a universe.

RHEA: Now, we talk about many universes. Just so you understand, all of them are unobservable except ours. This is where we take off and transgress science by lightyears.

CALLISTUS: Indulge me in this Gedanken Experiment.

RHEA: Saying it in German makes you feel like you're Einstein.

ARIEL: You need to work on the hair.

CALLISTUS: Let's take the cosmological constant. Its natural value might be about 10^{120} times the present one. But are we likely to observe a Universe with such a huge Λ?

ARIEL: Such a Universe would immediately blow up and inflate way.

RHEA: No galaxies, no stars, no life. Except we will never know if such universes exist.

CALLISTUS: I will agree with Rhea that this is not a scientific argument in that it cannot be tested by measurements. It's a last resort to explain the low cosmological constant.

RHEA: Until we find a better explanation.

CALLISTUS: But the same could be said about the fine structure constant. 1/137 is just about right, but above 1/80 or below 1/180 would not allow life due to proton decay.[19]

RHEA: I think this argument works better for restaurants.

ARIEL: I'll explain this to the new owner of Thirteen. The anthropic principle of restaurants asserts –

CALLISTUS: That they need to serve good food.

After the first stars exploded, the intergalactic medium was enriched with metals. This is clear from theory, and it is seen in spectra of high-redshift galaxies and quasars (although with a significant scatter). Metallicity of galaxies is a function of redshift, stellar mass and star formation rate. For example, if the stellar mass of a system is larger, we may expect that a larger amount of stars have also exploded already and injected their metals into an interstellar medium. This is indeed seen in Fig. 7.8 for redshifts $z \simeq 0.1$–2.3 from a study by Zahid *et al.* (2013) (left panel) for redshifts $z \simeq 3$–11 from a study by Sarkar *et al.* (2025) (right panel). As a very rough average,

---

[19]Barrow & O'Toole (2001).

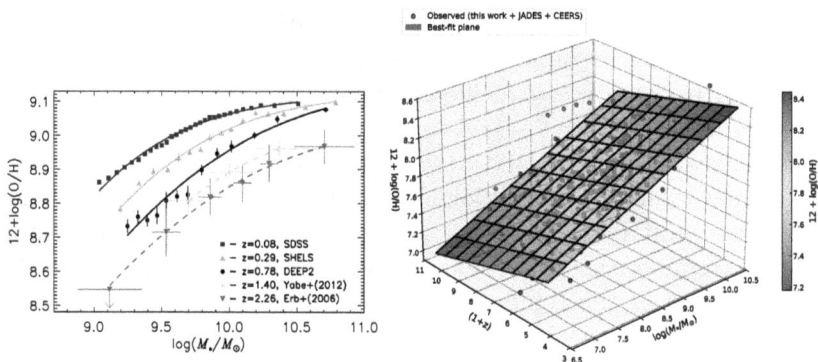

Fig. 7.8. Metallicity of galaxies as a function of stellar mass and redshift. The figures were taken from the studies by Zahid *et al.* (2013) (left panel) and Sarkar *et al.* (2025) (right panel).

by the redshift $z \sim 3$, the mean metallicity was about 10% of the solar value.

7.2.1 *Thermal instability*

To analyse the possible states of the interstellar gas media in galaxies, it is necessary to start from its thermal state or from the question of whether the gas is stable or not stable. Thermal instability means that the heating or cooling of the gas by little results in continuing heating or cooling. Thus, the corresponding instability criteria should depend on the specific heating and cooling processes. When studying energy balance in gaseous media, a so-called cooling rate is introduced. A simple example of such a function is $\dot{Q} = A\rho T^{\alpha} - H$. Here, $\dot{Q}$ is energy change per unit gas mass (let us say that per kg or per whatever unit), the first term on the right side characterises cooling and the second term heating, assumed to be constant here. When cooling dominates, $\dot{Q} > 0$.

Several processes may contribute to heating and/or cooling: thermal conductivity, convection, radiation, cosmic rays, etc. In astrophysics, thermal conductivity can usually be ignored, with the exception of the inner regions of white dwarfs and shock wave fronts. In many cases, convection can also be ignored. Heating due to cosmic rays is important, in several cases (in a dense gas cloud, for example), it can be assumed to be constant. The most important and complicated is energy transfer via radiation. Many different processes take part in radiation: recombination, free–free transitions, collisional excitation, etc. Different processes dominate at different temperatures.

Radiative cooling involves two particles; thus, per unit volume, radiative cooling is proportional to the density squared, or when taking per unit mass, it is $\sim\rho$. As different processes dominate at different temperatures, the corresponding cooling term should also be a function of the temperature $f(T)$. Thus, we may take as the condition for stationarity

$$\rho f(T) - H = 0.$$

Although the real calculations can be rather complicated, in principle, this relation gives us some stationary temperature and pressure:

$$T_{\text{stats}} = F(\rho), \quad p_{\text{stats}} \sim \rho T_{\text{stats}} = \rho F(\rho).$$

Detailed calculations show that for interstellar gas in galaxies, the derived functions $p_{\text{stats}}(\rho)$ are not monotonic functions. For regions of interesting pressures, several density and temperature values often correspond to a single pressure value. Thus, there should be regions where

$$\frac{dp}{d\rho} < 0,$$

which means instability (when density increases, the pressure decreases — this causes a further increase in the density). This is illustrated in Fig. 7.9 taken from an excellent classical textbook by Zeldovich & Novikov (1975, Fig. 45)

For example, in the Milky Way, three stationary states of the gas exist side by side: cold gas ($T \sim 10^2$ K), warm gas ($T \sim 10^4$ K) and even warmer gas. When we follow the time evolution (including a nonlinear phase), we derive that in the initially uniform gas density distribution, denser gas clouds form where the temperature is lower. In the remaining regions, density is correspondingly smaller and temperature higher. This state is already thermally stable. In denser gas clouds, the density is about a 100 times higher than initially. Instability can be triggered by some shock waves or collision of some gas flows.

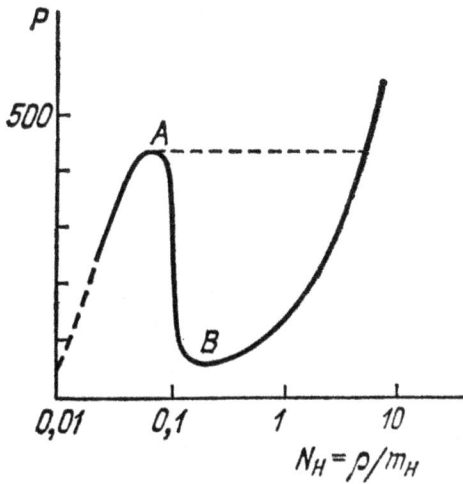

Fig. 7.9. Pressure of the gas P as a function of the number density of atoms N calculated from the stationarity condition above. The gas at densities between points A and B is unstable (Zeldovich & Novikov, 1975). The units in the axis were not indicated in the textbook; probably, they are CGS units.

When magnetic fields are present, the overall picture is similar. Instead of a single gas compression, there can be a series of compression. However, the details depend on the angle between the magnetic field lines and gas flow velocities.

To summarise, the existence of dense and colder gas clouds is the natural state of interstellar gas in galaxies.

7.2.2 *Molecular clouds*

Now, the gravitational instability enters the stage. In some clouds, density starts to grow due to gravitational instability. Although it may be intuitively unclear, the density increase is accompanied by a drop in temperature — cooling due to collisions and radiation can be surprisingly efficient. We know about these and subsequent processes from hydrodynamical simulations.

Efficient coolers are heavier elements, molecules and dust. Heavier elements have several fine and hyperfine transitions.[20] Molecules,

---

[20]Fine-structure energy states are due to interactions between the magnetic moments of spins of electrons and the orbital angular momentum of electrons.

especially CO molecules, have rotational energy states. In both cases, gas atoms or hydrogen molecules with even small thermal velocities collide with atoms or molecules and are able to excite them, losing, therefore, energy. Excited atoms or molecules soon return to the ground state by emitting photons. Photons escape the gas. Collisional excitation involves two atoms/molecules being, thus, proportional to the square of the density. Hence, cooling is more efficient in denser regions.

Dust is also a good cooler. Gas atoms or molecules collide with dust grains and give their energy to the dust grain. Due to this, dust heats a little and its thermal emission increases. Again, photons escape the gas.

If the cooling is fast, the gas cloud contracts and becomes denser and denser. Not only contraction increases density. Gas clouds also accrete mass from their surrounding regions. The evolution of gas clouds is also influenced by the stability properties of the overall gas disk (Toomre instability, Toomre (1964)). As a result of these processes, molecular gas clouds with a variety of masses form. Clouds consist in a significant fraction of H_2 and CO molecules because their temperatures are very low, 10–20 K.

Usually, two basic kinds of clouds with different substructures are distinguished. Giant molecular clouds have typical masses of 10^5–$10^7\, M_\odot$, and dimensions of 30–200 pc. Simply molecular clouds have typical masses of 10^2–$10^4\, M_\odot$, and dimensions of 10–20 pc. These clouds are surrounded by warmer (8000 K) atomic medium. Molecular clouds consist of many substructures, being rather inhomogeneous in this way — their density distribution is not smooth but is fragmented into filaments, walls, clumps, etc. Walls and filaments themselves consist of smaller clumps within which even denser cores form. Clumps have typical sizes of $\sim$1 pc and cores of $\sim$0.1 pc. An overview of the gas cloud structures and substructures is given by Ballesteros-Paredes *et al.* (2020).

The most important structural elements of a molecular cloud are filaments. The filamentary inner structure of molecular clouds was firmly confirmed by *Herschel Space Observatory* observations. *Herschel* observations revealed that there are filaments in all scales,

---

Hyperfine-structure states are due to interactions between the nuclear magnetic moment and the total magnetic moment of electrons.

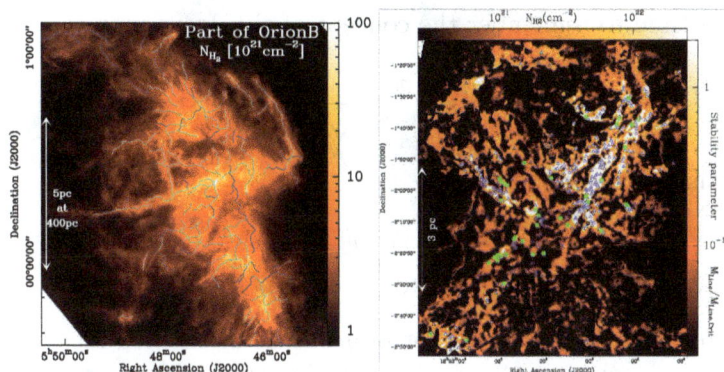

Fig. 7.10. Filamentary structure of molecular clouds revealed by *Herschel Space Observatory* observations. Left: Filaments in a part of the Orion molecular cloud (Arzoumanian *et al.*, 2019), right: protostars (green) and prestellar bound cores (blue) in the Aquila star formation complex (André *et al.*, 2010). Reproduced with permission © ESO.

0.5–100 pc (André *et al.*, 2010; Arzoumanian *et al.*, 2019). The mean width of filaments is surprisingly stable, 0.1–0.4 pc, but it depends on the definition of the width. Filaments fragment into protostars; where filaments intersect, star groups form, see Fig. 7.10.

Clumps and cores are not stationary but increase their masses by accreting gas in their surrounding environment. As a rule, accretion is along the filaments. As gas velocities between neighbouring layers may differ, Kelvin–Helmholtz instability appears, causing their connecting surface to fluctuate.[21] Simulations show that Kelvin–Helmholtz instability is rather important; due to accretion, turbulent velocities are maintained; formed star clusters are also dynamically non-stationary. As newly formed star clusters may contain thousands of stars, several stars can be ejected from the cluster, and the initial cluster may spread over the volumes of a few tens of parsecs.

In addition to accretion, the evolution of molecular clouds is influenced by large-scale gas flows, by moving through spiral arms, collisions with other clouds, passing of shock waves, and large-scale instabilities.

---

[21] The reader may google the words "atmosphere + clouds + Kelvin–Helmholtz" and look at the beautiful pictures of the instability of clouds.

It is time to remember the concept of the Jeans mass

$$M_J = \frac{\pi^{5/2}}{6} \frac{c_s^3}{\sqrt{\rho} \, G^{3/2}}$$

or by using gas temperature

$$M_J = \frac{\pi^{5/2}}{6} \frac{1}{\sqrt{\rho}} \left(\frac{7k_B T}{3\mu m_p G} \right)^{3/2}.$$

Here, instead of m_H, we introduced the often used mean atomic weight μm_p being $\mu \simeq 2.3$ in the case of a usual molecular cloud. The relation assumes adiabatic diatomic gas (multiplier 7/3). In the case of monoatomic gas, the multiplier is 5/3; in the case of isothermal gas, there is no multiplier. If the cooling time (or thermal adjustment time) is short, it is justified to use the isothermal approximation. In this case, for an average giant molecular gas cloud ($T = 15\,\mathrm{K}$, $\mu = 2.3$, $n = 10^9\,\mathrm{m}^{-3}$), the Jeans mass is $M_J \sim 30\,M_\odot$; in the case of dense clumps of these clouds ($T = 10\,\mathrm{K}$, $n = 10^{11}\,\mathrm{m}^{-3}$), the Jeans mass is $M_J \sim 1\,M_\odot$.

As we know, the reality may be always more complicated, and the derived quantities should be taken as rough estimates only. We ignored magnetic fields and turbulences. To take these quantities also into account, the effective sound speed consists of three components: the thermal contribution (the "usual" sound speed), the contribution due to magnetic fields and the contribution due to turbulences

$$c_{\mathrm{eff},s}^2 = v_T^2 + v_A^2 + v_t^2,$$

respectively. In fact, we have added the corresponding pressures here. Thermal sound speed for a gas at $T = 15\,\mathrm{K}$ is $v_T = 0.35\,\mathrm{km/s}$. In the Milky Way, gas disk Alfvén velocity due to magnetic field pressure is $v_A = 1.5\,\mathrm{km/s}$ (Heiles & Troland, 2005). Turbulent velocities are not well known; often, values $v_t = 6$ km/s are used. We see that, in fact, turbulent velocities dominate, and Jeans mass can be an order of magnitude larger. But still, as the giant molecular clouds have masses of $10^6\,M_\odot$, we may conclude that the giant molecular clouds can be unstable and contract.

Despite instability, the giant molecular clouds do not contract as a whole to form stars. As we mentioned already, they have inner filaments, clumps and cores. Numerical simulations and observations indicate that this kind of inner structure results from small-scale shock waves within the clouds. Some of these structures remain gravitationally bound and may collapse if their mass exceeds the Jeans mass. We also need to take into account that during the contraction, Jeans mass changes. If the Jeans mass of a core (or a clump) decreases to, e.g., half of the original mass, the core (or clump) may split into two independent parts. This is the process of fragmentation (see Fig. 7.10). The smallest protostellar cores from simulations have masses of only $\sim 0.001\, M_\odot$.

Let us look at the mass distribution function of giant molecular clouds and molecular clouds. In this function, the number of clouds per unit of the logarithm of mass is usually parametrised as

$$\frac{dN}{d\log M} \sim M^{-\gamma},$$

where γ is a parameter. Although derived from observations, the values of γ differ; it seems to be clear that they are too small compared with the value from the Salpeter mass function for stars ($\gamma = 1.35$, see the following section). The mass function of dense cores of molecular clouds has $\gamma = 1...2$, but the range of masses studied is too small to draw some conclusions. The most promising is the study of the line mass function of filaments (line mass is the mass per unit length of the filament). Here, observations of filaments give values $\gamma = 1.4 - 1.6$. We do not know with certainty, but this may be a clue to the initial mass function of stars.

7.2.3 *Initial mass function*

Initial mass function characterises the distribution of stellar masses at their birth. Usually, a unit volume is used, and in this case, the number of stars in this volume, dN, with masses between m and $m + dm$, is

$$\frac{dN}{dm} = A\xi(m)$$

and $\xi(m)$ is just the initial mass function (IMF). A is a normalising constant. Sometimes, instead of using dm, the logarithmic mass range is used:

$$\frac{dN}{d\log m}.$$

It is easy to see that when IMF is assumed to be a power law:

$$\frac{dN}{dm} \sim m^{-\alpha} \quad \text{or} \quad \frac{dN}{d\log m} \sim m^{-\gamma},$$

and there is a simple correspondence between them $\alpha = 1 + \gamma$.

The first determination of the IMF was done by Salpeter (1955), who calculated the IMF of main-sequence stars in the solar neighbourhood in the range of masses 0.4–$10\,\mathrm{M}_\odot$ and derived $\gamma = 1.35$. As Salpeter mentioned, the main uncertainties are due to the small volume considered and the conversion from present-day visual luminosities to bolometric luminosities, then to present-day masses, and thereafter to initial masses.

IMF is an essential input characteristic of several fields in galactic and extragalactic astronomy. For example, it is necessary to know the initial mass function in all calculations involving the usage of the chemical evolution models of stellar systems. It is essential to know it as precisely as possible, but it is not easy to derive it with sufficient precision. In addition, the natural question arises: how universal is IMF? At present, it is assumed to be constant in space and time. Clearly, it is a simplification only. We know that when the first stars were born, their masses were very different from those of today. IMF may be a function of redshift, environment and dynamical processes in galaxies. As we lack specific information about it, the IMF is assumed to be some fixed function at present.

There are three different IMF forms used most widely today, namely the Scalo, the Kroupa and the Chabrier IMFs. The first two of them are segmented power-law functions, and the third one has a somewhat different analytical form. The Scalo (1986) IMF is

$$\xi(m) = \begin{cases} m^{-1.8} & \text{for } 0.2 \le m < 1, \\ m^{-3.25} & \text{for } 1 \le m < 10, \\ 0.16\,m^{-2.45} & \text{for } m \ge 10. \end{cases}$$

The Kroupa (2002) IMF is

$$\xi(m) = \begin{cases} 0.0375\,m^{-1.3} & \text{for } 0.08 \le m < 0.5, \\ 0.0187\,m^{-2.3} & \text{for } 0.5 \le m < 1, \\ 0.0187\,m^{-2.3} \text{ or } m^{-2.7} & \text{for } m \ge 1. \end{cases}$$

We see that Kroupa gives two different possibilities for the largest mass range. Although the value -2.3 is usually used, even a steeper decrease of the IMF for the largest masses was derived (Mor *et al.* 2017, 2018). Kroupa extends his IMF even further to the low-mass region for brown dwarfs with masses $0.01 \leq m < 0.08 \, M_\odot$ (see details from the original paper).

Chabrier (2005) IMF for disk stars is

$$
\xi(m) = \begin{cases} 0.093 \, \dfrac{1}{m} \exp \left\{ -1.653 \left[\log \left(\dfrac{m}{0.2} \right) \right]^2 \right\} & \text{for } m < 1, \\ 0.041 \, m^{-2.35} & \text{for } m \geq 1. \end{cases}
$$

In Fig. 7.11, a comparison of the slopes of the IMFs by Salpeter, Kroupa and Chabrier as a function of stellar masses is given together with corresponding observational data from different studies (Hennebelle & Grudić, 2024).

Normalising constant of these IMFs can be calculated in a different way, depending on the purpose of using the IMF. For example, if the aim is to construct a mock sample of local stellar content, it would be straightforward to demand that the mass-weighted integral of the IMF should give the local stellar mass density

$$
A \int_{m_{\text{low}}}^{m_{\text{high}}} m \, \xi(m) \, dm = \rho_{\text{loc}},
$$

where m_{low} is the lower limit of masses and m_{high} is the higher limit of masses. In the case of chemical evolution models of stellar systems, the integral is often defined to be

$$
A \int_{m_{\text{low}}}^{m_{\text{high}}} m \, \xi(m) \, dm = 1.
$$

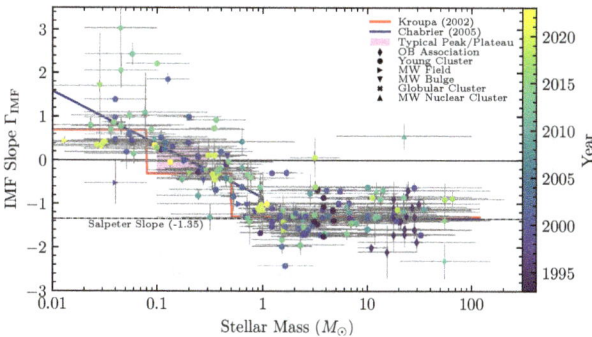

Fig. 7.11. Slopes of the IMFs by Salpeter, Kroupa and Chabrier as a function of stellar masses. For the references plotted in the figure, see the original paper (Hennebelle & Grudić, 2024).

In usual applications, $m_{low} = 0.08\,M_\odot$ and $m_{high} = 120\,M_\odot$ or $150\,M_\odot$ are used. The high-mass end value is not critical because even if there exist higher-mass stars, their number is very small.

In cosmology and in studying the constituents of the large-scale structure elements of the Universe, we are usually interested in the global, i.e. averaged over a galaxy IMF. Unfortunately, the determinations of the IMF are based on much smaller scale averaging of stellar data. We know from the previous section that molecular cloud masses and fragmentation depend on both local and galaxy-wide processes. One can say roughly that the high-mass part of the IMF depends quite a lot on the accretion of gas to the protostar and is thus clearly a function of the star formation processes of a galaxy. It was pointed out by Kroupa & Weidner (2003) and Yan *et al.* (2017) that the slope of the high-mass end of the galaxy-wide IMF is different from the one accepted in a usual stellar cluster-based IMF and indeed depends on the SFR of a galaxy (see Fig. 7.12). The smaller

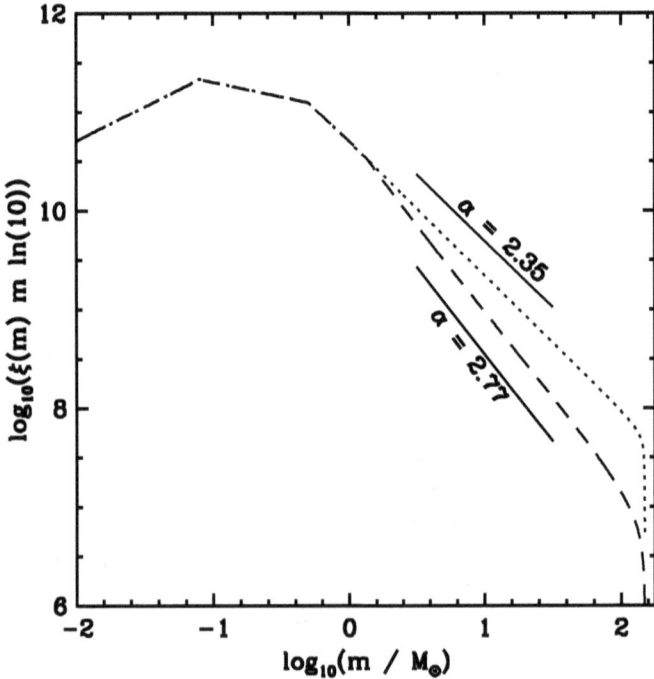

Fig. 7.12. High-mass end slopes of the usual IMFs (dotted line and $\alpha = 2.35$) and galaxy-wide IMF (dashed line and $\alpha = 2.77$) as a function of stellar masses (Kroupa & Weidner, 2003).

mass part of the IMF depends more on the local fragmentation processes, but this part is a function of metallicity (Yan *et al.*, 2024) and, thus, can be a function of galactic properties.

Clearly, when using the Salpeter IMF (it is also still used), the number of low-mass stars is higher than in other cases. This results in higher mass-to-light ratios when using chemical evolution models. When using the Chabrier or Kroupa IMFs, the calculated mass-to-light ratios are by a factor of 1.5–2 smaller than for the Salpeter IMF, respectively.

7.3   Gas inflows and outflows in galaxies

Galaxies are not isolated objects. As we know, they are often members of groups and clusters. Even when there are not seen other nearby galaxies, they are surrounded by the environment of both the dark matter and the baryonic matter.

7.3.1   *Gas in and around galaxies: Current situation*

Let us map the current situation of the galactic-scale gas distribution first. According to their temperature, we distinguish here four or five different phases of the gas at present.

Starting from the coldest, there is a *cold neutral gas* with temperatures $T < 10^4$ K lying approximately within or a little farther of the optical dimensions of galaxies.[22] In the MW galaxy, two basic, approximately disk-like subcomponents with temperatures of about 50 K and 8000 K can be distinguished. In addition, it includes high-velocity clouds and the Magellanic stream quite far above the galactic plane. Of course, finer divisions are possible.

The next two phases are the *lukewarm and warm gas* with a temperature range of $T \sim 10^4$–10^6 K. They can be handled together, but the distinction between them is usually at about 10^5 K. This gas is seen in UV absorption lines and is responsible for the Lyα forest in quasars. The gas lies within about the virial radius of the galaxy and is considered the dominant component of the circumgalactic media.

---

[22]In cosmology when describing the cosmic web, gas with $T < 10^5$ K is often called as cold gas.

Then, there is a *hot gas* with $T > 10^6$ K. The lower temperature end of this phase is in non-virialised distant regions around the galaxies and is often called WHIM (warm-hot intergalactic gas) and we know about it from soft X-ray emission of galaxies. The hotter part $(T > 10^7$ K) is virialised gas in X-ray haloes of galactic clusters.

These multiple phases can exist side by side in the outer parts of galaxies. In MW, for example, high-velocity clouds move within a hotter circumgalactic medium, as can be seen from their head–tail structure (Brüns *et al.*, 2000). The inner structure of high-velocity clouds themselves shows a two-phase structure.

The physical properties of these gas phases can be derived by the ionisation states and line widths of different chemical elements, for example, by comparing O II versus O V versus O VII line widths. Higher ionisation states of a chemical element demand higher temperatures, and thus, the referred sequence corresponds to an increasing gas temperature and decreasing density sequence. Due to the evolution of cooling and heating functions over time, the balance between these phases and their relative amounts change over time. For example, in Fig. 7.13, the overall relative distribution of the lukewarm and warm gas as a function of redshift according to simulations by Cen & Chisari (2011) is plotted.

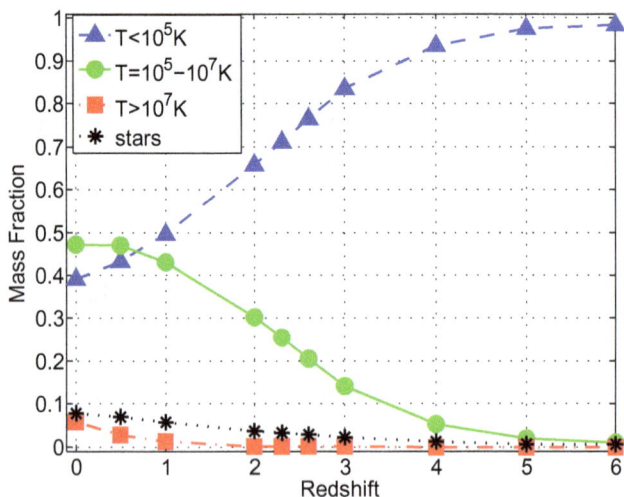

Fig. 7.13. Evolution of the mass fraction of three gas phases with redshift (Cen & Chisari, 2011).

Gaseous media within the optical dimensions of the galaxy is usually called the interstellar medium (ISM). Gas within the surrounding DM halo of the galaxy is usually called the circumgalactic medium (CGM). Gas outside the galaxy is an intergalactic medium (IGM). Excellent recent reviews about the properties of the gas are given by Tumlinson *et al.* (2017) and Faucher-Giguère & Oh (2023), although these reviews are concentrated on the CGM.

Our understanding of the evolution of gas in galaxies has changed in the past few decades or so. Galaxies and gas within them are not a kind of isolated systems. For galaxies, it has been known for a long time, but the same is true for gas. Galaxies, together with their surrounding gas, belong to the cosmic web, and their evolution is quite a lot regulated by the processes determined by the web. Galaxies exchange gas with the surrounding environment.

7.3.2 *Basic physical characteristics of the gas*

7.3.2.1 *Virial temperature of the gas*

Before moving on, let us first introduce an important characteristic of the gas in DM haloes: the virial temperature of the gas. For stellar systems, a conception of the virial balance of the system is often used. In an equilibrium system, the kinetic energy K and the potential energy U are balanced by the virial equation $2K = -U$ (U itself is negative). The virial conception can also be used for the gas in dark matter haloes. The kinetic energy is now the internal energy of the gas and can be expressed as

$$K = \frac{3}{2}k_B T\, N = \frac{3}{2}\frac{k_B T M_{\text{gas}}}{\mu m_p},$$

where N was the number of atoms. The potential energy for a uniform cloud can be expressed as

$$U = -\frac{3GM_{\text{gas}}M}{5R_g},$$

where R_g is defined just in a way the relation holds and M is the dark halo mass. The virial temperature can be defined by demanding that the virial theorem $2K = -U$ holds, and we have

$$T_{\text{vir}} = \frac{\mu m_p G M}{5 k_B R_g}.$$

Introducing v_c as the circular velocity at R_g ($v_c^2 = GM/R_g$), we have

$$T_{\text{vir}} = \frac{\mu m_p}{5k_B} v_c^2.$$

We assumed here that the gas is in equilibrium with the dark halo, and its virial temperature is determined by the mass of the dark halo. For a typical halo with a circular velocity at the virial radius of 160 km/s (typical velocity for the MW and the Andromeda galaxy in their outskirts), this gives $T_{\text{vir}} \sim 10^6$ K.

Just for future reference, for fully ionised gas with today's mean metallicity $\mu = 0.60$ and in the adiabatic case, the sound speed in units of km/s is $c_s \simeq 0.152\sqrt{T}$, where temperature T is in Kelvin.

7.3.2.2 *Cooling and dynamical timescales*

Next, let us derive convenient (thus simplified) expressions for two important timescales of gaseous media: the cooling time t_{cool} and dynamical time t_{dyn}.

The internal energy of a gas per unit volume can be written as

$$\mathcal{E}_{\text{int}} = \frac{E_{\text{int}}}{V} = \frac{3}{2}\frac{Nk_BT}{V} = \frac{3}{2}nk_BT,$$

where N is the number of gas particles and n is their concentration. Cooling time is usually defined as

$$t_{\text{cool}} \equiv \frac{\mathcal{E}}{|d\mathcal{E}/dt|}.$$

Now, we need to express how internal energy changes over time

$$\frac{d\mathcal{E}}{dt} = \frac{dE}{V\,dt} = \rho\frac{dE}{M\,dt} = \rho\dot{Q}.$$

Here, $\dot{Q}$ is the overall cooling rate (heating minus cooling) introduced in Section 7.2.1. At present, we are interested in cooling and assume that there is no heating. We also assume that we can ignore the thermal conduction, and thus, cooling is due to radiative losses. Usually, to describe just cooling, the cooling function $\Lambda(T)$ is used via relation $\rho\dot{Q}_{\text{cool}} = -n^2\Lambda(T)$ (in the literature the definition of the cooling function can differ). In this simplified approach, under the concentration n, the concentration of hydrogen atoms n_H is understood.[23] For the cooling time, we have thus

$$t_{\text{cool}} = \frac{3k_B T}{2n\Lambda(T)}.$$

The dynamical time (often also called free-fall time) is defined as the time when the test mass, released from the rest at an arbitrary radius within a sphere of constant density, reaches the centre of the sphere. This time is (see, e.g., Binney & Tremaine, 2008, Section 2.2)

$$t_{\text{dyn}} = \sqrt{\frac{3\pi}{16G\rho}}.$$

For inhomogeneous spheres, the density is the mean density of the matter within the sphere.

7.3.2.3 *Radiative cooling*

Gas particles can lose their kinetic energy by emitting photons, i.e. via radiation. This may happen with several physical processes, and as a rule, different processes dominate at different temperatures.

At the highest temperatures, e.g., in massive haloes with $T_{\text{vir}} \geq 10^7$ K, when gas atoms are mostly fully ionised, free electrons and ions interact with each other. As a rule, electrons, being much lighter than ions, are deflected from their path, meaning they are accelerated and accelerated charged particles emit radiation. Free electrons remain free after radiation, and transitions responsible for this emission are called free–free transitions or bremsstrahlung.

When an atom is only partially ionised, an initially free electron may emit radiation when it recombines with the ion; as a result, a neutral or less ionised atom forms. Initially, the free electron becomes bound with the atom/ion, and transitions responsible for this emission are called free-bound transitions.

When an atom is not ionised but due to collisions only in an excited state, the electron from the higher energy level, as a rule,

---

[23]In the case of neutral gas with solar abundance $n \simeq 1.10 n_H$; in the case of completely ionised gas with solar abundance $n \simeq 2.30 n_H$ (Ryden & Pogge, 2021, Section 1).

quite quickly moves to the lower energy state, and one or several photons will be emitted. Transitions responsible for this emission are called bound–bound transitions.

On average, this list of transitions is also the sequence from higher to lower temperatures.

Different chemical elements have different ionisation and excitation energies, so cooling rates also depend strongly on the chemical composition of the gas. For most elements, the structure of energy states is rather rich with basic states, fine structure states and hyperfine structure states (molecules have also rotational and vibrational states). All this makes the forms of cooling functions rather complicated.

As an example, in Fig. 7.14, the cooling functions calculated by Gnat & Sternberg (2007) for the low-density radiative cooling gas are given. The upper panel of the figure is for equilibrium, and the lower panel is for the non-equilibrium ionisation states, i.e. when the cooling times are longer or shorter compared to the recombination times of various ions. Gas is assumed to be dust-free without external radiation. In the upper panel, the dominant cooling elements at various temperatures are given near the curves. Cooling by elements other than C, N, O, Mg, Si, S, and Fe is negligible in these conditions.

7.3.2.4 *Basic gas heating processes*

Heating means energy transfer to the gaseous environment. First heating process is photoionisation or photoheating. Stars emit photons in a wide energy range, including energies exceeding the ionisation potential of atoms — they photoionise atoms and ions. To ionise, the photon energy must exceed the binding energy of an atom or ion. In the context of galaxy formation, photons should be in the UV wavelength range. These photons are produced by massive stars and quasars, forming a UV background.

If an atom/ion absorbs a high-energy photon, the escaping electron acquires kinetic energy $K = h\nu - E_{\text{ion}}$, where E_{ion} is the ionisation energy. This free electron collides with other particles and transfers its energy to these particles, i.e. heats them. For photoheating to be effective, there must be a sufficient number of atoms/ions to ionise. As an example, in the case of hydrogen, photoionisation is effective for temperatures $<10^5$ K, as for higher temperatures, all the hydrogen atoms are already ionised.

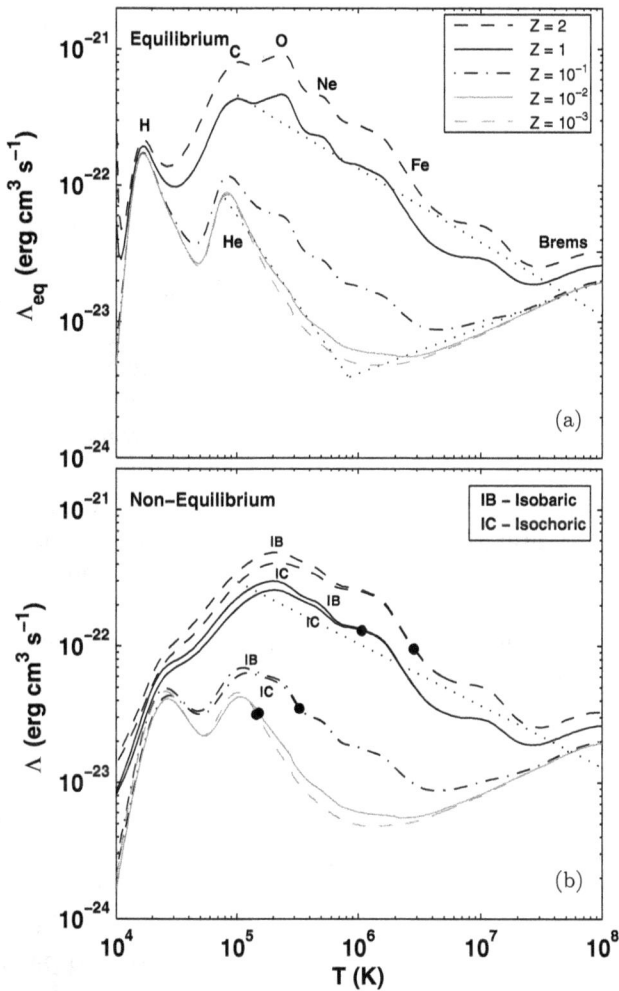

Fig. 7.14. Cooling functions as a function of the temperature for different metallicities Z. In the upper panel, cooling functions for the equilibrium ion abundances are given. The dominant cooling elements at various temperatures are given near the curves. In the lower panel, the cooling functions for the non-equilibrium ion abundances are given. For each Z, two curves are displayed: the isobaric (constant pressure) and the isochoric (constant density) (Gnat & Sternberg, 2007).

The photoheating rate per unit volume is proportional to the intensity of the ionising radiation field, the density of the particles that can be ionised and the ionising cross-section. In addition to atoms/ions, photons can free different amounts of electrons also from dust grains. Small grains have larger photoionisation efficiency and can contribute significantly to net heating.

The ionising radiation field (ionising background) is a complicated function of emissivity and optical depth of the medium. Both of these parameters, being in fact, also complicated functions, depend on wavelength and redshift.

In addition to photons, gaseous medium can be *heated by particles*. The first particles that come to mind are cosmic rays. Cosmic rays are high-energy charged particles (protons, electrons, nuclei). About 1% of them are electrons, and 90% are protons. The remaining are nuclei of heavier elements. Usually, only the low-energy cosmic rays are sufficiently abundant to contribute significantly to ionisation. A single cosmic ray particle can ionise several atoms, and the process is most effective in dense gas, e.g., in dense molecular clouds.

The third heating mechanism is *heating by shock waves*. But this deserves a special subsection.

7.3.2.5 *Shock waves*

Let us discuss first about what the shock waves are. If a gas in steady motion receives a slight perturbation at any point, the effect of the perturbation is subsequently propagated through the gas with the velocity of sound. Thus, relative to a fixed system of coordinates, the rate of propagation of disturbance consists of two parts: (1) the perturbation is carried along by the gas flow and (2) it is propagated relative to the gas with the sound speed in that medium. These sound waves carry information about the perturbation. If gas velocity is smaller than the sound speed, information about the disturbance is carried away by sound waves to everywhere in the gas. But if the gas flow velocity is larger than the speed of sound, the situation changes radically. Information about the disturbance can propagate only downstream of the flow. Medium ahead of the disturbance has no information about the arrival of the disturbance and is not able to react and accommodate to it. As a result, there is a discontinuity between the regions behind and ahead of the disturbance. This discontinuity is called a shock wave. The discontinuity is very thin, and one may take the typical thickness of the order of few free path lengths of gas particles.

Shock waves arise in several situations. In our context, they form when some "objects" flow supersonically through a gas. Under the "object", we also mean the situation when a cloud of cold gas moves in a hot environment or a gas cloud collides with another gas cloud, etc.

Shock waves are characterised by the Mach number defined as the ratio of the gas flow (or an object) velocity v to the velocity of the sound c_S:

$$\mathcal{M} \equiv v/c_S.$$

If $\mathcal{M} > 1$, we talk about the supersonic motion and formation of shock waves; in the opposite case, there is subsonic motion. Sound speed in cosmic gas is, as a rule, quite low, in adiabatic case $c_s \simeq 0.152\sqrt{T}\,\mathrm{km/s}$, i.e. even for gas with $T \sim 10^6\,\mathrm{K}$, the sound speed is only $\sim 150\,\mathrm{km/s}$ and all the motions exceeding this cause shock waves.

Let us discuss now about how the gas properties change due to shock waves. Let the characteristics of undisturbed gas be p_1, ρ_1, T_1, c_1 (pressure, density, temperature and sound speed, respectively), and the shock wave propagates in this environment with the speed of $v_{\mathrm{sh}} > c_1$ (see Fig. 7.15). Behind the shock wave, the characteristics of the gas are p_2, ρ_2, T_2, c_2. In fact, the shock wave is just the boundary between these two gas media. Usually, the analysis is continued in the reference system where the shock wave is at rest. In this case, the undisturbed gas flows towards the discontinuity at velocity $v_1 = |v_{\mathrm{sh}}|$. Away from the discontinuity, its velocity becomes v_2. Instead of sound speeds, we use corresponding Mach numbers $\mathcal{M}_1$ and $\mathcal{M}_2$, where $\mathcal{M}_1 > 1$ or even $\mathcal{M}_1 \gg 1$.

The behaviour of the gas on passing through the shock wave is described by a set of conservation relations: the mass conservation, the energy conservation and the momentum conservation, called the Rankine–Hugoniot jump conditions. After some algebraic manipulations with these conditions, it can be quite easily derived how the

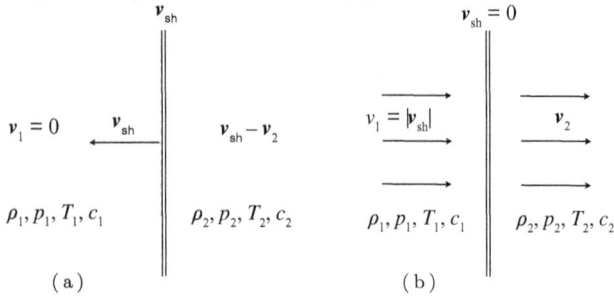

Fig. 7.15. (a) A shock wave propagates in undisturbed gas in left direction with the velocity v_{sh}. (b) In the reference frame where the shock wave is at rest, the undisturbed gas flows towards the discontinuity at velocity $v_1 = |v_{\text{sh}}|$.

density, pressure and temperature of the gas changes after the shock passes it:

$$\frac{\rho_2}{\rho_1} = \left[\frac{1}{\mathcal{M}_1^2} + \frac{\gamma - 1}{\gamma + 1}\left(1 - \frac{1}{\mathcal{M}_1^2}\right)\right]^{-1},$$

$$\frac{p_2}{p_1} = \frac{2\gamma}{\gamma + 1}\mathcal{M}_1^2 - \frac{\gamma - 1}{\gamma + 1},$$

$$\frac{T_2}{T_1} = \frac{\gamma - 1}{\gamma + 1}\left[\frac{2}{\gamma + 1}\left(\gamma\mathcal{M}_1^2 - \frac{1}{\mathcal{M}_1^2}\right) + \frac{4\gamma}{\gamma - 1} - \frac{\gamma - 1}{\gamma + 1}\right].$$

In case of adiabatic strong shock, $\gamma = 5/3$, $\mathcal{M}_1 \gg 1$, we have much simpler relations

$$\frac{\rho_2}{\rho_1} = 4,$$

$$\frac{p_2}{p_1} = \frac{5}{4}\mathcal{M}_1^2,$$

$$\frac{T_2}{T_1} = \frac{5}{16}\mathcal{M}_1^2.$$

Thus, shock compresses and heats the gas. During the formation of a DM halo, merging and accretion of subhaloes create shocks. Although the DM is collisionless, both the already collapsed part of the DM halo and accreting subhaloes contain gas. As a rule, subhaloes accrete with supersonic velocities, and the halo gas is heated.

For illustration of how shock front reveals itself in observations, let us look at the Chandra X-ray satellite observations of a merging

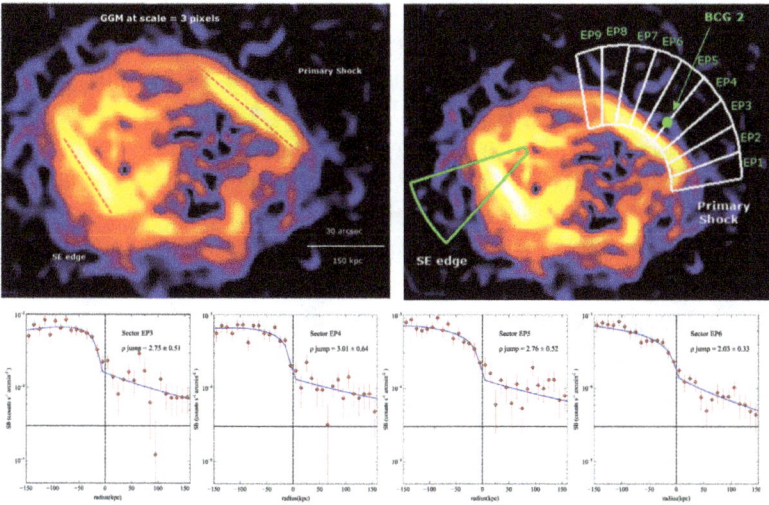

Fig. 7.16. Top left panel: Image of a part of the cluster in the 0.5–7.0 keV range. The red dashed lines highlight the two surfaces with pronounced intensity gradients; top right panel: sectors to measure density jump across the Primary Shock assuming elliptical geometry of the shock; bottom panel: a selection of sectors fitted with the broken power-law density model (blue) and giving the density jump. See details from the original paper (Diwanji *et al.*, 2024).

galaxy cluster SPT-CLJ 2031-4037 at $z = 0.34$ by Diwanji *et al.* (2024) given in Fig. 7.16. X-ray data confirm two shock fronts by directly measuring the temperature jump of gas across the surface brightness edges. Average density jump in the binned sectors is $\rho = 3.04 \pm 0.36$ giving the corresponding Mach number $\mathcal{M} = 3.09^{+0.75}_{-0.43}$.

7.3.3 *Gas accretion*

In the case of accretion, gas flows from the more distant intergalactic medium to the circumgalactic medium and thereafter to the interstellar medium. As a rule, the temperature of the infalling gas is colder than the gas in the dark matter haloes (we are not looking at galaxies in the hot cluster gas environment at present). The same is true for the corresponding sound speeds. The sound speed is a function of the temperature, and thus, the sound speed is smaller in the infalling gas than in the halo gas. Thus, in most cases, gas accretes with supersonic velocities and shock wave fronts form in accreted gas. As the shock front passes the accreting gas, the gas behind the shock front is compressed and heated. Density and temperature increases were described just earlier as the Rankine–Hugoniot conditions. In principle, the gas heated by shock wave then cools until a (new) thermal equilibrium is reached. But if the gas accretion continues and cooling is slow, the hot gaseous halo forms within the dark matter halo of the galaxy. In this way, a significant fraction of the kinetic energy of the infalling gas can be thermalised.

In a more general case, the result of the gas accretion depends on the balance between the cooling time t_{cool} and dynamical time t_{dyn}. If $t_{cool} \ll t_{dyn}$, the gas cools quickly, and nothing prevents it from collapsing to the centre within the free-fall time. In the opposite case, if $t_{cool} \gg t_{dyn}$, the gas remains in approximate equilibrium with the dark halo. Whether the gas remains hot or still slowly cools depends on whether the cooling time is shorter or longer than the Hubble time. If the cooling time is shorter than the Hubble time (inner regions of the galaxy), the gas slowly and smoothly accretes to the centre of the galaxy. In the other case, it remains hot in the halo.

We saw that the dynamical time is $t_{dyn} \sim 1/\sqrt{\rho} \sim 1/\sqrt{n}$, the cooling time in case of radiative cooling is $t_{cool} \sim 1/n$. We see that cooling dominates (time is shorter) in denser systems or in denser regions of a system. Thus, the central regions of a galaxy may contain cooled gas, while the outer, less dense regions have hot, virialised gas. Specific numerical values depend on cooling details in the case of radiative cooling, e.g., on the metallicity of the gas.

Let us describe the shock heating of the gas in more detail. In addition to the general considerations above, also from the simulations (see, e.g., Birnboim & Dekel, 2003), we know that it is very important to take into account the radiative cooling of the compressed and heated gas. Without radiative cooling, in the adiabatic case, the shock forms always at about the virial radius of the galaxy. In time, this limit gradually propagates outwards but approximately continues to coincide with the virial radius. But when radiative cooling is taken into account, at first, the formation of stable shocks takes a time of nearly 4 Gyrs. During that time, due to cooling, gas collapses freely to the halo centre, more precisely up to the disk, being stopped by the angular momentum of the disk. Then, a stable shock forms at the edge of the disk and starts to propagate outwards. The shock does not reach the virial radius of the halo and, after some time, starts to fall again. Thus, in the case of radiative cooling, the shock never expands beyond the virial radius; it remains at temperatures

$T \sim 10^4$–10^5 K and the accreted gas is a resource for star formation. This is called the cold-mode accretion.

The details of the described results depend on the halo mass. Birnboim & Dekel (2003) found that virial shocks, i.e. shocks at the virial radius called the stable hot-mode accretion, form only in massive haloes, and the just described cold-mode accretion occurs in smaller masses haloes. The distinction can be taken approximately at $M \sim (3\text{–}5) \times 10^{11} \, M_\odot$, but it depends on the metallicity of the gas as the cooling is more efficient in higher metallicity gas. Thus, we have the hot-mode and cold-mode accretion. Indeed, simulations by Kereš *et al.* (2005) have shown that the histogram of maximum temperatures in accretion is clearly bimodal with a similar distinction line at $M \simeq 3 \times 10^{11} \, M_\odot$.

Cold and hot modes also differ in how the accretion looks like. Simulations by Kereš *et al.* (2005) indicated that cold accretion is often directed along filaments called cold streams. In hot mode, the sound speed of hot gas is rather high, and pressure waves can smooth out density fluctuations. Thus, in hot mode, accretion is quasi-spherical. The cold mode dominates at high redshifts and in today's low-density regions and the hot mode in groups and clus-ter environments at low redshifts. Even when accretion shock occurs close to the virial radius, the cold mode via filaments can be active at high redshifts.

In Fig. 7.17, the evolution of the same galaxy is in the cold-mode accretion and the hot-mode accretion (Kereš *et al.*, 2005). The first one corresponds to the situation at $z = 5.52$ when the galaxy mass is $M = 2.6 \times 10^{11} \, M_\odot$, the second one when redshift is already $z = 3.24$ and the galaxy mass has reached $M = 1.3 \times 10^{12} \, M_\odot$. The gas is represented by green dots, the remaining particles are colour-coded by overdensity from the darkest blue (lowest) to the white (highest). Box sizes are $4 \, R_{\text{vir}}$. It is seen that cold accretion (left panel) is clearly directed along the intersecting filaments, allowing the galaxy to accrete gas from large distances. In the hot-mode stage (right panel), the accreted gas is coming from the inner region of the hot quasi-spherical halo.

Cold-mode accretion is more rapid than the hot-mode one, and a significant amount of the gas is not shock-heated to the virial temperatures and remains cold (Brooks *et al.*, 2009). In this way, cold gas can still accrete as filaments within an overall spherical hot gas halo. This is illustrated in Fig. 7.18 where results of the

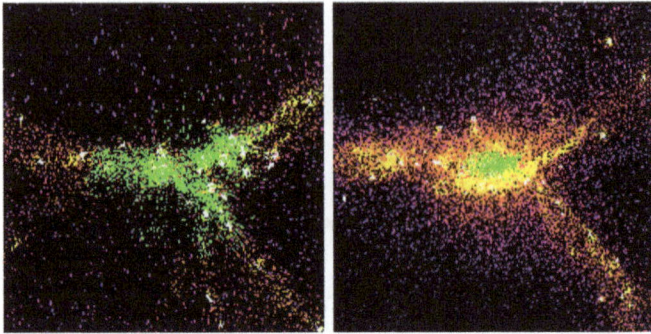

Fig. 7.17. The evolution of the same galaxy from the cold-mode to the hot-mode accretion. The cold mode (left panel) corresponds to the situation at $z = 5.52$ when the galaxy mass is $M = 2.6 \times 10^{11}\,M_\odot$, the hot mode (right panel) when redshift is already $z = 3.24$ and the galaxy mass has reached $M = 1.3 \times 10^{12}\,M_\odot$. The gas is represented by green dots, the remaining particles are colour-coded by overdensity from the darkest blue (lowest) to the white (highest). Box sizes are $4\,R_{\mathrm{vir}}$ (Kereš *et al.*, 2005).

high-resolution hydrodynamical simulation show two $10^{12}\,M_\odot$ haloes at $z = 2$ (Nelson *et al.*, 2016). The simulation h2L11 illustrates a more or less round hot halo. The viral shock at $\sim 1.25\,r_{\mathrm{vir}}$ is seen as a quite abrupt increase in the temperature. The simulation h4L11 results in a more disturbed halo. No clear shock boundary is seen. Near the centre, there is a merger of two subhaloes. In both runs, filamentary inflows across the virial radius are seen. When approaching the centre, filamentary gas is heated.

Even in the case of a hot galaxy cluster environment, simulations by Sharma *et al.* (2012) show that within this hot halo gas, dense and cold filaments condense out, falling eventually towards the centre of a galaxy. Condensation starts when the overdense cold gas blob cools below the temperature $10^7\,$K. Then, the cooling gas, cooling rate becomes faster than the turbulent mixing rate, and the blob quickly cools to a thermally stable temperature of $10^4\,$K. Near the filamented blob, secondary cold filaments form; see Fig. 7.19.

After some processing in galaxies by star formation and subsequent evolution, the recycled gas flows back (at least in part) to the circumgalactic medium. Probably, the flow back does not extend to the intergalactic media. Of course, during this process, gas's physical properties change. Taking into account the outflows, the just

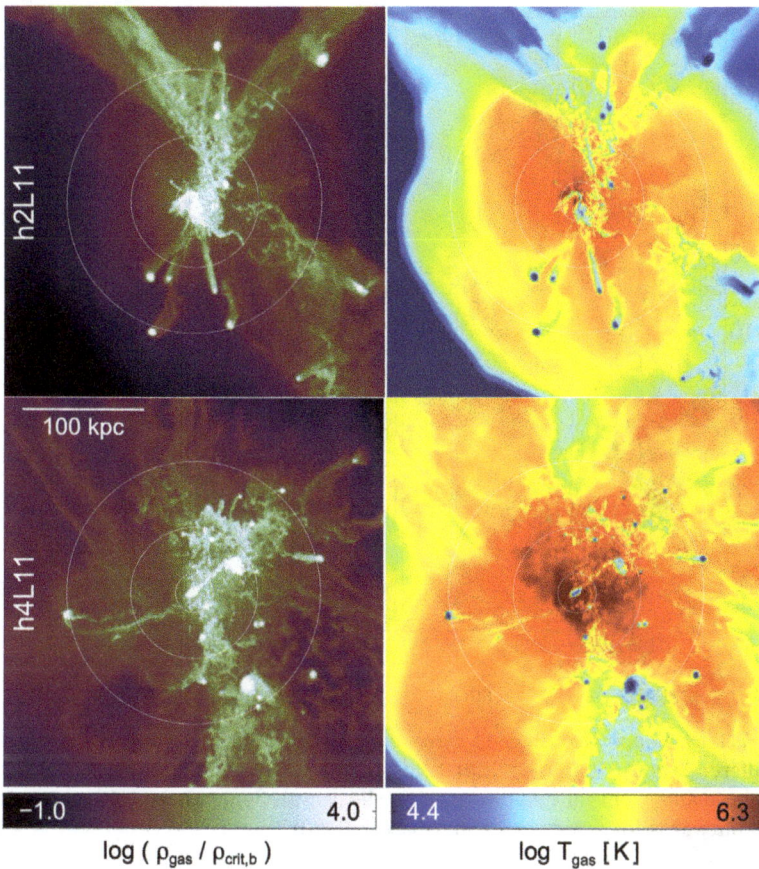

Fig. 7.18. The inner halo structure from the two simulated haloes at $z = 2$. The mass-weighted projections of the gas densities and temperatures are shown. The outer white circle is at the radius r_{vir} (Nelson *et al.*, 2016).

described accretion processes change. For example, when ignoring the feedback and averaging over all galaxies, Benson & Bower (2011) derived from their simulations that at high redshifts, about 90% of baryons accrete in cold mode. But when feedback is included, the cold mode is clearly less apparent.

Further complications to evaluate the relative importance of the cold and hot accretion result from the complicated structure of the whole intergalactic and circumgalactic media. We started this section by describing the multi-phase nature of the gas. Due to this, the

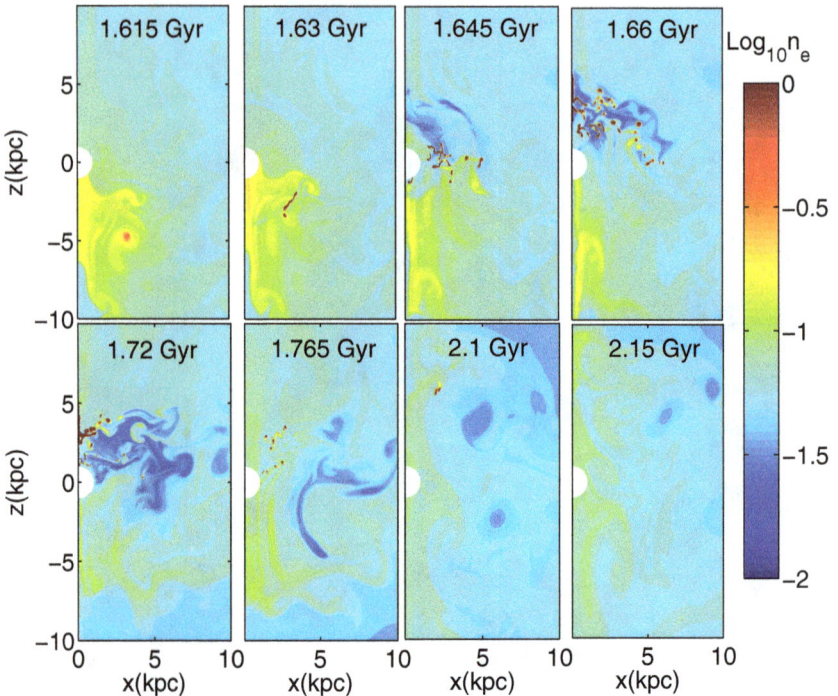

Fig. 7.19. Snapshots of the electron density in the inner 10 kpc of a simulation at different times. It is seen that cold and dense filaments condense out of the hot phase and eventually fall in towards the centre. Some of the cold gas can also move outwards for some time (Sharma *et al.*, 2012).

balance between the cooling and dynamical times, t_{cool} and t_{dyn}, can be different in different regions.

7.3.4 *Gas inflows and outflows*

Necessity of the accretion and outflow circle roots in star formation timescale. When using, for example, a current SFR and the amount of available gas in a typical galaxy, it can be easily shown that the gas will be exhausted rather soon. Let us take the MW with the current SFR $2.0\,M_\odot\mathrm{yr}^{-1}$ (Elia *et al.*, 2022) and disk gas mass $1.2 \times 10^{10}\,M_\odot$. We see that the disk gas should be exhausted within 6 Gyrs. This is a very conservative estimate, as it assumes the same SFR also during the formation period of the MW. For other galaxies, gas exhaustion timescales for the so-called closed box models (i.e. without inflows

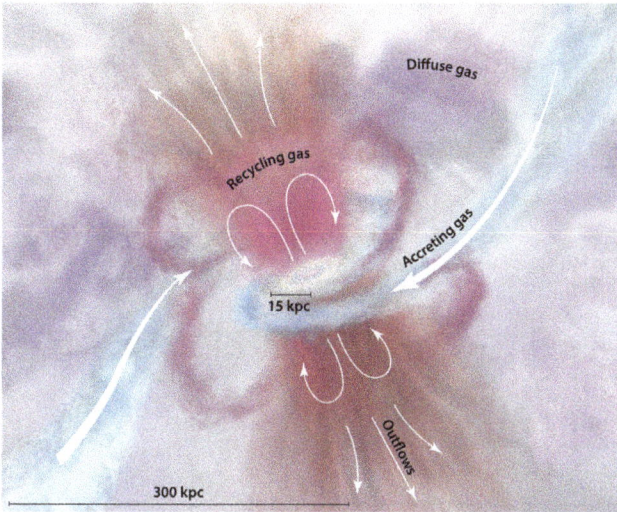

Fig. 7.20. Illustration of the inflows and outflows of galactic and circumgalactic gas (Tumlinson *et al.*, 2017). Used with permission of *Annual Reviews* from Tumlinson *et al.*, 2017. Permission conveyed through Copyright Clearence Center, Inc.

and outflows) are also 2–6 Gyrs. Thus, there is a need for an external gas supply. Schematically, the process is illustrated in Fig. 7.20.

Taking into account the inflow-outflow circle, the accretion of both the fresh, as a rule, pristine gas from intergalactic space and the processed gas from the circumgalactic medium (from the space within the virial radius of a galaxy) occurs. In Fig. 7.21, the observed surface density distribution of different gas components within a virial radius of a galaxy is seen (Tumlinson *et al.*, 2017). For a reference, the NFW profile for $M_{\mathrm{DM}} = 2 \times 10^{12}\,\mathrm{M}_\odot$ is given in black. Colour coding is in the top of the figure. Gas surface densities are from different neutral and ionised elements tracers, see the details and references from the original paper (Tumlinson *et al.*, 2017). Circumgalactic medium at low redshifts has a diverse kinematic structure. Unfortunately, as the observations are only along the line of sight, the kinematics and positions of gas clouds are degenerate.

Observations of HI clouds (Lyman limit systems) at different redshifts seem to indicate that galaxies with rather different masses have large reservoirs of circumgalactic gas. For very low-mass galaxies, the observations are not so firm, but a general trend can be seen. It is interesting that this gas is also observed in significant quantities even

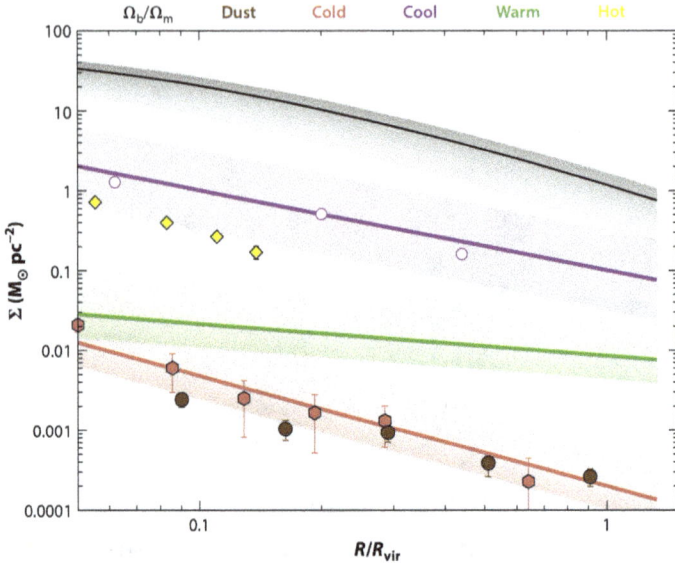

Fig. 7.21. Circumgalactic media mass densities as a function of galactocentric radius in the virial radius units. The black line is the NFW profile for the dark halo with $M_{DM} = 2 \times 10^{12} \, M_\odot$ (Tumlinson *et al.*, 2017). Used with permission of *Annual Reviews* from Tumlinson *et al.*, 2017, Permission conveyed through Copyright Clearence Center, Inc.

around passive early-type galaxies without clear star formation. The gas is colder ($T \leq 10^5 \, K$) than the virial temperature of their host galaxies (Thom *et al.*, 2012). Thus, the existence of the circumgalactic gas does not necessarily imply its falling to the galaxy.

Rubin *et al.* (2012) discovered partially enriched cool gas inflows of $0.2 - 3 \, M_\odot$/yr in four isolated star-forming galaxies at intermediate redshifts. These are disk-like galaxies about the MW stellar masses. In the MW, we see gas accretion as there are high-velocity clouds with such blueshifted velocities that they will reach the galaxy disk within $10^7 - 10^8$ years and cause mass inflow rate $0.1 - 0.4 \, M_\odot$ (Putman *et al.*, 2012). But this is less than the current SFR of the MW. The cold gas inflow reservoir is bound to the gravitational potential of the galaxy. In addition to the HVC-like accretion, there is also a smooth accretion of ionised gas.

To replenish observed inflows, there should also be outflows. Outflows are also necessary to explain the observed metal lines in gas haloes around star-forming galaxies. This is confirmed both by observations of local and intermediate redshift galaxies and simulations. Statistics of outflows indicate preferredly biconical outflows perpendicular to the galactic disk. This direction preference disappears at the heights of $\sim$60–80 kpc, but outflows reach $\sim$100 kpc scales. It may be that a large part of the circumgalactic medium is made from outflows (Tumlinson *et al.*, 2017). Outflows are caused by stellar winds and supernova explosions. As a rule, stellar winds do not have enough energy to escape galactic haloes, and this gas remains in a circumgalactic environment, loses its energy and finally reenters the galaxy.

Interplay between the accretion flows and outflows is beautifully illustrated in a simulation by Anglés-Alcázar *et al.* (2017), where the evolution of different types of circumgalactic gas flows is shown (see Fig. 7.22). The central galaxy has at $z = 0$ dark halo mass

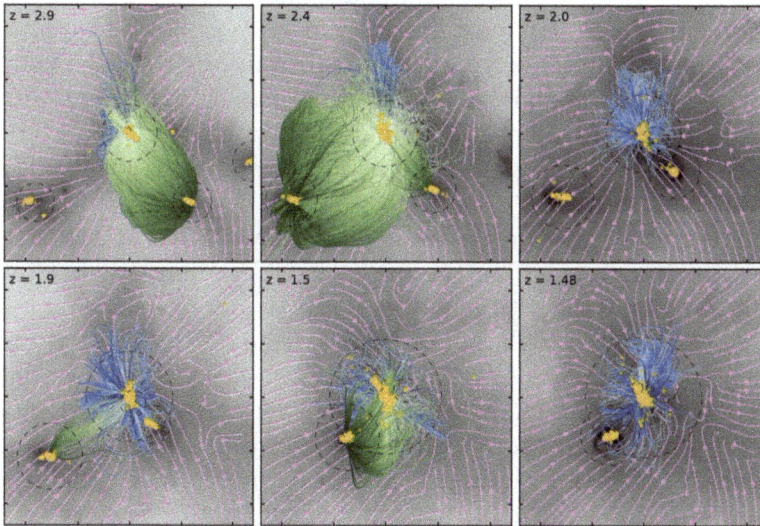

Fig. 7.22. Simulation of the evolution of different types of circumgalactic gas flows near the main galaxy. The background grayscale is the projected gas density in a cube with a side of 240 kpc Tick marks indicate 50 kpc physical length. Orange stars show the location of randomly chosen star particles. Purple lines indicate projected mean mass-weighted steamlines of fresh gas. Blue lines denote future trajectories of gas outflows from the central galaxy that will accrete later back. Green lines are future trajectories of gas removed from another galaxy that will accrete to the central galaxy as part of intergalactic gas transfer. The black dashed lines show the virial radii of identified dark haloes (Anglés-Alcázar *et al.*, 2017).

of 1.4×10^{11} M$_\odot$. Only six snapshots at different redshifts are given here. The background grayscale is the projected gas density in a cube with a side of 240 kpc (tick marks are to 50 kpc physical length). Orange stars show the location of randomly chosen star particles. Purple lines indicate projected mean mass-weighted steamlines of fresh gas. Blue lines are future trajectories of gas outflows from the central galaxy that will accrete later back. Green lines are future trajectories of gas removed from another galaxy that will accrete to the central galaxy as a part of intergalactic gas transfer. The black dashed lines show the virial radii of identified dark haloes. At all redshifts, gas flows from the intergalactic medium onto the central galaxies. However, at later redshifts, some of the gas is ejected back into the intergalactic space and, at least in part, is accreted later again. The simulation also illustrates the role of satellite galaxies. It is seen that the satellites are strongly influenced by the central galaxy. Green lines dominate in several snapshots! The flows in opposite directions are seen. At redshifts 2.9 and 2.4, gas flows to the main galaxy from the intergalactic space and from other smaller haloes. Outflows do not dominate. At $z = 2$, fresh gas from intergalactic space continues to flow to the main galaxy, but outflows start to dominate. These processes, more or less periodically, may continue several times.

So, there is a circle of inflows and outflows. Even at higher redshifts, those accreted from the intergalactic medium gas is already somewhat enriched with metals (Pop III stars!). This recycling started early in the evolution of galaxies, and all the star formation models should account for it. The inflowing or outflowing gas has a range of metallicities as the gas can be in various enrichment stages. Outflows due to supernova explosions can escape the circumgalactic environment and reach intergalactic space. But also this enriched gas may later re-enter the galactic environment.

However, as it often is, the inflow–outflow mechanism does not always work. There is a known problem — why does star formation in some galaxies quench? We described that the dark mass haloes and their outer regions around galaxies contain a lot of gas. The amount of this gas is sufficient to feed galaxies for a rather long time. In some galaxies, it is indeed so, but in some galaxies, it is not. As we know, the gas falls to the halo via cold accretion filaments and, in this way, feeds star formation. But in massive galaxies, the circumgalactic medium pressure is so high that it prevents the falling of cold gas.

In massive galaxies, the falling gas is heated by virial shocks, and it remains quite hot, being not able to form stars. For this reason, the massive galaxies are, as a rule, passive. (We do not discuss here additional environmental processes, such as ram pressure. This is the topic of other chapters.)

7.4 Supernovae explosions and chemical evolution of galaxies

7.4.1 *Supernovae explosions*

In the case of the Milky Way, the phenomenon of occasionally extremely bright new stars appearing in the night sky was probably first registered by ancient Chinese observers. A nice overview of these records was published by Clark & Stephenson (1976). Some of these records note that these "new stars" were seen for quite a long time even in the daytime. The discovery of the first extragalactic supernova (as we know them nowadays) was made in 1885 in the old Tartu (Dorpat) Observatory, Estonia, by Ernst Hartwig (Hartwig, 1885; de Vaucouleurs & Corwin, 1985). This was the SN 1885A in the Andromeda galaxy.

Understanding the supernova (SN) explosion mechanisms seems to be an old story by now. Indeed, the phenomena of SNe as a specific physical phenomenon was understood already in 1934 by Walter Baade and Fritz Zwicky (Baade & Zwicky, 1934), and a path to explosion of the *core-collapse supernovae* was proposed in 1941 by George Gamow and Mario Schoenberg (Gamow & Schoenberg, 1941). However, it was only in 1987 that the explosion of the supernova SN 1987A in the Large Magellanic Cloud allowed the observers to really measure the resulting neutrino flux from the explosion and confirm that our understanding is correct. Still, to talk about a flux was somewhat overstating as, in reality, it was detected in a total of 25 antineutrinos in three different neutrino observatories. Still, the observations confirmed the theoretical models and neutrino astronomy started. A path to the explosion of the *thermonuclear supernovae* roots from the calculation of the limiting mass of white dwarfs by Subrahmanyan Chandrasekhar (Chandrasekhar, 1931) and resulting from this a possibility of the supernovae explosions by Fred Hoyle and William Fowler (Hoyle & Fowler, 1960).

Although there are many slightly different SN explosion mechanisms, usually two major mechanisms are reviewed, as referred to above: the core-collapse supernovae and thermonuclear supernovae. The first one is related to the final stages of the evolution of massive, above $\sim$8–10 $M_\odot$ stars, and the second one is related to a white dwarf approaching and exceeding the Chandrasekhar limit. The second one is also known as the type Ia supernova.

We do not review the evolution stages of stars or the formation of chemical elements here; we are interested in how these elements enrich the ISM in SN explosions. We start from times when, in the case of the core-collapse SN, the progenitor massive star already contains an iron core surrounded by shells of most chemical elements up to iron; the nuclei produced by the helium-capturing process dominate. Due to extremely high temperature in the centre, $T_c \sim 10^{11}$ K, the iron and other heavy nuclei start to photodissociate into helium nuclei and neutrons, and then into protons and neutrons, and thereafter due to β-decay (the electron capture) into neutrinos and neutrons. As a result, the core of a star consists of pure neutrons and neutrinos. Neutrinos are also surrounding the core. As the electrons are gone, the pressure support due to electrons is gone as well, and the free-fall collapse of the core follows. This collapse is stopped when nuclear matter reaches the nuclear density, and a repulsive strong interaction force stops the collapse. Due to a momentary overcompression of the matter, the core bounces back and drives an outward-moving shock wave into the still-infalling outer core.

For a long time, there was an understanding that just this shock wave created the SN explosion. However, recent theoretical models clearly show that the energy of the shock is lost due to the creation of a burst of neutrinos and due to the accreting matter with still-running photodissociate processes from the outer core of the star. The shock is stalled and cannot create an explosion. Thus, a puzzle was that we see supernovae explosions, but theoretically, they cannot exist! A possible solution was found only recently with the help of very resource-demanding numerical calculations, see, e.g., Janka *et al.* (2016), Burrows *et al.* (2020) and Burrows & Vartanyan (2021). It was necessary to develop 3D models by taking into account hydrodynamic instabilities, turbulences, multi-dimensional motions, radiation, etc. Only within the past 10 years was some progress achieved.

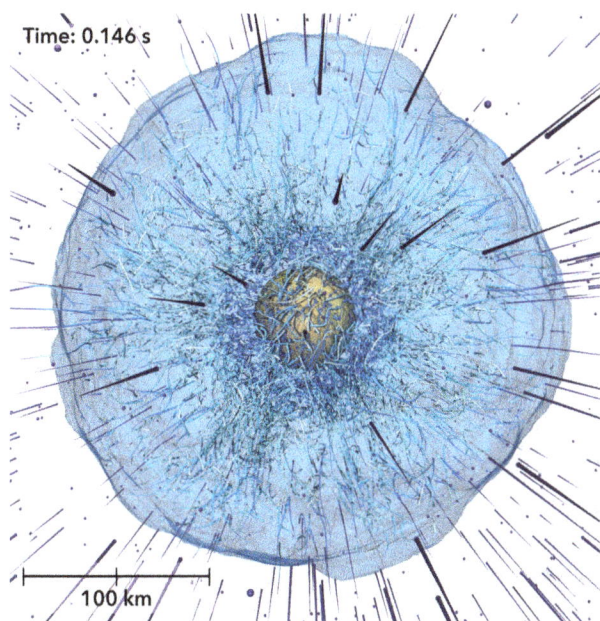

Fig. 7.23. Core-collapse SN just before the explosion. The shock wave is at about 150 km. The proto-neutron star is the inner ball at the centre surrounded by swirling turbulent matter. The trajectories depict the recent 5 ms in the positions of the individual accreting matter elements (Burrows & Vartanyan, 2021).

According to the current (but still uncertain) understanding, the explosion results from the critical combination of several processes. Outside of the central core (the proto-neutron star), heating by neutrinos[24] creates growing turbulence[25] and corresponding additional pressure in the still-accreting matter and moves the stalled shock region outwards into less dense regions, see Fig. 7.23 where the turbulences and instabilities are illustrated in the region just before the explosion in the region between the central core and stalled shock. When all these growing processes reach a critical stage, an explosion follows.

---

[24]Heating by neutrinos is caused by processes $\nu_e + n \rightarrow e^- + p$, $\bar{\nu}_e + p \rightarrow e^+ + n$ and by scattering.

[25]As in the case of water boiling by heating from downside.

The second explosion mechanism, the type Ia SN, occurs when the mass of the carbon–oxygen white dwarf reaches the Chandrasekhar limit of $\sim$1.37 M$\odot$. When this happens, the carbon starts to be ignited in the centre. At first, the reaction energy is carried successfully outwards by convection and radiated away in the form of neutrinos. This quasi-silent stage may last a few 100 years. But then, the reaction intensity reaches the stage when it is not possible to transport the created energy sufficiently efficiently away anymore. The core matter of the white dwarf is degenerate, and its temperature starts simply to increase without the pressure regulating it. This runaway increase in temperature increases the nuclear reactions, and within a few seconds, it culminates in the explosion which completely destroys the white dwarf. It is like the thermonuclear explosion. A typical SN of this kind produces $\sim$0.7 M$\odot$ of iron.

The basic processes of the thermonuclear SN are more or less well understood, although they are not so well understood when calculating the maximum brightness of Ia SNe from the theory, allowing for the calibration of the distance scale. The main theoretical problem is how the reaction front spreads outwards or in another way when the slow-burning changes to detonation (when the front spread speed increases the sound speed). As usual, the calculations are complicated by turbulence and instabilities (mainly the Rayleigh–Taylor instability), see, e.g., Fig. 7.24 (Röpke *et al.*, 2007).

Taking together both types, the SNe are a major source of chemical elements. Averaging over the universe, the core-collapse SNe produce mainly elements from oxygen to rubidium, and the thermonuclear SNe produce mainly elements from silicon to zink. Elements from oxygen to aluminium and from gallium to rubidium are produced only by core-collapse SNe. We note that there are certain differences in producing different elements. Let us compare, for example, the relative amount of [Mg/Fe]. Magnesium is produced in core-collapse SN only, and iron is produced in both the core-collapse and thermonuclear SNe. If we look at core-collapse SNe, the elements with a certain [Mg/Fe] ratio are produced. But if we look at both kinds of SNe together, compared with the core-collapsed only, then some additional iron will be produced but not magnesium. Thus, the [Mg/Fe] ratio in this case is smaller. Taking into account that core-collapse SNe are from massive (and thus young) stars, specific element ratios may give us information on recent star formation processes.

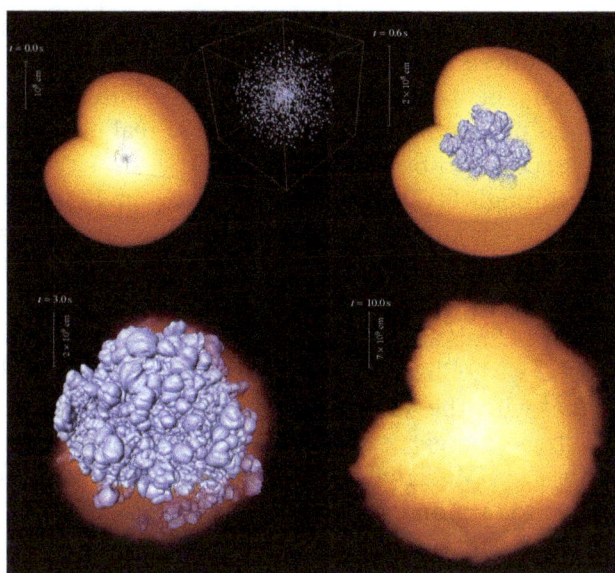

Fig. 7.24. Evolution of the thermonuclear supernova explosion simulation. The upper left panel shows the initial setup and the close-up illustrates the chosen flame ignition configuration. The next two panels show the propagation of the turbulent flame through the white dwarf. Initial kernels deform due to Rayleigh–Taylor instabilities, grow and merge. In subsequent evolution, buoyancy instability dominates in larger scales and turbulent cascades in smaller scales. The density structure of the remnant is shown in the lower right panel (Röpke *et al.*, 2007).

7.4.2 *Merger of neutron stars*

The discovery of gravitational waves (GW) from the merger of two neutron stars (the GW170817 event) with the subsequent registration of gamma-ray emission (GRB170817A event) and thereafter X-ray, radio, optical and near-infrared emission demanded a revision of the heavy element synthesis theory.

According to current understanding, the nuclei of heavy elements are formed primarily by neutron capture reactions. Two separate processes are distinguished, the slow *s*-process and the rapid *r*-process. In the first case, the timescale between the successive neutron captures is longer than the beta decay timescale; in the second case, it is shorter. As a rule, the *r*-process produces heavier elements than the *s*-process. It is well established that *s*-processes occur in the ordinary evolution of massive stars. Sources of *r*-process element production were more debatable, but as a rule, it was assumed that these elements were produced in SN explosions. There were theories that mergers of binary neutron stars can also contribute to the production of r-process elements, but due to a lack of real merger data, this channel was assumed to be insufficient.

The breakthrough GW signal was detected on 17 August, 2017, by Advanced LIGO and Advanced Virgo interferometers. About 1.7 s later, the Fermi Gamma-ray Space Telescope and INTEGRAL detected a γ-ray burst. About 11 hours later, the optical counterpart of the merger was first seen with the Swope Telescope, 9 days later, the X-ray signal with Chandra X-ray Observatory and 16 days later the radio signal with the VLA Observatory. It was derived that the host galaxy of the merger is NGC 4993, about 41 Mpc away from us. Analysis of the GW signal allowed us to derive the masses of the merging neutron stars and rough estimates of their spins. Gamma rays gave the burst total energy and peak luminosity, but of course, some assumptions about the geometry of the burst were needed (e.g., isotropy versus anisotropy, jets). An important additional parameter was the time delay of the gamma burst. Optical observations gave light curves of the emission from near-UV to near-IR, spectral features, expansion velocities of the optical photosphere of the fireball and its cooling rates. Thus, it was quite a lot of initial data to start modelling what precisely happens.

According to models, the GW170817 resulted in the ejection of $\sim 0.05\,M_\odot$ of r-process elements. The current estimate is that the binary neutron star merger rate in an MW-like galaxy is $37^{+27}_{-11}\,\mathrm{Myr}^{-1}$ (Pol *et al.*, 2020). Although the numbers are model-dependent, one can conclude that the merger of neutron stars should be considered as at least one source of r-process nucleosynthesis. A detailed overview of the analysis of the GW170817 event is given by Kasen *et al.* (2017); Smartt *et al.* (2017), Margutti & Chornock (2021). After the GW170817/GRB170817A event, several other events were studied, e.g., the GRB230307A burst with the JWST observations (Levan *et al.*, 2024).

An obvious question arises: Can neutron star mergers alone explain the r-process enrichment in the Milky Way? This was the title of the recent paper by Kobayashi *et al.* (2023). An important issue here is that there is a long time delay between the SN explosion (i.e. neutron star formation) and the merger of neutron stars. Thus, when looking at the amount of r-process enhancement of, e.g., [Eu/Fe] in most metal-poor stars in the Milky Way, is it possible that a sufficient number of formed neutron stars had time to merge in order to produce these elements for these stars? Using the abundances of oxygen, iron and europium, Kobayashi *et al.* (2023) concluded that time delays in neutron star merger models are too long to fully account for r-elements. But the mergers between black holes and neutron stars have a better chance. Unfortunately, at present, there are significant uncertainties in binary evolution models.

7.4.3 *Production of chemical elements*

CALLISTUS: How was spring break?

ARIEL: Ugh, working away on my project.

CALLISTUS: Right. As expected.

RHEA: I was expecting "Dancing like we are made of starlight."

ARIEL: Makes no sense. We are made of stars, aren't we?

CALLISTUS: As you Millenials say, *complicatus*.

ARIEL: Gen Z. And Harry Potter fans might say that otherwise, "it's complicated".

CALLISTUS: Never mind. Our body has a lot of hydrogen. Most of it was created during the first few minutes after the Big Bang.

ARIEL: But the rest of it comes from supernova explosions, doesn't it?

RHEA: Nucleosynthesis is still an intense area of research. Sure, most elements necessary for life were created in stars, so it's still true that part of us have been in a supernova explosion parsecs away from the solar system.

CALLISTUS: Your carbon and nitrogen might have been released by a moribund low-mass star withering away. Most of your body is water, and its oxygen most likely comes from a massive star explosion.

RHEA: The diversity and complexity of these processes are astounding, ranging from cosmic ray fission through exploding white dwarfs to merging neutron stars.

CALLISTUS: Gold is expensive because you have to bang two neutron stars together to make it. Diamond, on the other hand, is just carbon.

RHEA: It comes from ordinary stars, but it shines!

ARIEL: Let's agree that they are both better than Bitcoin.

After about 1 s of the Big Bang, the neutron ratio to protons was 1:6 (which is a result of a little difference in their masses). Soon, these baryons formed the nuclei of deuterium, lithium, and helium isotopes. In addition, a lot of protons remained (free neutrons decayed). In essence, the Universe consisted mainly of hydrogen and helium nuclei — for every helium nucleus, there were about 12 hydrogen nuclei with a very small amount of deuterium and lithium.

After the star formation started, the creation of heavier elements started. As the most massive stars evolved faster, at first, the ISM was enriched by core-collapse SNe and by neutron star merger. Of course, there was a time delay between these two enrichment channels. Unfortunately, we do not know how long it takes neutron stars to merge and how many neutron stars there may be. Still, it is clear that the next generation of stars forms from the enriched media. This subsequent enrichment can be imagined as the spiralling enrichment model, see Fig. 7.25.

Massive stars ($>8\,M_\odot$) enrich the ISM with metals first via stellar winds. For smaller mass stars, the contribution of winds is rather small (few per cent), but for larger masses, it can be up to 60–80% of the total elected mass. Next, the massive stars enrich the ISM via SN explosions. SNe explosions enrich the surrounding media in three ways. First, the matter surrounding the neutron core (C, O, Mg, etc.) is ejected to interstellar medium; second, the shock wave drives subsequent nuclear reactions, e.g., most of ejected Fe is produced just in the shock wave; third, when the expelled matter reaches the surrounding gas, it creates again a shock wave being responsible for the acceleration of particles to cosmic rays which can split heavier nuclei to the lighter ones, e.g., Li, Be, B, and in lesser extent also O, C.

When the mass of the progenitor of the core-collapse SN exceeds a certain limit, instead of a neutron star, a black hole may form, and the subsequent matter ejection will be much smaller. It is not quite clear what the limit is. It can be $\sim 40\,M_\odot$, but it can be even as small as $\sim 20\,M_\odot$. Thus, the progenitor's mass range to explode as a core-collapse SN may be quite limited, and it is essential to know it.

Low-mass stars start to lose their mass by stellar winds when reaching the asymptotic giant branch. Their inner structure is largely convective, and they enrich the ISM with elements up to oxygen and due to neutron capture process also with elements heavier than Fe. They eject about 60–80% of their initial masses and end up

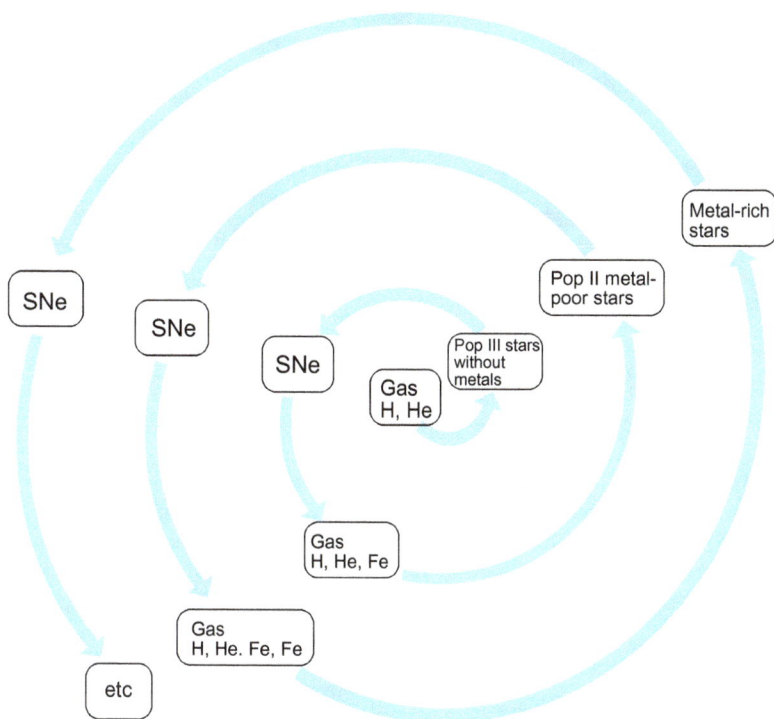

Fig. 7.25. The circle of stellar evolution and gas enrichment. Interstellar gas increasingly enriched in heavy elements is labelled "H, He, Fe" and "H, He, Fe, Fe". The symbol "SNe" designates the enrichment of gas with heavy elements in all processes: supernova explosions, neutron star collisions, and the final stages of stellar evolution by stellar winds.

as the white dwarf. Some of the end products, the accreting or merging white dwarfs, explode as thermonuclear supernovae. During the explosion, nearly all the mass will be converted into iron-peak elements, with 50% Fe, 11% Si, 10% O, etc.

An uncertainty here is how often we may expect white dwarfs to merge. Estimates are uncertain; fortunately, there is hope that the next-generation LISA gravitational wave space interferometer, which is planned to be launched in 2035, will also be able to register gravitational waves from white dwarf mergers.

At present, about 2% of the baryonic mass in the Universe is in the form of heavier elements or metals, as astronomers say. A summary of the chemical element production is seen in Fig. 7.26 where the

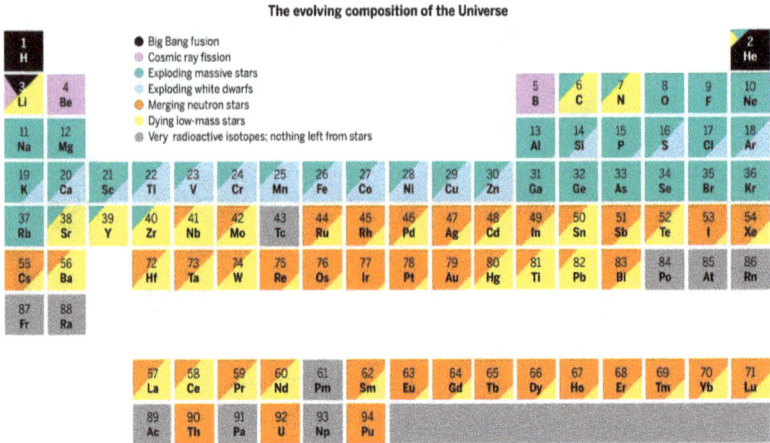

Fig. 7.26. Sources of chemical element production. The coloured areas within each element correspond to the contribution of various processes listed in the legend in the upper left region of the figure. Clearly, merging neutron stars changed our understanding of the production of heavy elements. Figure created by Jennifer Johnson, The Ohio State University, USA. *Credits*: ESA/NASA/AASNova.

relative contribution from various element synthesis sources are given (see details in the work of Johnson, 2019). The figure is composed of the time of the solar system formation, i.e. for about 4.5 Gyrs lookback time.

7.4.4   *Chemical evolution of galaxies*

Knowing the initial mass function, chemical enrichment, gas inflows and outflows, we can start to construct a chemical evolution model of a galaxy.

7.4.4.1   *System of equations of chemical evolution*

The initial mass function $\xi(m)$ (IMF), introduced in Section 7.2.3, was defined in a way that it will give us the number of stars in a unit volume, dN, with masses between m and $m + dm$, i.e.

$$\frac{dN}{dm} \sim \xi(m).$$

The normalising constant here can be calculated in several ways, depending on how we intend to use the IMF. In chemical evolution calculations, IMF is usually normalised in a way that

$$\int_{m_l}^{m_h} m\xi(m)\,dm = 1$$

(masses are measured in Solar mass units). Here, m_l is the lowest stellar mass and m_h the highest stellar mass.

Once a star with the mass m reaches its evolution endpoint after a time τ_m, a remnant is left behind: a white dwarf, a neutron star or a black hole. Let the mass of the remnant be m_r. For stars of different masses, it is different. The rest of the mass $(m - m_r)$ is rejected away into the ISM in the form of gas enriched by different metals. If we consider the formation of a single population of stars, formed at the same time $t = 0$, we can write that the fraction in mass returned back to the ISM by the time (t) is

$$R(t) = \int_{m_t}^{m_h} \xi(m)(m - m_r)dm.$$

Here, m_t is the lowest mass that ended its life by the time t. This function is zero for $t = 0$ (no stars have ended their life, or $m_t = m_h$).

The next important function is the stellar yield, Y_i, indicating the mass fraction of a star ejected to the ISM due to stellar evolution in the form of the element i. This is surely a function of the stellar mass m and can be written as

$$Y_i = \frac{m - m_r(m)}{m} Z_{i,0} + y_i(m),$$

where $Z_{i,0}$ is the mass fraction of element i being present at the time of formation of the star and y_i is the mass fraction of newly synthesised yield. For a given element, y_i can be positive or negative as the elements can be synthesised or destroyed. From the theory of stellar evolution, these functions, $Z_{i,0}$ and y_i, can be calculated with the help of knowledge of nuclear physics and stellar structure. We described this a little in the previous section.

The next important function is the star formation rate (SFR) $\psi(t)$, which is defined as how much gas mass is turned to stellar mass per unit of time, i.e. the Schmidt law (Schmidt, 1959)

$$\psi(t) \equiv \frac{d\rho_s}{dt} = k\rho_g^\alpha(t).$$

For the constants here, the value $\alpha = 1.3$–1.9 was derived by Bacchini *et al.* (2019), the value of k depends on a specific galaxy group and gas. When written for surface densities, it is called the Kennicutt–Schmidt law (Kennicutt, 1998). In the case of chemical evolution models, several approximations are usually used, such as the Kennicutt–Schmidt law for molecular gas. In SFR, it is assumed that the spatial density of newly born stars is according to the assumed IMF.

Now, we can start to write a basic system on the chemical evolution equations. Let the mass in stars be M_s, the mass in gas be M_g, the total baryonic mass is $M = M_s + M_g$, and the metallicity of the gas be Z.[26] The evolution of the mass components are (see, e.g., Ferreras, 2019, Ch. 6.3)

$$\left. \begin{aligned} \frac{dM}{dt} &= f(t) - o(t), \\ \frac{dM_s}{dt} &= \psi(t) - E(t), \\ \frac{dM_g}{dt} &= -\psi(t) + E(t) + f(t) - o(t). \end{aligned} \right\} \qquad (7.1)$$

The first equation is the time evolution of the baryonic mass, where the functions $f(t)$ and $o(t)$ are the gas inflow and outflow rates, respectively. The second equation is the stellar mass evolution. Stellar mass increases due to SFR $\psi(t)$ but decreases because, during the evolution, stars lose their mass due to stellar winds and, at the final stage, due to explosions. This evolutionary mass loss rate, or the mass return rate to the gas phase, is characterised by the function $E(t)$. The third equation concerns gas mass change. Now, the first and the second terms obviously have opposite signs compared with the previous equation, and gas inflows and outflows are also added.

Let us look at the function $E(t)$. At some time t_*, the stellar total mass increases with the rate $\psi(t_*)$. We need to split this stellar mass change into stars having different masses, according to IMF, i.e. instead of using simply $\psi(t_*)$, we have

$$E(t) \rightarrow \int_{m_l}^{m_h} \psi(t_*) m \, \xi(m) \, dm.$$

---

[26] Z is usually understood as the contribution of elements heavier than He, weighted by mass.

This concerns star formation time t_*. We are interested in (i) the present time t and (ii) only these stars which have ended their life by the time t. This means that, first, the function ψ should be taken at the argument $(t - \tau_m)$ and second, in the integral, the lower limit must be m_t. Thus, we have

$$E(t) \to \int_{m_t}^{m_h} \psi(t - \tau_m) m \, \xi(m) \, dm.$$

Now, the last step. To have the function $E(t)$, we are not interested in stellar masses but the mass they eject in the form of the gas to the ISM. Thus, instead of m, we need to write $(m - m_r)$. Now, we have the expression to the function $E(t)$

$$E(t) = \int_{m_t}^{m_h} \psi(t - \tau_m)(m - m_r) \, \xi(m) \, dm.$$

The evolution of the metal content is similar to the third gas equation

$$\frac{d(Z M_g)}{dt} = -Z\psi(t) + E_Z(t) + Z_f f(t) - Z o(t),$$

with the corresponding mass return rate

$$E_Z(t) = \int_{m_t}^{m_h} \psi(t - \tau_m) \left[(m - m_r) Z(t - \tau_m) + m y_i \right] \xi(m) \, dm.$$

In square parenthesis, there is, in fact, the stellar yield for the total metallicity Z. The equation can also be written for individual elements i.

The solution of the system of equations (7.1) to the full extent is very complicated undoubtedly. Therefore, in many cases, simplifications are used. In the instantaneous recycling approximation, it is assumed that the return of metals into the ISM takes place just at the moment when the star is born, i.e. no evolution time delay is considered. Indeed, when one is interested in the overall evolution of an old galaxy, the evolution time of a considerable amount of stars is less than the age of the galaxy, and one can ignore the metal production time delay. In this approximation, $\tau_m = 0$ and $m_t = m_l$ and the function $E(t)$ simply is

$$E(t) = \psi(t) \int_{m_l}^{m_h} (m - m_r)\,\xi(m)\,dm.$$

The integral can be calculated separately; we introduced it earlier in this section as $R(t)$. The second evolution equation is now in a simple form

$$\frac{dM_s}{dt} = (1 - R)\psi(t).$$

This is an example of how the equations simplify.

An extreme simplification is the closed box model with no inflow and outflow and with $M_s = Z_i = 0$ at $t = 0$. In addition, the instantaneous recycling approximation is added to this model.

7.4.4.2 *Simulation of stellar populations*

When the aim is to calculate the time evolution of the integrated light of a galaxy or a part of the galaxy results of the detailed models described above should be integrated over stars. These models are called stellar population synthesis models.

As a starting point, let us look at a simple stellar population (SSP) implying that we assume all the stars formed simultaneously as an instantaneous burst of star formation and with the same metallicity Z. The metallicity can be handled as a free parameter. Calculations are usually made for a few different values of Z.

The monochromatic flux of an SSP at a time t is integral to the fluxes of all stars, i.e. integral over stellar masses m

$$F_\lambda(t, Z) = \int_{m_l}^{m_h} F_\lambda(m, t, Z)\,\xi(m)\,dm.$$

The whole spectrum and, hence, all the colour indices we are interested in can be calculated then easily. Thus, we need to know monochromatic fluxes as a function of stellar masses at different times and metallicities $F_\lambda(m, t, Z)$ or formulated in a different way; we need to know the isochrones and the library of stellar spectra.

The term "isochrone" means that time is held to be a constant. In stellar evolution, isochrones are curves in Hertzsprung–Russell (HR) diagram joining the positions of stars with different masses. These curves are calculated theoretically on the basis of stellar structure models to the best of the our knowledge. An output of stellar structure models is a library of stellar evolution tracks which can be converted to isochrones. Calculation of stellar structure models is a computationally time-consuming task and, unfortunately, we may not know, even theoretically, all the aspects of stellar evolution. Due to this, several isochrone libraries exist.

Two approaches are used to have a library of stellar spectra (and, as usual, their combinations). One approach is to have a large collection of observed stellar spectra, the second approach is to calculate stellar spectra from theoretical stellar structure models. Both have their advances and limitations. Observed spectra are, as a rule, obviously the correct spectra, and this is an advantage. A limitation is that we do not have them in a sufficient number for all masses and spectral regions (for example, spectra of faint low-mass stars in the near-IR region demand satellite observations). In the case of theoretical spectra, an advantage is that we can calculate them for all wavelengths; a limitation is that our theoretical models may not be sufficiently correct.

But let us assume that we have the library of isochrones and the library of stellar spectra. In this case, we can fix t and Z, and from the library of isochrones, we can see all the evolutionary stages of stars with given masses. Then we take their spectra $F(m, t, Z)$ from the library of stellar spectra and calculate the spectra of the whole SSP $F(t, Z)$. An example of SSP spectra calculated by Bruzual & Charlot (2003) is seen in Fig. 7.27. Calculated time evolution of colour indices and mass-to-light ratios for different metallicities are given in Fig. 7.28.

A step further from the SSP is to model composite stellar populations. Here, stars are born at different times and can have different metallicities. The first change is characterised by introducing a time-dependent star formation rate $\psi(t)$, the second change by simply a time-dependent metallicity $Z(t)$. Metallicity is not anymore simply a parameter. The monochromatic flux at a time t is now an integral over SSP parameters, more precisely, over metallicity and again time. The metallicity is not simply a function of time, but a more complicated function as at every moment, there can exist simultaneously different metallicities. If we denote the metallicity

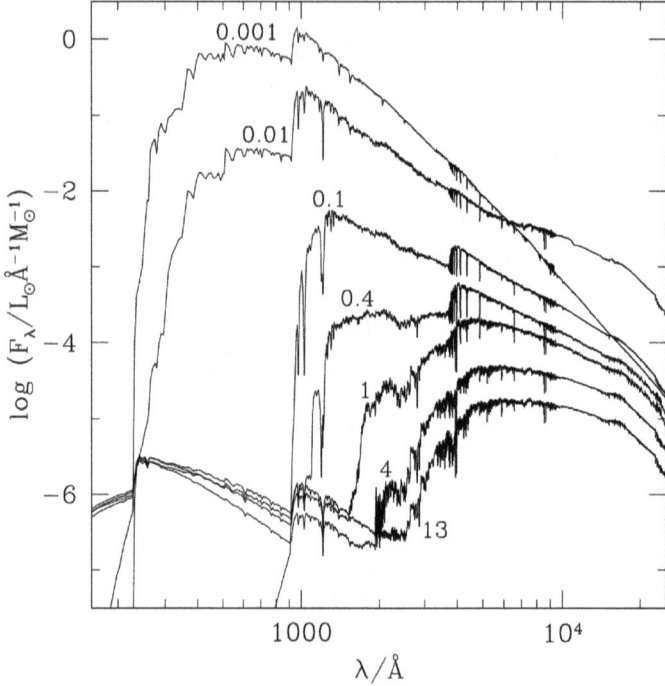

Fig. 7.27. Spectral evolution of the standard SSP model for different ages. Solar metallicity is assumed (Bruzual & Charlot, 2003).

function as $P(Z,t)$, we may write

$$F_\lambda(t) = \int_0^t dt_* \int_0^{Z_m} dZ \, F_{\lambda,SSP}(t_*,Z) \, \psi(t - t_*) \, P(Z, t - t_*).$$

Here, t_* is the age of SSP, i.e. $t_* = 0$ means zero age or current time. Integration over t_* from 0 to t with function arguments $(t - t_*)$ mean that we integrate from the beginning of star formation to the end of star formation. The expression can be simplified by assuming still a fixed constant metallicity. In this case

$$F_\lambda(t, Z) = \int_0^t dt_* \, F_{\lambda,SSP}(t_*,Z) \, \psi(t - t_*).$$

Conventional assumptions for SFR is exponentially declining SFR and delayed SFR.

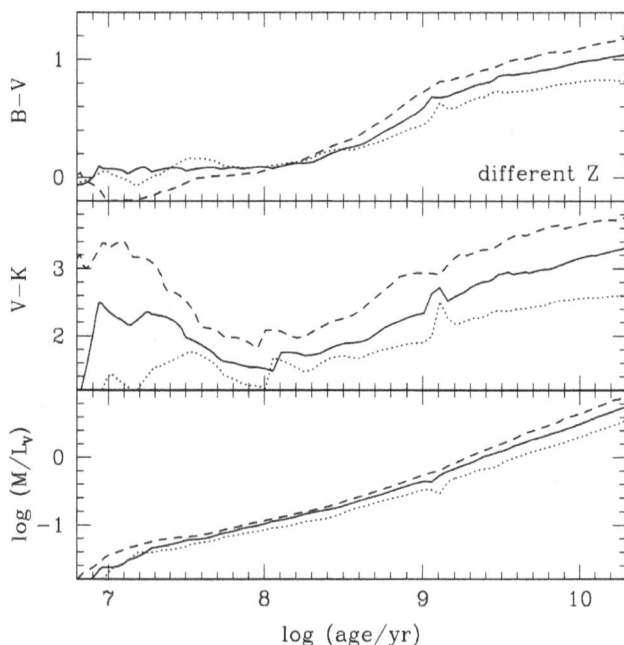

Fig. 7.28. Evolution of $B - V$ and $V - K$ colours and stellar mass-to-light ratio M/L_V of SPP for different metallicities, $Z = 0.004$ (dotted line), the solar metallicity $Z = 0.02$ (solid line) and $Z = 0.05$ (dashed line) (Bruzual & Charlot, 2003).

In most advanced models all the aspects of chemical evolution discussed above should be taken into account – gas inflows and outflows, different aspects and possible bursts of star formation, abundances of different elements, etc.

7.4.4.3 *Comparison of simulated populations with observations*

Some additional aspects should be taken into account when comparing the results of calculations with observations.

First, we like to mention dust extinction.[27] Extinction due to dust absorption in our Galaxy is quite simple to take into account.

----

[27] Although the terms of extinction and attenuation are sometimes used interchangeably, they are different terms, see Salim & Narayanan (2020).

There are standard absorption maps, and for known celestial coordinates, these maps give us the amount of absorption at various wavelengths. Perhaps, most often, the Schlegel *et al.* (1998) and Schlafly & Finkbeiner (2011) maps are used. It is more complicated to consider the intrinsic absorption, or attenuation, in a galaxy, see a review by Salim & Narayanan (2020). For this, we need to know how dust and stars are distributed in a galaxy. As a rule, one can assume the dust to be a thin layer at the galactic symmetry plane, although this is, again, clearly a simplification. For a single galaxy, it is possible to analyse the distribution of certain objects/stars over the visible image of the galaxy and, in this way, to estimate the intrinsic extinction (see, e.g., Tempel *et al.*, 2010). For statistical analysis of a large number of objects, there are only approximate empirical relations: attenuation as a function of galactic stellar mass, specific SFR, axis ratio, metallicity, etc. (see Salim & Narayanan, 2020).

Second, the so-called K and E corrections. These corrections are important when we need to compare the spectra of distant galaxies with the models. Due to the expansion of the Universe, emitted from the distant universe, photons are redshifted. In fact, the whole spectrum of a distant galaxy is redshifted and it is necessary to correct the observed spectrum from the redshift or to apply the redshift to our computed spectrum. This is called the K correction. E correction is related to the time evolution of the observed object and its spectrum. The spectrum emitted by a high-redshift galaxy is emitted a long time ago. We need to take into account that the time t in all previous formulae is a time in past. Even in cases with no obvious star formation, we need to take into account the passive evolution of stellar populations.

7.5 Supermassive black holes in galactic centres

7.5.1 *Direct observations of the supermassive black holes in the centres of galaxies*

Supermassive black holes (SMBHs) at the centres of galaxies are now recognised as the normal components of galaxies. The most direct evidence for the presence of SMBHs at the centres of galaxies was found in the giant elliptical galaxy M87 and in our Milky Way

April 11

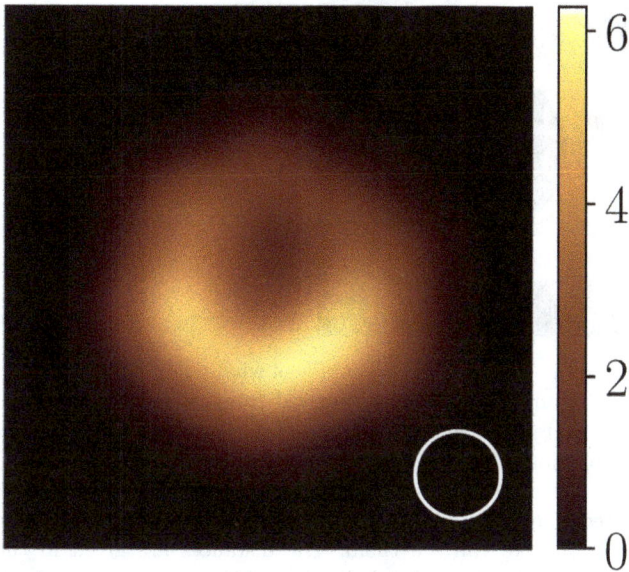

Fig. 7.29. Image of the photon capture region around the SMBH at the centre of giant elliptical galaxy M87 derived with the Event Horizon Telescope. Colour code is the brightness temperature in units of 10^9 K, and the white circle in the lower-right corner is the beam size. The size of the picture is about 120 μas. (Event Horizon Telescope Collaboration *et al.*, 2019). CC BY 3.0.

galaxy. It was a public-wide press conference event when, in 2019, the Event Horizon Telescope team announced the direct observations of registered emission near the event horizon in the galaxy M87. Simultaneous measurements were done at a wavelength of 1.3 mm, with several telescopes being away from each other up to about 10 000 km and working as interferometers. That was nearly an impossible mission to fit the individual wavefronts from such distant telescopes. But it was done, and the famous images were derived, see, e.g., Fig. 7.29.

Initially, there were two candidates for the observations: the central black holes of the Milky Way and the galaxy M87. For both of them, nearly the same angular resolution was necessary, so in this context, there was no difference. For the first object, M87 was selected because the emission from the centre of the MW was expected to be somewhat more time-variable.

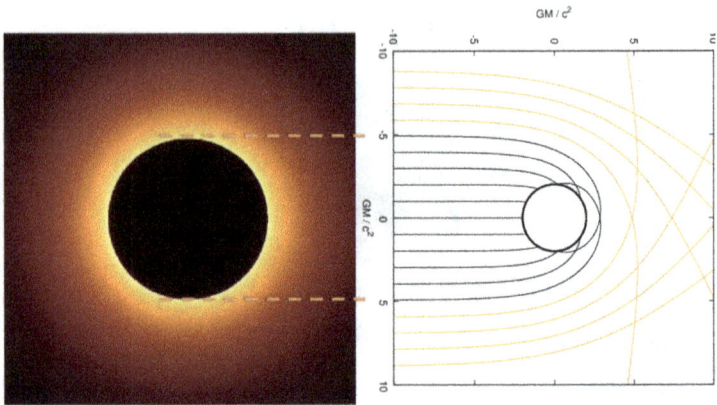

Fig. 7.30. In the left panel is an example of what a non-rotating black hole looks like, and this is an idealised image derived from the Event Horizon Telescope. Shadow is created at the photon capture radius $R_c = \frac{3\sqrt{3}}{2} R_S$. In the right-hand panel, orbits of photons coming from infinity from the left side with various impact parameters are given. Photons with black trajectories are absorbed by the black hole (captured photons), while photons with orange trajectories distances are not captured, and we can see them. The distances are given in the so-called gravitational radius units, being half of the Schwarzschild radius. The central black ring corresponds to the Schwarzschild radius.

Source: OdysseyEdu.worldpress.com.

The radius of the emission ring in Fig. 7.29 is 21 μas corresponding at the distance of the galaxy to 5.2×10^{10} km. The ring corresponds to the 1.3-mm synchrotron radiation from photons at the so-called photon capture radius R_c of the black hole being directly related to the Schwarzschild radius R_S of the black hole $R_c = \frac{3\sqrt{3}}{2} R_S$ (see Fig. 7.30). From this formula, we derive that $R_S = 2 \times 10^{10}$ km giving for the mass of the black hole $M_\bullet = 6.77 \times 10^9 M_\odot$. Precise modelling gives the value $(6.6 \pm 0.9) \times 10^9 M_\odot$. This is the first direct observation of the "shadow" of a black hole. From modelling the asymmetry of the brightness between the lower and upper parts of the ring, it was possible to derive constraints to the rotation of the black hole.

Soon, a similar study was carried out for the SMBH at the centre of the Milky Way galaxy (Event Horizon Telescope Collaboration *et al.*, 2022), see one of the derived images in Fig. 7.31. The emission ring has a radius of 26 μas, giving the black hole mass $M_\bullet = (4.0^{+1.1}_{-0.6}) \times 10^6 M_\odot$.

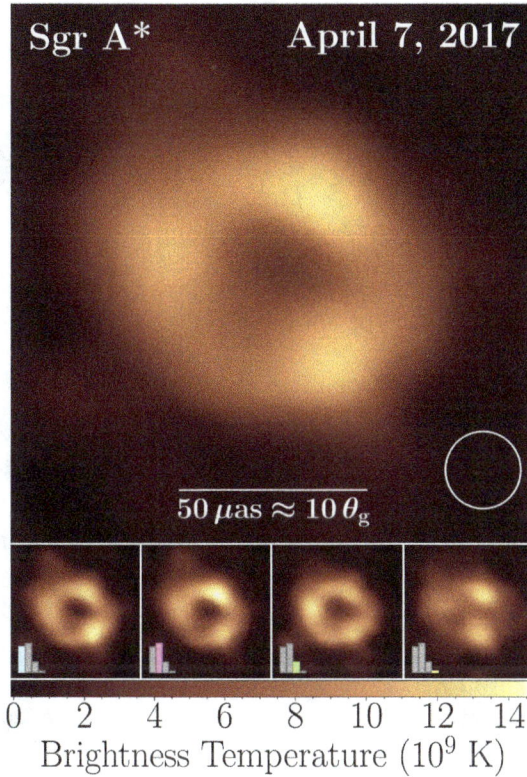

Fig. 7.31. Image of the photon capture region around the SMBH at the centre of the Milky Way galaxy derived with the Event Horizon Telescope. Colour code is the brightness temperature in units of 10^9 K, and the white circle in the lower-right corner is the beam size (Event Horizon Telescope Collaboration *et al.*, 2022). CC By 4.0.

The value of the SMBH mass at the centre of the MW was not a surprise. Within the statistical errors, it coincided well with the value calculated from the orbits of several stars. A compact star cluster surrounds the central SMBH of MW, and the motions of these stars around the SMBH can be used to determine the mass of the BH. Eisenhauer *et al.* (2005) and Genzel *et al.* (2010) used adaptive optics near-IR observations with the ESO VLT Telescope to study motions of stars around the central source Sgr A*. Authors combined the new radial velocities with astrometry to derive three-dimensional stellar orbits for six of these "S stars" in the central 0.5 arcsec, see Fig. 7.32,

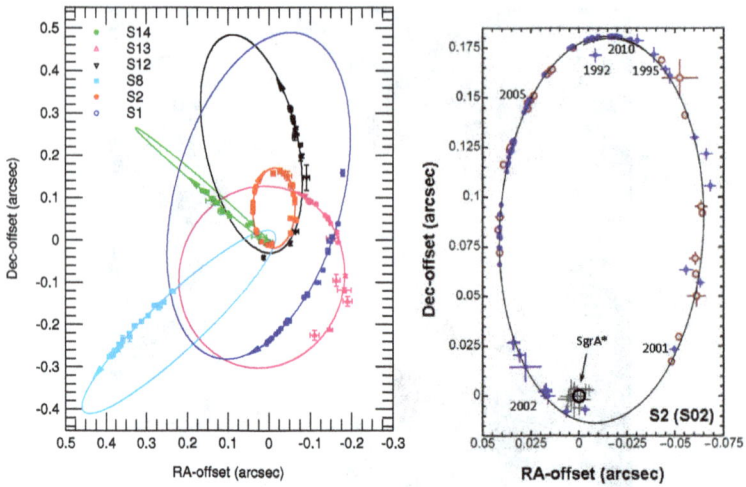

Fig. 7.32. Left: Projection on the sky of the six stars near the central black hole. The various colour curves are the result of the best global fit to the spatial and radial velocity data of S1, S2, S8, S12, S13, and S14 (Eisenhauer *et al.*, 2005). Right: Orbit of the star S2 (S02) on the sky. Blue, filled circles are the ESO NTT/VLT observations, red circles are from Keck data. It is seen that Sgr A* is not at the focus of the ellipse. The difference between the location of the focus of the apparent ellipse and the true focus allows us to determine the spatial orientation of the orbit and the central BH mass (Genzel *et al.*, 2010). Reprinted figure with permission from Genzel *et al.*, 2010. Copyright 2010 from American Physical Society.

left panel. The authors updated the estimate of the distance to the Galactic centre from the S2 orbit fit, $R_0 = 7.62 \pm 0.32\,\text{kpc}$, which yields a central mass value of $(3.61 \pm 0.32) \times 10^6\,M_\odot$. An independent determination of the mass of the MW central SMBH was made by Ghez *et al.* (2008), who used observations with the Keck 10-m Telescope to independently measure the positions of these stars. The most valuable star in the study of both groups was S2 (Ghez group designated it as S02), revolving with a period of 15.8 years, allowing us to follow the whole orbital period. The pericentre of the orbit of S2 is only about 125 au $= 1400\,R_S$ (see Fig. 7.32 right panel). Fits of the Keplerian orbit model to these datasets result in estimates of the black hole's mass, $(4.1 \pm 0.6) \times 10^6\,M_\odot$, and distance of $8.0 \pm 0.6\,\text{kpc}$, in good agreement with earlier estimates. Quite justly due to this study, the leaders of both teams, Reinhard Genzel and Andrea Ghez, were awarded the Nobel Prize in Physics in 2020.

7.5.2 Statistical properties of SMBHs

In addition to the two galaxies referred above, there are indirect hints that at the centres of many galaxies, there are very localised mass concentrations believed to be SMBHs. The Hubble Space Telescope revolutionised SMBH research and it was found that stellar and/or gas velocity dispersions at the centres of many galaxies increase; it was possible to model such an increase by assuming that there is a point mass at the centre of these galaxies. The number of these galaxies is sufficient to study possible statistical correlations related with the masses of the SMBHs.

7.5.2.1 SMBH mass as a function of $M_{\rm sph}$

First and perhaps most important was the discovery of a tight correlation between SMBH mass $M_\bullet$ and the velocity dispersion σ of the spheroidal (the bulge) component of the host galaxy. Together with similar correlations with bulge luminosity and mass, this led to the belief that SMBHs and bulges coevolve by regulating each other's growth. Gebhardt *et al.* (2000) used a sample of 26 galaxies to find a correlation between SMBH masses, $M_\bullet$, and the luminosity of the bulge, L_B, or luminosity-weighted aperture effective star velocity dispersion, σ_e. The sample includes our Galaxy (MW), Andromeda galaxy M31 and its companion M32, and central galaxies of several groups and clusters, including the central galaxy M87 of the Virgo cluster. Figure 7.33 shows these correlations. Green squares denote galaxies with maser detections, red triangles are from gas kinematics, and blue circles are from stellar kinematics. Solid and dotted lines are the best-fit correlations and their 68% confidence bands. The correlation with the velocity dispersion is very clear. In this dataset, MW, M32 and NGC 7457 have the smallest dispersions, $\sigma_e = 67-75\,\mathrm{km/s}$, and SMBH masses, $M_\bullet = (3-4)\times 10^6\,\mathrm{M_\odot}$. The largest SMBH masses are about $M_\bullet \sim 2\times 10^9\,\mathrm{M_\odot}$. Thus, the relation covers nearly three orders of magnitude in SMBH masses. The correlation remains nearly unchanged in later studies; see a review by Graham (2016).

During subsequent years, the dataset was enlarged, and the sample by Kormendy & Ho (2013) contained nearly 90 galaxies with direct BH mass determinations already. The tight correlation between SMBH mass and velocity dispersion strongly suggests a

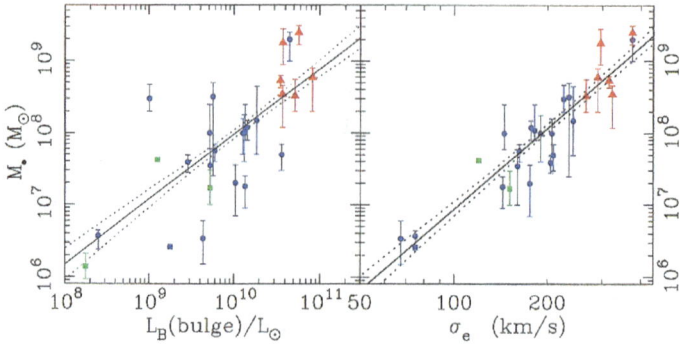

Fig. 7.33. Black hole mass versus bulge luminosity (left) and the luminosity weighted aperture dispersion within the effective radius (right) (Gebhardt *et al.*, 2000).

causal connection between the formation and evolution of the central black holes and bulges. A causal connection between the AGNs and SMBHs was indicated already by Rees (1984a). Thus, it is natural to assume that bulges, black holes, and quasars formed, grew, or (in case of quasars) turned on as parts of the same process due to the collapse or merger of smaller systems, bulges in particular. This process can provide a rich fuel supply to a centrally located black hole.

Next, by using a careful decomposition of the near-IR surface brightness distribution of a sample of 66 galaxies and avoiding pseudobulges, the correlation between the SMBH and the bulge mass was studied by Savorgnan *et al.* (2016), see Fig. 7.34. In this analysis, two different relations were derived between the central BH masses $M_\bullet$ and the stellar masses of spheroids $M_{\rm sph}$. The bulges of early-type galaxies follow $M_\bullet \sim M_{\rm sph}^{1.04\pm0.10}$ relation consistent with a dry-merger formation scenario. The bulges of late-type galaxies follow a steeper $M_\bullet \sim M_{\rm sph}^{2-3}$ relation, indicating that gas-rich processes feed the BHs more efficiently than the host bulge growth in stellar mass. In a later study, Davis *et al.* (2019a) derived, on average, power 2.45 for late-type galaxies.

7.5.2.2 $M_{\rm BH}$ as a function of stellar masses of galaxies

Heckman & Best (2014) studied the coevolution of the stellar masses of galaxies and SMBH masses in the nearby universe. Black hole

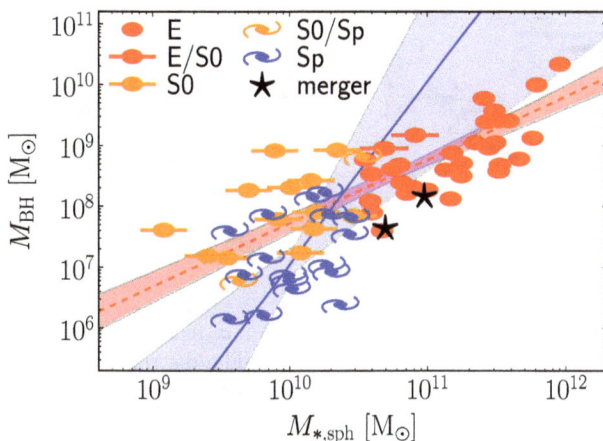

Fig. 7.34. SMBH masses plotted against spheroid stellar masses. Symbols are coded according to the galaxy morphological type (see the legend). The red dashed line indicates the linear regression for the bulges of 45 early-type galaxies, the solid blue line shows a regression for the bulges of 17 late-type galaxies. Shaded areas correspond to $1\,\sigma$ uncertainties (Savorgnan *et al.*, 2016).

masses were estimated for Type 1 AGNs from velocity dispersions of Broad Line Regions and for Type 2 AGNs directly through the dynamical modelling of spatially resolved kinematics of stars or gas. First, again, in the left panel of Fig. 7.35, the relation between $M_{\bullet}$ and velocity dispersions σ is given. Both AGNs and quiescent galaxies are consistent with the relation derived earlier (shown by the solid line). This relation was then used to study the correlation with the whole stellar masses of galaxies in the SDSS main sample: the right panel of Fig. 7.35 (as mentioned, $M_{\bullet}$ were derived from the $M_{\bullet}$ - σ relation, shown in the left panel). The greyscale indicates the volume-weighted distribution of all galaxies, with each lighter colour band indicating a factor of 2 increases. It is clearly seen that the black hole mass is not a fixed fraction of the total stellar mass — the spread is quite wide. This is also true for AGNs: the blue and red contours show the volume-weighted distributions of high ($\geq 1\%$; mostly radiative mode) and low ($<1\%$; mostly jet mode) Eddington-fraction AGN, with contours spaced by a factor of 2.

As there was a tight correlation for the bulges but clearly larger scatter for the total stellar mass, it is natural to ask: Is the correlation valid also for disks? Kormendy *et al.* (2011) analysed this question

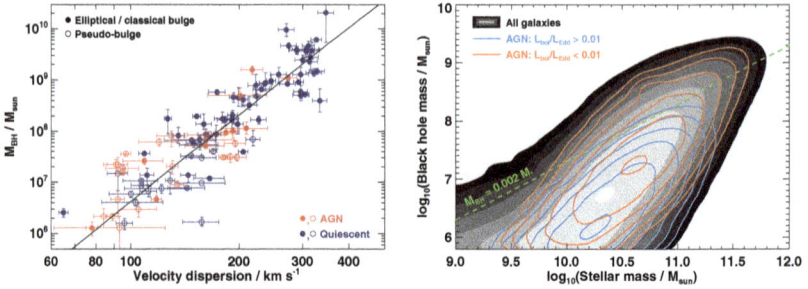

Fig. 7.35. Left: The black hole mass M_{BH}–σ relation of local galaxies with direct black hole mass measurements. Right: The distribution of galaxies in the SDSS main galaxy sample on the stellar mass versus black hole mass plane (Heckman & Best, 2014). Used with permission of *Annual Reviews* from Heckman and Best, 2014, Permission conveyed through Copyright Clearence Center, Inc.

and found that SMBHs do not correlate with galaxy disks. This can be a reason for the significant scatter in the case of the total stellar masses of SDSS galaxies.

7.5.2.3 *Black hole masses as a function of M_{DM}*

Now, let us describe the correlation between masses of DM haloes and SMBHs. Usually, in these studies, the maximum rotational velocity v_{max} is used as a measure of the DM halo's mass. Correlation studies for the possible $M_{\bullet}$–v_{max} relations have been done for more than two decades with various galaxy samples. The published results are quite diverse (see Graham, 2016; Davis *et al.*, 2019a). For example, Kormendy *et al.* (2011) came to the conclusion that the masses of SMBHs do not correlate with the masses of DM haloes surrounding galaxies: masses of DM haloes determine the rotation velocity in the outer, flattened parts of galaxies, but masses of SMBHs determine near-central velocity dispersions. On the other hand, in several studies, a correlation was found with slopes from shallow to steep.

Davis *et al.* (2019b), using the sample of 48 spiral galaxies with directly measured BH masses, studied the correlation between the black hole masses and v_{max} (and other correlations) and derived quite a tight relation, steeper than the previously reported one, $M_{\bullet} \sim v_{max}^{10.62\pm1.37} \sim M_{DM}^{4.35\pm0.66}$, see Fig. 7.36. As the authors mention, the reason why, in earlier studies, a shallower relation was found is the assumption that the relation between $M_{\bullet}$ and $M_{*,tot}$ was assumed nearly linear. An approximately linear relation makes

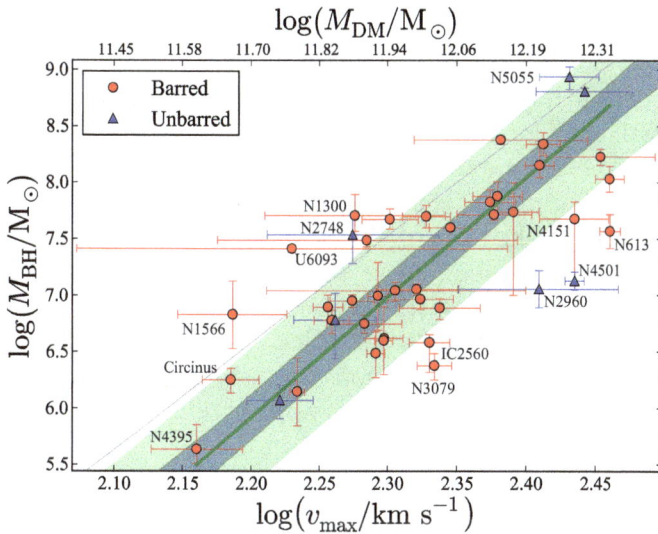

Fig. 7.36. Black hole mass versus maximum rotational velocity (and DM halo mass) for 42 galaxies (Davis *et al.*, 2019b).

sense for galaxies formed from the result of many dry mergers and, thus, for early-type galaxies. For spiral galaxies, this is clearly not the case. A similar correlation for 25 spiral galaxies was found by Smith *et al.* (2021). Although Marasco *et al.* (2021) confirmed in their study a clear correlation between the $M_\bullet - M_{DM}$, they found that the slope in this relation is not uniform but shows break at $M_{DM} \sim 10^{12}\,M_\odot$ or $M_\bullet \sim 10^7 - 10^8\,M_\odot$. The authors mention that masses of DM haloes were calculated more carefully than in previous studies, using detailed modelling of rotation curves or kinematics of globular cluster subsystems. For the region of DM halo, masses studied by Davis *et al.* (2019b); Smith *et al.* (2021) also Marasco *et al.* (2021) derived quite steep relation. At higher masses, the relation becomes shallower.

Steep relation $M_\bullet$ vs M_{DM} implies that BHs grow much more rapidly relative to their hosts. Again, a possible reason is additional gas mass accretion. In addition, it means that the rôle of DM haloes is quite significant, influencing most of the observable properties of hosts.

The results of correlations and reasons of such correlations are not clear. To summarise, "the fundamental physical connection between black hole and bulge growth still awaits discovery" (Graham, 2016).

7.5.2.4   *Summary of statistics*

To understand the BH–galaxy coevolution, the relationship between
DM and stellar masses of galaxies is to be understood. This relation
was investigated by Kormendy & Ho (2013) and is summarised in
Fig. 7.37. The left panel shows with black points the field galaxy
baryonic mass function fitted by solid curve with the Schechter
(1976). Upper dashed curve shows the mass function of cold DM
haloes, and lower dashed curve shows the expected baryonic mass
function for the cosmological baryon fraction of $\sim 1/6$. We see that
the baryon content of galaxies (black points) is lower than the
expected function. The deviations are larger in the low-mass and
high-mass regions. The largest baryon fraction is in haloes of mass
$M_{\mathrm{DM}} \sim 10^{12}\,\mathrm{M}_\odot$. This is clearly seen in the right panel of Fig. 7.37,
which shows directly the baryon fraction $M_\star/M_{\mathrm{DM}}$ as function of the
halo DM mass M_{DM}.

   The reason of the lower baryonic content in lower and higher
masses (again the left panel) are probably stellar winds and AGN
feedback: smaller galaxies miss more baryons because they were
ejected by supernova (SN) driven winds (Dekel & Silk, 1986), in high-
mass region (from $M_{\mathrm{DM}} \sim 10^{12}\,\mathrm{M}_\odot$), baryon content is progressively
lower as more baryons are kept suspended in hot gas by a combi-
nation of AGN feedback and cosmological gas infall. We described

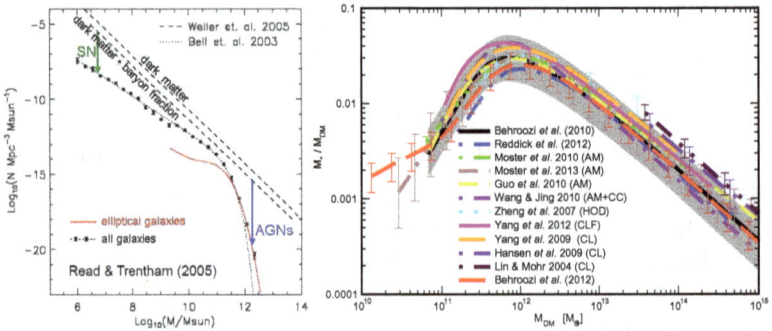

Fig. 7.37.   Left: The total field galaxy baryonic mass function (black points)
and Schechter function fit (upper dotted curve) compared to the mass spectrum of
cold DM haloes from numerical simulations, and DM mass function multiplied by
the universal baryon fraction of 0.163 (lower dashed curve). Right: The baryon
fraction $M_\star/M_{\mathrm{DM}}$ as function of the halo DM mass M_{DM} (Kormendy & Ho,
2013). Used with permission of *Annual Reviews* from Kormendy & Ho, 2013.
Permission conveyed through Copyright Clearence Center, Inc.

these processes earlier in this chapter but not in the context of black hole growth.

7.5.3 *A pathway from black holes to SMBHs*

Inflows and outflows of gas, being natural processes in galaxy evolution, also contribute to the growth of BHs. In a review paper about the assembly of the first massive black holes, Inayoshi *et al.* (2020) referred to four possible formation pathways of the initial seed BHs. These pathways are illustrated in Fig. 7.38.

The pathways are as follows:

(1) BHs with masses $M_{seed} \approx 10^{1-2} \, M_\odot$ formed as the remnants of massive Pop III stars after they end their life and grow via mergers.
(2) BHs of masses $M_{seed} \approx 10^{5-6} \, M_\odot$ formed in cooling gaseous haloes with virial temperatures $T_{vir} \sim 10^4 \, K$ under peculiar conditions such as strong 11.2–13.6 eV radiation field, suppressing H_2 cooling, and/or rapid merger history of DM haloes, and/or high-velocity collisions of baryon-DM haloes.
(3) BHs with masses $M_{seed} \approx 10^{3-4} \, M_\odot$ may form promptly through stellar or smaller BH mergers in the cores of dense stellar clusters.
(4) Stellar mass BHs formed at the centres of dense pristine gas clouds as in the case of the scenario (1) and can grow via

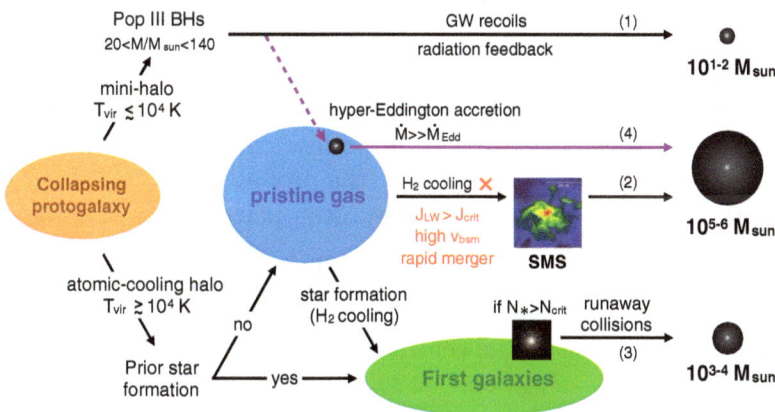

Fig. 7.38. Formation pathways of seed BHs in early protogalaxies (Inayoshi *et al.*, 2020). Used with permission of *Annual Reviews* from Inayoshi *et al.*, 2020. Permission conveyed through Copyright Clearence Center, Inc.

hyper-Eddington accretion ($\dot{M}_{\text{seed}} \gg \dot{M}_{\text{Edd}}$) reaching masses of $M_{\text{seed}} \approx 10^{5-6} \, \text{M}_\odot$.

The initial mass function of BH seeds probably covers the whole range of masses from $M_{\text{seed}} \approx 10 \, \text{M}_\odot$ to $10^6 \, \text{M}_\odot$ with a maximum at smaller masses $\approx 10^{1-2} \, \text{M}_\odot$.

Whatever the origin of the seeds of SMBHs would be, it is necessary to grow their masses by many orders of magnitude. Thus, one must ask whether there is an Eddington limit for accretion (Inayoshi *et al.*, 2020). When the gas accretes onto the BH, a lot of radiation is emitted. In fact, the BH accretion is perhaps the most efficient energy creation mechanism: in gas accretion, about 10–30% of the rest-mass energy of the gas is emitted via radiation, in the case of nuclear reactions, less than 1 % of the rest mass energy is released. Emitted photons scatter from electrons creating the outward radiation pressure.

The Eddington limiting luminosity is defined as the situation when the outward radiation pressure balances the inward gravitation force. Outward energy flux from a point source with the luminosity L at radius r is $F = L/(4\pi r^2)$. As the relation between the momentum and energy of the photon is $p = E/c$, the momentum flux or radiation pressure is $P_{\text{rad}} = L/(4\pi r^2 c)$. The force the electron feels from the radiation is the radiation pressure multiplied by the scattering Thomson cross-section σ_e:

$$F_{\text{rad}} = \frac{\sigma_e L}{4\pi r^2 c}.$$

Protons also feel radiation pressure, but the scattering cross-section for protons is much smaller, and we may ignore it. However, protons and electrons are electromagnetically tied, and when calculating the inward gravitational force, only the proton's larger mass counts. Gravitational force to the electron–proton pair is

$$F_{\text{grav}} \simeq \frac{GMm_p}{r^2},$$

where M is the mass of the central body (BH). For the limiting case of stability, $F_{\text{grav}} = F_{\text{rad}}$, and we have

$$L = \frac{4\pi c G M m_p}{\sigma_e}.$$

This is the Eddington luminosity $L_{\rm Edd}$. Let the efficiency of the accretion be η. In this case, if the mass accretion rate is $\dot{m}$, the luminosity created by this is $L = \eta \dot{m} c^2$. For the Eddington accretion rate, we have the BH mass increase rate

$$\frac{dM}{dt} = (1 - \eta)\dot{m} = \frac{(1 - \eta)L_E}{\eta c^2} = \frac{(1 - \eta)}{\eta}\frac{4\pi G m_p}{\eta \sigma_e c}M.$$

This is a simple equation giving after the integration the accretion time to reach from a seed mass $M_{\rm seed}$ to a final mass $M_{\bullet}$:

$$t = A \ln\left(\frac{M_{\bullet}}{M_{\rm seed}}\right),$$

where A takes together all the constants in the previous formula and for $\eta = 0.1$ is 5.03×10^7 yr.[28] The time to grow from $M_{\rm seed} = 100\,M_{\odot}$ to $M_{\bullet} = 10^9\,M_{\odot}$ will be 0.8 Gyr. Thus, there will be a question as to, how it is possible for BHs to reach the masses $\sim 10^9\,M_{\odot}$ by the time of redshifts $z \sim 6$–7 when the age of the Universe is nearly the same value or even less? Although we know now that there are large gas reservoirs outside the visible dimensions of galaxies directing gas to the central regions of galaxies as cold streams, sustaining continuous Eddington level accretion is challenging over a long time. A natural question arises: Can the accretion exceed the Eddington limit significantly?

First, observations seem to support it — the examples of SS 433, super-luminous stars like η Carinae, etc. To have a super-Eddington accretion, it is argued that radiation photons can be trapped in slim but optically thick accreting matter disk, and much of them advected back to the BH. This may happen if the gas inflow speed is faster than the outward photon diffusion speed. As a result, the radiative efficiency η drops and BH masses grow faster. High-resolution simulations suggest that, indeed, radiation may decrease due to the formation of high-density gas clumps; rapid gas accretion occurs in the disk region, and the radiation emerges towards the polar conical regions. Although accretion details depend on gas geometry and super-Eddington accretion does not occur continuously but in a series

---

[28] Here, we equalised the number of electrons and protons, i.e. ignored the helium content of the gas. But the correction would be small.

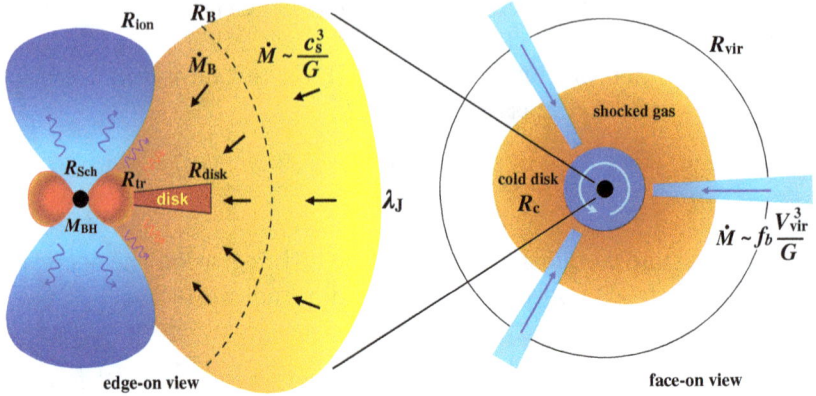

Fig. 7.39. Schematic illustration of the accretion flow onto a massive BH with a mass of $M_\star$ at a rate significantly exceeding $\dot{M}_{\mathrm{Edd}}$ (edge-on view, left) and on the early protogalaxy that hosts the accreting BH (face-on view; right) (Inayoshi *et al.*, 2020). Used with permission of *Annual Reviews* from Inayoshi *et al.*, 2020. Permission conveyed through Copyright Clearence Center, Inc.

of short periods, it is often sufficient to explain the necessary BH growth rate in several cases. A schematic illustration of the hyper-Eddington accretion onto a massive BH is shown in Fig. 7.39.

In the last paragraph of the review article, Inayoshi *et al.* (2020) concluded: *In order to probe the specific seed models, it will be necessary to detect BHs with masses of $\leq 10^5\,\mathrm{M}_\odot$ at redshifts $z \geq 10$, because the newly born BHs will likely lose the memory of their birth by the time they grow well above $\sim 10^5\,\mathrm{M}_\odot$ and are incorporated into more massive host galaxies.*

JWST surveys of several fields in 2023–2024 have revealed a surprisingly large number of red point-like sources called as 'little red dots' (LRDs) (see, e.g., Gentile *et al.*, 2024; Greene *et al.*, 2024; Kokorev *et al.*, 2024; Matthee *et al.*, 2024), see Fig. 7.40. These objects have redshifts $9 < z < 4$, they are compact and most of them have broad H_α and H_β lines with widths of 1200–3700 km/s. The data indicate that they can be dust-reddened AGNs with the central BH masses of $M_\bullet \sim 10^5 - 10^9\,\mathrm{M}_\odot$. Spectral properties of these compact sources can be explained by super-Eddington accretion. Although a larger statistics of LRDs are needed (as always!) even at present, a surprising finding is a relatively large number of these objects: the number of LRDs exceeds about 100 times the number of comparable faintest UV-selected quasars. Still, the number density of

Fig. 7.40. False colour JWST images of three Balmer Hα emitters at $z \sim 5$ little red galaxies (LRDs) from the study by Matthee *et al.* (2024, Fig. 1). Balmer Hα emitters stand out as red point sources (Matthee *et al.*, 2024).

LRDs is approximately consistent with the major merger statistics (Greene *et al.*, 2024), indicating the relationship between AGN triggering, merging and reddening of LRDs.

In some cases, host galaxies of LRDs are partially seen, allowing us to study hosts of distant AGNs. Hosts are rather compact star-forming galaxies, the spectral and photometric analysis indicate rather large BH to stellar mass ratios $M_\bullet/M_*$. Although a comparison can be somewhat biased due to sample selection, this deserves attention and further study (Greene *et al.*, 2024). It may be that to explain all the variety of LRD properties, self-interacting DM models are needed.

7.6 Observing the galaxy evolution with redshift

7.6.1 *Historical notes*

Eggen *et al.* (1962) were the first to show that it is possible to study galactic evolution using stellar abundances and kinematics. They inferred that as the metal-poor stars reside in a more or less spheroidal (stellar) halo with randomly oriented and quite elliptical orbits, the Galaxy was created during a rapid collapse of a relatively uniform, isolated protogalactic cloud. This simple picture was later challenged by Sandage & Eggen (1969) and Sandage (1970), who showed that old galactic and globular clusters have a large spread of ages.

Essential steps in understanding the dynamical evolution of galaxies were made by Lynden-Bell (1967, 1969). In the first of these papers, Lynden-Bell studied the statistical mechanics of violent relaxation to explain a smooth form of elliptical galaxies where a simple two-body relaxation time was very long. We described this process in Section 7.1.3.1 in relation to the formation of dark haloes. But the mechanism is also important in the case of merging galaxies and to explain cuspy profiles of DM haloes, such as the NFW profile (8.2) and the Einasto profile (2.4). In the second paper, Lynden-Bell suggested that galactic nuclei can be collapsed quasars.

The first steps in understanding the physical evolution of galaxies were made by Tinsley (1968), who used stellar evolutionary tracks to calculate models of the evolution of stellar populations of galaxies. The essential results of this study were confirmed by Einasto (1972), see Chapter 2.

7.6.2 *A time machine of distant galaxies*

Further progress in studying the evolution and formation of galaxies was made using modern large telescopes, which allowed us to study the physical properties of more and more distant and, hence, younger and younger galaxies. A very interesting summary of such efforts was given by Ellis (2022). Reviews of efforts to understand the formation and evolution of young galaxies were given by Stark (2016) and Robertson (2022). In the following, we give a summary of these efforts.

The overview of cosmic history is shown in Fig. 7.41. After the Big Bang (time $t = 0$), the Universe expanded and cooled until electrons

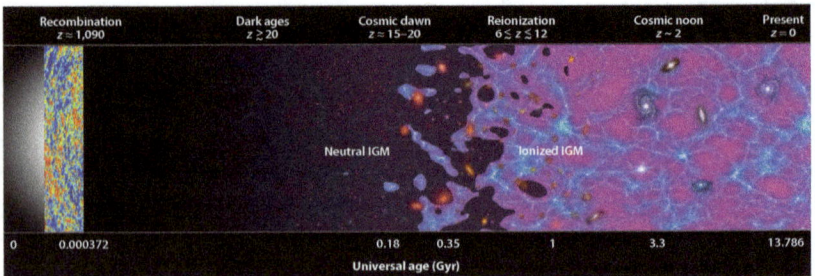

Fig. 7.41. Overview of universal history highlighting the epoch of reionisation (Robertson, 2022). Used with permission of *Annual Reviews* from Robertson, 2022. Permission conveyed through Copyright Clearence Center, Inc.

could recombine with protons to form neutral hydrogen and some helium. This recombination occurred at time $t = 372,000$ years at redshift $z = 1,090$. During the Dark Ages, the Universe remained dim and neutral, until the first stars formed at the highest peaks of the density distribution at cosmic epoch $z \approx 30$ or even earlier. This was the Cosmic Dawn. The formation of these Pop III stars terminated the cosmic dark ages and initiated the gradual enrichment of the Universe with metals. Dwarf galaxies formed in lower-mass haloes containing gas that has been polluted with earlier generations of stars. The next generations of stars contained some metals and formed Pop II. Over the next billion years, galaxies accreted rapidly cold gas, growing considerably in mass. The Lyman continuum radiation released from these early stars, galaxies and their active nuclei ionised their intergalactic surroundings, leading to cosmic reionisation: over the period of time $0.35 \leq t \leq 1$ Gyr (of $12 \leq z \leq 6$), the hydrogen of the intergalactic medium is transformed into an ionised state. Then a period of intensive galaxy formation followed: the star formation peaked at Cosmic Noon at epoch $t \approx 3.3$ Gyr ($z \approx 2$). We know this from the observations and the laws of physics.

The Hubble Deep Field (HDF, 1996), the Hubble Deep Field South (2000), followed by the Hubble Ultra Deep Field (HUDF, 2004) and the Hubble eXtreme Deep Field (2012) allowed us to follow the evolution of galaxies up to the time $t \sim 0.5$ Gyrs from the Big Bang ($z = 10$). All these fields complement each other, as they have different areas covered and different depths. The redshifts (and ages) of the most distant objects in these deep fields are only photometric and not spectroscopic. The detailed properties of the most distant spectroscopically confirmed galaxies remain to be unveiled with the JWST, and it remains to the future years (we described the current state in Section 7.1.4). In observations, the filters used are usually in optical, near-IR and mid-IR wavelengths corresponding to the UV emission in rest frame of these galaxies, implying that the results may be biased to star-forming galaxies.

An early picture taken with the JWST is presented in Fig. 7.42. This image was first presented at the US President's White House briefing on 11 July 2022. It shows the capability of the JWST to observe faint galaxies. The faintest galaxies visible in this photo have redshift about 10, and their age is 13.1 billion years. These dwarf galaxies are only visible in infrared light, to which JWST is sensitive.

Fig. 7.42. Photo taken with the JWST space telescope of the galaxy cluster SMACS 0723. The cluster of galaxies SMACS 0723 visible in the foreground is four billion light years away from us at redshift 0.39. The clusters DM halo distorts the light from distant galaxies, making them appear to be curved (NASA archive).

An easy and efficient way to find an initial sample of distant galaxy candidates is to search the Lyman-break objects, also called a "drop-out" technique. Most, if not all, of the distant galaxies contain a lot of neutral hydrogen. The transitions of the neutral hydrogen atom are as a rule, from the ground level to the excited levels, starting from the absorbed energy 10.2 eV or the photon wavelength 121.6 nm and thereon. Thus, in a galaxy, a lot of the photons with $\lambda < 121.6$ nm will be absorbed by hydrogen atoms. Rather soon, they will be emitted back but in an arbitrary direction, and the number of photons in the line-of-sight direction will be much smaller. The absorption is continuous after wavelengths shorter than the Lyman limit of 91.2 nm. Thus, at wavelengths shorter than 121.6 nm, the continuum of stellar emission is depressed, and at $\lambda < 91.2$ nm, it is depressed even more. These wavelength cutoffs are in the rest frame of a galaxy. For distant galaxies, the Lyman break wavelengths are shifted by the galaxy's redshift. For example, if we note that the brightness of a galaxy decreases at about 600 nm, i.e. it is seen in the V-band but not in B-band, we may guess its redshift is about $z \sim 4$. This technique can be used to select high redshift candidate galaxies using multiband images of certain fields in the sky and for follow-up detailed spectroscopic studies.

7.6.3 Properties of young high-redshift galaxies

7.6.3.1 Morphological types

Perhaps, the first statistical study, how the morphological types of galaxies evolve with redshift was done by Abraham *et al.* (1996), who used HDF images of faint galaxies, detected in the HDF in the range of infrared magnitudes $21 < I < 25$ mag. Magnitudes and morphological types were estimated independently from visual appearance by Richard Ellis and Sidney van den Bergh and by an automated classification algorithm. At faint magnitudes $I \geq 22$, there was an overabundance of irregular/peculiar/merging systems of all magnitudes. At the limit of $I = 25$ mag, these systems represent 30–40% of all galaxies. Authors suggest that a significant fraction of these systems may lie beyond the redshift $z \simeq 2.5$.

This was confirmed later on the basis of deeper photometric and spectroscopic redshift surveys by, e.g., Conselice *et al.* (2008) from the analysis of the HUDF survey, Mortlock *et al.* (2013) from the CANDELS survey, and Talia *et al.* (2014) from the GMASS and CANDELS surveys (see, e.g., Figs. 7.43 and 7.44). These observations also support earlier suggestions that at higher redshifts, there is an increasing number of non-regular clumped galaxies, and the conventional Hubble classification system no longer adequately describes the structural characteristics for a significant fraction of galactic and possibly proto-galactic systems.

7.6.3.2 Dimensions and stellar masses

It is natural to suppose that the sizes of galaxies increase with their stellar mass. According to our current understanding, it is also natural to suppose that galaxy masses increase with time. We do not look at this in more detail.

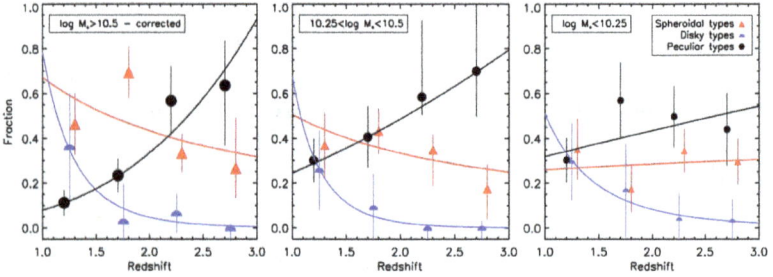

Fig. 7.43. The galaxy-type fraction split by mass as a function of redshift. Types are corrected to account for the fact that the misclassification due to image quality will depend on galaxy brightness (i.e. stellar mass) (Mortlock *et al.*, 2013).

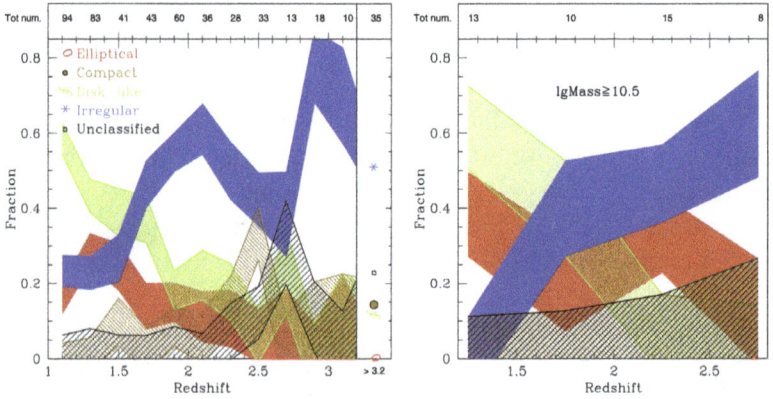

Fig. 7.44. The relative fraction of the galaxy types as a function of redshift. The left plot is for the whole sample, and the right plot is for a mass complete sample. The total number of galaxies in each redshift bin is indicated in the upper part of the plots (Talia *et al.*, 2014). Reproduced with permission © ESO.

Due to the difficulty of observations, the redshift samples of higher-mass galaxies are used to study size evolution in time. Observations indicate that galaxy sizes increase with time for redshifts $0 < z < 3$: distant galaxies are more compact than local galaxies with the same stellar mass. In the case of larger galaxies (stellar masses $\log M_* > 11.3$), Mowla *et al.* (2019) found that there is a power-law dependence of their median size with redshift $\langle r_{\mathrm{eff}} \rangle = 13.4(1 + z)^{-0.95}$ kpc. No statistical difference was found between the star-forming and quiescent galaxies.

The power-law fit

$$r_{\rm eff} \sim (1+z)^\alpha$$

was also found to be adequate in other studies. Trujillo *et al.* (2007) and Buitrago *et al.* (2008) analysed separately spheroidal and disklike galaxies and derived in both cases the power-law fits the size distribution, but with different α values, -1.48 and -0.82, respectively, indicating that spheroids evolve faster (see Fig. 7.45).

For masses, Mowla *et al.* (2019) found that $r_{\rm eff} = A \times m_*^\alpha$, where $r_{\rm eff}$ is in kpc and m_* is the stellar mass in units of $5 \times 10^{10}\,{\rm M_\odot}$. Both the parameters in this relation, A and α, are functions of redshift: $\log A = -0.25 \log(1+z) + 0.80$ and $\alpha = -0.13 \log(1+z) + 0.27$. With time, a clear increase in sizes is seen for more massive galaxies (Fig. 7.46); for smaller masses, the increase is quite small.

7.6.3.3 *Mergers*

This increase in the size of galaxies with time is ascribed to galaxy mergers and the accretion of gas. The relative importance of these factors is not clear yet. Merging of galaxies, both the stellar and DM components, is a key aspect of galaxy formation models in cold DM cosmology. In simulations, it is relatively simple to follow the merger history of a galaxy; in observations, it is far from an easy task.

An overview of merger analysis is given by Conselice (2014, Sec. 4.3). Usually, the asymmetry index, the clumpiness index and the concentration index are calculated for galaxies, as it is natural to assume that concentration is important to understand the relative importance of smooth monolithic collapse or merger build-up of a galaxy; asymmetries and clumps dominantly arise during the process of mergers (Conselice, 2003), see Fig 7.47. The asymmetry index A measures the similarity of a galaxy to the same galaxy but rotated by 180° relative to its centre. The clumpiness index S compares the image of a galaxy with the somewhat convolved Gaussian filter image, i.e. with the smoothed image. The concentration index C is the ratio of two radii, containing 50% and 80% of galactic light. In addition, these CAS parameters (as they are called), other parameters, characterising, e.g., starburst properties, are added. For example, the Gini

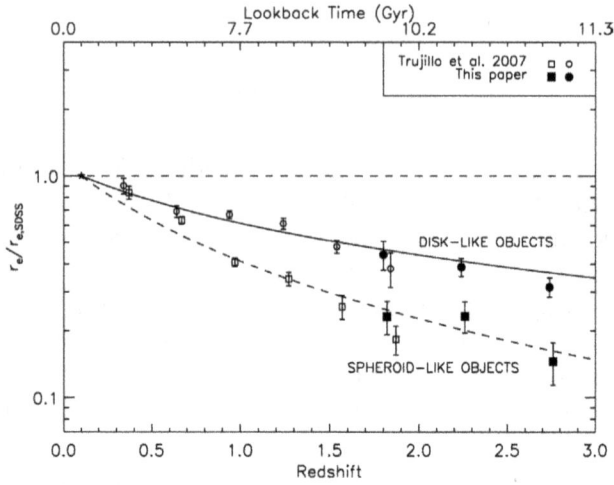

Fig. 7.45. Effective size evolution of spheroidal and disk-like galaxies relative to corresponding local SDSS galaxies as a function of redshift. Galaxies were separated to disk-like and spheroidal types by the profile Sersic index value $n = 2.5$ (Buitrago *et al.*, 2008).

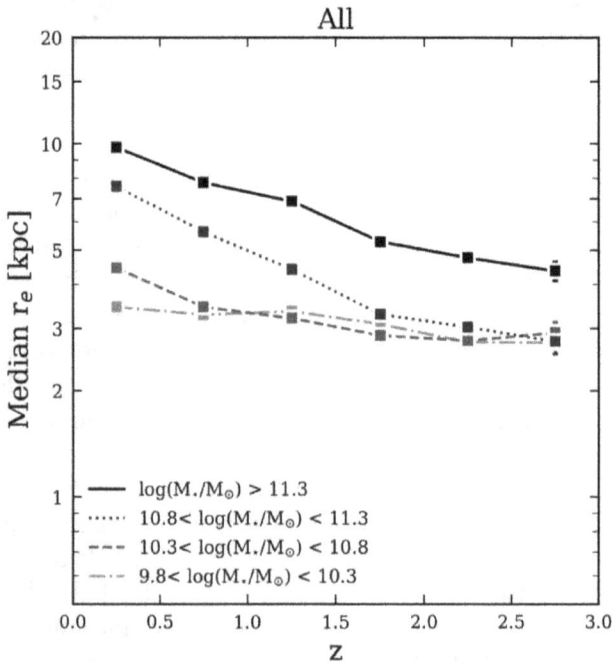

Fig. 7.46. Median sizes of galaxies as a function of redshift for different stellar mass ranges (Mowla *et al.*, 2019).

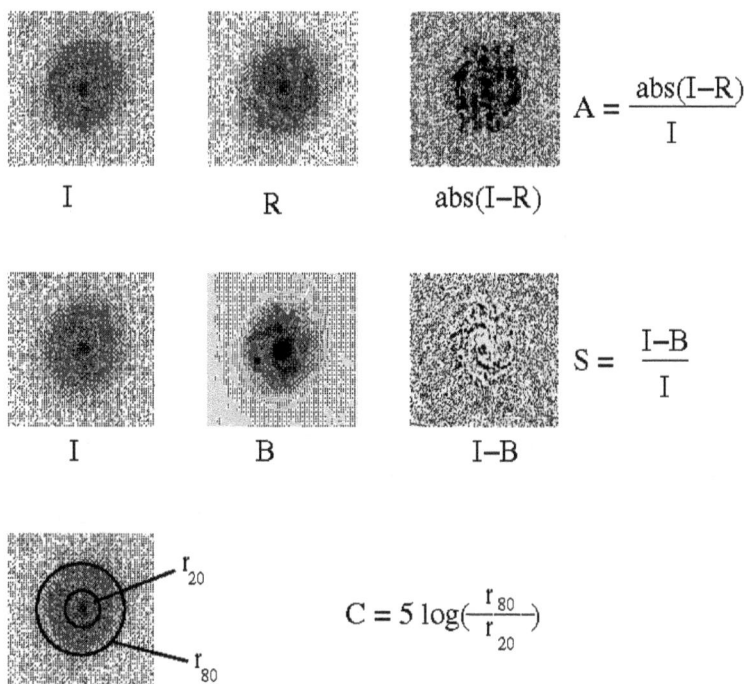

Fig. 7.47. Graphical representation of how three parameters, asymmetry (A), clumpiness (S), and concentration (C), are measured: I is the original galaxy image, R is the this image rotated by $180°$, B is the smoothed by a Gaussian image (Conselice, 2003).

coefficient G compares how the light distribution fluctuates between pixels.[29]

By comparing the space of these parameters with galactic images, Conselice (2003) determined the criteria to find the merged galaxies automatically. The method was later modified by Lotz *et al.* (2008) and Freeman *et al.* (2013).

The merger fraction f_m of galaxies is defined as the ratio of merged galaxies to the total number of galaxies:

$$f_m(M_*, z) = \frac{N_m}{N_T}.$$

---

[29]The Gini coefficient was introduced in economics and estimates the income equality/inequality distribution within a population.

The redshift evolution of the merger fraction is characterised by the power law (Conselice, 2014)

$$f_{\mathrm{m}}(z) = f_0(1+z)^m$$

or by the combined power law and exponent

$$f_{\mathrm{m}}(z) = \alpha(1+z)^m \exp[\beta(1+z)].$$

According to the first formula, there is a monotonic increase with redshift; according to the second formula, a peak redshift exists. As a result of analysis and comparison of various studies, Conselice (2014) concluded that the merger history tends to peak at $z = 2.5$ for massive galaxies $M_* > 10^{10}\,\mathrm{M}_\odot$ at values of $f_{\mathrm{m}} \sim 0.3$–0.4 and decline at lower redshifts. The power m varies significantly in different studies. In studies using redshifts up to $z \sim 1$, the power is 2–2.7, in the case of higher redshifts, it is higher, being 3–4 for massive galaxies $M_* > 10^{10}\,\mathrm{M}_\odot$ and of 1–2 for lower mass galaxies.

The other important merger characteristic is the merger rate. The merger rate per galaxy Γ is the average time between mergers for a galaxy with given properties (e.g a stellar mass range) within a given redshift range. According to Conselice (2014), the average time between mergers increased nearly 10 times from $z \sim 2.8$ to 0.5. Calculated from these quantities, Conselice *et al.* (2013) derived that during $z = 1$–3, about half of the initial stellar mass at $z = 3$ was added due to mergers. Comparing this to the observed SFRs, Conselice *et al.* (2013) concluded that for sustaining the observed SFR, it is necessary that gas accretion from the intergalactic medium should be of approximately the same amount.

At even higher redshifts, detecting CAS characteristics and identifying merged galaxies are more complicated. By studying the SFR in close pairs of galaxies based on JWST data, Duan *et al.* (2025) derived the merger rate for redshifts from $z \sim 5$ to $z \sim 10$ shown together with previous studies in Fig. 7.48. Note that in this figure, the merger rate is $\mathcal{R} = \Gamma^{-1}$ with the dimension of Gyr^{-1}. JWST data suggest that at early epochs at $z > 6$, the merger rate is saturated, and the power-law exponent describes the evolution better. The short average time between mergers, 0.2–0.5 Gyr, observed during the Universe's first 1 Gyr indicate a rapid transformation of young galaxy properties.

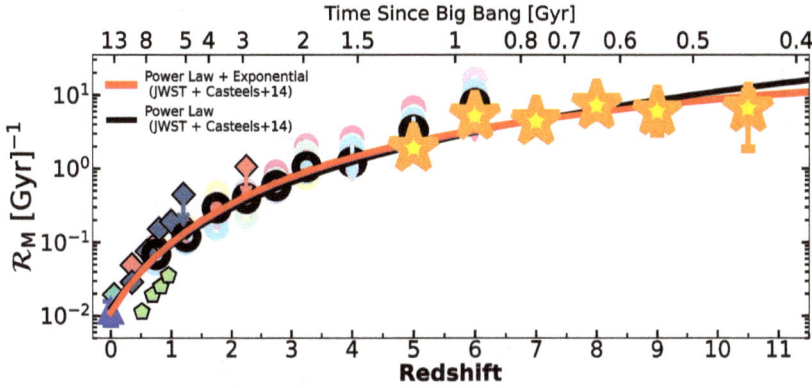

Fig. 7.48. Evolution of the galaxy merger rate with redshift (Duan *et al.*, 2025). High-redshift major merger rates from JWST results are represented by yellow-orange stars; for other designations in the figure, see the original paper (Duan *et al.*, 2025).

7.6.3.4 *Luminosity function and star formation rate*

One of the principal descriptive functions of galaxies is the luminosity function (LF). The LF is often described by the Schechter (1976) function:

$$\phi(L) = (\phi^*/L^*)(L/L^*)^\alpha \exp\left(-L/L^*\right)\mathrm{d}(L/L^*), \qquad (7.2)$$

where the parameter ϕ^* describes the intensity galaxy formation, α is the exponent at low luminosities $(L/L^*) \ll 1$, and L^* (or the respective absolute magnitude M^*) is the characteristic luminosity of galaxies. Another possibility to describe the LF is to use the double power law:

$$\phi(L) = (\phi^*/L^*)(L/L^*)^\alpha (1 + (L/L^*)^\gamma)^{(\delta-\alpha)/\gamma}\mathrm{d}(L/L^*), \qquad (7.3)$$

where α is the exponent at low luminosities $(L/L^*) \ll 1$, δ is the exponent at high luminosities $(L/L^*) \gg 1$, γ is a parameter that determines the speed of transition between the two power laws, and L^* is the characteristic luminosity of the transition similar to the characteristic luminosity of the Schechter function.

A review of recent determinations of LFs of young galaxies was given by Robertson (2022). We show in Fig. 7.49 both LFs of galaxies during the cosmic reionisation period at redshifts $z = 6$–10, the Schecter law by solid lines and the double power law by dotted and

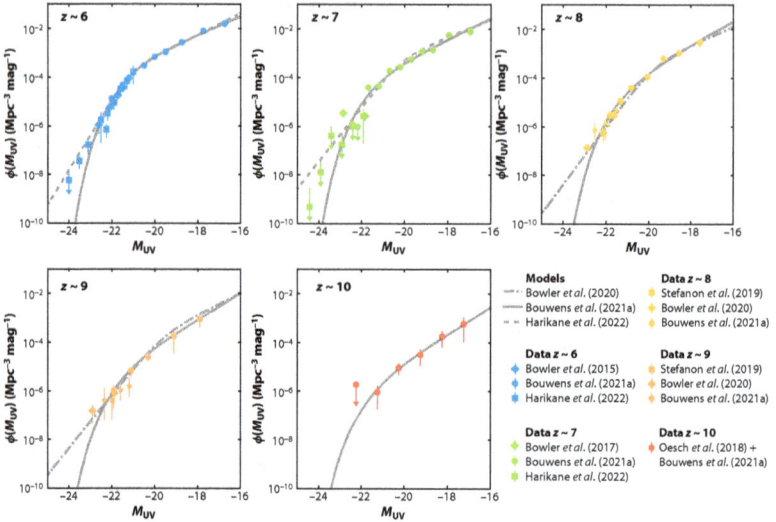

Fig. 7.49. Rest-frame luminosity function of galaxies during cosmic reionisation at redshifts $z = 6$ (blue), 7 (green), 8 (yellow), 9 (orange), and 10 (red) (Robertson, 2022). Used with permission of *Annual Reviews* from Robertson, 2022. Permission conveyed through Copyright Clearence Center, Inc.

dashed lines. Over most redshifts, the double-power-law function fits data better than the Schechter function.

The sharp UV luminosity cutoff at the bright end of the LF can be regulated by a combination of physical processes we noted earlier in this chapter, including the inefficiency of gas cooling in massive haloes, active galactic nucleus heating, but also the increased dust content of the most luminous galaxies. At highest redshifts, these processes may not yet be effective in the dark matter haloes that are present. If this is the case already at $z \simeq 6$–7, the bright end of the LF may not exhibit an exponential cutoff as dictated by the Schechter function and is better represented by the double-power-law function. For precise analysis, the data is scanty.

Figure 7.50 shows the SFR at redshifts $z \sim 4$–10 from rest-frame UV-selected galaxies and theoretical models. The best agreement between the theoretical models and the observed cosmic SFR history occurs for models with the star formation efficiency as a function of halo properties being roughly constant with redshift. In these models, the SFR density does decline rapidly with redshift in accordance with the abundance of dark matter haloes (Robertson, 2022). At lower redshifts, the SFR starts to decrease. This is seen from the

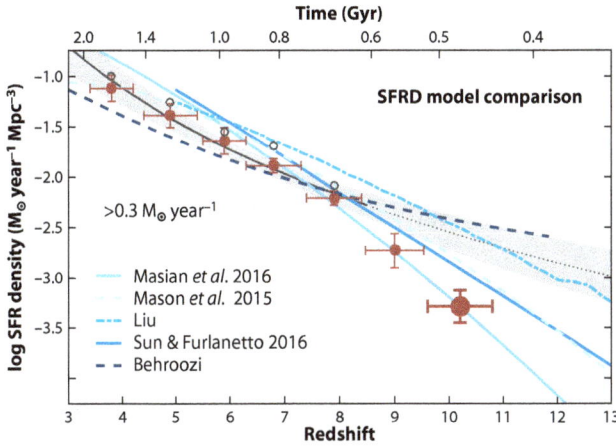

Fig. 7.50. Evolution of the SFR density inferred from the evolving UV luminosity function (Robertson, 2022). Used with permission of *Annual Reviews* from Robertson, 2022. Permission conveyed through Copyright Clearence Center, Inc.

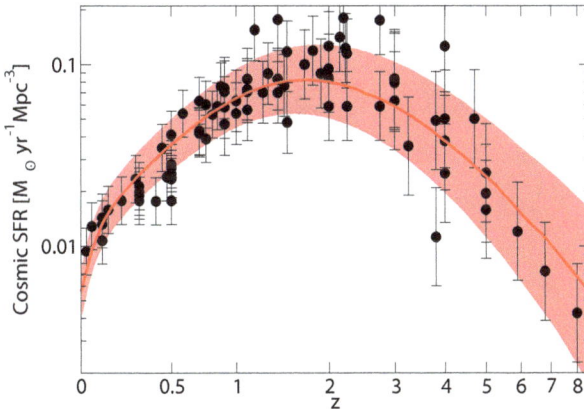

Fig. 7.51. Observational constraints on the cosmic star formation rate (black points) compared to the best-fit model (red solid line) and the posterior 1σ distribution (red-shaded region) (Behroozi *et al.*, 2013). Used with permission of *Annual Reviews* from Robertson, 2022. Permission conveyed through Copyright Clearence Center, Inc.

study by Behroozi *et al.* (2013) determined the average star formation history from $z = 0$ to $z = 8$, as shown in Fig. 7.51. SFR has a clear maximum at redshift $z \approx 2$ and decreases about 10-fold, both at $z = 0$ and $z = 8$.

Chapter 8

Structure and Evolution of Galaxies in the Cosmic Web

In Chapter 7, we described various physical processes in galaxies and in the cosmic web. Now, we discuss how these processes influence the structure and evolution of galaxies and the cosmic web. We start with the discussion of the formation and evolution of galactic populations. Thereafter, we discuss the formation and evolution of galaxies and systems of galaxies. We use as examples our own Galaxy, Andromeda galaxy, Local Group and Local Supercluster. In these nearby systems, physical and dynamical processes can be studied in great detail. Most important of these processes are merging of minor systems to form larger ones. The merging effect depends on the scale of the system from galaxies to superclusters.

8.1 Formation and evolution of galactic populations

The presence of galaxies of various morphological types was known long ago. As discussed in Chapter 2, differences in morphological types can be explained by the presence in galaxies of stellar populations of different spatial distribution, kinematics, ages and compositions. Differences of galaxies of various types are well seen in the distribution of systems of galaxies in the absolute magnitude versus central surface brightness diagram, as presented in Fig. 8.1. The figure demonstrates that in the high central surface density range, giant and dwarf elliptical galaxies and globular clusters occupy well-separated regions. Galactic disks, spheroidal galaxies and

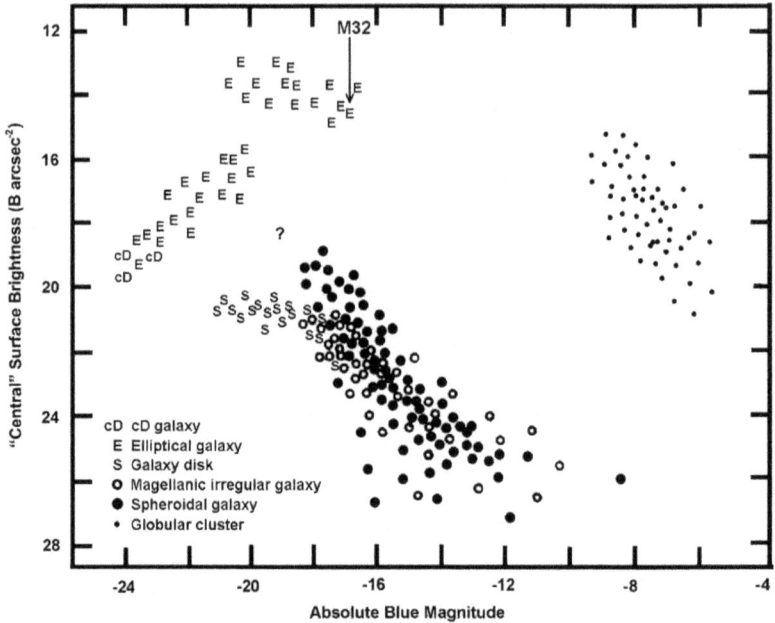

Fig. 8.1. Schematic illustration of the differences of stellar systems in the distribution of the absolute magnitude versus central surface diagram. Various symbols denote spheroidal and irregular galaxies, galaxy disks, globular clusters, elliptical and cD galaxies. Surface brightnesses apply to the main bodies of the galaxies (Kormendy *et al.*, 2009).

Magellanic irregular galaxies occupy in the diagram the low central surface density area, very different from populations in the high surface density area. These differences hint to various physical processes in the formation of various galaxies.

We discuss first general dynamical processes of the evolution of the Universe. Next, we describe the processes influencing the evolution of various stellar populations: stellar haloes, globular clusters, bulges and disks. Subsequently, we consider the structure and evolution of galaxies of different morphological types.

8.1.1 *Dynamical processes in the evolving Universe*

ARIEL: I feel so existential today.

RHEA: How so?

ARIEL: An asteroid that has a 3% chance of hitting Earth–

CALLISTUS: Hannibal, ante portas![1] It immediately was downgraded to a negligible likelihood.

RHEA: We'll be alright. It wouldn't have destroyed Earth anyway.

CALLISTUS: And isn't it great that we detect these things in advance? We might be able to do something about it.

RHEA: Send in Bruce Willis!

ARIEL: Armageddon! But hear me out. The asteroid made me think that we might survive an asteroid hit, an AI apocalypse, or even a nuclear war. I mean humanity. Life. But what about a supernova exploding nearby?

RHEA: Did you hit a stumbling block with your project?

ARIEL: What's the point of my project if we can be wiped out by a supernova?

RHEA: Come on, what are the odds of that?

ARIEL: If we are made of stars, one has had to explode nearby to create the elements crucial to our life. It happened before, it can happen again. Maybe now...

CALLISTUS: Right. Let's take this seriously for one second.

RHEA: But it's only a distraction from their project, isn't it, Ariel?

ARIEL: How close would it have to be to wipe us out?

CALLISTUS: If it's within 10 parsecs, it's not gonna be pretty.

RHEA: In addition to light, it would produce an intense flux of photons in the gamma and X-ray regions.

ARIEL: Let's assume it is at 10 parsecs. If one went off today, we would only find out 32 years from now. Just when I'd be at the peak of my scientific career, perhaps describing this very same argument to a young graduate student and... BOOOM! Is it worth persevering through grad school for that?

RHEA: It could be worse. One could have gone off 30 years ago...

ARIEL: That would just about cut off my PhD!

RHEA: You will not have a PhD if you don't finish your project!

CALLISTUS: After noticing the explosion in the visible, the flux of gamma rays would reach us about 100 days later and last a year or so. The effect on the biosphere would be catastrophic.

RHEA: Then, a year or so later, the X-rays would reach us, lasting for about ten years.

ARIEL: It would be a boring horror movie. No final girl, nobody would survive. The ozone layer: destroyed. UV radiation reaching phytoplankton in the surface layer of water. I'm depressed just thinking about it.

CALLISTUS: Some life could still survive deep in the oceans for the next thousand years. I trust evolution.

---

[1] Hannibal (great danger) is at the door.

ARIEL: Life 2.0, the movie. Promise little, deliver nothing.

RHEA: Kinda like your project. What's worse, the coup de grace would arrive after a thousand years: cosmic rays would pepper us, or whatever's left, creating high-energy muons and harmful radiation for all life. Not much would survive that.

ARIEL: I conclude it's pointless to finish my project, huh?

RHEA: Even though something like this happened before, the likelihood of a supernova going off within 10 parsecs is extremely low.

CALLISTUS: Maybe once every few billion years at most. So, it might have happened in the early stages of life, but it's nothing to worry about now.

ARIEL: What about within 100 parsecs?

RHEA: Maybe once every few millions of years. Fossil radioactive isotopes suggest such an explosion 3.5 million years ago.

CALLISTUS: But that would not cause widespread catastrophe. We are still here. And we can likely survey all the stars within 10 parsecs in the near future and exclude even that infinitesimal probability.

ARIEL: Ugh, should I go back to worrying about the AI apocalypse, then?

CALLISTUS: I don't think so.

ARIEL: I guess climate change it is.

RHEA: Time to go back to worrying about your project.

CALLISTUS: Every science project is like a movie: if it is interesting at all, it has to have an "all is lost" moment.

RHEA: That's when you clench your teeth, work even harder and muddle through if need be.

CALLISTUS: Or find a creative solution.
[*Silence.*]

ARIEL: Aye-Aye Captain. Back to my project.

All massive bodies are gravitational attractors. Galaxy-type and larger attractors are of interest in cosmology. It is well known that smoothing affects the character of the high-density regions that are found. As discussed in Chapter 4, smoothing with a kernel of length $1 \ h^{-1}$ Mpc highlights ordinary galaxies together with their satellite systems similar to our Galaxy and M31. Smoothing with a kernel of length $4 \ h^{-1}$ Mpc finds high-density regions of the cosmic web that have an intermediate character between clusters and traditional superclusters, such as central regions of superclusters. The optimal smoothing length for finding superclusters is $8 \ h^{-1}$ Mpc; see Liivamägi *et al.* (2012).

The property of superclusters to act as great attractors was the basis of the suggestion by Tully *et al.* (2014), Pomarède *et al.* (2015) and Graziani *et al.* (2019) that superclusters should be defined on the basis of their dynamical effect on the cosmic environment, the basins of attraction (BoA) or cocoons (Einasto *et al.*, 2019b). Conventional superclusters form central high-density regions of BoAs. Neighbouring BoAs have common sidewalls. The BoA is the volume containing all galaxies whose flow lines converge at a given attractor, the local minimum of the gravitational potential. In this method, Universe is divided into BoAs — cells of dynamical influence. Cells of dynamical influence are regions around superclusters, from which superclusters collect their matter.

Elements of the cosmic web evolve with time. Clusters of galaxies grow by merging of smaller clusters and by infall of non-clustered matter, filaments merge, and voids became emptier. Superclusters also change, their sizes shrink in comoving coordinates, and masses grow by infall and merging. Similar general visual appearance of the density fields at very early and present epochs as well as the percolation analysis suggest that supercluster embryos were created very early. This result is not surprising, already Kofman & Shandarin (1988) demonstrated that the whole present-day structure is seen in the initial fluctuation distribution. The development of the density field in the early phase is well described by the Zeldovich (1970) approximation and its extension, the adhesion model by Kofman & Shandarin (1988), see Fig. 4.1.

Galaxy and supercluster embryos were created by high peaks of the initial field. The initial velocity field around peaks is almost laminar. Pichon *et al.* (2011) and Dubois *et al.* (2012, 2014) showed that a significant fraction of cold gas falls towards haloes along filaments, which are oriented nearly radially to the centres of high-redshift rare massive haloes, see Fig. 8.2. This process rapidly increases the mass of the central halo. We may conclude that depending on the height of the initial density peak, in this way, embryos of galaxies, galaxy clusters, and superclusters cores were created. However, the further evolution of superclusters differs from the evolution of galaxies and galaxy clusters. Galaxies and ordinary clusters of galaxy are local attractors and collect additional matter from their local environment. Superclusters are global attractors and collect matter from a much larger environment.

Clusters move together with filaments in the large potential well of superclusters. The simultaneous movement of clusters with their surrounding filament follows from the simple fact that the filamentary character of the cosmic web is preserved at the present epoch. If clusters had high peculiar velocities with respect to the surrounding filaments, the filamentary character of the web would be destroyed during the evolution. The laminar character of the velocity field is explained by the presence of the DE, as suggested already by Teerikorpi *et al.* (2005) and Sandage *et al.* (2010). Very rich clusters are central clusters of superclusters, and they are connected with other structures by many filaments. The central clusters of these superclusters lie at minima of potential wells created by respective superclusters. They are fed by filaments from several sides and are suitable locations for cluster merging, which means that small clusters fall onto the central cluster along filaments surrounding the central cluster. Or in other terms, the clustering is essentially a regular flow of smaller systems towards larger ones, not a completely random process.

8.1.2 *Dynamical processes in galactic population*

One of the dynamical processes, which influences the formation and evolution of galactic populations, is the merging of galaxies. As suggested by Toomre (1977), the merging is the dominant process in forming elliptical galaxies. During the merger, stars and dark matter in each galaxy become affected by the approaching galaxy. In late stages of the merger, the gravitational potential is quickly changing. Star orbits in both merging galaxies are greatly altered. This process is called "violent relaxation" studied in detail by Lynden-Bell (1967). Numerical simulations by van Albada (1982), Angulo *et al.* (2017) and Nipoti (2017) show that during the collapse phase of galaxy evolution with many merging events, a density distribution with the de Vaucouleurs $r^{1/4}$ model is formed, as observed in spheroids of elliptical galaxies. This distribution is well represented by the Einasto profile (2.4) with parameter $N = 4$ and approximately with the NFW profile (8.2).

Galaxy mergers depend on parameters and relative motions of both merging galaxies. Most often are minor mergers where one of the galaxies is significantly larger than the other. This process is often called "galactic cannibalism", where the major galaxy is absorbing most of the gas and stars of the minor galaxy. If two merging gas-rich galaxies are approximately of the same size, a large amount of star formation occurs, which quickly fades and produces a elliptical galaxy. Such events are called wet mergers. Dry mergers are mergers between gas-poor galaxies. These mergers do not greatly affect the morphological type of galaxies, but increase the stellar mass of the major galaxy.

The second important dynamical process is the slow inflow of matter to galaxies. Dekel *et al.* (2009) used hydrodynamical simulations in a ΛCDM Universe to find the role of cold gas streams to the evolution of galaxies. The simulation is in a box of comoving size $50\,h^{-1}$ Mpc with $1,024^3$ particles, where more than 100 haloes of mass $\sim 10^{12} M_\odot$ allow us to get a large statistical sample. Authors found that hot gaseous medium in haloes of virial masses $M_v \geq 7 \times 10^{11} M_\odot$ at earlier epochs $z \simeq 2$ forms penetrating cold streams towards the halo centre. Left panel of Fig. 8.2 shows the distribution of the entropy $\log K = \log(T/\rho^{2/3})$ in a typical halo. Here, the temperature and gas density are in units of the virial temperature and mean density within the halo virial radius R_v. The narrow streams are of much lower temperature and entropy by more than three orders of magnitude. The boundaries between the streams and the surrounding hot medium within the virial radius are sharp. This picture is supported by observational data, which show that rich

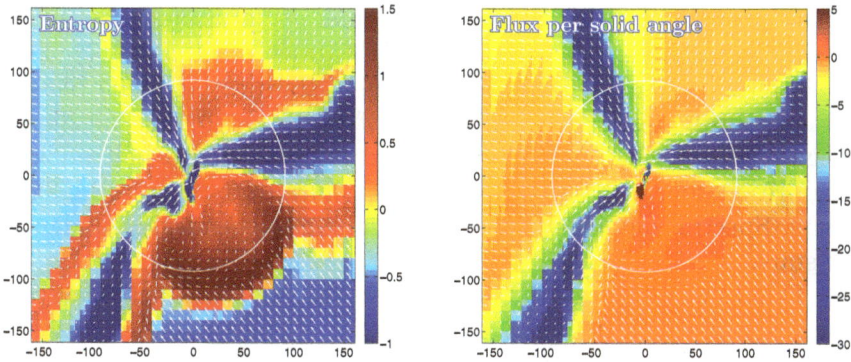

Fig. 8.2. The distribution of gas in a halo of the hydrodynamical simulation in a thin equatorial slice. The left panel shows the entropy $\log K = \log(T/\rho^{2/3})$ in units of the virial quantities, and the right panel gives the radial flux $\dot{m} = r^2 \rho v_r$ in units $M_\odot \text{yr}^{-1} \text{rad}^{-2}$. The circle marks the virial radius (Dekel *et al.*, 2009). Used with permission of *Nature* from Dekel *et al.* (2009). Permission conveyed through Copyright Clearence Center, Inc.

clusters of galaxies are centres of multiple filaments directed towards the central cluster, see Fig. 8.38.

The arrows in the both panels mark the velocity field projected on the slice plane. The flux colour map in the right panel shows the flow rate per solid angle, $\dot{m} = r^2 \rho v_r$. The flux inward is almost exclusively channelled through the narrow streams, typically involving 95% of the total inflow rate. The opening angle of a typical stream is 20–30°, so the streams cover a total area a few percent of the sphere. A comparison with the observed abundance of star-forming galaxies implies that most of the input gas must rapidly convert to stars. One-third of the stream mass is in gas clumps leading to mergers of mass ratio greater than 1:10, and the rest is in smoother flows. Three-quarters of the galaxies forming stars at a given rate are fed by smooth streams.

8.1.3 *Evolution of galactic disks and stellar haloes*

Galactic disks and stellar haloes are most prominent populations of Milky Way-type galaxies. The classical view of the structure and evolution of galaxies was discussed in Chapter 2. In the last 20 years, our understanding of the structure and evolution of galactic populations has improved dramatically. This field of astronomy is known as Galactic Archaeology. Most important new kinematical, spatial and spectroscopic data came from the second data release of the Gaia Mission (Gaia DR2) (Gaia Collaboration *et al.*, 2018) and from large spectroscopic surveys such as APOGEE (Majewski *et al.*, 2017). The Gaia DR2 has sampled six million F-G-K stars with full 6D phase-space coordinates. APOGEE is one of the programs in the Sloan Digital Sky Survey III (SDSS-III) and has collected a half million high-resolution, high signal-to-noise ratio infrared spectra for 146,000 stars. Spectral data allow us to determine the chemical abundance of stars.

Thanks to these new datasets, a new, more mature picture of the evolution of the Milky Way-type galaxies is emerging. A review of the structure and evolution of galaxy based on Gaia data is given by Helmi (2020). New data suggest that two of the galactic populations, the stellar halo and disk, are closely linked. The structure of both populations is formed via multiple mergers. Most mergers were small, but include several large ones, such as the Helmi streams, Sequoia, and Thamnos events, and the Magellanic and Sagittarius streams.

Stars retain memory of their origin when they move. During merger events, a star cluster or dwarf galaxy gets torn apart by the tidal forces of a larger system like the Milky Way, and the stripped stars continue to follow similar trajectories as their progenitor system. This implies that if the Milky Way type halo is the result of the mergers of many different objects, their stars should follow streams that cross the whole Galaxy. Using numerical simulations of merger events, Helmi & White (1999) found that if the stellar halo had been built via mergers, approximately 500 streams would be expected in the halo near the Sun. The distribution on the sky of the currently known streams and spatial substructures is shown in Fig. 8.3. Thin streams tell us the story of the destruction of less massive objects, smaller dwarf galaxies, open and globular clusters.

Stars belonging to different subpopulations and streams can be selected using the Toomre diagram, as shown in left panel of Fig. 8.4, and the diagram energy E versus z-angular momentum L_z in the right panel. Toomre diagram is similar to the Strömberg diagram, as

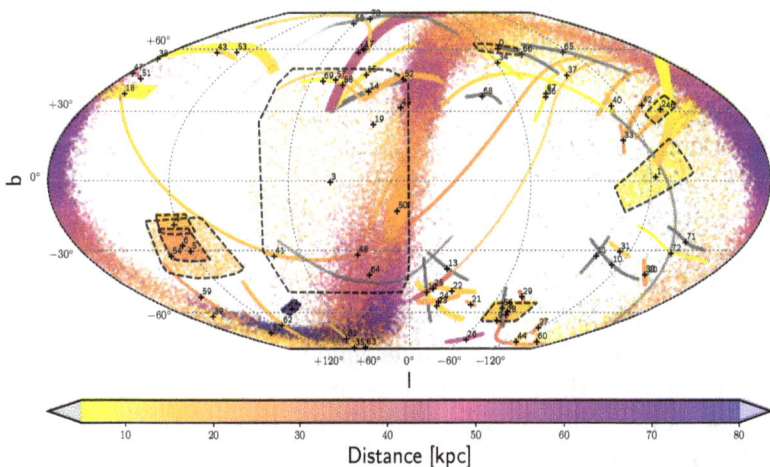

Fig. 8.3. Sky distribution of currently known spatially coherent streams and overdensities (indicated as regions delimited by dashed lines) (Helmi, 2020). Used with permission of *Annual Reviews* from Helmi (2020). Permission conveyed through Copyright Clearence Center, Inc.

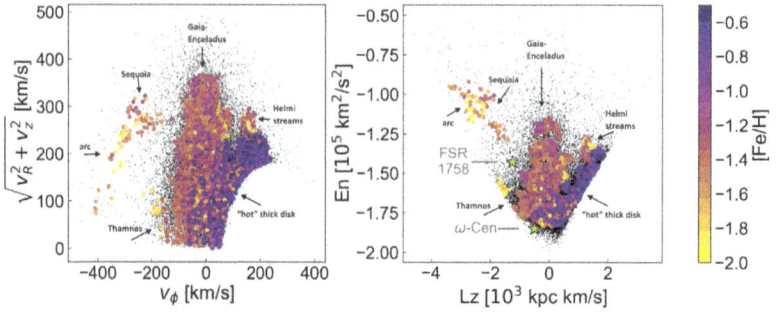

Fig. 8.4. Distribution of halo stars selected kinematically, within 3 kpc from the Gaia RVS sample extended with radial velocities from APOGEE, RAVE, and LAMOST with metallicities from the latter. The left panel shows their distribution in the Toomre diagram, while on the right energy, E and z-angular momentum L_z are plotted (Helmi, 2020). Used with permission of *Annual Reviews* from Helmi (2020). Permission conveyed through Copyright Clearence Center, Inc.

shown in Fig. 2.7, but instead of the mean full velocity dispersion, it uses the velocity dispersion $\sigma = \sqrt{V_R^2 + V_z^2}$ versus the rotation velocity $V_\phi = V_\theta - V_0$, where V_θ is the observed velocity in the direction of galactic rotation, and V_0 is the galactic rotation velocity (the local standard of rest). In Fig. 8.4, stars are colour-coded by [Fe/H], when they were identified in clumps in the space of E, L_z, eccentricity, and [Fe/H], the remaining stars are shown as black dots. The arrows indicate the approximate location of the various groups. Comparison of their extent in E, L_z, to N-body simulations suggest that the stellar mass of the Thamnos stream is $M_\star \approx 6 \times 10^5$ M$_\odot$ and that of the Sequioa stream is $M_\star \approx 10^7$ M$_\odot$.

New data suggest that the structure of the disk and halo of the Milky Way was mostly influenced by a merger of a Magellanic cloud-type satellite 10 Gyr ago, named Gaia–Enceladus (G–E). Helmi *et al.* (2018) demonstrated that the inner halo is dominated by debris from the G–E. Stars of the halo originated from a "heated" thick disk and from debris from the G–E satellite. The merger with G–E triggered star formation in the early Milky Way, leading to the appearance of the thick disk. The stars that originated in G–E form streams and slightly retrograde and elongated trajectories. With an estimated mass ratio of four to one, the merger of the Milky Way with G–E must have led to the dynamical heating of the precursor of the galactic thick disk, thus contributing to the formation of this component approximately 10 billion years ago. These findings are in line with

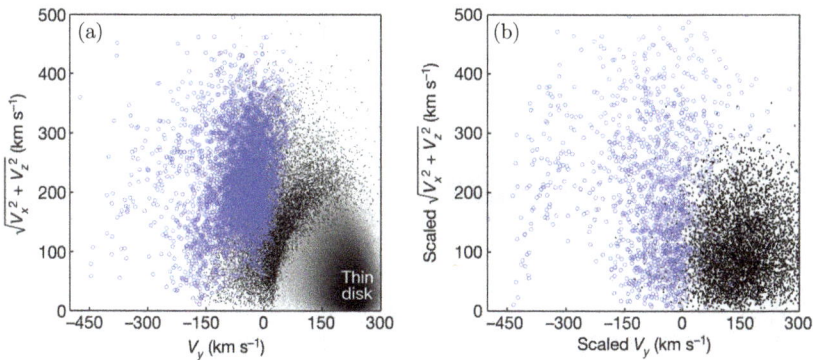

Fig. 8.5. Measured velocity distribution of stars in the solar vicinity compared with merger simulation results. (a) Velocities of stars in the disk are plotted with grey density contours and halo stars are shown as black points. The blue points are part of stars in G–E. (b) Distribution of star particles in a small volume extracted from a simulation of the formation of a thick disk from the merger of a satellite galaxy (blue symbols) with a pre-existing disk (black points) at a mass ratio of 5:1 (Helmi *et al.*, 2018). Used with permission of *Nature* from Helmi *et al.* (2018). Permission conveyed through Copyright Clearence Center, Inc.

the results of galaxy formation simulations, which predict that the inner stellar halo should be dominated by debris from only a few massive progenitors. Left panel of Fig. 8.5 shows the velocity distribution of stars inside a volume with a radius of 2.5 kpc in the solar vicinity, as obtained from the Gaia data. Right panel of the figure shows the velocity distribution from a simulation of the formation of a thick disk via a 1/4 mass-ratio merge. The similarity between the graphs suggests that the retrograde structure could be largely made up of stars originating in an external galaxy that merged with the Milky Way.

Support for the merger hypothesis also comes from the chemical abundances of stars. In the left panel of Fig. 8.6, we plot the abundances [α/Fe] and [Fe/H] for a sample of nearby stars. α-elements are produced by massive stars that die fast as core-collapse SN (type II), while iron is also produced in thermonuclear (type Ia) supernovae explosions. Therefore, in a galaxy, [α/Fe] decreases with time as [Fe/H] increases. Figure 8.6 shows the well-known sequences defined by the thin and thick disks of the galaxy. The vast majority of the retrograde structure's stars (in blue) follow a well-defined separate sequence that extends from low to relatively high [Fe/H]. Right panel of Fig. 8.6 shows the Hertzsprung–Russell diagram (HRD) of halo

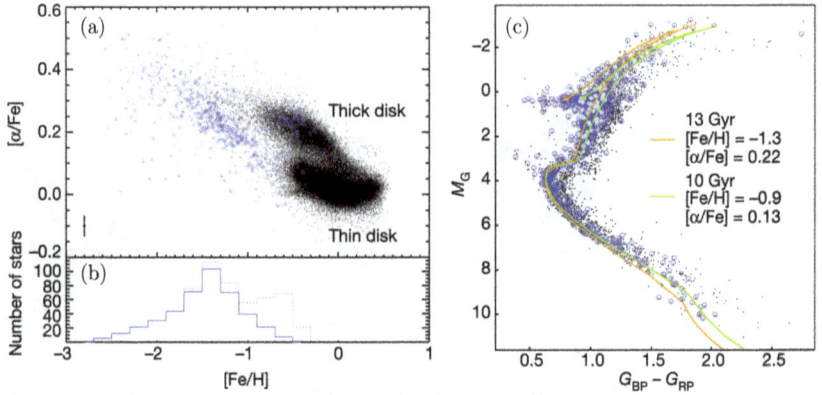

Fig. 8.6. Astrophysical properties of stars in G–E. (a) Chemical abundances for a sample of stars located within 5 kpc from the Sun. The blue circles correspond to 590 G–E stars. We note the clear separation between the thick disk and G–E stars. (b) The solid (dotted) histogram shows the metallicity distribution of the retrograde structure without (with) the subset of α-rich stars. The distribution, which peaks at [Fe/H]≈ -1.6, is reminiscent of the distribution of the stellar halo of the Galaxy. (c) The HRD for halo stars (black points). Dark blue symbols represent G–E stars and light blue symbols are those that are within 5 kpc of the Sun and with [α/Fe]$< -0.14 - 0.35$[Fe/H]. The superimposed isochrones (orange and green lines) show that an age range of 10–13 Gyr is compatible with the HRD of G–E (Helmi *et al.*, 2018). Used with permission of *Nature* from Helmi *et al.* (2018). Permission conveyed through Copyright Clearence Center, Inc.

stars for two ages, 10 and 13 Giga-years, and various abundances. The large metallicity spread of the stars of the retrograde structure implies that they did not form in a single burst in a low-mass system.

Bignone *et al.* (2019) identified a Milky Way analogue in the EAGLE suite of cosmological simulations. This simulated galaxy was selected because its merger history resembles that of the Milky Way with a stellar mass ratio ~ 0.2. The spheroidal component of the galaxy has dynamical properties similar to those of the Milky Way — a significant population of stars were on very eccentric orbits as a result of a merger with another system that was completed at red-shift $z \sim 1.2$. For each star particle, authors calculated the angular momentum L_z as well as the maximum angular momentum along the main axis of rotation, $L_{z,\max(E)}$, which defines the circularity $\epsilon = L_z/L_{z,\max(E)}$ of the star orbit.

Top panel of Fig. 8.7 shows the star formation history of the three components identified in the MW analogue at $z = 0$. A starburst

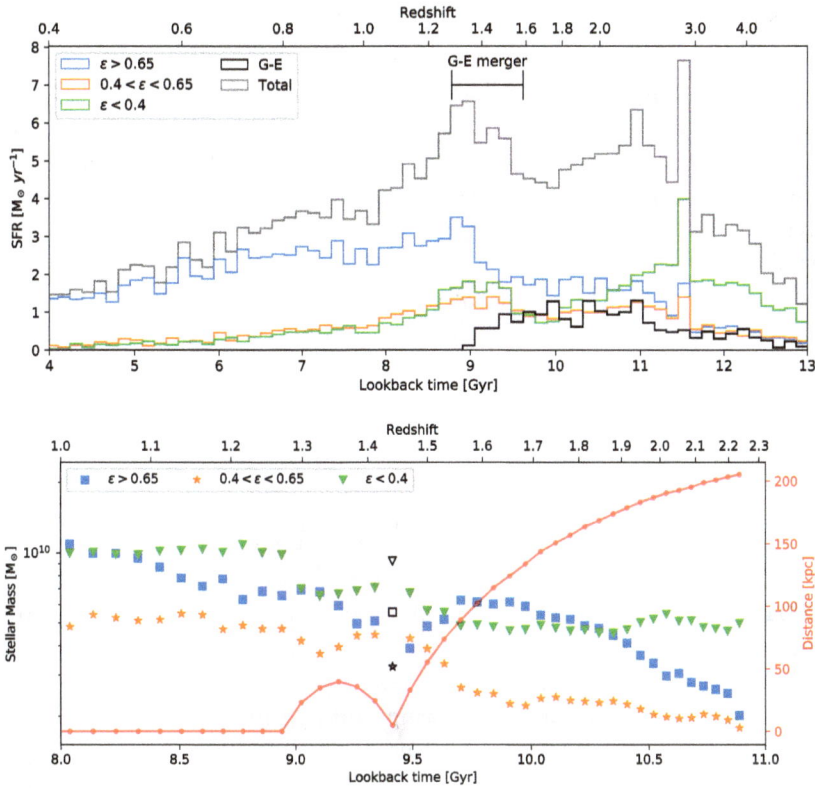

Fig. 8.7. Top: Star formation history of the host galaxy. In black, we show the star formation history of stellar particles accreted from the satellite. Bottom: Evolution of the stellar mass content with lookback time in the host galaxy for stellar particles in three circularity ranges (Bignone *et al.*, 2019).

coinciding with the time of the merger (shaded interval) can be seen for all stellar particles. Stars were associated to components on the basis of their circularity, with low circularity (green) representing the spheroid, intermediate (orange) — the thick disk and high circularity (blue) — the thin disk. Stars accreted from G–E analogue are show in black. The gray line shows the star formation history of all stellar particles. The thick disk in this simulation originates in part from a heated disk (star particles transferred from the thin disk and put on less circular orbits during the merger) and also from stars formed during the merger itself. The shaded region on the left corresponds to the G–E-type merger at $z \approx 1.2$. Note the increase in the SFR of all components during the G–E merger as well as in an earlier merger at redshift $z \approx 3$.

Bottom panel of Fig. 8.7 shows the evolution of the stellar mass with redshift. The stellar mass in particles with $\epsilon < 0.4$ presents a marked increase due to the accretion of stars from the satellite. The stellar mass with intermediate circularities ($0.4 < \epsilon < 0.65$) also presents an increase just before and during the merger that can be attributed to a thickening of the disk triggered by the merger. The shaded region marks the lookback times between 8.8 and 9.6 Gyr, when the merger is expected to have its most significant dynamical effects. The red line represents the distance between the centres of mass of the satellite and its host galaxy.

8.1.4 *Evolution of galactic bulges*

Bulges are dominant populations of elliptical galaxies and are present in disk-dominated galaxies. As discussed by Kormendy & Kennicutt (2004), the origin of bulges in giant elliptical galaxies is very different from the origin of bulges in dwarf elliptical galaxies and disk-dominated spheroidal galaxies. At early times, galactic evolution was dominated by hierarchical clustering and merging, processes that are violent and rapid. These processes formed bulges of giant elliptical galaxies. The other process is secular: the slow rearrangement of energy and mass that results from interactions involving collective phenomena, such as bars, oval disks, spiral structure, and triaxial dark haloes. This process forms bulges of dwarf elliptical galaxies and spheroidal galaxies. Kormendy & Kennicutt (2004) called these populations pseudobulges.

Various populations are in the dynamical equilibrium due to various processes. Globular clusters, open clusters and the compact nuclear star clusters in galaxies are supported by random motions. In the absence of any gas infall, they spread in three dimensions by outward energy transport. The mechanism is two-body relaxation, and the consequences are core collapse and the evaporation of the outer parts. A disk is supported by rotation, so evolution is by angular momentum transport. The "goal" is to minimise the total energy at fixed total angular momentum.

As discussed by Kormendy (2013), the slow secular evolution of galaxies of various types can be explained by various rotation

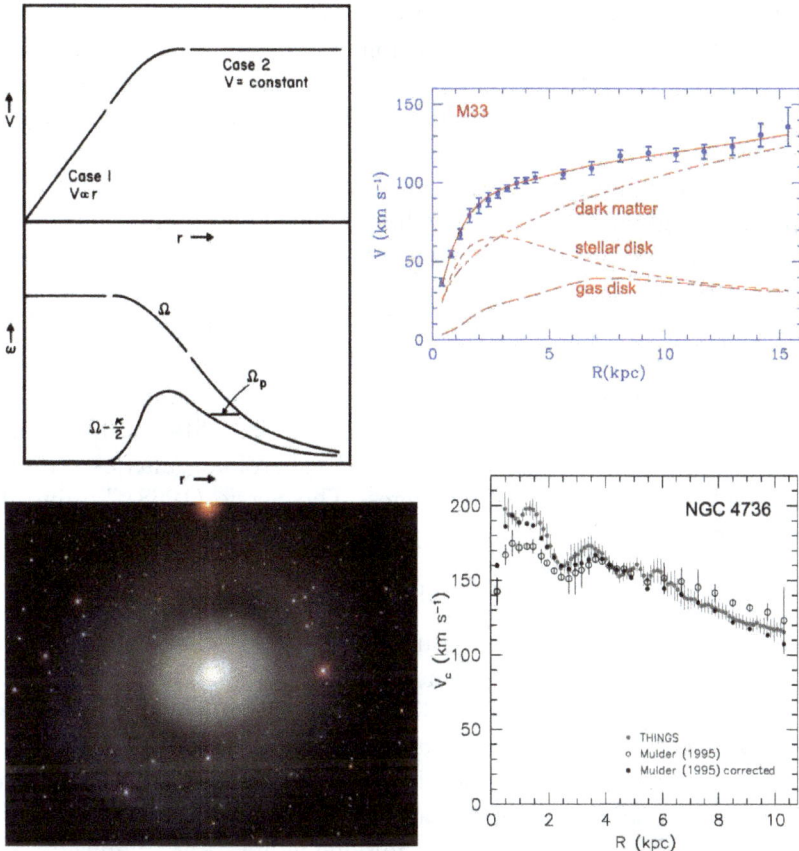

Fig. 8.8. Typical examples of spiral galaxies that do (NGC 4736, bottom) and do not (M33, top) have mass distributions that are conducive to secular evolution (Kormendy, 2013).

velocity laws. Suitable examples are galaxies NGC 4736 and M33, as shown in Fig. 8.8. The top-left panel of Fig. 8.8 shows for the M33 galaxy the rotation curve $V(r)$ and associated angular velocity curves $\Omega(r) = V(r)/r$ and $\Omega - k/2$, where k is the epicyclic frequency. In this galaxy, $V \sim r$ and $\Omega = $ const., thus the secular evolution is not energetically favourable. Even if $V(r)$ is turning downwards from $V \sim r$ towards $V = $ const., the decreases of $V(r)$ outwards is so slow that the secular evolution is disfavoured. This is the situation for galaxies like M33, whose rotation curve and its decomposition to populations are shown in the right top panel. M33 has neither a classical bulge nor a significant pseudobulge. The galaxy NGC 4736 (image at bottom left), has a rotation curve that decreases outwards with radius, see bottom-right panel. Disk spreading is energetically very favourable. The rotation curve of NGC 4736 suggests a mass distribution which supports the formation of the pseudobulge by secular evolution, and the galaxy M33 does not support it. NGC 4736 is a typical strongly oval galaxy, so it has an engine for secular evolution. It is a good example for secular evolution — an unbarred galaxy with a pseudobulge.

Kormendy *et al.* (2009) used the high spatial resolution of the Canada–France–Hawaii Telescope (CFHT) to obtain surface photometry of elliptical and spheroidal galaxies in the Virgo cluster. Authors showed that the Sersic (1968) functions fits the brightness profiles $I(r)$ of nearly all elliptical and spheroidal galaxies remarkably well over large dynamic ranges. The Sersic (1968) luminosity density profile is given by the expression

$$I(r) = I_0 \exp[(-r/r_0)^{(1/n)}], \tag{8.1}$$

where l_0 is the central luminosity density, r_0 is the scaling radius, and n is the Sersic index, which we denote as n_S. The Sersic profile is identical to the Einasto profile (2.4), but instead of the spatial density, it is applied to the projected (surface) density.

Kormendy *et al.* (2009) found three kinds of departures from the Sersic fit. All 10 Virgo ellipticals with total absolute magnitudes $M_V \leq -21.7$ have cuspy cores — "missing light" — at small radii. All 17 ellipticals with $-21.7 \leq M_V \leq -15.5$ do not have cores. Authors emphasise that cores of galaxies are scoured by binary black holes (BHs) formed in dissipationless ("dry") mergers. The third departure is observed in cD galaxies, which have extra light at large radii with respect to the outward extrapolation of a Sersic functions fitted to the inner profile.

These departures are well seen in Fig. 8.9, where we show the Sersic function fits for two galaxies, the central cluster galaxy NGC 4486 (M87), and the brightest cluster spheroidal galaxy NGC 4482. The galaxy M87 has two departures form the Sersic fit: at small radii, the luminosity is lower than the overall fit with index $n_S = 11.8$, an indication of the presence of a cuspy core with missing light. The large n_S value is due to the inclusion of a low-surface-brightness cD halo. If

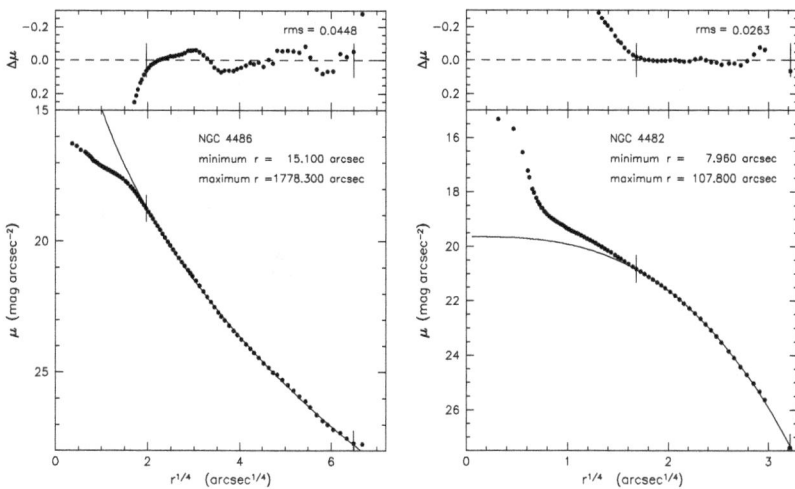

Fig. 8.9. Left: Sersic function fits to the major-axis profile of NGC 4486 (M87), the central galaxy of the Virgo cluster. Right: Sersic function fits to the major-axis profiles of NGC 4482, the brightest spheroidal galaxies in the Virgo cluster (Kormendy *et al.*, 2009).

this halo is excluded, the Sersic fit yields $n_S = 8.9$, a typical value for bright ellipticals. cD haloes are believed to consist of stars that were stripped from individual galaxies by multiple collisions, which are essential for central galaxies of clusters (as discussed in Chapter 2, all central galaxies of clusters of the main filament of the Perseus–Pisces supercluster are cD galaxies). Most ordinary ellipticals have Sersic index $n_S \approx 4$. Right panel of Fig. 8.9 presents the luminosity profile of the brightest spheroidal Virgo cluster galaxy NGC 4482. The Sersic index of the galaxy is $n_S = 1.4$, a typical value for the disk of a spheroidal galaxy. The inner profile shows the presence of extra light.

Figure 8.10 summarises the physical properties of the surface brightness profiles of gas-rich wet merger remnants and gas-poor or dry remerger remnants. In the left panel, the distribution of the surface luminosity density is given for a gas-rich merger remnant and in the right panel, the same is given for a gas-poor remerger remnant. The gas-rich merger remnant is a two-component system, with a Sersic-like violently relaxed envelope with characteristic Sersic index $n_S \approx 2.7 \pm 0.7$, and an inner dissipational/extra-light

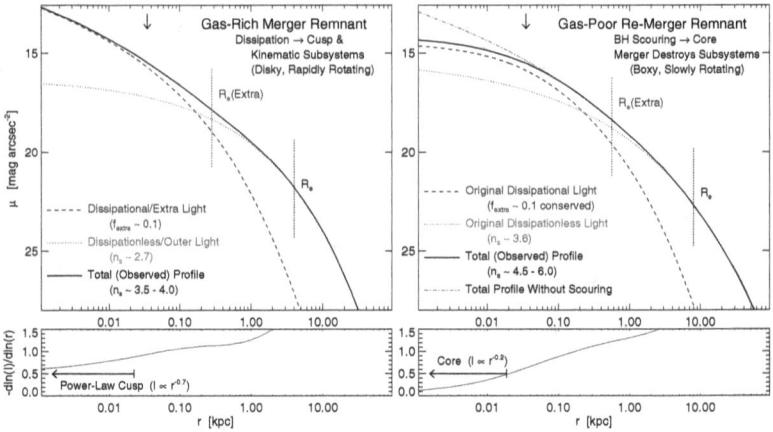

Fig. 8.10. Summary of the physical properties of spheroid profile shapes for a typical $\sim L^*$ elliptical, as formed in a major, gas-rich merger (left) and modified by a major gas-poor, spheroid–spheroid remerger (right) (Hopkins *et al.*, 2009).

component that dominates the light profile inside of $\sim 0.5 - 1$ kpc. The inward extrapolation of this extra light forms the central cusp, a power-law-like continuation with characteristic power-law slopes $d \ln I / d \ln r \approx -0.7 \pm 0.3$. Lower panels of the figure show the logarithmic derivative of the total (observed) profile. In the original merger, stars from the premerger disks are violently relaxed into an extended Sersic-like envelope of index $n_S \approx 2.7$. Conservation of phase-space density prevents this component alone from reaching the high densities of the observed ellipticals. Gas dissipation, however, yields a nuclear starburst, leaving a dense "extra-light" component on top of the outer component, dominating the profile at $r \lesssim 0.5 - 1$ kpc. The nuclear "cusp" is the inward continuation of this dissipational component. Together this yields a global profile with typical Sersic index $n_S \approx 4$ and densities of the observed ellipticals, which are much higher than their progenitor spirals.

In a dissipationless remerger, shown in the right panel of Fig. 8.10, both components are "puffed up" by a factor of ~ 2. The scattering of stars makes the envelope of the original dissipationless component broader, raising its Sersic index to $n_S \approx 3.6$ and leading to a total observed profile with Sersic index $n_S \approx 4.5 - 6$. There is no new dissipation, but the remnant of the original dissipational component continues to dominate the inner profile within ~ 1 kpc. With no scouring mechanism, the central slope would still eventually rise about as steeply as the progenitor cusp. Scouring, or scattering of stars from the central cusp by a nuclear black hole, evacuates $\sim M_{BH}$ worth of stars and flattens the central profile to form a nuclear "core". Although the apparent effects on the extra-light profile can be large, the total mass in the "scoured" region here (the "missing mass") is only $\sim 2\%$ of the extra-light mass, most of which is near $\sim 0.5 - 1$ kpc.

Kormendy & Bender (2019) investigated the structure and evolution of structural analogues of Milky Way galaxies NGC 4565 and NGC 5746. All three are giant SBb–SBbc galaxies with two pseudobulges, i.e. a compact, disky, star-forming pseudobulge embedded in a vertically thick, "red and dead" boxy pseudobulge, which really is a bar seen almost end-on. Authors used the Hobby–Eberly 10-m Telescope spectroscopy in McDonald Observatory to analyse the chemical composition and structure of NGC 4565 and NGC 5746 galaxies. These two galaxies as well as our MW are relatively mature in that most of their gas has been evacuated inside the inner ring. This has been interpreted as the signature of gas shocks that result in gas flow towards the centre. Authors determined chemical composition for NGC 4565 and NGC 5746, and calculated structural parameters of main components of all three galaxies. Circular velocities of main bodies of galaxies are $V_c = 220 - 311$ km/s, absolute magnitudes are $M_K = -23.7, -24.8, -25.3$, respectively. Nuclear clusters relative masses in units of the total stellar mass are of the order of 0.0004; relative stellar masses of disky and boxy pseudobulges are about 0.04 and 0.28, respectively. Scale heights of disky and boxy pseudobulges are about 90 pc and 500 pc, respectively. Authors emphasise that bar-driven secular evolution drives some disk gas towards the centre, where it forms stars and builds "disky pseudobulges".

8.1.5 *Formation and evolution of globular clusters*

Globular star clusters (GCs) are among the oldest luminous objects in the Universe. With typical masses 10^4–10^6 $M_\odot$ and compact sizes (half-light radii of a few pc), they are easily observable in external galaxies. Their masses are comparable to masses of dwarf galaxies, but in contrast to galaxies, they are compact stellar systems, as seen in Fig. 8.1, and there exist no intermediate stellar systems which could fill the gap between dwarf galaxies and GCs.

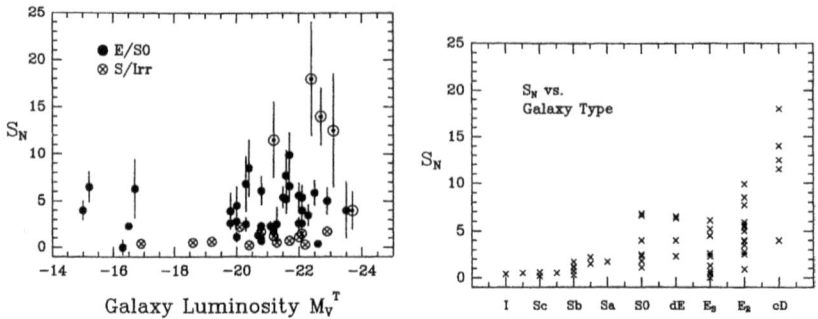

Fig. 8.11. Specific frequency S_N of globular cluster systems in galaxies of various luminosity and morphological type. In the left panel, S_N is plotted versus the luminosity of the parent galaxy. The five giant ellipticals at the centres of rich clusters (Virgo, Fornax, Hydra, Coma, A2199) are denoted by circled dots. In the right panel, S_N is plotted against the morphological type. Here, E_S and E_R refer to ellipticals in sparse and rich clusters. The last bin cD denotes five central ellipticals in clusters (Harris, 1991). Used with permission of *Annual Reviews* from Harris (1991). Permission conveyed through Copyright Clearence Center, Inc.

The number of GCs in galaxies is very different. To characterise the ability of galaxies to develop GCs, Harris & van den Bergh (1981) proposed to use specific frequency S_N defined as the number of GCs per unit luminosity (absolute magnitude $M_V = -15$). Harris (1991) collected available data on specific frequencies of galaxies of various morphological type and luminosity, results are shown in Fig. 8.11. We see that the specific frequency depends on the luminosity of the parent galaxy only moderately. Much stronger is the dependence on the morphological type. The highest specific frequency have central galaxies of rich clusters of galaxies of type cD.

GCs are found in all galaxies in the Local Universe and have old ages of $\sim 10\,\mathrm{Gyr}$. The origin of GCs is a major unsolved problem on the interface between star and galaxy formation. In part, our understanding of GC formation is limited because most GCs in the Universe must have formed at redshifts $z > 2$. At such distances, the sizes of star-forming giant molecular clouds (GMCs) are unresolved, implying that the physical processes leading to GC formation must be inferred indirectly. We discuss in the following some observational properties of GCs and thereafter models of GC formation.

The traditional concept of GCs is that they form a simple population of compact stellar systems, where all stars share the same age and abundances. New ground-based and HST observations indicate distinctive chemical anomalies that cause the complexity in colour–magnitude diagrams (CMDs). These star-to-star abundance variations within clusters are known as multiple populations (MPs). Stars in GCs can be divided to two main populations: (1) primordial or first population (1P) stars: stars having the same abundances as the field at the same metallicity [Fe/H] and (2) enriched or second population (2P) stars: stars showing enhanced N, Na, and Al and depleted C and O abundances with respect to the field at the same metallicity [Fe/H] (Bastian & Lardo, 2018).

Bastian & Lardo (2018) presented the CMD of one of the most massive Milky Way globular cluster NGC 2808, as shown in Fig. 8.12. The left panel shows the CMD of the central region of the GC. Note the distinct multiple red-giant-branch (RGB) stars and the highly structured horizontal branch. This complexity is due to light element abundance variations (He, C, N, and O) between cluster stars. The right panel shows two-colour "chromosome map", where the x-axis is mainly sensitive to variations in He(Y), while the y-axis is dominated by variations in $N(\Delta N)$. Stars above the dashed line are considered to be population 2P, whereas stars below the same line are

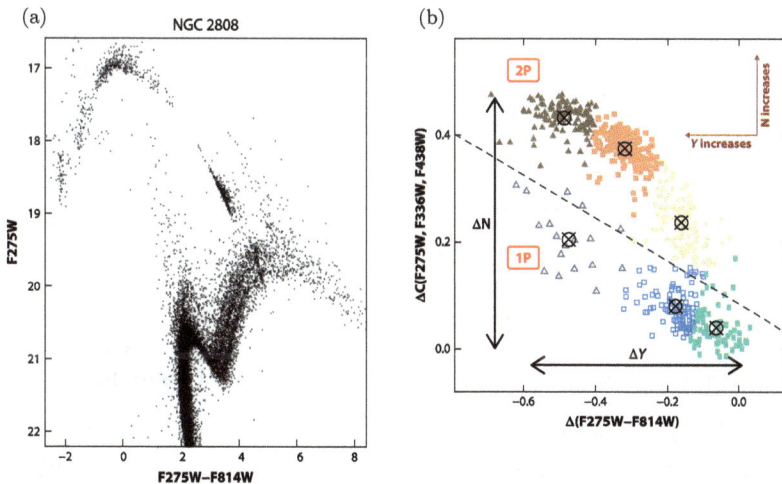

Fig. 8.12. (a) An HST UV–optical CMD of the central regions of NGC 2808; (b) a "chromosome map" of NGC 2808 for red giant branch (RGB) — relative positions of the stars on the RGB in different filter combinations that are sensitive to different abundance variations (Bastian & Lardo, 2018). Used with permission of *Annual Reviews* from Bastian & Lardo (2018). Permission conveyed through Copyright Clearence Center, Inc.

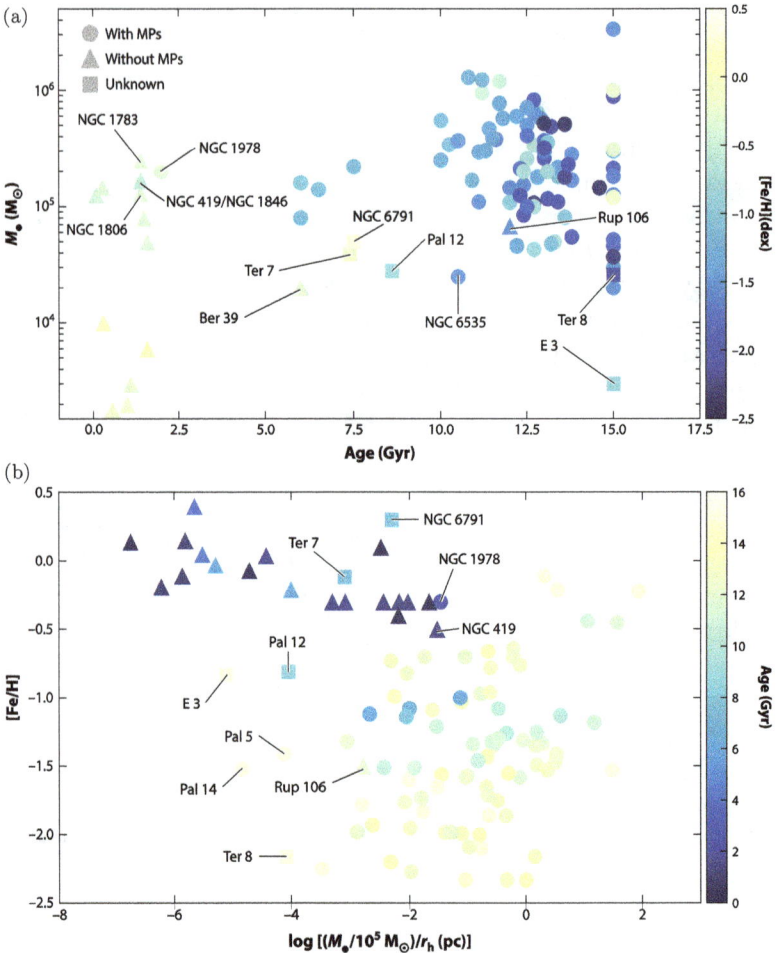

Fig. 8.13. Distribution of GCs with and without multiple populations (MPs) in (a) the age–mass plane and (b) the concentration–[Fe/H] plane (Bastian & Lardo, 2018). Used with permission of *Annual Reviews* from Bastian & Lardo (2018). Permission conveyed through Copyright Clearence Center, Inc.

population 1P. Note that both the 1P and 2P consist of three subpopulations with distinct chemistry. The cluster NGC 2808 belongs now to the Milky Way, although it was likely stolen from a dwarf galaxy that collided with the Milky Way. The summary of the analysis of the presence of multiple populations is given in Fig. 8.13. Distributions are given in age–mass and concentration–[Fe/H] planes. Circles denote clusters where MPs have been unambiguously detected, triangles show where they have not been detected, and squares show ambiguous cases. Some particularly interesting cases are labelled. An age of 15 Gyr has been assigned to clusters for which no age determination has been found in the literature. Whether or not a cluster hosts MPs depends on its mass (or density) as well as its age.

Kruijssen (2015) presented a two-phase model for the origin of GCs. In this model, early populations of stellar clusters form in the high-pressure thin disks of high-redshift ($z > 2$) galaxies. This is a rapid-disruption phase due to tidal perturbations from the dense interstellar medium. Thereafter, galaxy mergers associated with hierarchical galaxy formation redistribute the surviving, massive clusters into the galaxy haloes, where they remain until the present day (a slow-disruption phase due to tidal evaporation). The high galaxy merger rates of $z > 2$ galaxies allow these clusters to be "liberated" into the galaxy haloes before they are disrupted within the high-density disks. The timeline of this two-phase model of GC formation is presented in Fig. 8.14. At the time t_{form}, stars and young massive clusters (YMCs) are formed through the regular star formation process in the gas-rich and high-pressure disks seen at high redshift (see panel (a)). During their subsequent evolution within the host galaxy disk (Phase 1), the clusters undergo rapid disruption by tidal shocks for a total duration t_{disc} until at a time t_{merge}, the host

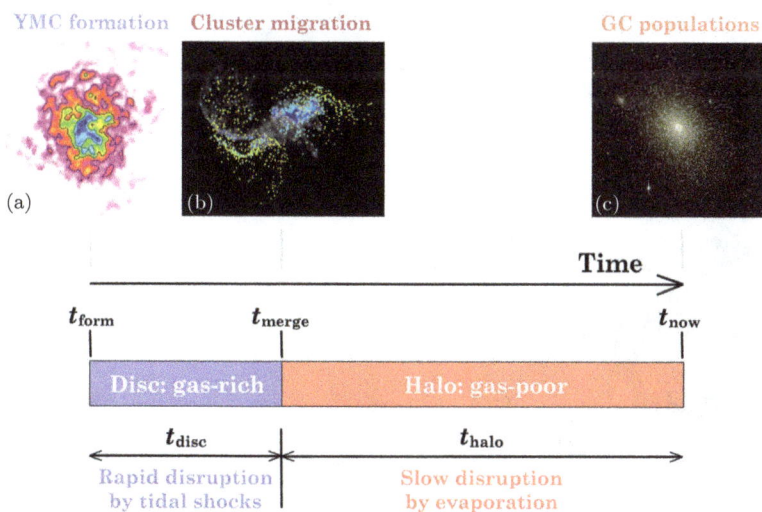

Fig. 8.14. Panel (a) shows the clumpy $z = 4.0$ galaxy, panel (b) shows the migration of star clusters in a galaxy merger, and panel (c) shows the GC population of the early type galaxy NGC 4365. (Kruijssen, 2015).

galaxy undergoes a merger and the clusters migrate into the galaxy halo (see panel (b)), thereby increasing their long-term survival chances and thus becoming GCs. The GCs subsequently undergo quiescent dynamical evolution in the galaxy halo (Phase 2), which is characterised by slow disruption by evaporation (see panel (c)). This phase lasts for a total duration t_{halo} until the present day t_{now}. Panel (a) shows the clumpy galaxy at redshift $z = 4.05$, panel (b) shows the migration of stellar clusters in a galaxy merger model, and panel (c) shows the GC population of the early-type galaxy NGC 4365.

Until recently, the formation and evolution of GCs was studied on galactic scale, and details of physical processes in the evolution of precursors of GCs, giant molecular clouds (GMCs) were not fully understood. Modern observational possibilities allow us to study these processes on pc scale. Kruijssen *et al.* (2019c) performed observations of the nearby almost face-on star-forming galaxy NGC300 to understand the physical processes during the evolution of GMCs. Authors used the Atacama Large Millimeter/submillimeter Array (ALMA) to observe with high resolution NGC300 carbon monoxide CO(1-0) emission, and the MPG/ESO 2.2-m telescope to map the hydrogen Hα emission. The Hα emission characterises star formation regions and the CO(1-0) emission GMCs. Authors derived CO-to-Hα flux ratios for increasing apertures, large aperture means the galactic average. Results of this analysis are shown in Fig. 8.15.

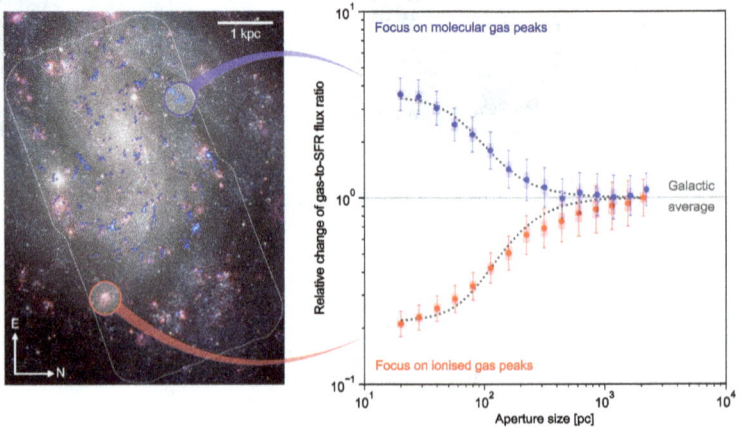

Fig. 8.15. Decorrelation of molecular gas and young stellar emission on sub-kpc scales (Kruijssen *et al.*, 2019c). Used with permission of *Nature* from Kruijssen *et al.* (2019c). Permission conveyed through Copyright Clearence Center, Inc.

The ALMA CO(1-0) map is shown in the left panel as a blue overlay on top of an optical composite image of NGC300 taken with the MPG/ESO 2.2-m telescope. The white contour indicates the extent of the CO map. Emission from young stars traced by Hα is shown in pink and emission from GMCs in blue. The change of the CO-to-Hα flux ratio relative to the galactic average is shown as a function of spatial scale in the right panel for apertures placed on CO emission peaks (top branch) and Hα emission peaks (bottom branch). The change of the CO-to-Hα flux ratio relative to the galactic average is shown as a function of spatial scale in the right panel for apertures placed on CO emission peaks (top branch) and Hα emission peaks (bottom branch). The scale bar shows the projected size scale uncorrected for inclination. The evolutionary timeline of the GMC lifecycle is indicated by the dotted lines. The left panel demonstrates that CO(1-0) and Hα emissions are rarely co-spatial — they do not trace each other on the cloud scale. This is quantified in the right panel, which shows the CO-to-Hα flux ratio as a function of the aperture size for apertures placed on either CO(1-0) or Hα emission peaks.

Kruijssen *et al.* (2019c) modelled how long GMCs live and form stars (t_{CO}), how long feedback takes to evacuate residual gas (t_{fb}, the time for which GMCs and HII regions coexist). The measured GMC lifetimes are spanning $t_{CO} = 9 - 18$ Myr, with a galactic average of $t_{CO} = 10.8 \pm 2$ Myr. The feedback timescale is $t_{fb} = 1.5 \pm 0.2$ Myr, the mean SFR $\epsilon_{sf} = 2.5\%$, and the characteristic region separation length 104 ± 20 pc. The timescale for depleting molecular gas is $t_{dep} \approx 2$ Gyr (Leroy *et al.*, 2008). The gas depletion time allows us to find the timescale over which gas is turned into stars, $t_{sf} = t_{dep} \times \epsilon_{sf} \approx 50$ Myr. These GMC lifetimes fall between the GMC crossing and gravitational free-fall times. The GMCs live and form stars for a dynamical timescale, after which they are dispersed relative to the new-born stellar population. The cloud-scale evolutionary cycling is illustrated in Fig. 8.16. The symbols show the observed relation between the total gas surface density (Σ), where the high-redshift sample is assumed to be fully molecular, and the SFR surface density (Σ_{SFR}) for galaxies in the Local Universe and at high redshift. Dotted lines represent constant gas depletion times, as indicated by the labels. The results of this work show that GMCs and star-forming regions move through this diagram. As a function of time, they increase their gas density, increase their SFR, expel gas through feedback, and eventually fade by stellar evolution, which is schematically illustrated by the red arrows.

Kruijssen *et al.* (2019a) presented cosmological simulations of Milky-Way-mass galaxies in the "MOdelling Star cluster population Assembly In Cosmological Simulations within EAGLE" (E-MOSAICS) project. Figure 8.17 shows the locations (yellow circles, with radii indicating the virial radii) of the 25 haloes in their parent EAGLE volume. The main panel shows the dark matter distribution of the EAGLE Recal simulation at $z = 0$, with the resimulated haloes marked with yellow circles, of which the sizes indicate the virial radii. Solid circles denote the two galaxies featured in the top right-hand panel, which shows gas density coloured by temperature, with red and white hues corresponding to $T = 10^5$ K and $T = 10^6$ K, respectively. The two middle right-hand panels zoom in on one galaxy (MW23) and show mock optical images, with the

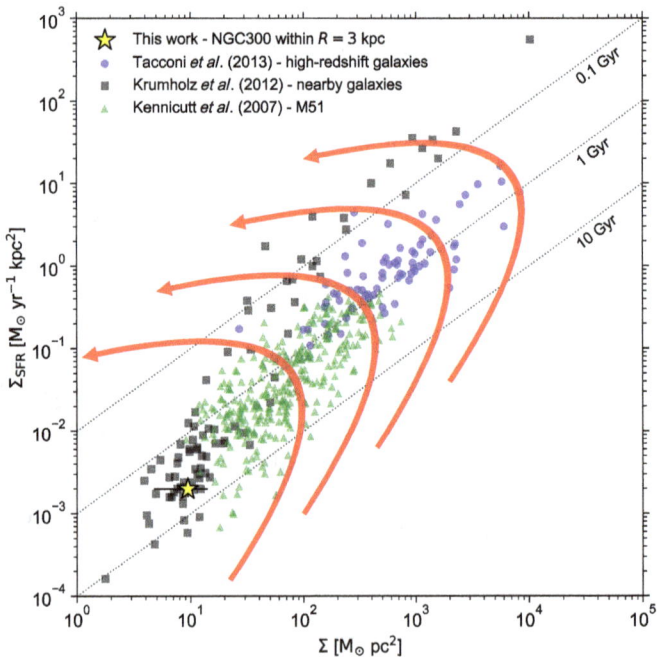

Fig. 8.16. Schematic illustration of cloud-scale evolutionary cycling in the $\Sigma - \Sigma_{\rm SFR}$ plane (Kruijssen *et al.*, 2019c). Used with permission of *Nature* from Kruijssen *et al.* (2019c). Permission conveyed through Copyright Clearence Center, Inc.

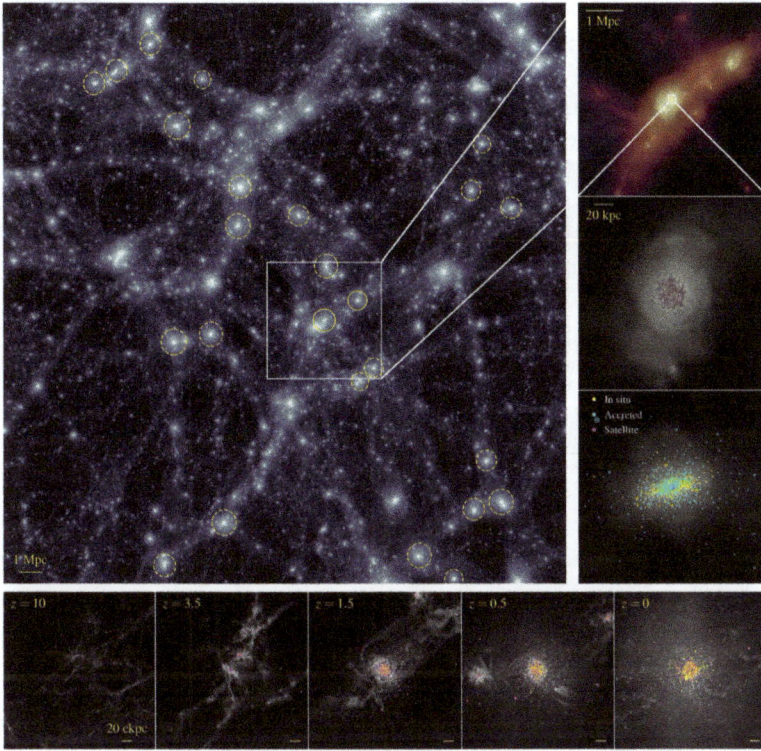

Fig. 8.17. Visualisation of the E-MOSAICS simulations to highlight all 25 Milky-Way-mass ($L^\star$) galaxies (Kruijssen *et al.*, 2019a).

bottom panel including massive ($M \geq 5 \times 10^4\,\mathrm{M}_\odot$) stellar clusters as dots, coloured by their origin — those that currently reside in a satellite are shown in magenta, those that formed in a satellite galaxy and were subsequently accreted are shown in cyan, whereas those formed in the main progenitor are shown in yellow. Some of the GCs formed *in situ* during the build-up of the main progenitor, whereas others formed *ex situ* in lower-mass satellites that have since been accreted. The bottom row visualises the assembly of the same galaxy and its cluster population, from $z = 10$ to $z = 0$, with gas shown in greyscale and coloured dots again representing the massive star clusters, this time coloured by their metallicities ($-2.5 \leq [Fe/H] \leq 0.5$). We see the formation of a number of satellites along filaments, who by merging form the final galaxy at redshift $z = 0$.

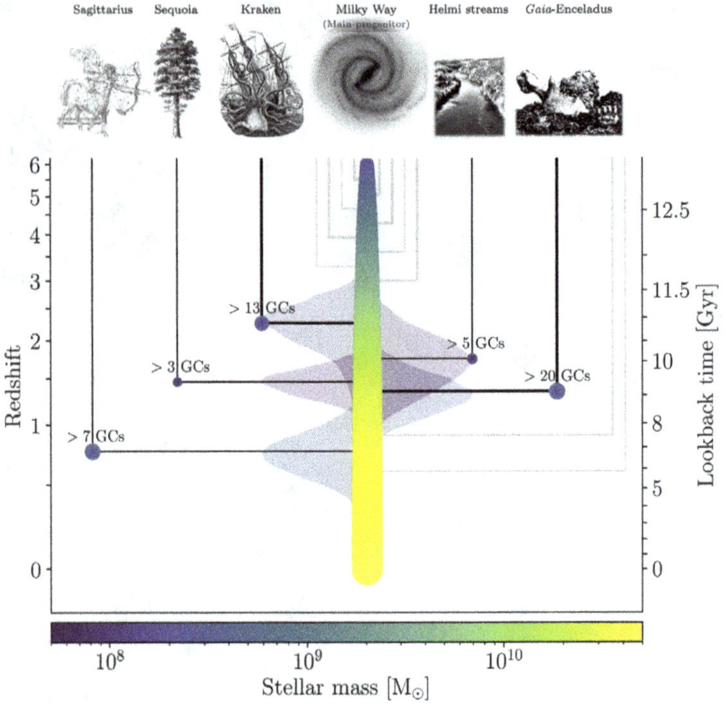

Fig. 8.18. Galaxy merger tree of the Milky Way inferred by applying the insights gained from the E-MOSAICS simulations to the Galactic GC population (Kruijssen *et al.*, 2020).

Kruijssen *et al.* (2020) used the information contained in the galactic GC population to quantify the properties of the satellite galaxies from which the Milky Way assembled. To achieve this, authors trained an artificial neural network on the E-MOSAICS cosmological simulations of the coformation and coevolution of GCs and their host galaxies. The network uses the ages, metallicities, and orbital properties of GCs that formed in the same progenitor galaxies to predict the stellar masses and accretion redshifts of these progenitors. Authors applied the network to Galactic GCs and found five progenitors associated with MW: G–E, the Helmi streams, Sequoia, Sagittarius, and the recently discovered "low-energy" GCs of the galaxy "Kraken". Figure 8.18 shows the merger tree of the MW. This figure summarises the main results presented in this paper. The main progenitor is denoted by the trunk of the tree, coloured by stellar mass (based on the reconstruction from Kruijssen *et al.* (2019b): see the colour bar. Black lines indicate the five identified (and likely most massive) satellites, with the shaded areas visualising the accretion redshifts. The coloured circles indicate the stellar masses at the time of accretion (both with their colours and sizes). The annotations list the minimum number of GCs brought in by each satellite. Light grey lines illustrate the global statistics of the Milky Way's merger tree inferred by Kruijssen *et al.* (2019b). Thin lines mark tiny mergers (with a mass ratio of $r_{M_\star} < 1/100$) and thick lines denote minor (or possibly major) mergers (with a mass ratio of $r_{M_\star} > 1/100$). This merger tree is consistent with the stellar mass growth history of the MW, the total number of mergers (N_{br}), the number of high-redshift mergers ($N_{N_{\mathrm{br}}, z>2} > 2$), and the numbers of tiny and minor mergers from Kruijssen *et al.* (2019b), as well as with the five identified satellite progenitors discussed in this work, including their stellar masses, accretion redshifts, and GC populations. Note that only progenitors with masses of $M_\star > 4.5 \times 10^6$ M$_\odot$ are included. From left to right, the six images at the top of the figure indicate the identified progenitors: Sagittarius, Sequoia, Kraken, the MWs Main progenitor, the progenitor of the Helmi streams and of the G–E.

8.1.6 *Galactic luminosity functions in various environments*

One of the aims of this book is to point out that many properties of galaxies — their morphologies, colours, gas contents, star formations, masses, dimensions, etc. — depend on whether they reside in the field, groups or clusters, and in a wider cosmological context in voids, filaments, walls, etc. Actually, the distribution of luminosities of galaxies is different for brightest cluster galaxies (BCG) and satellite galaxies. Luminosities of galaxies also depend on the global environment. These relations can be determined for nearby galaxies. Now, we describe galactic luminosity functions (LFs) in the context of the global environment.

Tempel *et al.* (2009) used the 2dFGRS to investigate LFs for galaxies in different environments. Figure 8.19 shows differential LFs in various environments — voids, filaments, superclusters (SC) and

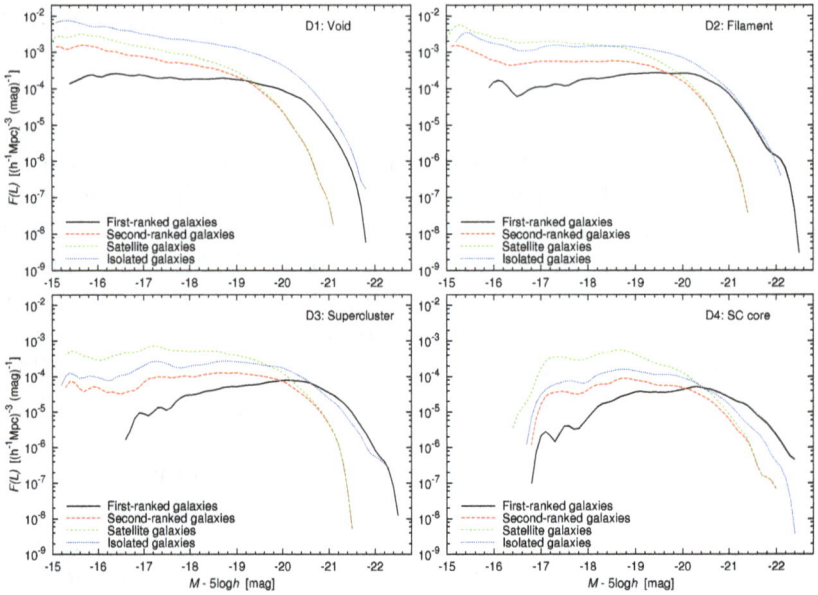

Fig. 8.19. Differential LFs in different environments and for different galaxy populations. Top-left panel: void environment; top-right panel: filament environment; bottom-left panel: supercluster environment; bottom-right panel: supercluster core environment. Solid line: first-ranked galaxies; dashed line: second-ranked galaxies; short-dashed line: satellite galaxies; dotted line: isolated galaxies (Tempel *et al.*, 2009).

SC cores. LFs were calculated for various galaxy populations — first-ranked galaxies, second-ranked galaxies, satellite galaxies, and isolated galaxies. Figure 8.19 (upper-left panel) shows that in voids, the bright end of LFs of all galaxy populations is shifted towards lower luminosities — in the void environment, first-ranked galaxies of groups are fainter than those in higher-density environments. The bright end of the LF for isolated galaxies in voids is comparable to that of first-ranked galaxies. The LFs of second-ranked galaxies and of all satellites are comparable, although there is a slight increase in the LF of satellite galaxies at the lowest luminosities. The LF of first-ranked galaxies, in contrary, has a plateau at the faint end without signs of increase.

In the filament environment, the bright ends of the LFs for first-ranked galaxies and for isolated galaxies are similar, while the brightest second-ranked galaxies and satellite galaxies are fainter. For a wide range of luminosities, the LF for first-ranked galaxies is slowly decreasing towards fainter luminosities, while LFs for other galaxy populations have a plateau.

The supercluster environment represents poor superclusters and the outskirt regions of rich superclusters. The figure shows that the first-ranked galaxies in the supercluster environment have luminosities comparable to those of the first-ranked galaxies in filaments, but the LFs for faint galaxies show a decrease towards the faint end. The LF for first-ranked galaxies in this region has a well-defined faint luminosity limit (approximately −17 mag), while the LFs for other galaxy populations extend to fainter luminosities.

The LFs for supercluster cores are shown in the lower-right panel of Fig. 8.19. We note the striking difference between the LFs in supercluster cores and the LFs in other environments: here, all LFs have a well-defined lower luminosity limit, about −17 mag, which for first-ranked galaxies was already seen in supercluster environment. Also, in supercluster cores, the brightest first-ranked galaxies are more luminous than the brightest first-ranked galaxies in other environments. In summary, the most dense environment (supercluster cores) is different from other environments: there are no very faint galaxies, and the brightest first-ranked galaxies are brighter than the first-ranked galaxies in lower-density environments. The lower luminosity limit is shifted to smaller luminosities if we move to less dense environments. The transition between different environments is smooth.

The overall shape of the LFs in Fig. 8.19 suggests that isolated galaxies may be a superposition of two populations: the bright end of their LF is close to that of the first-ranked galaxy LF and the faint end of the LF is similar to the LF of satellite galaxies. This is compatible with the assumption that the brightest isolated galaxies in a sample are actually the brightest galaxies of invisible groups.

In the supercluster core environment, the brightest isolated galaxies are fainter than the brightest first-ranked galaxy, but they are almost as bright as the second-ranked galaxies in this environment. Earlier, we showed that the second-ranked galaxies in high-density regions are similar to first-ranked galaxies in lower-density regions. In other words, second-ranked galaxies in supercluster core clusters can be considered as first-ranked galaxies of clusters before the last merger event.

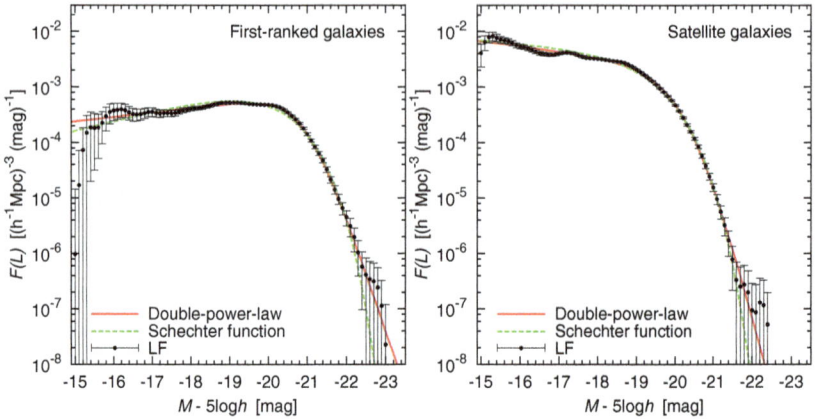

Fig. 8.20. Differential LFs for various galaxy populations: first-ranked and satellite galaxies. The points denote LFs using 2dF galaxy catalogue. Error bars are Poisson $1 - \sigma$ errors. The red solid line denotes the double-power-law and the green dashed line denotes the Schechter function (Tempel *et al.*, 2009).

Luminosities of galaxies also depend on the local environment — whether galaxies are central (first-ranked) in groups or satellites. Figure 8.20 presents the LFs for various galaxy populations: first-ranked and satellite galaxies. When calculating LFs, we have selected galaxies from all density regions. The LFs have a well-defined bend around $L^\star \simeq 10^{10}\, h^{-2} L_\odot$ ($M^\star - 5 \log h \simeq -20$). There are remarkable differences between faint regions of LFs of first-ranked and satellite galaxies: the LFs of satellite galaxies slightly increase by moving towards lower luminosities, but for the first-ranked galaxy sample, the LF is decreasing.

For all galaxy populations, we have fitted for LFs Schechter and double-power-law functions. In general, both functions give a pretty good fit. There is still one big difference between the Schechter and the double-power law: for most populations, the Schechter law predicts too few bright galaxies; the double-power-law gives a much better fit for the bright end of the LF and a better fit in the bend region.

Properties of the LF of various types of galaxies in different environments can be interpreted by differences of galaxy and group evolution. In supercluster cores, rich groups form through many mergers, thus the second-ranked galaxies have been the brightest galaxies of poorer groups before they have been absorbed into a larger group.

In lower-density environment, the merger rate is lower and groups of galaxies have been collected only from nearby regions through minor mergers and continuous infall of matter to galaxies, as suggested by White & Rees (1978).

8.1.7 *Evolution of galactic luminosity functions*

HST and JWST observations allowed us to find LFs of galaxies visible at early epochs and large distances. By default, these galaxies were most luminous at respective epochs. Most probably, these galaxies were the brightest galaxies of early groups and clusters.

The evolution of the UV LF of galaxies over the redshift ranges $4 < z < 10$ is shown in Fig. 8.21. The figure reveals a rapid decline in the abundance of UV luminous galaxies at $z > 4$. It shows that during the early period with redshifts from $z = 10$ to $z = 4$, the amplitude of the LF increases more than 10-fold, and the faint end slope of the LF decreases from $\alpha \approx -2.2$ to $\alpha \approx -1.6$. The evolution in the luminosity function at $z > 4$ can be explained by a rapid decline in

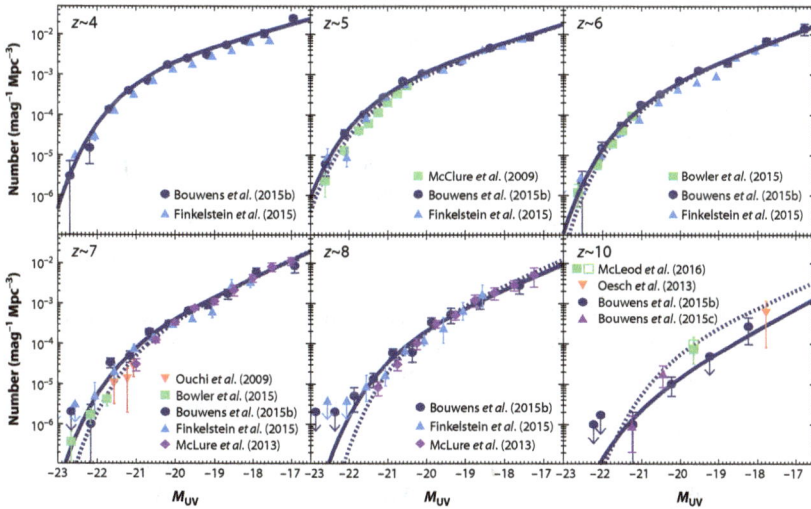

Fig. 8.21. Evolution in the rest-frame ultraviolet (UV) luminosity function of UV-continuum selected dropouts over the redshift range $4 < z < 10$. All data points have been adjusted to a cosmology with $\Omega_m = 0.3$, $\Omega_\Lambda = 0.7$ and $H_0 = 70\,\mathrm{km/s/Mpc}$ (Stark, 2016). Used with permission of *Annual Reviews* from Stark (2016). Permission conveyed through Copyright Clearence Center, Inc.

the characteristic density of galaxies and a steepening of the faint-end slope. The change in the amplitude causes the decrease in the space density of galaxies seen in Fig. 7.49, and the evolution in the faint end slope causes the decline to be less rapid at low luminosities.

8.2   Structure and evolution of local galaxies

RHEA: Andromeda and the 36 unruly dwarfs.

ARIEL: That's a terrible title. Since when is 7 not enough? And why are they unruly? I'd go with Milky Way and the 14 dwarfs. A lot punchier.

CALLISTUS: Both of these fairy tales are spellbinding.

ARIEL: What's wrong with the poor dwarfs?

CALLISTUS: M31 or Andromeda is the closest galaxy to our home, the Milky Way, less than an Mpc away. Despite being our neighbour, a beautiful whirling spiral, its satellites are quite different from ours.[2]

ARIEL: They are named like the infinite sequels of a boring Hollywood movie, Andromeda I, II, all the way to XXXIII –

RHEA: Better than Deadpool and Wolverine III ...

ARIEL: How about Guardians of the Galaxy I, II, etc.?

CALLISTUS: You can take it up with the International Astronomical Union.

ARIEL: Sure. But what makes them the Unruly 36? Which is, by the way, a catchy title.

RHEA: And all the stars aligned... Most satellites reside in a plane, the Great Plane of Andromeda.

ARIEL: Wait, I heard that most of the Milky Way satellites reside in a plane as well. I'm tempted to call it the Great Plane of the Milky Way, the IAU permitting.

CALLISTUS: Some say that such an alignment is inconsistent with our present cosmological theories: simulations are unlikely to produce such confinement in a plane.

RHEA: I'd argue, along with many others, that these results strongly favour MOND.

CALLISTUS: Except, a recent paper[3] claims it's temporary: their velocity field suggests the Milky Way satellites will spread out in the near future. We just happen to see them aligned today by sheer luck.

---

[2]Savino *et al.* (2025).
[3]Sawala *et al.* (2023).

ARIEL: Lucky to be alive today. However, given the alignment of the M31 satellites, I have my doubts. Moreover, I still don't get the unruly part because, frankly, being in a plane is pretty "ruly".

CALLISTUS: Carpe diem.[4] Let's get to the unruliness. While the Milky Way satellites stopped forming stars a while ago, the Andromeda satellites still do. Even if some of M31's companions have fallen in only relatively recently, the observed star formation activity is difficult to explain based on the most recent simulations.

ARIEL: Ugh, star-forming is unruly. I guess that's in the eye of the beholder.

CALLISTUS: We expected these satellites to quiet down in a few gigayears after arriving.

RHEA: Satellite galaxies, while predicted by the concordance model and confirmed in simulations, are a result of complex non-linear and baryonic processes. Our present simulations cannot perfectly track that "sub-grid physics". It's still an art rather than science. Yet, the diversity and the difference between the Milky Way and M31 satellites are fascinating. We need more observations... and theories.

CALLISTUS: Something left for you, Ariel.

ARIEL: I guess these fairy tales do not have an ending yet, happy or otherwise. But they're all the more fascinating.

In this section, we discuss the structure and evolution of several relatively isolated galaxies. Galaxies with stellar masses comparable to the Milky Way ($M_\star \sim 6 \times 10^{10}$ $M_\odot$) host the majority of the stellar mass in the present-day Universe. Most of these galaxies are isolated in the sense that within their DM haloes, there are no other galaxies of the same luminosity. For this reason, their evolution is determined only by local processes within their DM haloes and least influenced by external factors. In contrast, galaxies in groups, clusters and central regions of superclusters are also influenced by external processes. We start our discussion with the analysis of the structure and evolution of our Galaxy and continue with the analysis of the Andromeda galaxy M31 as best examples of single isolated galaxies. We note that very detailed information of Milky Way disk, bulge and globular clusters and their evolution was discussed in the previous section.

---

[4]Seize the day.

8.2.1 *Structure and evolution of Milky Way*

The Milky Way is a fairly typical disk galaxy of estimated stellar mass $\sim 5 \times 10^{10} M_\odot$, which implies a luminosity close to the characteristic value $L^\star$ of the galaxy luminosity function. The Milky Way has several visible populations: a thin disk, thick disk, bulge/bar and a stellar halo. Each of these populations has individual characteristics. Their stars differ in their age, chemical distributions and kinematics, as shown in the Strömberg diagram in Fig. 2.7. This implies that the populations are truly physically distinct. Their constituent stars inform us about the various processes that are important in the build-up of a galaxy throughout its life.

Our Galaxy differs from other galaxies in one important aspect: our position inside the Galaxy does not allow us to measure some important parameters, characterising the Galaxy — the distance of the Sun from its centre, R_0, and the rotation velocity, V_0, at this distance. Both parameters are members of the set of Galactic parameters. Since these parameters enter in all dynamical models of the Galaxy, it is reasonable to agree with their values. To achieve this goal, the Committee 33 of the IAU formed in 1982 a working group to discuss the problem. The report of this group was published by Kerr & Lynden-Bell (1986). Authors discussed all recent determinations of these constants and obtained the following mean values: $R_0 = 8.5 \pm 1.1$ kpc and $V_0 = 222 \pm 20$ km/s. Authors suggested the use of the following as IAU standard values: $R_0 = 8.5$ kpc, $V_0 = 220$ km/s.

Rotation data of various stellar and gas populations have been used to construct a dynamical model of the Galaxy. One of the first models, which takes into account the presence of a dark halo in galaxies, was made by Haud & Einasto (1989). The model has following populations: flat population, disk (both are ring-like, since there exist no young stars in the central region of MW), bulge, stellar halo, and DM halo (corona), see Fig. 8.22. For all populations, the generalised exponential density function, Eq. (2.4), was used. More recent studies by Sofue & Rubin (2001) and Sofue (2017) have reached more distant regions in good agreement with data used by Haud & Einasto (1989).

Gaia Collaboration *et al.* (2021) focused on several aspects of the Galaxy that coexist in the anticentre and that will help us towards

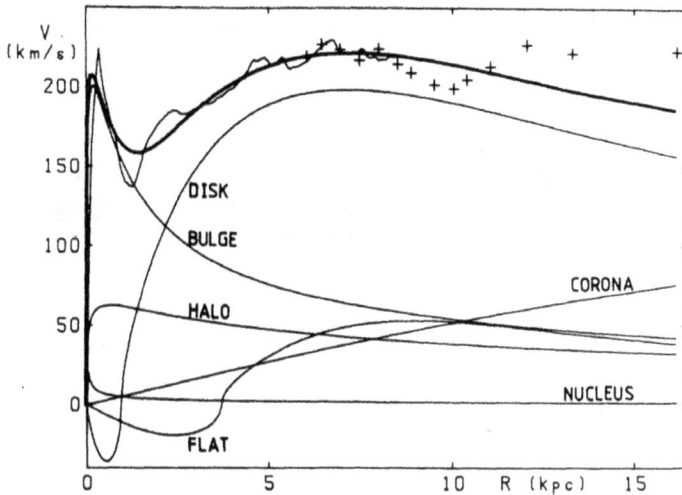

Fig. 8.22. Rotation curve of our Galaxy. Wavy curve: HI observations, crosses: HII observations, thick line: model. Thin lines represent the contributions of populations to the rotation curve (Haud & Einasto, 1989).

answering a single question: how the Galaxy appears today and how it became like this. Authors selected stars in the anticentre region based on their positions in the HRD, constructed on the basis of photometric and colour data. These populations are shown in Fig. 8.23: EYP: extremely young massive stars with ages $\tau \leq 0.2$ Gyr, YP: young main sequence with ages $0.2 \leq \tau \leq 2$ Gyr, IP: intermediate main sequence stars with ages $2 \leq \tau \leq 8$ Gyr, OP: old main sequence stars, RG: red giants, and RC: red clump stars; here τ is the age of stars in Gyr.

The Early Third Data Release (EDR3) of the Gaia mission allowed us to study the kinematics and morphological properties of much fainter stars than previous releases. One of the areas of interest is the anticentre region. The anticentre is a meeting point of several distinct components of the Galaxy (the disk, the halo) and possibly hosts ancient and recently disrupted stellar systems of extragalactic origin. The anticentre is also an excellent window to the dynamics and the past of the MW since, due to the lower gravitational potential, any perturbation on the disk would cause more significant deformations than in the inner disk. Rotation and velocity dispersion profiles of selected populations are shown in Fig. 8.24. We see that

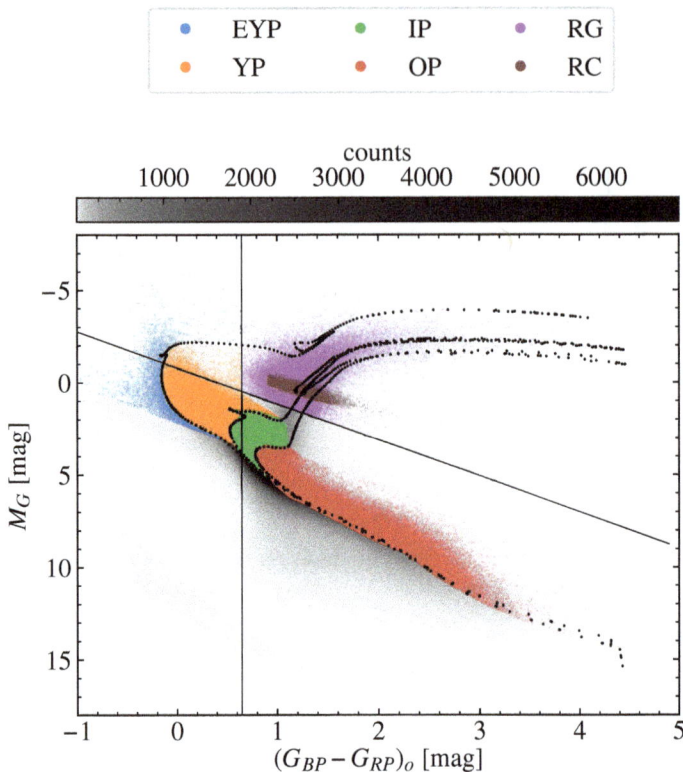

Fig. 8.23. HR diagram of the anticentre region and different selected populations. The diagram is shown for sources with available photometry (G, BP, RP) and extinction data and absolute magnitudes. We overplot three stellar isochrones with $[M/H] \approx 0$ for the ages of 0.2, 2 and 8 Gyr, a line at colour $BP - RP = 0.65$ and a diagonal line following the extinction slope used for the selection of populations (Gaia Collaboration *et al.*, 2021).

the youngest population EYP has the fastest rotation and smallest velocity dispersion profile, and the old main sequence population OP has the slowest rotation and highest velocity dispersion profile. This is consistent with the expectation that younger stars rotate as fast as the cold interstellar gas, thus at velocities closer to the true circular velocity of the MW. Globally, all the rotation curves decline for $R \geq 9.5$ kpc and show a bump at around 10 kpc. Beyond this, the curve of YP stars is flat, while those of older stars decrease again. The second important result is — both the rotation velocity and the

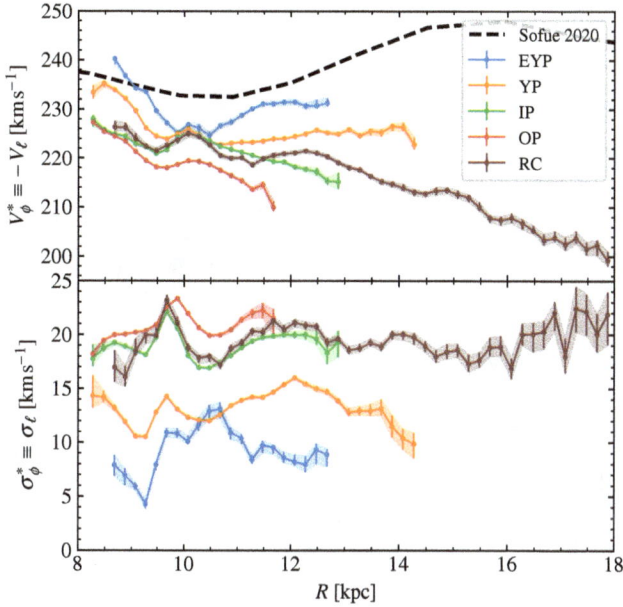

Fig. 8.24. Top: Rotation velocity profiles in the anticentre of the populations EYP, YP, IP, OP and RC as a function of the distance R from the centre. The rotation curve by Sofue is plotted for comparison. Bottom: Same as top but for the velocity dispersions (Gaia Collaboration *et al.*, 2021).

velocity dispersion profiles have small oscillations that correlate to each other. This is a possible signature of multiple merger effects.

The influence of an early merger was confirmed by the Apache Point Observatory Galactic Evolution Experiment (APOGEE). Mackereth *et al.* (2019) investigated the origin of accreted stellar halo populations in the Milky Way using APOGEE and Gaia observations, and the EAGLE simulations. APOGEE observations were part of SDSS-III program to sample all major populations of Milky Way stars with the Sloan 2.5-m telescope. Spectra of stars were made with high resolution, which allowed us to determine the chemical compositions of stars with high precision. Among all α elements available from APOGEE spectra, Mg is the one for which abundances are the most reliable in the metal-poor regime due to the number of available lines and their strength at low metallicity. Therefore, Mackereth *et al.* (2019) decided to look how stellar populations are distributed in the Mg–Fe plane as a function of orbital eccentricity ϵ. Figure 8.25 shows

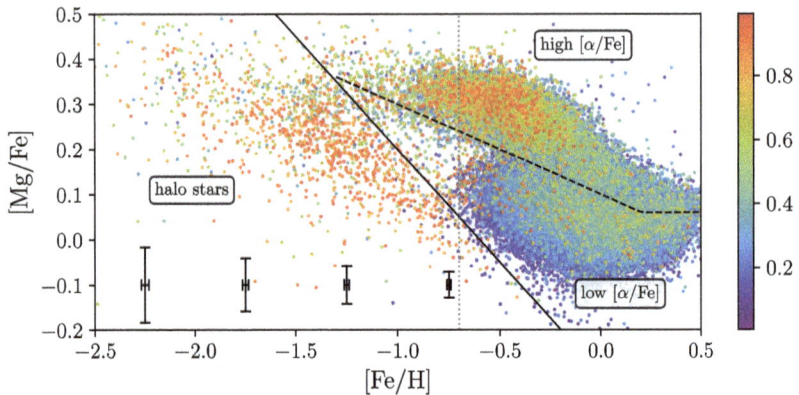

Fig. 8.25. The [Mg/Fe]–[Fe/H] plane in APOGEE coloured by orbital eccentricity ϵ. The points are plotted in a way that the highest ϵ populations stand out. A population extends from ([Fe/H],[Mg/Fe]) $\sim (-2.0, 0.3)$ to $\sim (-1.0, 0.1)$ that consist mainly of stars on highly eccentric orbits, with a distinct element abundance pattern to that of the Galactic disc (at [Fe/H]> -0.7) (Mackereth *et al.*, 2019).

the [Mg/Fe]–[Fe/H] distribution of APOGEE stars, where symbols are colour-coded according to each star orbital eccentricity. High-eccentricity group extends between ([Fe/H], [Mg/Fe]) $\sim (-2.0, 0.3)$ to $\sim (-0.7, -0.1)$, this population consists mainly of halo stars. These stars have chemical compositions that are characteristic of those seen in massive dwarf galaxy satellites of the Milky Way today, suggesting that this population is likely the progeny of a single, massive accretion event which occurred early in the history of the Milky Way. Low Fe/H stars form groups in high-α/Fe and low-α/Fe disk populations, having high and low eccentricities, respectively. They are likely the result from a mixture of different origins, including the remnants of less massive accretion events, *in situ* star formation, disk heating, and some contamination from the high-eccentricity (ϵ) population.

8.2.2 *Structure and evolution of Andromeda galaxy M31*

ARIEL: I don't understand where the NFW profile comes from.

RHEA: The simple answer is that it comes from simulations.

ARIEL: And you consider that understanding?

CALLISTUS:　The NFW profile comes from cosmological N-body simulations where particles interact with Newtonian gravity in an expanding universe governed by the Friedmann equation.

RHEA:　Typically, the outcome of fitting halo profiles is a double power law; the most popular among them is the NFW profile, but some simulations better fit with the Moore profile.

ARIEL:　Why do I feel like something is missing? There are many ways to find haloes and multiple parameter choices when setting up and running the simulation. Some of those factors might be responsible for the different profiles. Also, individual haloes vary, so filtering your dataset will influence the final results.

CALLISTUS:　A fitting formula from simulations is a different kind of understanding than what we are used to in classical physics. But it is an understanding. Intelligo me intelligere.[5] Perhaps our understanding of understanding is expanding.

RHEA:　Like the universe. Some haloes evolve in large tidal fields and are distorted. Some of them are not even spherical. But on average, they reflect the spherical symmetry of the Newtonian potential.

ARIEL:　That's a great point. Punctus greatus, as they say in Rome.

CALLISTUS:　That'd be magnum punctum.

ARIEL:　Not only is Newtonian gravity isotropic, but so are the initial conditions, at least statistically. Anisotropies should average out in large samples. But the rest is still a numerical result: Can I trust it without equations derived from clear underlying principles and physics?

CALLISTUS:　Oh, you want an analytical understanding. Sometimes, you solve equations numerically. Does only an analytical solution count as understanding?

RHEA:　What is an analytical solution? $\sin(x)$ is a function that has a name, but in the end, we have to compute it numerically. What if I name the solution of my equation the Rhea function?

CALLISTUS:　In our next paper! For the NFW profile, there is a paper[6] where they derive an analytical formula based on statistical physics. They assume spherical symmetry and Newtonian gravity and use a grand canonical ensemble to show that there is a maximum entropy solution if the central potential is limited.

ARIEL:　Do they derive the NSW profile?

CALLISTUS:　Their analytical expression is different but almost indistinguishable if you plot it against the NFW fit.

ARIEL:　Why limit the central potential?

---

[5]I understand that I understand.

[6]Carron & Szapudi (2013).

CALLISTUS: That's a mathematical condition for a solution to exist. Nevertheless, it is reasonable to assume that some additional physics would prevent infinite potential wells like black hole formation or baryonic processes.

RHEA: If you doubt that simulations constitute understanding, what about AI? It's all the rage in our field. And I imagine in every other field.

ARIEL: Just because an AI can predict something, I will not call it understanding.

CALLISTUS: Some would argue that it's enough if the AI understands it since we created the AI... but I tend to agree with you. At the same time, the AI impact on understanding is still fluid and up for debate.

RHEA: Our concept of understanding evolved for millennia since the first Greek philosophers. Relatively recently, quantum mechanics and then computers revolutionised what we call understanding. I have a practical view: if I can predict something, I feel I understand it, be it analytical calculations, computer simulations, or even AI.

CALLISTUS: Many physicists would see it that way. But, remember that Feynman famously said that nobody understands quantum mechanics.

ARIEL: I'll quote those words on my comprehensive exam. It's like a get-out-of-jail card.

CALLISTUS: I think he meant we can't have an intuitive understanding similar to mechanics, which relies on everyday experience. We might even be born with some intuitive understanding of basic geometry and mechanics, at least according to Pinker. I view understanding as radical compression. The more we can compress data while preserving its essence, the more we understand it.

RHEA: So, an MP3 music file represents an understanding of music over a WAV file?

CALLISTUS: Of course, not. It's only a small amount of compression. Sheet music certainly is a more massive compression, and it can express harmony, counterpoint and rhythm better than a WAV or MP3 file.

ARIEL: Sheet music is a lossy compression. It might say, rallentando, but it will not track the difference between Glenn Gould, Horowitz, or Lang Lang slowing down, their touch and dynamics.

CALLISTUS: Which brings us to the proverbial spherical cow.

RHEA: I prefer point-like cows.

ARIEL: The Cow, that's cinema *verité*. My favourite. Yet, the interplanetary orbit of point-like cows and Teslas are exactly the same. Although the latter is more dangerous right now.

CALLISTUS: Exactly, we can compress those orbits into a few orbital elements regardless of the details.

RHEA: And predict them with reasonable accuracy. I don't care if that prediction is analytical, numerical, or AI.

CALLISTUS: Let's agree that there are different ways and levels of understanding. An explainable AI can be easily understood by humans. Black box AI might use opaque algorithms that are difficult or impossible to understand. For me, the latter is far from understanding, even if it's capable of prediction.

ARIEL: I agree that there is a difference. But some phenomena might be predicted by black box AI where all other more transparent methods fail.

RHEA: Exactly. We cannot give that up.

CALLISTUS: Just like the NFW profile. We haven't found a profound understanding of it over decades. There may be none beyond maximum entropy. Yet, it helps us understand the universe since it compresses the results of many studies and thousands of CPU hours into a simple formula.

ARIEL: Hmm.

RHEA: Shake it off. Maybe you just have to get used to it.

ARIEL: I hear you, and I'm working on it.

The essential parameter of the Andromeda galaxy M31 is its distance. The first reliable distance to M31 was estimated with a dynamical method by Öpik (1922): 450 kpc. Modern data are based on Hubble Space Telescope survey of M31 satellite galaxies by Savino *et al.* (2022), which identified about 4700 RR Lyrae stars and determined their light curves. For the M31 distance, authors found 776 ± 21 kpc in good agreement with distance found from cepheid variables.

The second essential information of M31 is its velocity vector with respect to the Milky Way, which allows us to find the mass of the Local Group and the future merging of M31 and Milky Way. This problem was investigated by Kahn & Woltjer (1959), who found that M31 and Galaxy approach each other with a velocity of 125 km/s. If these galaxies form a physical unit, they must have performed the larger part of the orbit around their centre of mass during the existence of the Universe, about 10^{10} years. From these data, authors concluded that the total mass of the Local Group is $M_{LG} \geq 1.8 \times 10^{12} \, M_\odot$. This problem was recently analysed by van der Marel *et al.* (2012), who used HST proper motion measurements of stars in M31. They found the radial velocity of M31 with respect to our Milky Way, $V_{\text{rad,M31}} = -109.3 \pm 4.4\,\text{km/s}$, and the tangential velocity of $V_{\text{tan,M31}} = 17.0\,\text{km/s}$. Hence, the velocity vector of M31 is statistically consistent with a radial (head-on collision) orbit towards the Milky Way. Authors derived the estimate of the Local Group mass, $M_{\text{LG}} = (3.17 \pm 0.57) \times 10^{12}\,M_{\odot}$.

The third essential information of M31 is its decomposition to homogeneous populations. The first dynamical model of M31 which included dark matter halo was calculated by Tenjes *et al.* (1994), see Fig. 2.16. This model included the following stellar populations: core, bulge, disk, flat and stellar halo. Data on rotation velocities and multicolour imaging accumulated rapidly, which allowed Tamm *et al.* (2012) to calculate a new dynamical model of M31, as shown in Figs. 8.26 and 8.27. This model is based on a variety of data: photometric images of M31 in SDSS colours u, g, r, i, z and Spitzer MIPS camera infrared images near $3.5\,\mu\text{m}$. The M31 images in various colours were pixelised, the final mosaic was resampled to 3.96 arcsec per pixel. Additional information was the spectral energy distribution (SED) of models by Blanton & Roweis (2007) called B07. This model provides five composite spectra corresponding to an extremely young population B07-1 of age 0.7 Gyr, [Fe/H]=0.40 and mass-to-light ratio in g-colour $M_{\text{tot}}/L_g = 0.76$, a very old population B07-5 of age about 12 Gyr, [Fe/H]≈ 0.01, and $M_{\text{tot}}/L_g \approx 8$, and three intermediate populations, among them B07-3 with age 0.4–1 Gyr, [Fe/H]=0.05 and $M_{\text{tot}}/L_g = 0.47$, and B07-4 with age 7–12 Gyr, [Fe/H]=0.03, and $M_{\text{tot}}/L_g = 5$. It is shown that a linear combination of spectra of these populations can adequately describe the spectral energy distribution of most of the galaxies.

Figure 8.26 presents examples of the observed and modelled SEDs for a random pixel in the bulge region (top panel) and in a young disk region (bottom panel). The sizes of the data points indicate the photometric uncertainties of each measurement. In most pixels, the oldest model population (B07-4) alone provides a good representation of the observed SED. In the young disc regions, the stellar populations are more diverse: the B07 model populations 1, 4 and 5 contribute 24.44%, 75.52%, and 0.04% of the mass, respectively. The corresponding SEDs are weighted according to the mass fraction. The stellar mass corresponding to each model spectrum was derived for each imaging pixel. This gives the two-dimensional stellar mass distribution in M31. The distribution appears featureless and regular,

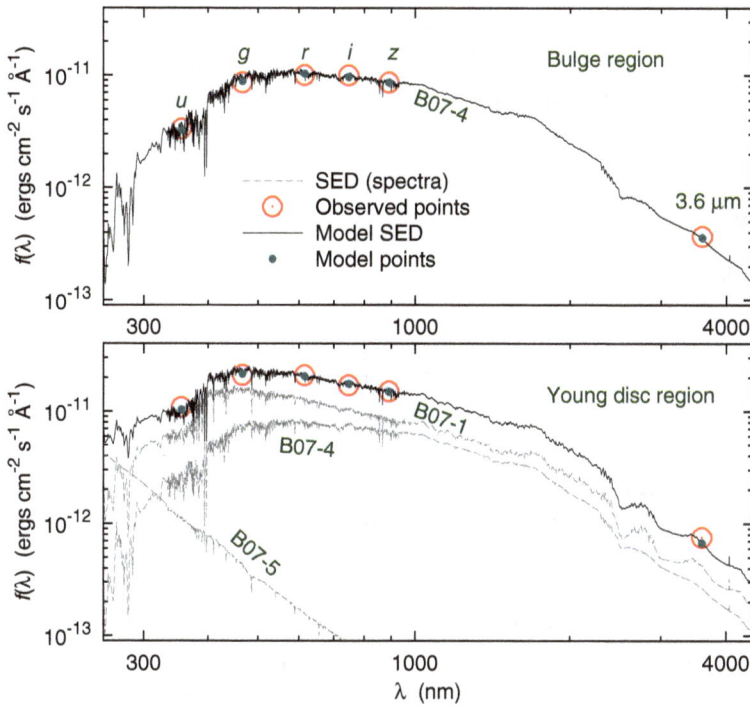

Fig. 8.26. Examples of the observed (large circles) and modelled (lines) SED for a random pixel in the bulge region (upper panel) and the young disc region (lower panel). The model values corresponding to each filter are also shown (small data points) (Tamm *et al.*, 2012).

resulting from the intensive smoothing, but it also indicates that the galaxy is generally undisturbed.

The final step in the calculation of the dynamical model of M31 is the estimate of masses of the DM halo and of the superposition stellar components, nucleus, bulge, disk, and halo, taking into account the observed rotation curve of M31. Tamm *et al.* (2012) used for stellar components the Einasto profile, Eq. (2.4). Parameters of stellar components are given in Table 8.1. Here, q is the axial ratio of equidensity surfaces and ρ_c is the central density. Masses and mass–light ratios correspond to the B07 model, i.e. the lower limits; the upper limits (from the maximum-stellar model) are 1.5 times higher in each case. Parameters were found by an automatic fitting procedure. Mass-to-light ratios of components are in good agreement with

Table 8.1. Parameters of stellar components.

Component	a_c [kpc]	q	N	ρ_c [$M_\odot\,pc^{-3}$]	M_{comp} [$10^{10}M_\odot$]	M/L_g [$M_\odot/L_\odot$]
Nucleus	0.0234	0.99	4.0	$1.713 \cdot 10^0$	0.008	4.44
Bulge	1.155	0.72	2.7	$9.201 \cdot 10^{-1}$	3.1	5.34
Disc	10.67	0.17	1.2	$1.307 \cdot 10^{-2}$	5.6	5.23
Young disk	11.83	0.01	0.2	$1.179 \cdot 10^{-2}$	0.1	1.23
Stellar halo	12.22	0.50	8.669	$4.459 \cdot 10^{-4}$	1.3	6.19

earlier estimations, only the M/L_g for the stellar halo is much higher than expected for an old population, dominated by metal-poor stars like RR Lyrae variables and globular clusters. In globular clusters, the mass can be found directly from the velocity dispersion of stars, which yields for the mass-to-light ratio $M/L_B = 2.1 \pm 0.6$ $M_\odot/L_\odot$ (Tenjes *et al.*, 1994), about three times less than accepted by the fitting procedure.

The gravitational potential of a galaxy is traced by the rotation curve. To match the calculated rotation curve with the observed one, the contributions of DM must be added to the stellar mass model of the galaxy. Tamm *et al.* (2012) used for DM the Einasto profile as well as some other popular profiles. Most frequently used DM profile is the Navarro *et al.* (1997) (hereafter NFW) profile:

$$\rho_{NFW}(r) = \frac{\rho_c}{\left(\frac{r}{r_c}\right)\left[1 + \left(\frac{r}{r_c}\right)\right]^2}, \tag{8.2}$$

where ρ_c is a density scale parameter. N-body simulations suggest steeply rising DM density towards the centre (therefore "cuspy" haloes) as the Moore *et al.* (1999) profile

$$\rho_{Moore}(r) = \frac{\rho_c}{\left(\frac{r}{r_c}\right)^{1.5}\left[1 + \left(\frac{r}{r_c}\right)^{1.5}\right]} \tag{8.3}$$

and the Burkert (1995) profile

$$\rho_{Burkert}(r) = \frac{\rho_c}{\left(1 + \frac{r}{r_c}\right)\left[1 + \left(\frac{r}{r_c}\right)^2\right]}. \tag{8.4}$$

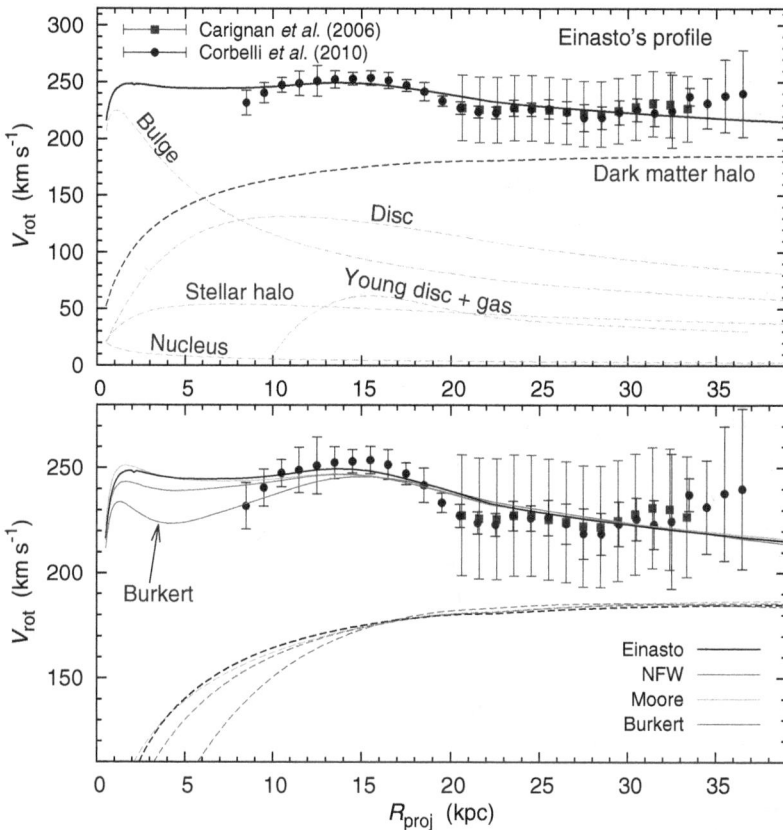

Fig. 8.27. Upper panel: The observed rotation curve (data points with error bars) overplotted with the model (solid line). Contributions of each component are also shown (dashed lines). The model corresponds to the B07 stellar mass estimates and the Einasto distribution for the DM density. Lower panel: The same stellar model with four different DM density distributions. For clarity, only the total rotation curves and the DM contributions are shown (Tamm *et al.*, 2012).

Each of the four DM distributions was used in combination with the stellar components as given in Table 8.1 to calculate the gravitational potential of the galactic model and the corresponding rotation curve. The upper panel of Fig. 8.27 presents the observed rotation curve, overplotted with the curve derived from the B07 stellar masses and the Einasto DM halo. Contributions of each stellar component and the DM halo are also shown. In the lower panel, model rotation curves for all the four DM models are presented. It is seen that within the observed range of the rotation curve, differences between different DM profiles are negligible. From 7 kpc inwards along the major axis, outside the range of observations, the model with the Burkert DM profile starts to deviate from the other models.

Wang *et al.* (2020) performed very high-resolution numerical simulations in a cold dark matter-dominated ΛCDM universe and determined DM halo profiles of haloes over the mass range from Earth mass to rich galaxy cluster mass. Authors demonstrated that halo profiles can be described by the NFW and Einasto profiles, and for Einasto profile, authors accepted the shape parameter $N = 6.25$. Over the whole halo mass range, the NFW profile describes the halo profile with 10% accuracy, the Einasto profile is better and fits to a few per cent accuracy.

Kalirai *et al.* (2010) presented Keck spectroscopic observations of hundreds of individual stars of the M31 dwarf spheroidal (dSph) galaxies as a part of the SPLASH Survey (Spectroscopic and Photometric Landscape of Andromeda's Stellar Halo). This sample was compared with the Milky Way dwarf spheroidal galaxies. This survey allowed us to find internal kinematics, chemical abundances and masses of dSphs. Main results of the survey are presented in Fig. 8.28, where properties of M31 dSphs are compared to Milky Way dSphs of the same luminosity. The left panel illustrates the luminosity–metallicity relation, where we find excellent overlap between the M31 and Milky Way satellites, suggesting that these systems have had similar chemical evolution histories. The middle-left panel illustrates the luminosity–size relation, where the size is represented by the two-dimensional elliptical half-light radius. The masses and mass-to-light ratios of the M31 dSphs are illustrated in the two right panels. The overall trend of increasing metallicity radius and mass with increasing luminosity and decreasing total mass-to-light ratio with increasing luminosity is similar for the dSphs for both galaxies, M31 and Milky Way. Also, we see that spheroids of Milky Way are systematically less luminous. Major differences among the population of satellites in the two systems have long been known. The most luminous Milky Way satellites, the Small and Large Magellanic Clouds, are irregular-type galaxies, of which M31 has none. The most luminous M31 satellites are the dwarf ellipticals M32, M110, NGC 147,

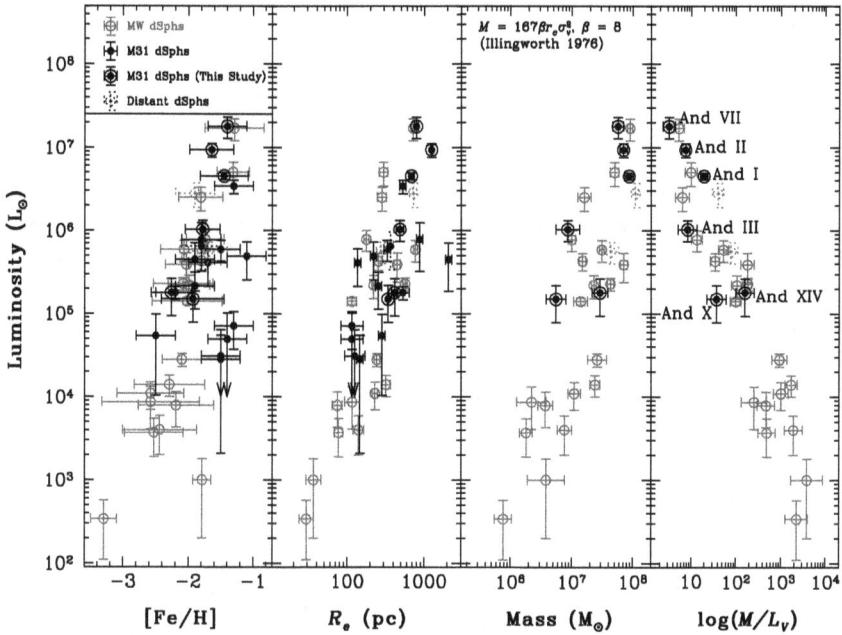

Fig. 8.28. Properties of M31 dSphs are compared to Milky Way dSphs of the same luminosity. The figure has four panels, where the luminosity in solar units, $L_\odot$, is presented as a function of abundance [Fe/H], size R_e in parsecs, mass in solar units, and mass-to-light ratio M/L_V in solar units. Satellites of M31 are shown with black symbols, Milky Way satellites with gray symbols (Kalirai *et al.*, 2010).

NGC 187, and NGC 205, of which the Milky Way has none. The overall population of globular clusters in M31 outnumbers the Milky Way by several times.

Various data indicate to accretion events and a wide range of merger histories in the history of the MW-like galaxies 11–8 Gyr ago (Bell *et al.*, 2017; Di Matteo *et al.*, 2019; Haywood *et al.*, 2016). Observations suggest that all isolated MW class galaxies are surrounded by dwarf satellites (Einasto *et al.*, 1974b; Haywood *et al.*, 2016; Bellstedt *et al.*, 2021). Such galaxies are fed by dwarf satellites inside dark matter haloes to which they belong as well as by pristine gas and dark matter from filaments. Initially, all galaxies are star-forming. During the evolution in a fraction of galaxies, star formation is suppressed, and galaxies become quenched. Depending on the character of seeds, and in the absence of the influence of other similar galaxies, a certain distribution of star-forming and passive galaxies at present epoch forms.

It is possible to trace this growth history via the properties of a galaxy's stellar halo. To get necessary information, the Pan-Andromeda Archaeological Survey (PAndAS) was initiated. The survey was conducted with the Canada–France–Hawaii 4-m Telescope near the summit of the Mauna Kea mountain. In total, there were 2985 exposes contributing to the PAndAS program, containing images of stars and galaxies of infrared magnitude $i \leq 25.5$. Figure 8.29 shows the map of the surveyed region around the traditional M31 galaxy. We see companion galaxies M32, M33, NGC 205, NGC 185, NGC 147 as well numerous streams and faint dwarf

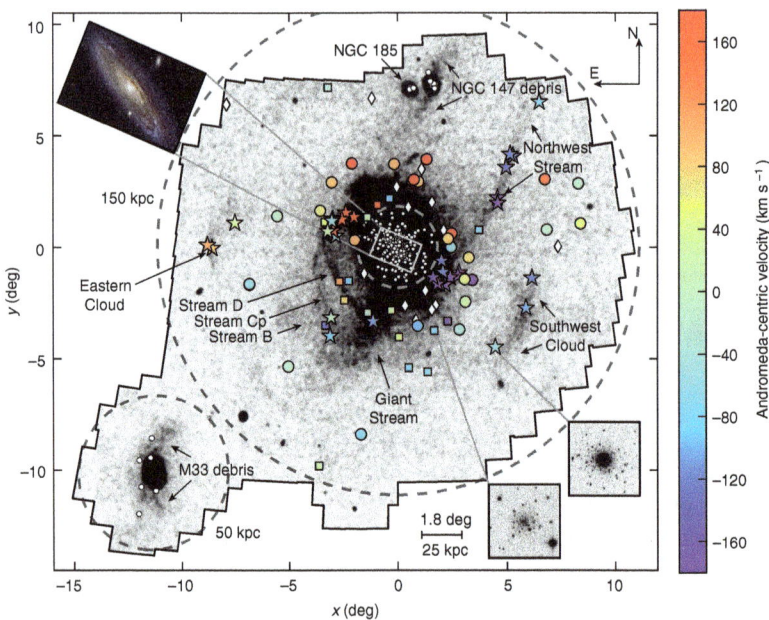

Fig. 8.29. Map showing the distribution of metal-poor red-giant stars in Andromeda's halo. The density has smoothed with a Gaussian kernel. Members of Andromeda's globular cluster system are marked with filled symbols. Those outside 25 kpc are coloured according to their line-of-sight velocities (where available) and have symbols according to their classification (circles for GC-Non, stars for GC-Sub, squares for ambiguous and diamonds for no data). Dashed circles represent projected galactocentric radii of 25 kpc and 150 kpc, respectively (Mackey *et al.*, 2019). Used with permission of *Nature* from Mackey *et al.* (2019). Permission conveyed through Copyright Clearence Center, Inc.

galaxies. The survey detected M31 globular clusters, physically associated with underlying halo substructures (a subgroup referred to as "GC-Sub"), while other clusters exhibit no association with substructure ("GC-Non") and are plausibly part of the smooth halo (Mackey *et al.*, 2019). Authors analysed kinematical properties of clusters of both subgroups. The analysis showed that the globular clusters have two distinct populations rotating perpendicular to each other. The rotation axis for the population associated with the smooth halo is aligned with the rotation axis for the plane of dwarf galaxies that encircles M31. Mackey *et al.* (2019) interpret these separate cluster populations as arising from two major accretion epochs, probably separated by billions of years, which formed two distinct populations of globular clusters.

Savino *et al.* (2022, 2023) made a survey of ultra-faint dwarf (UFD) satellites on M31, based on deep colour–magnitude diagrams from Hubble Space Telescope. This showed that a half of stellar mass of these galaxies was formed by $z = 5$, i.e. about 12.6 Gyr ago, similar to UFDs of the Milky Way. About 10–40% of their stellar mass was formed at later times at $2 < z < 3$. The faintest MW UFDs of mass $M_\star(z = 5) \leq 5 \times 10^4 \, M_\odot$ are likely quenched at reionisation epoch, whereas more massive M31 UFDs of masses $M_\star(z = 5) \geq 5 \times 10^5 \, M_\odot$ have their star formation only suppressed by reionisation and quench at a later time.

8.2.3 Structure and evolution of galaxies in the Local Universe

Classical observations including Hubble Space Telescope and Sloan Digital Sky Survey Telescope yield one spectrum per galaxy. This is adequate to determine the redshift of the galaxy and its chemical composition. However, it yields no information on the spatial structure and kinematics of the galaxy. Many elliptical galaxies and bulges of spiral galaxies are triaxial and consist of multiple kinematical components. This makes two-dimensional integral-field spectroscopy of galaxies essential for determining the internal structure of these systems and for understanding their formation and evolution. To achieve this goal, the Spectrographic Areal Unit for Research Optical Nebulae (SAURON) project was developed in a collaborative effort between groups at Observatoire de Lyon, Leiden Observatory, and the University of Durham. The SAURON spectrograph consists of an array of 1500 square lenslets, each 1.35 mm on side, and is optimised for wavelengths of 4500–7000 Å, with spectral resolution 2.8–3.6 Å. The SAURON spectrograph is mounted on the William Herschel 4.2-m Telescope at the Roque de los Muchachos Observatory in Canary Islands (de Zeeuw *et al.*, 2002). Early results of the SAURON project include the study of kinematics of nearby galaxies NGC 3384, NGC 4365, NGC 5813 and M32, whose stellar velocities and velocity dispersions were determined to a precision of 6 km/s. The kinematics of M32 requires the presence of a central black hole of mass $(3.4 \pm 0.7) \times 10^6$ $M_\odot$ (van der Marel *et al.*, 1998; de Zeeuw *et al.*, 2002).

The next stage of spatial spectroscopy is the ATLAS^{3D} project, which includes, in addition to the SAURON spectrograph, deep imaging with MegaCam CCD imaging facility on the Canada–France–Hawaii 3.6-m Telescope, Westerbork Radio Synthesis Telescope, and millimetre observations with the IRAM 30-m Radio Telescope. The goal of the project is to combine a multi-wavelength survey of a complete sample of early-type galaxies (ETGs) within the Local 42 Mpc volume with numerical simulations and semianalytic modelling of galaxy formation (Cappellari *et al.*, 2011a, 2011b). The parent sample of the ATLAS^{3D} survey consists of 871 ETGs within the volume: distance $D < 42$ Mpc, declination $|\delta - 29°| < 35°$, galactic latitude $b > 15°$. From this parent sample, a volume-limited sample of 260 galaxies was selected brighter than $M_K < -21.5$ mag (stellar masses $M_\star \geq 6 \times 10^9 M_\odot$). The project aims to quantify the global stellar kinematics and dynamics to characterise ETGs and relate this to their formation and evolution. Combining the stellar population diagnostics from the multi-wavelength coverage, the ATLAS^{3D} project will derive the star formation and the mass-assembly history of early-type galaxies. For all 260 ATLAS^{3D} project galaxies, direct images as well as spatial spectroscopy were obtained.

A striking impression derived from the inspection of the images of the galaxies is the structural similarity between many nearly edge-on fast-rotator galaxies and spiral galaxies in the parent sample. For every fast rotator ETG that is known to be close to edge-on, one can find a corresponding spiral galaxy with the same general shape, except for the presence of a prominent dust lane. This is in agreement with the classification scheme proposed by van den Bergh

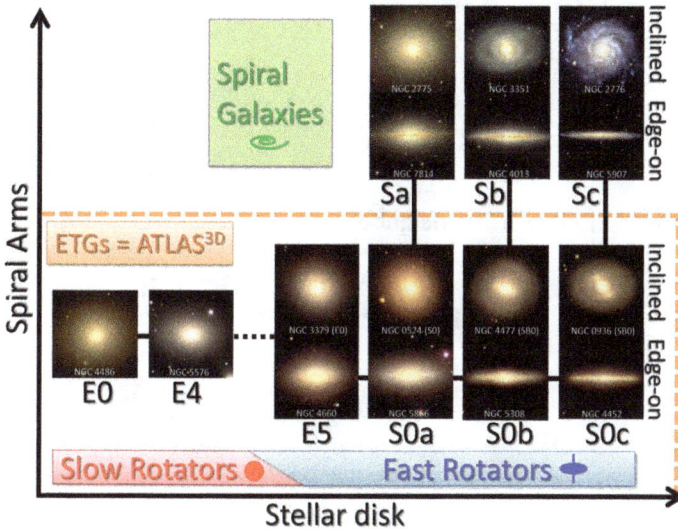

Fig. 8.30. Morphology of nearby galaxies from the ATLAS^{3D} parent sample. The volume-limited sample consists of spiral galaxies (70%), fast rotators ETGs (25%) and slow rotators ETGs (5%). The ATLAS^{3D} sample consists of the ETGs only, classified according to the absence of spiral arms or an extended dust lane. The edge-on fast rotators appear morphologically equivalent to S0s or to flat ellipticals with disky isophotes (Cappellari *et al.*, 2011b).

(1976) if one associates fast rotators with S0 galaxies. The new classification of galaxies is presented in Fig. 8.30. The fast rotators can be recognised from integral-field kinematics even when they are nearly face-on. They form a parallel sequence to spiral galaxies as already emphasised for S0 galaxies by van den Bergh (1976), who proposed the above distinction into S0a–S0c. Fast rotators are intrinsically flatter than $\epsilon = 0.4$ and span the same full range of shapes as spiral galaxies, including very thin disks. However, very few Sa have spheroids as large as those of E(d) galaxies. The slow rotators are rounder than $\epsilon = 0.4$. The ellipticity distribution in the outer parts of the galaxies in our sample is characterised by a roughly constant fraction of galaxies up to ellipticities $\epsilon \approx 0.75$. Under the reasonable assumption of random orientations for the galaxies in our sample, this indicates that most of the galaxies, even when they appear round in projection, must possess quite flat disks.

In Fig. 8.31, we show the kinematic morphology–density relation for fast and slow rotator ETGs and spiral galaxies using the volume

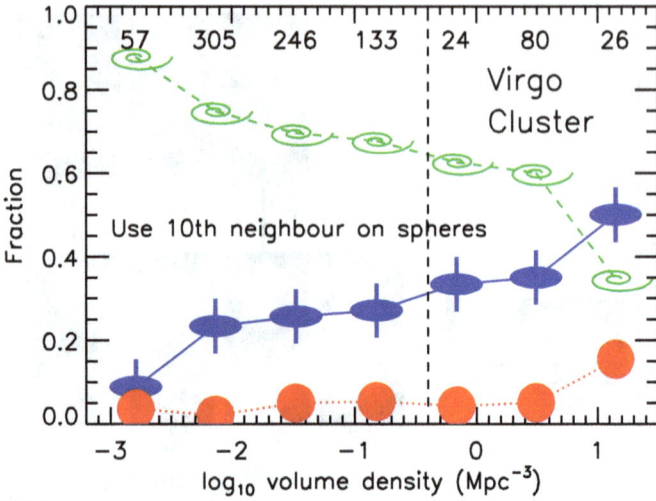

Fig. 8.31. The morphology–density relation for fast rotators (blue ellipse with vertical axis), slow rotators (red filled circle) and spiral galaxies (green spiral) (Cappellari *et al.*, 2011b).

density estimator. The dashed vertical line indicates an approximate separation between the density of galaxies inside/outside Virgo. The numbers above the symbols represent the number of galaxies included in each of the seven density bins. Cappellari *et al.* (2011b) found a clear trend for the spiral fraction to gradually decrease with environmental density while the fraction of ETGs correspondingly increases. This trend continues smoothly over four orders of magnitude in density and does not flatten out even at the lowest densities.

The connection between bulge mass and galaxy properties and the close link between ETGs and spiral galaxies are illustrated in Fig. 8.32, which includes the location of the spiral galaxies of the ATLAS^{3D} parent sample together with the ETGs. The masses of spiral galaxies were estimated, assuming a fixed $M/L_K = 0.8$ $M_\odot/L_\odot$. The figure shows that ETG properties vary smoothly with those of spiral galaxies. Galaxies with negligible bulges are almost invariably star-forming and classified as late spirals. Galaxies with intermediate bulges can still form stars and be classified as early spirals, or can be fast rotator ETGs. The galaxies with the most massive bulges, as indicated by their largest concentration or velocity

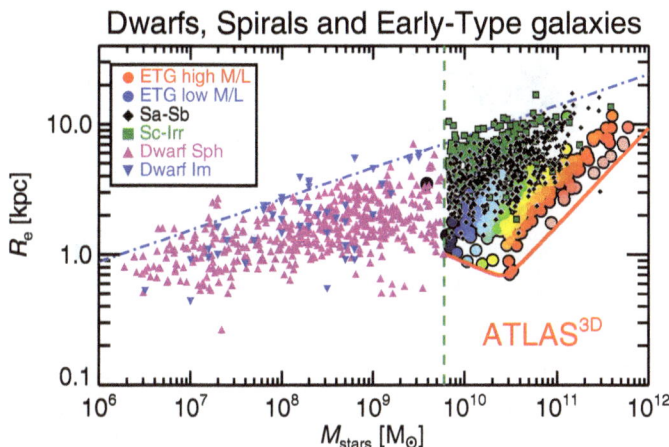

Fig. 8.32. The mass–size distribution for dwarfs, spiral galaxies and ETGs. Dwarf spheroids and irregulars are only shown below the mass limit of the volume-limited ATLAS^{3D} survey (vertical dashed line) (Cappellari *et al.*, 2013).

dispersions, are invariably ETGs. They have the largest M/L, the reddest colours, smallest $H\beta$, and lowest molecular gas fraction, but are still flat in their outer parts, indicating that they have disks and generally still rotate fast. Also included in Fig. 8.32 are dwarf ETGs (spheroidals) and dwarf irregulars (Im) galaxies. The plot shows that the spheroidal galaxies lie along the continuation of the mass–size relation for spiral galaxies and naturally connect to the region of fast rotator ETGs with the lowest M/L, young ages and small bulges.

An even better spatial spectroscopic resolution can be obtained with the Multi Unit Spectroscopic Explorer (MUSE) spectrograph installed at the Nesmyth focus of the fourth unit of the ESO Very Large Telescope (VLT) at Paranal Observatory, Chile. MUSE covers the wavelength range of 4750–9350 Å with an average spectral resolution of 2.5 Å, has a spatial resolution of 0.2 arcsec/pixel, and covers a field of view of 1×1 arcmin^2. Bidaran *et al.* (2020, 2022, 2023) used VLT/MUSE integral-field unit spectra to derive kinematics and specific angular momentum profiles of infalling dE galaxies on to the Virgo cluster. For the galaxy VCC0170, the direct image and kinematic maps are shown in Fig. 8.33. The detection in central part of VCC0170, nebular emission lines is consistent with

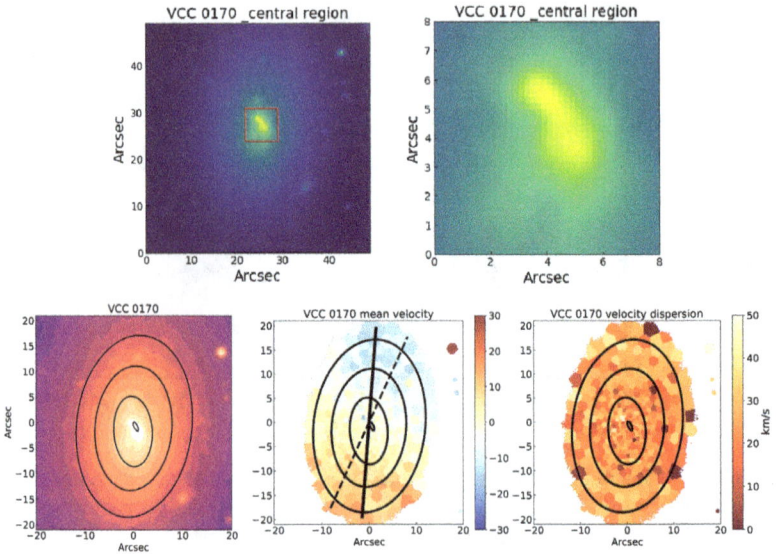

Fig. 8.33. VCC0170 and its central gas component. Top panel: The left picture is the MUSE stacked image, where the central region is marked with a red box, 8×8 arcsec in size. In the right-hand panel, a zoomed-in image of this region is provided, where the irregular shape of the core can easily be recognised. Bottom panel: The kinematic maps of the stellar component in VCC0170. The left-hand panel shows the MUSE stacked image while the rotation velocity and velocity dispersion of stellar component are plotted in the middle and right-hand panels, respectively (Bidaran *et al.*, 2020).

its blue colour, which is likely due to the presence of young massive stars. Bidaran *et al.* (2020) measured the gas kinematics for VCC0170 inside the marked region in Fig. 8.33 by fitting a Gaussian function to the $H\alpha$ line for each spaxel with signal-to-noise ratio SNR ≥ 3. Bidaran *et al.* (2022) used MUSE spectra to investigate how pre-processing and accretion on to a galaxy cluster affect the integrated stellar population properties of dwarf early-type galaxies (dEs). Authors analysed a sample of nine dEs with stellar masses of $\sim 10^9 \, M_\odot$, which were accreted ~ 2–3 Gyr ago on to the Virgo cluster as members of a massive galaxy group. These results suggest that the stellar population properties of low-mass galaxies may be the result of the combined effect of pre-processing in galaxy groups and environmental processes (such as ram-pressure triggering star formation) acting during the early phases of accretion on to a cluster.

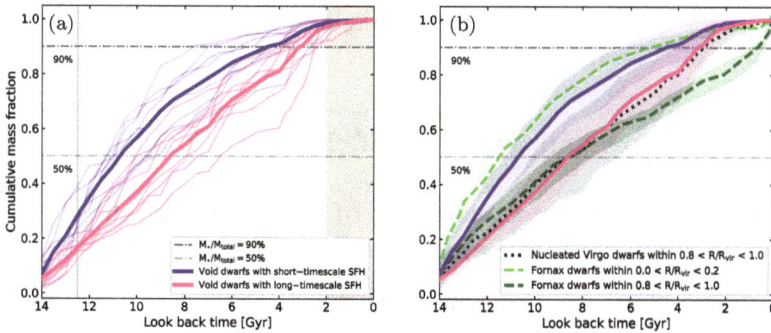

Fig. 8.34. Panel A: The mass assembly history of isolated dwarf galaxies as a function of look back time. Panel B: Average profiles of isolated and quenched dwarf galaxies (in purple and pink, as in panel A) and their range of error (shaded area) are now compared with the average profiles of their counterparts in the core and outskirts of the Fornax cluster (in light and dark green, respectively) (Bidaran *et al.*, 2025).

Bidaran *et al.* (2025) analysed the mass assembly history of isolated dwarf galaxies as a function of look back time, as presented in Fig. 8.34. Horizontal dashed lines from bottom to top in both panels represent the 50% and 90% cumulative stellar mass fractions, respectively. In panel A, the brown shaded area highlights the last 2 Gyr. We see that in all isolated dwarf galaxies, there was no additional mass assembly during the last 2 Gyr. The vertical solid line in this panel marks the age of 12.5 Gyr, which is the criteria authors adopted to classify galaxies into short- and long-timescale star formation histories. For comparison, the authors calculated the mass assembly history of Fornax dwarf galaxies and nucleated Virgo dwarfs, as shown in the right panel of the figure. Virgo dwarf galaxies hosting a nuclear star cluster (NSC) are located at large cluster-centric distances (whose average cumulative mass fractions are shown as dotted black line), exhibiting a mass assembly history akin to isolated dwarf galaxies with beginning of star formation at epoch <12.5 Gyr. The analysis of the light distribution of the isolated and quenched dwarf galaxies shows that all of them are NSC hosts. NSCs are very dense and massive stellar assemblies frequently found in the centre of galaxies with $M_\star \sim 10^8$–$10^{10} M_\odot$. The NSC formation in galaxies could be a consequence of *in-situ* star formation, where the compression of gas in their gravitational potential well, particularly in its central few parsecs, can trigger episodic and intense star formation bursts and subsequent formation of a central NSC.

8.3   Structure and evolution of clusters of galaxies

8.3.1   *Structure and evolution of the Virgo cluster and the Local Supercluster*

The Virgo cluster is the central cluster of the Local Supercluster, the nearest supercluster. The main body of the Local Supercluster is a flattened disk, which allows us to determine supergalactic coordinates. We discuss here first the properties of the Virgo cluster and thereafter the properties of the whole Local Supercluster.

Kashibadze *et al.* (2020) studied the structure and dynamics of the Virgo cluster using most recent data. Authors found for the cluster the following parameters, assuming Hubble parameter $H_0 = 73 \, \mathrm{km/s}$ per h^{-1} Mpc: distance $D = 16.5 \pm 0.2 \, \mathrm{Mpc}$, core (virial) radius $R_c = 1.7 \, \mathrm{Mpc}$, radius of the zero-velocity surface $R_0 = 7.15 \pm 0.15 \, \mathrm{Mpc}$, virial mass $M_{\mathrm{vir}} = (6.3 \pm 0.9) \times 10^{14} \, M_\odot$, total mass inside the zero velocity radius R_0 $M_T = (7.2 \pm 0.9) \times 10^{14} \, M_\odot$, luminosities in B and K colours, $L_B = (1.8 \pm 0.2) \times 10^{12} \, L_\odot$ and $L_K = (8.6 \pm 1.1) \times 10^{12} \, L_\odot$. The agreement of virial and external mass estimates suggests that the wide outskirts of the Virgo cluster between the core and the zero-velocity radii do not contain significant amounts of dark matter.

Giant elliptical galaxies at the centres of galaxy clusters called brightest cluster galaxies (BCGs) are extreme systems that provide stringent tests for theories of galaxy formation and evolution. Current models of BCG formation predict dualistic histories where most of the stellar mass is formed very early in the highest density peaks of the dark matter distribution, while the assembly of additional mass occurs later through the merging of smaller galaxies that are already old. The late assembly of BCGs can be considered as examples of the picture of two-phase assembly for massive galaxies, where a seed galaxy grows gradually through the accretion through minor mergers.

The nearest giant elliptical galaxy is M87 in the Virgo cluster. It is very different from central galaxies of small groups, such as Milky Way and M31 in the Local Group. Virgo cluster and Local Group present extreme cases of galaxy systems of various richness, both in the number of member galaxies and in the mass and luminosity of its BGCs. Both systems consist of several subgroups and are in the process of forming. Galaxies of the Virgo cluster form several subgroups: M87 (NGC 4486) — the brightest in subgroup Virgo A and the dynamical centre of the cluster, M49 (NGC 4472) — the brightest in subgroup Virgo B, and M60 (NGC 4649) — the brightest in subgroup C, see Fig. 8.35. The main concentration A is made up of predominantly early-type centre galaxies and has a large velocity dispersion, $\sigma_v \approx 750\,\text{km/s}$. The second concentration B near M49 contains mainly late type galaxies and has a smaller velocity dispersion, $\sigma_v \approx 400\,\text{km/s}$. The mean radial velocity of B is lower than that of A by $\Delta v \approx 100\,\text{km/s}$. This difference suggests that the cluster B is falling towards A from behind.

Virgo clusters of galaxies are too distant to see individual stars. However, they contain globular clusters (GC), which can be used to trace the history of cluster formation and evolution. The estimated number of GCs in M87 is about 13,000 and that in M49 about 6000 compared to about 200 in M31 and even less in Milky Way. This shows that the formation and evolution of BCGs in rich clusters is very different from the formation and evolution of central galaxies of small groups.

Strader *et al.* (2011) and Romanowsky *et al.* (2012) made a combined photometric and spectroscopic study of the GC system around M87. A view of the M87 in V-band image of the region around M87 is presented in Fig. 8.36. Here, we see the enormous image of M87, its close satellite galaxies and GCs as small dots. Romanowsky *et al.* (2012) obtained high-quality redshifts and colours for 488 GCs around M87 up to distance of 200 kpc. Authors found the signatures of two substructures in position–velocity phase space. One is a cold stream associated with a stellar filament in the outer halo; the other is a large shell-like pattern in the inner halo that implies an accretion event. The cold outer stream is consistent with a dwarf galaxy accretion event. The shell and stream GC support a scenario where the extended stellar envelope of M87 has been built up by a steady rain of material that continues until the present day. The shell of GCs has a chevron-like shape in phase space that resembles expectations for a disrupted infalling system.

Figure 8.36 shows that there are five low-luminosity ellipticals within a 30–80 kpc projected radius of M87: NGC 4486A, NGC 4486B, NGC 4476, NGC 4478, and IC 3443. Velocity dispersions

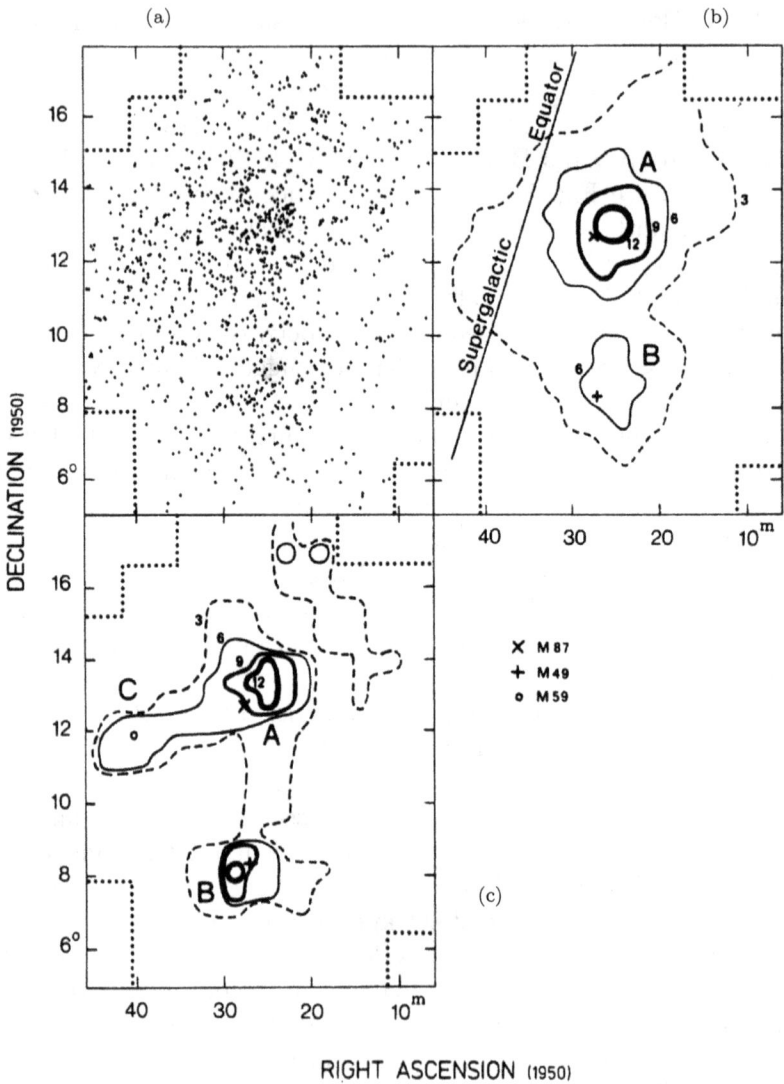

Fig. 8.35. General outline of the Virgo cluster based on all cluster members listed in the Virgo Cluster Catalogue. (a) Plot of all members. (b) Isopleths based on number counts in cells of size $0°.5 \times 0°.5$, smoothed by averaging over $1°.5 \times 1°.5$. The contours are labelled in units of galaxies per cell, i.e. number of galaxies per 0.25 square degree. (c) Same as (b) but with luminosity-weighted isopleths labelled in units of 10^{10} $L_\odot$ per deg^2. The positions of M87, M49, and M59, which are associated with the three major clumps A (M87 cluster), B (M49 cluster), and C are indicated. For orientation, the position of the supergalactic equator is given in (b) (Binggeli *et al.*, 1987).

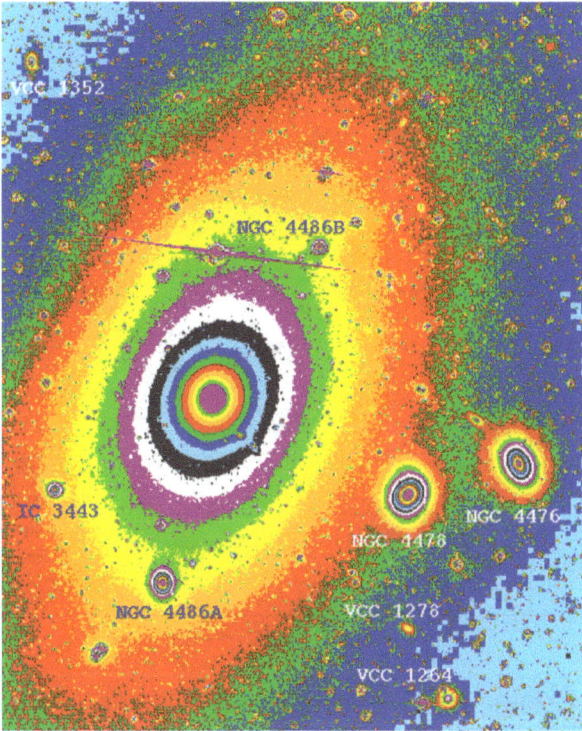

Fig. 8.36. A view of the M87 V-band image, showing a $23' \times 24'$ (110×120 kpc) region. Some small galaxies in the field are labelled. The shading is logarithmically spaced by surface brightness and arbitrarily coloured in order to enhance any isophote twisting (e.g., in the case of NGC 4478). The most probable shell GCs have positions spread along the outermost, dark-green and dark-blue isophotes, and preferentially in the northern half of the galaxy (Romanowsky *et al.*, 2012).

of NGC 4486A and NGC 4486B suggest that they originally hosted $\sim 70-200$ and $\sim 100-500$ GCs, respectively. Presently, NGC 4486A and NGC 4486B have no GCs. For NGC 4478 and NGC 4486B, the dynamical friction timescales are about 1–3 Gyr, suggesting that these galaxies are in the final stages of orbital decay towards the centre of M87.

Observations of galaxy mergers inside galaxy clusters suggest high star formation rates in the ejected tidal tails and in developing new dwarf galaxies. Ivleva *et al.* (2024) performed numerical simulations to investigate the effect of merging in the development of tidal dwarfs. Authors found several tidal dwarf galaxies per merger, which survive

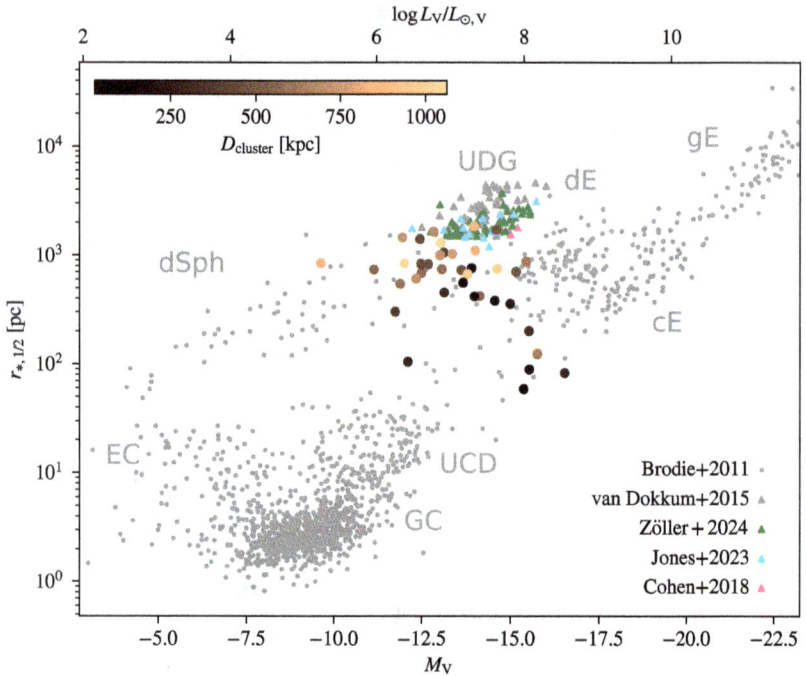

Fig. 8.37. Half-light radii versus visual band luminosities for various object types. Gray circles show stellar systems observed in the Local Universe including giant, compact, and dwarf ellipticals (gE, cE, and dE), dwarf spheroidals (dSph), ultra-compact dwarfs (UCD) as well as globular and extended clusters (GC and EC). Triangles indicate ultra-diffuse galaxies (UDG): UDGs inside the Coma cluster in gray and green: gas-rich field UDGs in light blue; and finally NGC 1052-DF2 and DF4 in light red. Simulated systems are coloured according to their radial distance from the cluster centre (Ivleva *et al.*, 2024).

longer than 1 Gyr after the merger event. Exposed to ram pressure, these gas-dominated dwarf galaxies exhibit high star formation rates while also losing gas to the environment. About 4 Gyr after the merger event, authors find several dwarf galaxies in two of the tested scenarios. The other stripped tidal dwarf galaxies either evaporate in the cluster environment due to their low initial mass or are disrupted as soon as they reach the cluster centre. Figure 8.37 shows positions of simulated dwarf galaxies in the diagram half-light radius $r_{\star 1/2}$ as a function of visual luminosity M_V. Here, in addition to simulated dwarf galaxies, also giant ellipticals gE, dwarf ellipticals dE, compact ellipticals cE, ultra compact dwarfs UCD, dwarf spheroids dSph, globular clusters GC and extended clusters EC are shown. The figure shows that all these systems of galaxies occupy relatively compact regions in the diagram, which suggests different origin and evolution. Simulated dwarf galaxies are plotted with coloured symbols, and they occupy mainly the region of UDGs.

Central galaxies of subgroups are supergiant ellipticals with extended cD-type envelopes and supermassive black holes (SMBH) at centres. The masses of SMBHs of these galaxies are $(5.5 \pm 0.4) \times 10^9 \, M_\odot$ in M87 (Simon *et al.*, 2024) and $(4.5 \pm 1) \times 10^9 \, M_\odot$ in M60 (Shen & Gebhardt, 2010). These galaxies are also prominent radio and X-ray sources, which suggest the presence of relativistic gas jets from nuclei.

Wood *et al.* (2017) used Chandra observations of hot gas to study the kinematics of the M60 cluster. This cluster is located 1 Mpc east of M87. Chandra X-ray images show a sharp leading edge in the surface brightness in the direction of M87, characteristic of a merger cold front due to M60s motion through the Virgo intergalactic medium. Authors found the infall velocity $v_{\text{tran}} = 1012 \pm 190 \, \text{km/s}$, which places an upper bound on the time of pericentre passage of about 0.95 Gyr. We can conclude that subgroups B and C are in the process of merging to form a larger single cluster. Earlier, these subgroups were enhancements inside galaxy and cluster filaments surrounding the Virgo cluster, see Fig. 8.38.

Tully & Fisher (1978) and Tully (1982) investigated the spatial three-dimensional distribution of galaxies in the nearby Universe. Results of this study demonstrated that the Virgo cluster is surrounded by a system of filaments as a spider with multiple legs, see Fig. 8.38. The filamentary system of galaxies around the Virgo cluster has two important properties: filaments are very thin and are directed towards the Virgo cluster.

Tully (1982) and Aaronson *et al.* (1982) studied the velocity field of galaxies around the Virgo cluster and found evidence that surrounding galaxies are moving towards the Virgo cluster. The infall of galaxies to the Virgo cluster was studied by Karachentsev & Nasonova (2010) using most recent data on distances and velocities of galaxies. Figure 8.39 shows the radial velocity versus the distance from Virgo centre relation for galaxies in the Virgo cluster

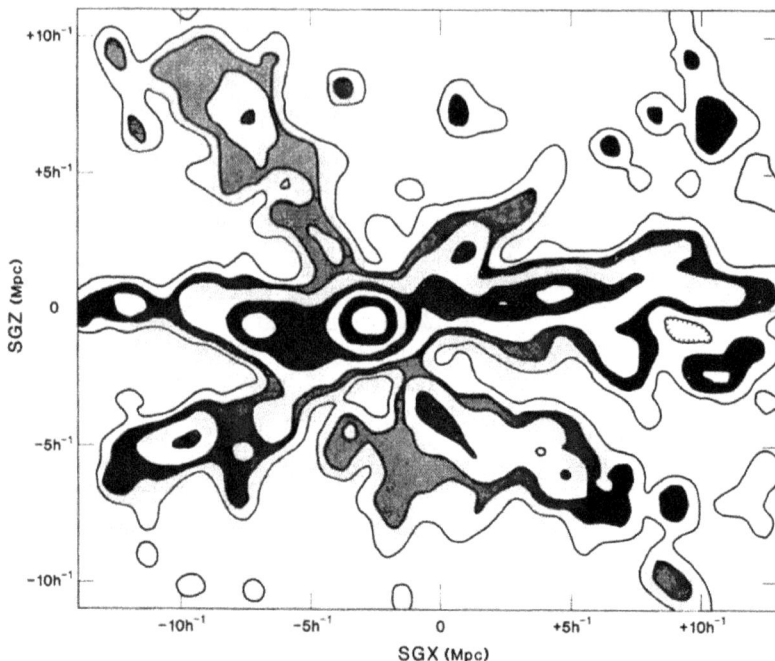

Fig. 8.38. Contour map of the distribution of all galaxies near the Virgo cluster projected onto the supergalactic SGX–SGZ plane. Surface densities have been computed by smoothing with a Gaussian with dispersion $0.5\,h^{-1}$ Mpc. Contours are on a logarithmic scale. The lowest contour corresponds to surface densities of 0.5 galaxies $(h^{-1}\text{ Mpc})^{-3}$. Successive contours correspond to levels of 1, 2, 4, 8, 16, and 32 galaxies $(h^{-1}\text{ Mpc})^{-3}$ (Tully, 1982).

region with respect to the Local Group centroid. Galaxy samples with distances derived by different methods are marked by different symbols. The inclined line traces the Hubble relation with the global Hubble parameter $H_0 = 72$ km/s per h^{-1} Mpc. The vertical dashed lines outline the virial zone. Two S-shaped lines correspond to a Hubble flow perturbed by virial masses of 2.7×10^{14} M$_\odot$ (dotted) and 8.9×10^{14} M$_\odot$ (solid) as the limiting cases within the confidence range. The top panel shows the Virgo cluster core within angular distance $\theta < 6°$. S-shaped lines indicate the expected infall at $\theta = 0°$. The bottom panel presents wider surroundings with $6° < \theta < 15°$, where the S-shaped lines indicate the infall at $\theta = 6°$. The typical distance error bars for each dataset are shown.

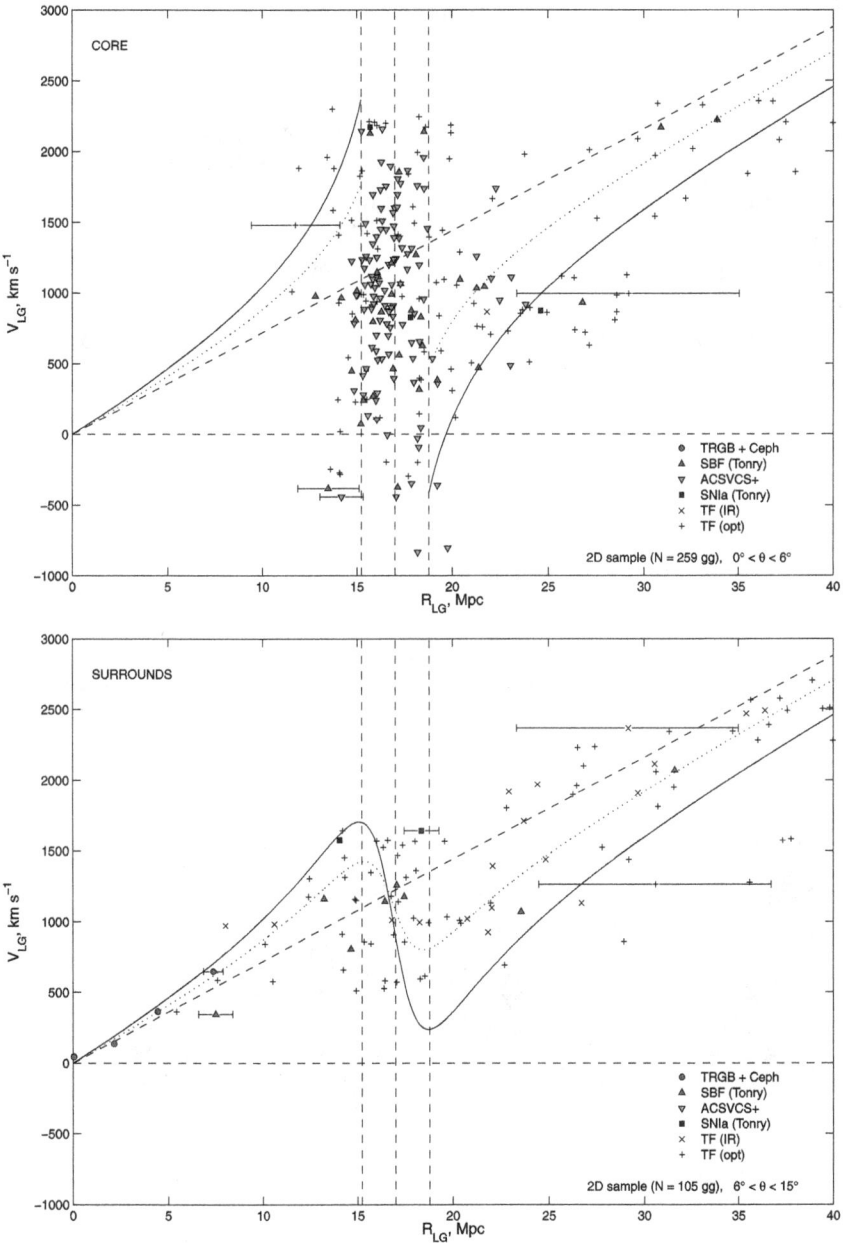

Fig. 8.39.   The radial velocity versus distance relation for galaxies in the Virgo cluster region. See text for details (Karachentsev & Nasonova, 2010).

8.3.2 *Structure and evolution of colliding clusters*

All gravitating bodies are attractors and massive systems of galaxies are great attractors. The Virgo cluster is an example of great attractors. It is surrounded by filaments directed towards the central Virgo cluster. As discussed above, the flow of matter towards the central cluster occurs along the filaments — subgroups B and C show signs of movement towards the main subgroup A. This phenomenon is one of the main properties of superclusters. However, the internal structure of subgroups A, B and C is not seen very clearly. We may ask the question: Can we find somewhere a more clear evidence for colliding clusters?

Given sufficient time galaxies, hot plasma and DM of clusters acquire similar, centrally symmetrical spatial distributions, tracing the common gravitational potential. During a collision of two clusters, galaxies (and DM) as collisionless particles (dust) of the minor cluster move through the main cluster undisturbed. The fluid-like intracluster plasma experience ram pressure and is spatially decoupled from galaxies and DM.

The most studied case of colliding clusters is the merging cluster 1E0657-56, as shown in Fig. 8.40. The optical image of the cluster was obtained with the 6.5-m Magellan telescope located at the Las Campanas Observatory in Chile, the X-ray image was made with the

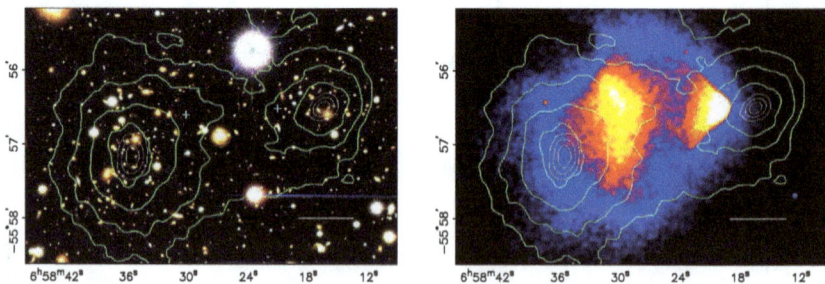

Fig. 8.40. Left: Colour image from the Magellan images of the merging cluster 1E0657-56, with the white bar indicating 200 kpc at the distance of the cluster. Right: 500 ks Chandra image of the cluster. Shown in green contours in both panels are the weak-lensing reconstructions of the DM density. The blue plus signs show the locations of the centres used to measure the masses of the plasma clouds (Clowe *et al.*, 2006a).

Chandra X-ray Observatory. The map of gravitational potential was constructed using weak gravitational lensing of background galaxies in Hubble Space Telescope images. Two clusters of galaxies are clearly visible. In this system, a massive subcluster or the "bullet" has moved through the main cluster on a trajectory nearly exactly in the plane of the sky (Markevitch *et al.*, 2002).

The parameters of subclusters have been estimated from weak lensing studies by Clowe *et al.* (2004, 2006b) from strong lensing data by Bradač *et al.* (2006) and from hydrodynamical simulations by Springel & Farrar (2007). According to these data, the masses are $M_{200} \simeq 1.5 \times 10^{15} \, M_\odot$ and $M_{200} \simeq 1.5 \times 10^{14} \, M_\odot$ for the main cluster and bullet, respectively, radii $r200 = 2136 \, \mathrm{kpc}$ and $r200 = 995 \, \mathrm{kpc}$, and concentration indexes $c = 1.94$ and $c = 7.12$ for the main cluster and bullet. Here, the concentration index c is defined as the ratio of radii enclosing 50% and 90% fluxes (Springel & Farrar, 2007). The shock Mach number of the bullet is $\mathcal{M} = 3.0 \pm 0.4$, and the shock velocity $v_s = 4740 \pm 630 \, \mathrm{km/s}$.

It has been assumed that this shock velocity is equal to the subcluster's relative velocity with respect to the parent cluster. This velocity is much higher than expected in the ΛCDM cosmology. To understand the dynamics of the bullet cluster, Springel & Farrar (2007) made hydrodynamical simulations of the evolution of the system. The model shows that despite a shock speed of 4740 km/s, the subcluster's mass centroid is moving only with $\sim$2700 km/s in the rest frame of the parent cluster. The difference arises due to a gravitationally induced inflow velocity of the gas ahead of the shock towards the bullet, which amounts to $\sim$1100 km/s.

The relative velocity of the bullet cluster $\sim$2700 km/s is expected for a satellite cluster moving in the gravitational potential field of a massive supercluster with central cluster of mass $M_{200} \simeq 1.5 \times 10^{15} \, M_\odot$. The mass of the Virgo cluster is much smaller, $M_{\mathrm{vir}} = 6.3 \times 10^{14} \, M_\odot$, and relative velocities of subclusters of the Virgo cluster are lower, about $\sim$1000 for the infalling cluster M60.

The bullet cluster is not the only known example of merging clusters. Merten *et al.* (2011) analysed another example of merging clusters — Abell 2744. The total mass of this cluster is $(1.8 \pm 0.4) \times 10^{15} \, M_\odot$, about the same as the mass of the bullet cluster. Here, we are witness of a multiple merger system.

8.3.3 Structure and evolution of the galaxy clusters of various richness

Einasto *et al.* (2024, 2022a) studied morphological properties of galaxies located in groups and clusters, determined by Tempel *et al.* (2014b) and based on SDSS data release 10 (DR10) data. They divided groups and galaxies into classes using three properties. The first property is the group luminosity: single galaxies, low-luminosity groups and high-luminosity groups. Single galaxies are lowest luminosity groups with no neighbours in the SDSS sample, their companions are of the type of dwarf satellites of the Milky Way and M31. High-luminosity groups are actually clusters of galaxies. The second property is the character of galaxies in groups: brightest group galaxies (BGGs) and satellite galaxies. The third property is the star-formation capacity: galaxies were divided according to star-forming intensity to passives and actives, using the index $D_n(4000)$: passive galaxies with old stellar populations with $D_n(4000) \geq 1.75$, and young star forming galaxies with $D_n(4000) < 1.75$ (this is a mixed population and includes quenched galaxies with no active star formation, red and blue star-forming galaxies, and recently quenched galaxies).

To study the dependence of morphological properties of galaxies on the global environment, authors used the luminosity density field smoothed with 8 h^{-1} Mpc B_3 spline kernel, see Eq. (4.5). Galaxies and groups with $D8 < 1$ ($D8.01$) in mean density units belong to cosmic voids, objects with $1 \leq D8 < 2$ ($D8.12$) and so on belong to regions of supercluster dynamical influence (cocoons) of corresponding global density, and systems with $D8 > 5$ are members of superclusters, i.e. dense parts of supercluster BoAs. Authors calculated for clusters of various richness distributions of stellar masses $\log M^*$, stellar velocity dispersions σ^*, $D_n(4000)$ indexes and concentration indexes C.

In Fig. 8.41, we present the distributions of stellar masses of passive galaxies of various types (Einasto *et al.*, 2022a). Data are given for low-luminosity and high-luminosity groups separately for satellites and brightest group galaxies (BGGs). Curves of different colours are for galaxies in various global environments $D8$. The distribution of stellar masses of isolated galaxies is shown as a solid red curve. The figure demonstrates several important properties of the morphology

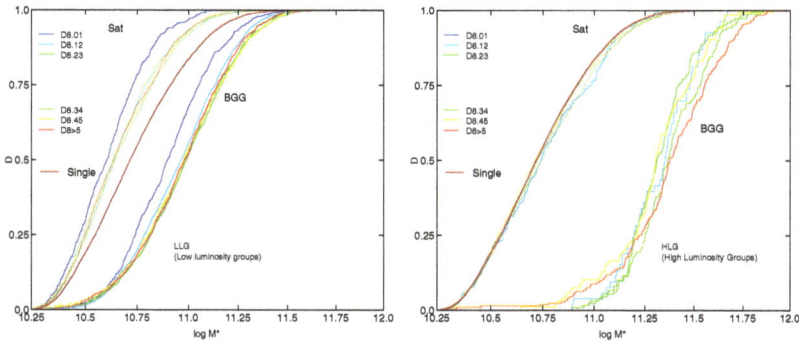

Fig. 8.41. Distribution of stellar mass $\log M^*$ of passive galaxies in regions of various global densities D8 for satellite galaxies and BGGs. The left panel denotes low-luminosity groups, which the right panel high-luminosity groups. Solid red curve in both panels shows the distribution of stellar mass for isolated galaxies (Einasto *et al.*, 2022a).

distribution of galaxies. First of all, we see that the largest effect have two characteristics — whether the galaxy is a satellite or the brightest in a group and the type of the system the galaxy belongs to — whether it is an isolated galaxy, or a member of a poor or a rich group. The influence of the global environment, characterised by the D8 parameter, has much smaller effect on the morphology of galaxies in groups.

Of special interest is the study of isolated galaxies, as the evolution of these galaxies is least influenced by external factors. Observations suggest that all isolated MW class galaxies are surrounded by dwarf satellites as discussed above. Such galaxies are fed by dwarf satellites inside dark matter haloes to which they belong as well as by pristine gas and dark matter from filaments. Initially, all galaxies are star-forming. During the evolution in a fraction of galaxies, star formation is suppressed, and galaxies become quenched. Depending on the character of seeds, and in the absence of the influence of other similar galaxies, a certain distribution of star-forming and passive galaxies at present epoch forms, see the left region in Fig. 8.42.

Figure 8.41 shows that the distribution of morphological properties of isolated galaxies and satellites of low-luminosity groups as well high-luminosity groups are very close; the curve for single galaxies

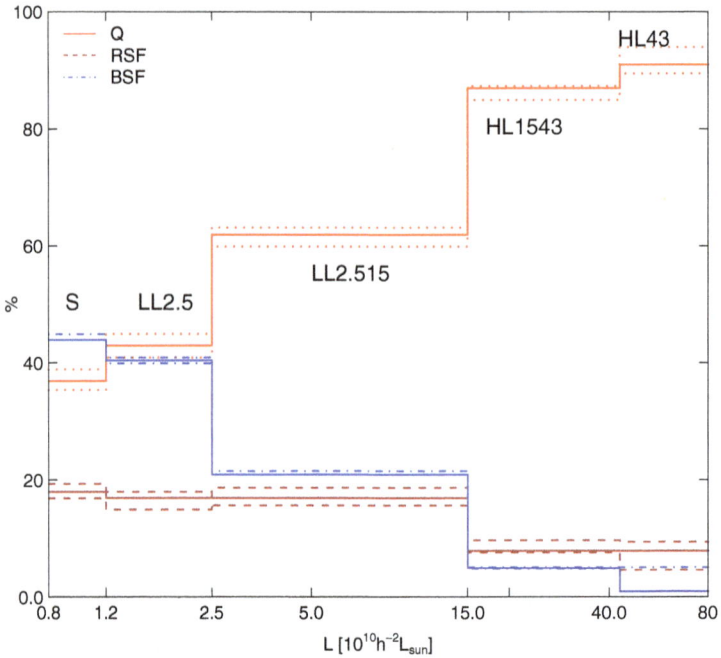

Fig. 8.42. Fractions of groups with BGGs of different star formation properties versus group luminosity for groups from different luminosity classes and for single galaxies. Red line shows the fraction of quenched BGGs and dark red dashed line shows the fraction of RSF BGGs. Blue line shows the fraction of BSF BGGs. Dotted, dashed, and dot-dashed lines show the changes in fractions if we use SFR limits $\log \mathrm{SFR} = -0.6$ and $\log \mathrm{SFR} \leq -0.4$ (Einasto *et al.*, 2024).

is plotted in both panels with a red line. This similarity suggests that evolutionary processes in satellite galaxies are similar to evolutionary processes in isolated galaxies, i.e. the infall does not change significantly the properties of infalling galaxies. Efficiency of star formation quenching of satellite galaxies as they fall into the potential well of the main galaxy (in the case of single galaxies) or to the group (for satellite galaxies in groups) depends on their orbital parameters. Infalling galaxies oscillate around the dynamical centre and preserve their internal structure. Typically, it takes more than one peripassage to quench a satellite galaxy. Only rarely the impact is directed exactly towards the central galaxies in a group, which leads to the merger of both galaxies and changes in morphological properties of central galaxies.

The global environment, characterised by the $D8$ index, has much smaller influence on the morphology of galaxies. Blue lines, which correspond to the global density interval $D8 < 1$ (region $D8.01$), lie higher than all other lines corresponding to intervals $D8 > 1$. This small difference shows that the morphology of galaxies in the lowest global density region is somewhat different from the morphology of galaxies in higher-density regions. Similarity of morphological properties of galaxies in regions $D8.12 - D8.45$ shows that in these regions, physical processes, influencing morphological properties, are similar.

Isolated galaxies, low-luminosity groups and rich clusters also differ in their location in the cosmic web. Einasto *et al.* (2024) showed that while isolated galaxies and low-luminosity groups can be found everywhere in the cosmic web, rich clusters are located only in high-density regions, especially in superclusters and in high-density cores of superclusters, see Fig. 8.42. Galaxy clusters of the highest luminosity are all located in superclusters. Groups of galaxies move in the smoothed velocity field together with surrounding galaxies and neighbouring groups in the direction towards the nearest great attractor. Great attractors form minima of the gravitational potential field and are dynamical centres of superclusters marked by dominant clusters of galaxies (Dupuy *et al.*, 2019, 2020; Einasto *et al.*, 2021b; Tully *et al.*, 2014). Spheres of dynamical attraction of groups are larger than respective spheres of isolated galaxies, but smaller than spheres of dynamical attractions of rich clusters. Thus, dynamical interactions in small groups are restricted to close neighbours. These interactions led to merging of galaxies and inflow of surrounding galaxies and gas with small inflow velocities.

Figure 8.42 shows how fractions of quenched, red star-forming and blue star-forming galaxies change with the group luminosity. The lowest luminosity interval corresponds to isolated galaxies. We see that fractions of quenched, red star-forming and blue star-forming galaxies of isolated galaxies and low-luminosity groups with number of group galaxies up to three are identical, i.e. star formation properties do not change when adding to groups only a few members. With additional increasing luminosity and membership of groups, the fraction of quenched BCGs increases from 0.37 to 0.91, the fraction of blue star-forming BCGs decreases from 0.44 to 0.01, and the fraction of red star-forming BCGs decreases from 0.18 to 0.08.

8.4 Structure and evolution of superclusters of galaxies

The large-scale distribution of galaxies in the Universe is very complex. There exist density enhancements of different sizes and shapes, such as clusters of galaxies, filaments, walls, and low-density regions (voids) between high-density regions. Largest building blocks of the Universe are superclusters of galaxies. As discussed in Chapter 2, superclusters were defined by Abell (1958) and Abell *et al.* (1989) as clusters of rich clusters of galaxies, but they also contain poor Zwicky *et al.* (1968) clusters, galaxies and galaxy filaments, which link superclusters to a connected network called the cosmic web (Bond *et al.*, 1996).

The definition of superclusters is not very precise since they have no well-fixed boundaries. Most recent nearby supercluster catalogues are found using the luminosity density field, as done by Luparello *et al.* (2011) and Liivamägi *et al.* (2012). More distant supercluster catalogues up to $z = 1$ were found by Sankhyayan *et al.* (2023) and Chen *et al.* (2024) applying the friend-of-friend method. The property of superclusters to act as great attractors was the basis of the Tully *et al.* (2014) suggestion to define superclusters on the basis of their dynamical influence to the cosmic environment — BoA.

8.4.1 *Structure of clusters in supercluster filaments*

Virgo cluster and the merging cluster 1E0657-56 are central clusters of superclusters and lie deep at the bottom of supercluster gravitational potential wells. Most clusters of galaxies are ordinary members of superclusters, they lie in the middle of filaments, far from central clusters where several filaments cross. It is interesting to study the structure of ordinary clusters in filaments.

One of the well-known cluster filaments is the main ridge of the Perseus–Pisces supercluster (PPS), consisting of Abell clusters A347 and A262, and clusters around galaxies NGC 507, NGC 383 and NGC 315 in increasing distance from the main cluster of the PPS, Abell

426. All main galaxies of these clusters are of supergiant cD type with extended envelopes. These main galaxies are also active radio galaxies with extended radio lobes. For the NGC 315, ALMA observations yield the mass of the central block hole, $M_{BH} = (1.67 \pm 0.3) \times 10^9$ M$_\odot$ (Boizelle *et al.*, 2021), about three times less than the mass of the black hole in the Virgo cluster. Virial masses of other clusters in the PPS were found from the velocity dispersions of cluster member galaxies and are as follows: A262 – 1.2, N507 – 1.3, N383 – 0.2 in units of $10^{14} h$M$_\odot$ (Sakai *et al.*, 1994).

8.4.2 Structure and evolution of the A2142, Corona Borealis and BOSS Great Wall superclusters

The A2142 supercluster has the highest luminosity peak among superclusters in the Liivamägi *et al.* (2012) catalogue of SDSS superclusters. The A2142 supercluster consists of quite a spherical main body with outgoing straight filament-like tail (the longest straight structure discovered so far), as seen in Fig. 8.43, where we also show with dashed red line its basin of dynamical influence (BoA) (Einasto *et al.*, 2020b). Two high-density regions form a high-density core (HDC) of the supercluster, lower-density regions form the outskirts of the supercluster. Galaxies in short filaments are denoted with violet colour, and galaxies in long filaments with length $\geq 20\,h^{-1}$ Mpc are denoted with dark red colour. HDC of the supercluster is marked with black circle. Orange circles mark the location of merging groups which will separate from the supercluster in the future. Blue circle marks turnaround (T) region, and light blue circle shows border of the zero gravity (ZG) region where long filaments are detached from the supercluster. Future collapse (FC) region lies between these regions. Navy line denotes the supercluster axis. Orange dashed line shows the BoA boundaries.

Einasto *et al.* (2020b) calculated the density contrast $\Delta\rho = \rho/\rho_m$ as a function of the cluster-centric distance D_c from A2142 for various regions. For a spherical collapse model with standard cosmological parameters ($\Omega_m = 0.3, \Omega_\Lambda = 0.7$), characteristic density contrasts are as follows: $\Delta\rho = 360$ (virial), $\Delta\rho = 200$ (r200), $\Delta\rho = 13.1$ (turnaround, blue dashed line in Fig. 8.44), $\Delta\rho = 8.73$ (future collapse (FC), magenta dash-dotted line), and $\Delta\rho = 5.41$ (zero gravity (ZG), light blue solid line). For details, see Dünner *et al.* (2006) and

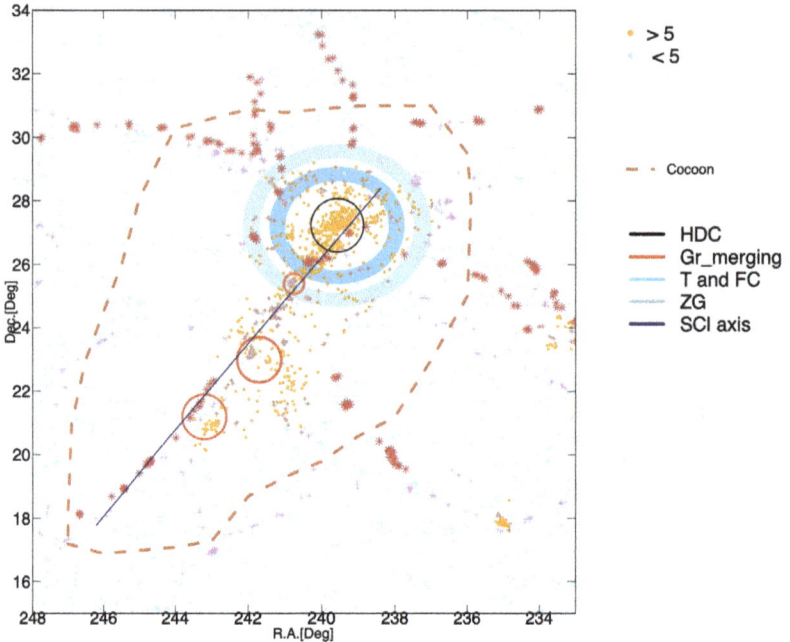

Fig. 8.43. Distribution of galaxies and filaments in the sky plane in and around the A2142 supercluster (Einasto *et al.*, 2020b).

Gramann *et al.* (2015). The connectivity of clusters is defined as the number of filaments towards the cluster. The connectivity of the cluster A2142 is 6–7, and 1–2 for poor clusters and groups far from the central cluster A2142. The supercluster main body is collapsing and groups in the HDC are falling into the central cluster A2142. Long filaments of galaxies and groups are detached from the central cluster at the turnaround region of the supercluster main body. These results suggest that the supercluster may split into several systems in the future.

Einasto *et al.* (2021d) made a similar analysis of the Corona Borealis supercluster, which has four rich clusters as its core region: A2061, A2062, A2065 and A2089. In their Figs. 13 and 14, authors presented density contrasts of clusters, which show that regions of different regimes (virial, r200, turnaround and future collapse) have decreasing cluster-centric distances for clusters A2065, A2061, A2089 and A2064. Distances between A2065 and other clusters are smaller than the radius of the future collapse region around A2065, thus

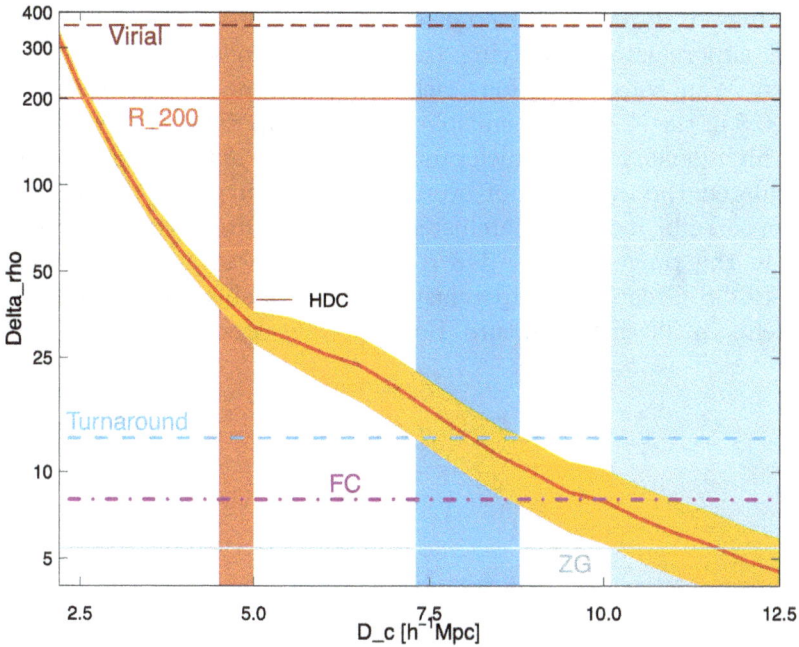

Fig. 8.44. Density contrast $\Delta\rho = \rho/\rho_m$ versus cluster-centric distance D_c for the A2142 supercluster main body (red line). Characteristic density contrasts are described in text. Tomato, blue, and light blue vertical areas mark borders of the HDC of the supercluster, turnaround region of the supercluster main body, and zero gravity region (Einasto *et al.*, 2020b).

A2089 lies within the turnaround region of A2065. The distance between the centres of A2065 and A2061 is approximately $9 \, h^{-1}$ Mpc within the future collapse radius of A2065. This suggests that all these clusters will merge into one collapsing structure in the distant future, forming one of the most massive collapsing structures in the Local Universe. The authors detected a distance minimum around rich clusters, which mark the border of the sphere of influence around clusters, within which all galaxies, groups, and filaments around clusters are infalling. Based on the spherical collapse model, they showed that these regions were at turnaround at redshift $z \approx 0.4$. This marks one important epoch in the evolution of rich clusters in superclusters. Note that at epochs between $z = 1$ and $z = 0.5$, evolutionary tracks of systems of galaxies have a break, see Figs. 6.2 and 6.3.

Einasto *et al.* (2022b) studied the evolution of the BOSS Great Wall superclusters, applying the spherical collapse model. BOSS Great Wall consists of four rich superclusters located at redshift $z = 0.5$ in the Northern hemisphere. These superclusters contain several high-density cores which probably collapse in the future. Authors calculated the evolution of overdensities at turnaround and future collapse radii for all superclusters. Radii of future collapse regions are in the range $R_{FC} \sim 4$–8 h^{-1} Mpc. The authors showed that the BOSS Great Wall superclusters will probably split into smaller systems in the distant future. However, this process is very slow.

Chapter 9

Structure and Evolution of the Ensemble of Superclusters and Voids

So far, we discussed the structure and evolution of various populations of the cosmic web: galaxies, clusters and superclusters of galaxies and cosmic voids. In this chapter, we discuss the structure and evolution of the whole ensemble of superclusters and voids. To select superclusters and voids, both the density as well as the velocity fields can be used. To follow the evolution, numerical simulations are needed. We start the analysis with the discussion of the structure and evolution of the Local Universe. Thereafter, we describe some geometrical properties of the cosmic web and of the whole ensemble of superclusters and voids, using spatial density and velocity data.

9.1   Structure and evolution of the Local Universe

The Local Universe is the volume around us which includes the Local Supercluster and other nearby superclusters and voids between them. This is the volume where observational data are most complete, and the structure and evolution can be studied in detail. One method to follow the evolution of the Local Universe is to apply constrained simulations, which allows us to study the evolution. We begin the discussion of the Local Universe with the analysis of planes of satellite galaxies, as demonstrated by the Magellanic stream and polar ring galaxies.

9.1.1 *Magellanic stream and polar ring galaxies*

Australian astronomers carried out with 18-m radio telescope measurements of neutral hydrogen in southern hemisphere and detected the Magellanic stream — a long filament of gas through the Magellanic clouds (Mathewson *et al.*, 1974), see Fig. 9.1. Lynden-Bell (1976), Mathewson & Schwarz (1976) and Mathewson *et al.* (1977) analysed the geometry, dynamics and origin of the Magellanic stream. Two options for the origin of the Magellanic stream were considered: (i) that the stream was pulled out of the Magellanic clouds by tidal forces produced by a close encounter with the galaxy and (ii) that the gas clouds of the stream are primordial and in the same orbit as the Magellanic clouds. Lynden-Bell (1976) preferred the first option and suggested that some dwarf galaxies and outlying galactic globular clusters as well as the Magellanic stream and some high-velocity H1 clouds (HVCs) were pulled out of the Magellanic clouds by strong tidal forces generated during close passages to the Galaxy. The map of HVCs in galactic coordinates with superposed Large and Small Magellanic clouds, and distance class XII globular clusters is presented in Fig. 9.1. The figure shows that there are actually three streams, two parallel streams are in the Northern Galactic hemisphere — the Draco–Ursa Minor stream and the second stream consisting mostly of HVCs and some distant globular clusters.

Lynden-Bell (1976) argued that the Magellanic stream is not a tail behind the Magellanic clouds but a tongue drawn forward and downward by the Galaxy. Magellanic clouds have not passed over the South Galactic Pole already but are instead heading towards it. The Draco–Ursa Minor stream in the Northern galactic hemisphere was probably torn off in an earlier encounter between the Greater Magellanic Galaxy (which includes both LMC and SMC) and the Milky Way. If so, the Magellanic clouds are on a bound orbit which is gradually coming closer to the Galaxy under the influence of tidal friction. If all the systems originated in the Greater Magellanic Galaxy which has been torn apart at least twice by tidal interaction, then the stellar populations of the dwarf spheroidals must be evolved samples of the stellar populations in the outer parts of the Magellanic clouds as they were long ago.

The plane of the Magellanic stream is inclined to the plane of the Galaxy by approximately 70°. It is interesting that satellites of

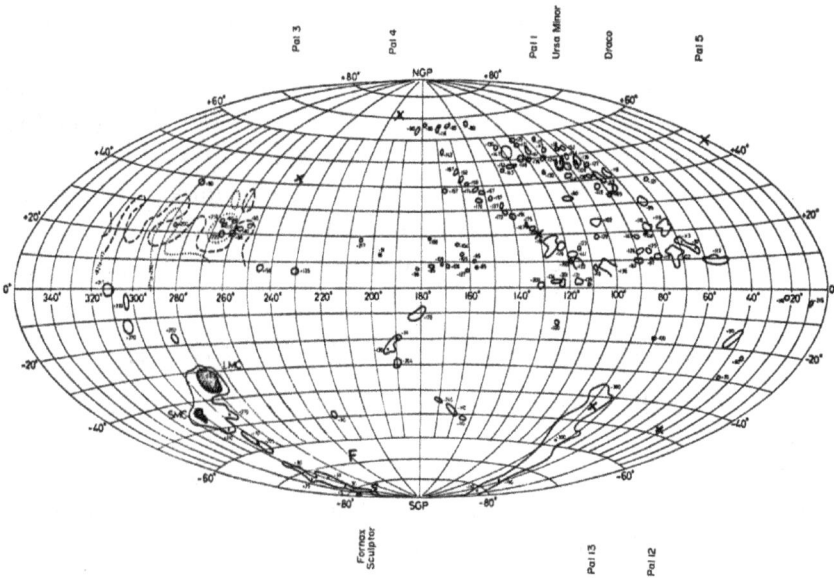

Fig. 9.1. A map of high velocity clouds in galactic coordinates with superposed Large and Small Magellanic Clouds, and U (Ursa Minor), D (Draco), S (Sculptor), F (Fornax) — distance class XII globular clusters (Lynden-Bell, 1976).

the Andromeda galaxy M31 form also a plane, highly inclined to the plane of M31. A similar phenomenon is observed in polar ring galaxies. Spindle-shaped S0 galaxies surrounded by severely tilted rings of luminous material have recently been found in increasing numbers (Schweizer *et al.*, 1983). The surrounding ring runs over the poles of the S0 disk. The ring motions suggest that a massive halo extends far beyond the S0 disk and that the halo is more nearly spherical than flat. The orbit of the ring offers the opportunity to probe the disk galaxy's gravitational field perpendicular to the plane. Schweizer *et al.* (1983) found that a few percent of all field S0s possess near-polar rings or disks. Authors suggested that these structures are due to a second event, most likely the transfer of mass from a companion galaxy during a close encounter.

Whitmore *et al.* (1990) showed that approximately 0.5% of all nearby S0 galaxies appear to have polar rings. Bournaud & Combes (2003) and Macciò *et al.* (2006) studied the origin of polar ring galaxies using numerical simulations including gas dynamics and star formation. Polar rings form naturally in a hierarchical Universe where most low mass galaxies are assembled through the accretion of cold gas infalling along megaparsec scale filamentary structures. Most of the gas is accreted in the warm phase galaxy mergers which lead to coalescence of the two haloes and central galaxies. The remnant is violently relaxating into a spheroidal system elongated in the plane of the merger. This is now the standard model for the formation of elliptical galaxies (Toomre & Toomre, 1972).

9.1.2 *Structure and evolution of the Local Group*

The plane of the Magellanic stream and satellite galaxies is oriented nearly perpendicular to the Galactic disk and its spatial extent encompasses the whole virial volume of the Galaxy. Satellite galaxies of the Andromeda galaxy M31 have a similar orientation. Pawlowski & Kroupa (2020) used Gaia DR2 data to determine orbits of 11 classical Milky Way satellite galaxies. Authors showed that these satellites are not only spatially but also kinematically correlated. Combining the proper motions increases the tension with ΛCDM cosmological expectations: only less than 0.1% of satellite systems in IllustrisTNG100-1 DM simulations contain orbital poles as closely aligned as observed.

However, earlier simulations of the Milky Way satellite systems were performed, ignoring the presence of close similar systems, the Andromeda galaxy with his satellite systems. To understand better the structure of Local Group galaxies, they must be simulated as one unit. Moreover, the problems arise when observations are confronted with predictions from DM-only simulations that treat the cosmic matter as a single collisionless fluid, which is a poor approximation on scales where baryonic processes are important. To avoid these limitations, Sawala *et al.* (2016) initiated the APOSTLE (A Project Of Simulating The Local Environment) simulations. Authors used the hydrodynamical code developed for the Evolution and Assembly of GaLaxies and their Environments (Eagle) project by Schaye *et al.* (2015). Authors selected 12 Local Group-like regions from a DM-only simulation of size $100^3\,\mathrm{Mpc}^3$ with 1620^3 particles. Each region contains a pair of haloes in the virial mass range $5 \times 10^{11}\mathrm{M}_\odot$ to $2.5 \times 10^{12}\mathrm{M}_\odot$, with median values of $1.4 \times 10^{12}\mathrm{M}_\odot$ for the primary (more massive) halo and $0.9 \times 10^{12} M_\odot$ for the secondary (less massive) halo of each pair. The combined median mass is $2.3 \times 10^{12} M_\odot$. The additional requirement was that the two haloes are separated by 800–200 kpc, approaching with radial velocity of 0–250 km/s and with tangential velocity below 100 km/s, to have no additional halo larger than the smaller of the pair within 2.5 Mpc and to be in environments with a relatively unperturbed Hubble flow out to 4 Mpc. The primary and secondary galaxies have in the mean 20 and 18 satellites more massive than $M_\star = 10^6 M_\odot$.

Each region was resimulated with the full Eagle hydrodynamical code, and surrounding regions were taken into account. At early times, the gas traces the filamentary DM structure, and stars begin to form in the most massive nodes of filaments. The first stars begin to form at $z \approx 17$ in the embryons of the pair of main LG galaxies, analogues to the Milky Way and M31. After the first stars have formed, feedback from star formation begins to blow out gas from the very low mass DM and gas haloes. At $z = 11.5$, reionisation heats the intergalactic gas and the gas already collapsed in haloes, quenching further gas cooling and accretion into small haloes. As a result, the formation of new galaxies is disrupted, until sufficiently massive haloes begin to form. Star formation begins anew in more massive haloes, and individual star-forming regions merge to assemble larger galaxies. As a result, the simulation contains tens of thousands of small DM substructures, but galaxies are forming only in the most massive haloes. This explains the "missing satellites" problem (Klypin *et al.*, 1999), and the difference of correlation functions of galaxies and DM in halo regions, as discussed in Chapter 6. Sawala *et al.* (2016, 2017, 2023) concluded that the orbital pole alignment is much more common than previously reported and that the orientation of Milky Way satellites are compatible with standard ΛCDM model expectations. Compared to corresponding DM-only simulations, the loss of baryons from subhaloes is due to the presence of baryons near the centre of the main halo. This reduces the number of subhaloes by a factor 2 to 4, independent of subhalo mass, but increasingly towards the main halo centre. The combined evidence from these model calculations suggest the formation of the Local Group with two large spiral galaxies, approximately 100 known satellite galaxies, and a vast number of completely dark substructures.

It is commonly believed that the Milky Way is approaching the neighbouring Andromeda galaxy, and in around five billion years, they merge and form a new giant elliptical galaxy. To check this prediction Sawala *et al.* (2024) studied the evolution of the Local Group using recent observations by the Gaia and Hubble space telescopes. Authors calculated orbits of MW and M31 galaxies, and studied the influence of their most massive satellites, LMC and M33. Both satellite galaxies change orbits of main galaxies. Sawala *et al.* (2024) concluded that with a probability of about 50%, there is no Milky Way — Andromeda merger during the next 10 billion years.

9.1.3 *A Council of Giants*

McCall (2014) analysed distances and luminosities of the brightest galaxies in the Local Volume and showed the presence of a Local Sheet which is both geometrically and dynamically distinct from the Local Group and the Local Supercluster. The sheet is inclined by 8° with respect to the Local Supercluster. A "Council of Giants" with a radius of 3.75 Mpc encompasses the Local Group, demarcating a clear upper limit to the realm of influence of the Local Group. Figure 9.2 presents the spatial distribution of a sample of nearby galaxies within the radius of 6.25 Mpc of the centre of the Council of Giants. The solid pink circle is the fit to the Council of Giants. The inner dashed pink circle marks the edge of the cylindrical realm of influence of the Local Group defined by density matching. The outer dashed pink circle marks the outer edge of the density-matched volume of the council. Dashed grey lines mark the intersections of the sheet with the Galactic and supergalactic planes. In the upper panel of Fig. 9.2, blue curves trace maxima in the potential surface described by the gravitational fields of the 14 giants in the Local Sheet.

9.1.4 *Constrained simulations*

Today, the most successful method to find the assembly history of the contemporary large-scale structure of the Universe is to apply constrained numerical N-body simulations. The region that is best studied and accessible observationally is the Local Universe, i.e. the Local Group and its immediate large-scale environment. The Constrained Local Universe Simulations (CLUES) project is dedicated to

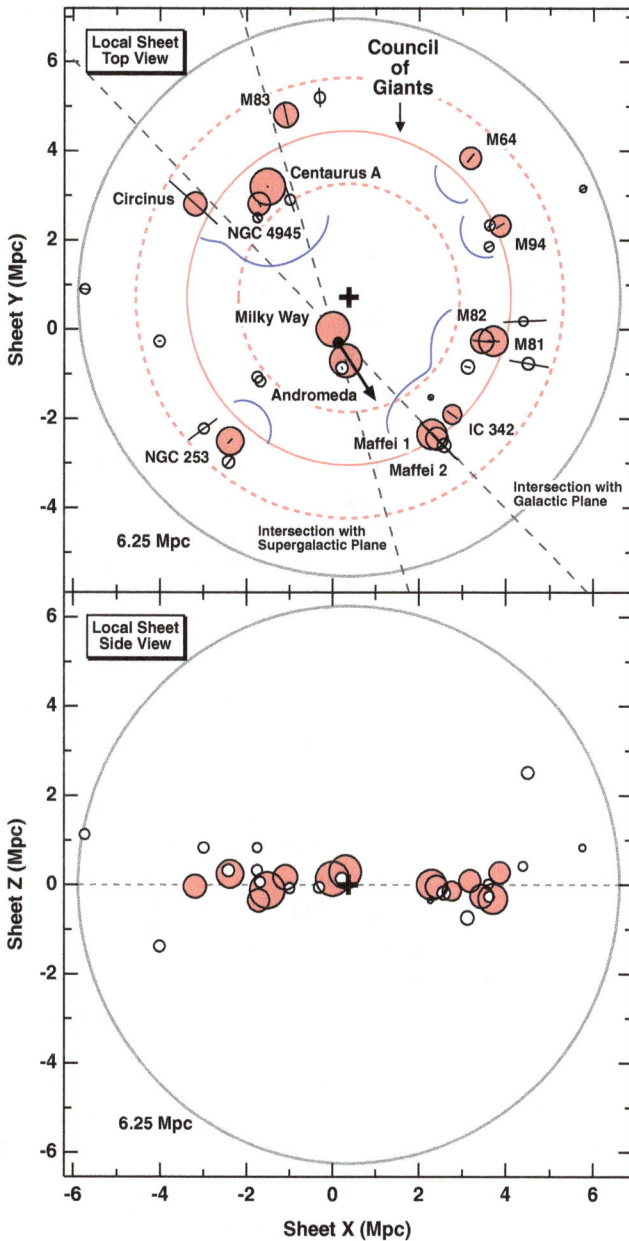

Fig. 9.2. The spatial distribution of sample galaxies within 6.25 Mpc of the centre of the Council of Giants. Shown are top and side views in a coordinate system with an x–y plane coincident with the mid-plane of the Local Sheet, which is displayed as a dashed grey line in the lower panel (McCall, 2014).

construct simulations that reproduce the Local Universe and its key ingredients, such as the Local Supercluster with the Virgo cluster, the Coma cluster and supercluster, the Great Attractor and the Perseus–Pisces supercluster. The main drawback of the CLUES simulation is that it does not directly constrain the structure on scales smaller than massive clusters. The reason is that the constrained realisation (CR) method is formulated within the formalism of Gaussian fields, assuming that the linear theory of density perturbations is valid on all scales. The data used as input are observed at the present time after having undergone the highly non-linear structure formation process. Thus, the linearity assumption is only valid down to a certain length scale.

To overcome these difficulties, Doumler *et al.* (2013) presented a method for the Lagrangian reconstruction of peculiar velocity data called reverse Zeldovich approximation (RZA). The RZA method consists of applying the Zeldovich approximation in reverse to galaxy positions and peculiar velocities to estimate the cosmic displacement field and the initial linear matter distribution from which the present-day Local Universe evolved. Halo peculiar velocities at $z = 0$ are close to the linear prediction of the Zeldovich approximation if a grouping is applied to the data to remove virial motions. The RZA is able to recover the correct initial positions of the velocity tracers with a median error of 5 h^{-1} Mpc for realistic sparse and noisy data.

Sorce *et al.* (2016a, 2016b) investigated the formation of the Virgo cluster using constrained simulations. To build initial conditions for dark matter numerical simulations, several techniques are needed assuming a prior cosmological ΛCDM model: (i) the Wiener filtering (WF) method to reconstruct the cosmic displacement field; (ii) the reverse Zeldovich approximation (RZA) method to relocate positions and velocities of particles; (iii) the constrained realisation (CR) of Gaussian field technique to produce overdensity fields constrained by observational data. Authors run 25 constrained simulations and 15 random simulations, all with resolution 512^3 particles in boxes of size 500 h^{-1} Mpc. In each constrained simulation, a unique DM halo could be identified as Virgo's counterpart. Simulation initial conditions follow observational constraints from the Cosmicflows-2 data by Tully *et al.* (2013). Figure 9.3 presents the reconstructed overdensity and velocity fields of the Local Universe. The left panel shows the observed Universe as reconstructed with the Wiener filter, and

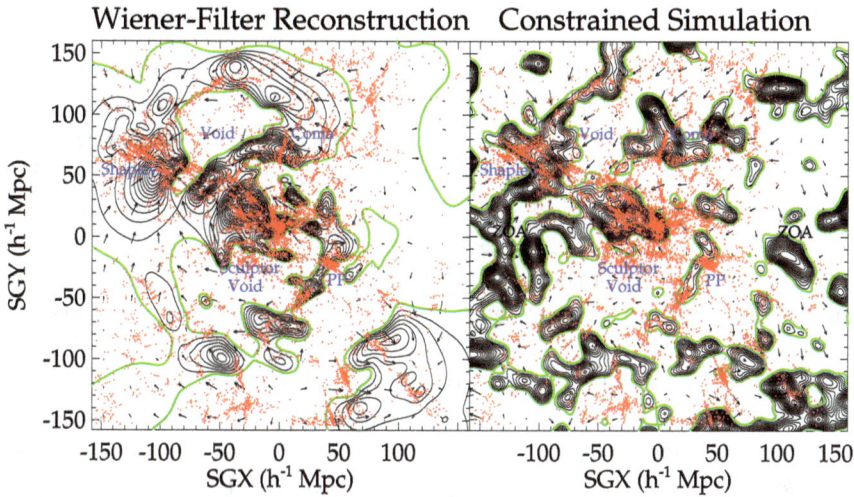

Fig. 9.3. Supergalactic plane of the reconstructed overdensity (contours) and velocity fields of the Local Universe obtained with the Wiener-filter technique (left) and of the simulated density (contours) and velocity fields of one realisation (right). The green colour stands for the mean density. Arrows represent velocity fields. Galaxies from the 2MASS redshift catalogue, in a $\pm 5\ h^{-1}$ Mpc thick slice, are superimposed as red dots (Sorce *et al.*, 2016b).

the right panel shows the constrained simulated Universe. Structures, voids and flows of the Local Universe are well recovered and simulated. A few of them are identified (blue names). While the Wiener filter reconstructs fairly well the Local Universe in the centre of the box, the simulation allows us to go farther in distances and deeper into the Zone of Avoidance and, more importantly, it supplies the whole density field (including non-linearities). Particles infalling on to Virgo haloes move along a preferred direction that is similar in all the simulations. This direction is along the filaments in which the Virgo cluster resides. This is in agreement with the general Cosmic Web analyses according to which the clusters accrete matter along their host filament, as suggested already by Shandarin & Klypin (1984) and Dekel *et al.* (2009), see Fig. 8.2 of Chapter 8. At around redshift 1, Virgo haloes have accreted 50% of their mass while an average random halo of the same mass has only accreted 30% of its mass. This suggests a relatively quiet merging rate for the Virgo cluster during the last 7 Gyr.

9.1.5 *Evolution of basins of dynamical influence*

Klypin *et al.* (2003) used constrained simulations to reproduce the large-scale density field with major nearby structures, including the Local Group, the Coma and Virgo clusters, the Great Attractor, the Perseus–Pisces, and the Local Supercluster. The structure of the Local Supercluster region (LSC: a 30 h^{-1} Mpc sphere around the Virgo cluster) and the Local Group environment is shown in Fig. 9.4. The fields were smoothed with the Gaussian filter of 1.4 h^{-1} Mpc smoothing length. The circle at the origin of the coordinates marks the position of the Milky Way galaxy. Points show dark matter (DM) particles (10% of all particles is shown) in a slice of 10 h^{-1} Mpc thickness centred on the supergalactic plane ($SGZ = 0$). Contours show the projected density in the slice: the thick contour corresponds to the average matter density of the Universe. The thin contours mark overdensities 2, 4, 6, and so on. Authors find that at the current epoch, most of the mass of the LSC is located in a filament roughly centred on the Virgo cluster and extending over $\sim 40\ h^{-1}$ Mpc. The

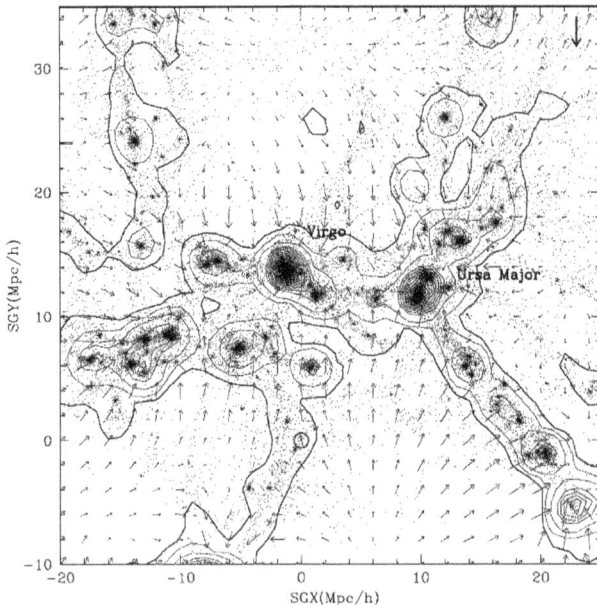

Fig. 9.4. Density and velocity fields in the 45 h^{-1} Mpc region around the Virgo cluster. The velocities are in the Virgo cluster rest frame (Klypin *et al.*, 2003).

simulated Local Group is located in an adjacent smaller filament, which is not a part of the main body of the LSC and has a peculiar velocity of $\approx 250\,\mathrm{km/s}$ towards the Virgo cluster.

As discussed in Chapter 8, basins of attractions (BoAs) are regions around superclusters, from which superclusters collect their matter. Hoffman *et al.* (2017) mapped the large-scale 3D velocity field using a Wiener filter reconstruction from the CosmicFlows-2 (CF2) dataset of peculiar velocities and identified the attractors and repellers that dominate the local dynamics. They showed that the local flow is dominated by a single attractor — associated with the Shapley concentration — and a single previously unidentified repeller. Streamlines of attractors and repellers are shown in Fig. 9.5. The knots and filaments of the velocity shear tensor web are shown for reference. The streamlines are seeded on a regular grid and are coloured according to the magnitude of the velocity. The flow stream lines clearly diverge from the repeller (Basin of Repulsion, BoR) and converge on the attractor. For the anti-flow, the divergence and convergence are switching roles.

Tully *et al.* (2019) used CosmicFlows-3 data to investigate the cosmography of the Local Void. This analysis confirmed the basic results of earlier studies. The database of directly measured distances of galaxies improved rapidly, and the current CosmicFlows-4 (CF4)

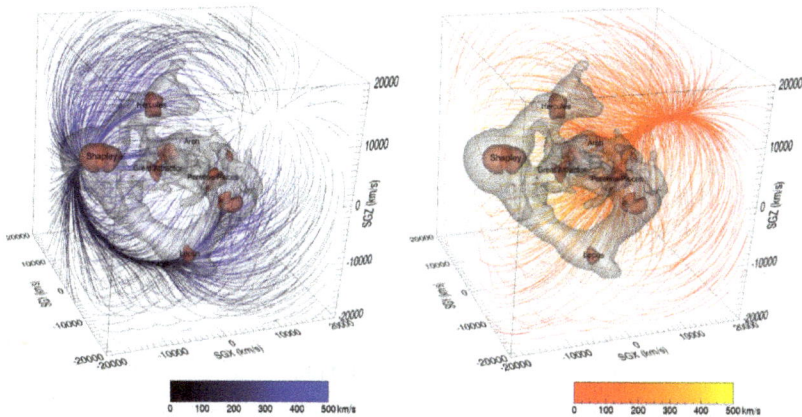

Fig. 9.5. A 3D view of the stream lines of the attractor flow field (in black-blue, left panel) and of the repeller flow (in yellow-red, right panel) (Hoffman *et al.*, 2017). Used with permission of *Nature Astronomy* from Hoffman *et al.* (2017). Permission conveyed through Copyright Clearence Center, Inc.

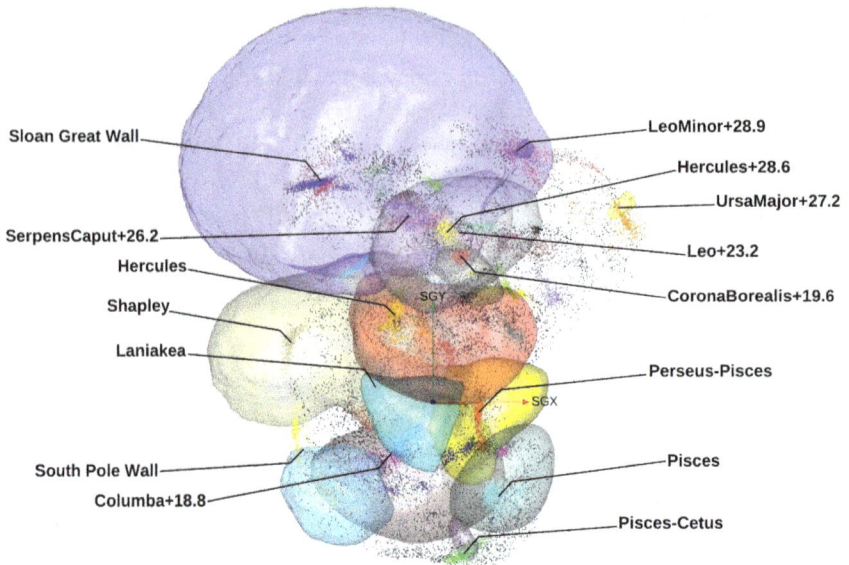

Fig. 9.6. Envelopes of major BoAs superimposed on the sinks. Names given to the major BoA are indicated. For previously unnamed objects, we use the convention "constellation+distance" in units of 1,000 km/s. Note that several structures can be found in the same constellation (e.g., Hercules) (Valade *et al.*, 2024).

dataset is about six times larger than the CosmicFlows-2 in terms of number of galaxies and is doubling its reach in the Northern hemisphere. This dataset allowed us to determine the 3D gravitational velocity field and bulk flow of galaxies in a much larger volume of the Local Universe. Figure 9.6 shows the visualisation of the BoA of nearby superclusters: Laniakea, Sloan Great Wall, Perseus–Pisces, Pisces, Pisces–Cetus, Hercules and Shapley (Valade *et al.*, 2024). The Coma supercluster is part of the Shapley basin. Each identified basin of attraction is annotated by the associated structure name. The coverage of CF4 allows us to see a few SDSS basins at much higher supergalactic Y values. One of these may be related to the A2142 supercluster.

The much richer CF4 dataset allowed us to calculate volumes of supercluster basins. Dupuy & Courtois (2023) found the following volumes for Laniakea, Apus, Hercules, Lepus, Perseus–Pisces and Shapley basins: 1.9, 9.5, 3.1, 8.1, 4.8, 7.9, in units of

10^6 $(h^{-1}$ Mpc$)^3$. Volumes of high-Y SDSS basins are much larger. The increase in volumes of distant basins is partly due to larger smoothing of the velocity field at high-Y, where the sample of galaxies with known distances and velocities is sparse. As shown in Fig. 9.17, a large smoothing can increase masses (and thus volumes) up to 100 times.

9.1.6 *Structure and future evolution of the Local Universe*

Araya-Melo *et al.* (2009) followed the evolution of bound objects from the present epoch up to a time in the far future of the Universe in a ΛCDM Universe. Authors used for simulations a 500 h^{-1} Mpc box with 512^3 DM particles and standard cosmology with $\Omega_m = 0.3$, $\Omega_\Lambda = 0.7$ and $h = 0.7$. Authors ran the simulation up to $a = 100$ and analysed the structure at epochs $a = 1$, 2, 5, 10. The main results can be summarised as follows: (i) The large-scale distribution of bound objects and superclusters in comoving space does not show any significant evolution between $a = 1$ and $a = 100$. (ii) The superclusters collapse between $a = 1$ and $a = 100$. While at the present epoch, they are embedded within the cosmic web, by $a = 100$, they have turned into isolated cosmic islands.

Seidel *et al.* (2024) investigated the future evolution of the Local Universe using the Simulation of the LOcal Web (SLOW) set of constrained simulations. This is the first simulation suite to combine constraints from the local velocity fields, a large volume and detailed galaxy evolution models, making it the ideal laboratory for studying local peculiarities. SLOW is made for a 500 h^{-1} Mpc large box and aims to simulate the Local Universe (Dolag *et al.*, 2023). Authors used various resolution levels to perform simulations with different levels of additional hydrodynamical and full galaxy formation physics. Using Cosmicflows-2 catalogue, the local velocity field was reconstructed using a Wiener filter technique by Zaroubi *et al.* (1995, 1999). The final initial conditions were obtained by applying the constrained realisation algorithm introduced by Hoffman & Ribak (1991) to the reconstructed velocities, for details of the simulations, see Seidel *et al.* (2024).

To analyse how the supercluster regions are collapsing in the future, authors performed *Clairvoyant* forward simulations of the

SLOW initial conditions set and let the DM particles evolve until a scale factor of $a = 1000$, equivalent to 119 Gyrs into the future, or equivalent cosmological "redshift" $z = -0.999$ (negative redshifts are not physically meaningful but serve as an illustrative parameter). This epoch corresponds to roughly $8t_H$ with $t_H = 1/H_0$ as the Hubble time and $h = 0.6777$. This specific time was chosen to ensure that the rapid expansion of the background due to dark energy domination has frozen the large-scale evolution.

Seidel *et al.* (2024) selected supercluster regions and member clusters from three observational supercluster catalogues and identified the corresponding regions in the Clairvoyant forward simulations by utilising the cross-matched most massive clusters in the simulation. General properties of six supercluster regions were found as follows: Shapley with maximum extent of the collapsed region at $z = 0$, $R_{max} = 15.95$ h^{-1} Mpc, Coma with $R_{max} = 13.93$ h^{-1} Mpc, Centaurus with $R_{max} = 12.71$ h^{-1} Mpc, Hercules with $R_{max} = 10.94$ h^{-1} Mpc, Perseus–Pisces with $R_{max} = 9.90$ h^{-1} Mpc, and Virgo with $R_{max} = 8.70$ h^{-1} Mpc. As an example of the reconstruction of the evolution, we show in Fig. 9.7 the Clairvoyant simulation of the Perseus–Pisces supercluster region. The left panel shows the

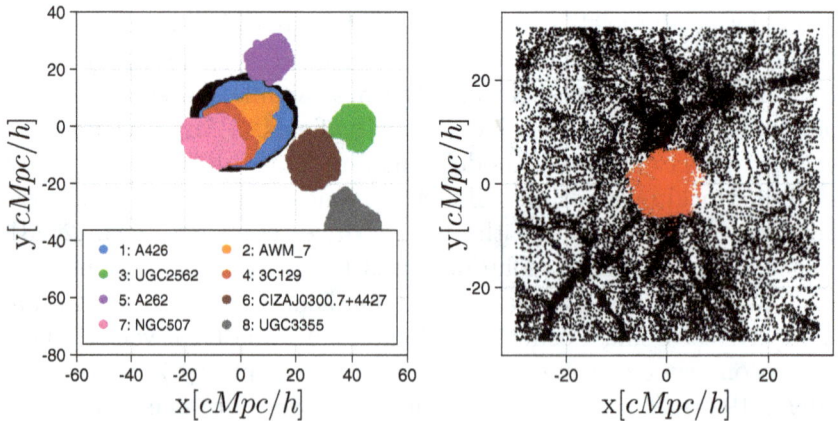

Fig. 9.7. The Perseus–Pisces supercluster region in the Clairvoyant simulation with the cross-identified counterparts: 1: Perseus (A426), 2: AWM 7, 3: UGC2562, 4: 3C129, 5: A262, 6: CIZAJ0300. 7+4427, 7: NGC507, 8: UGC3355. Left: Supercluster according to the N-Body simulation at $z = 120$. Right: The collapsed region (red) and its environment at the present epoch $z = 0$ (Seidel *et al.*, 2024).

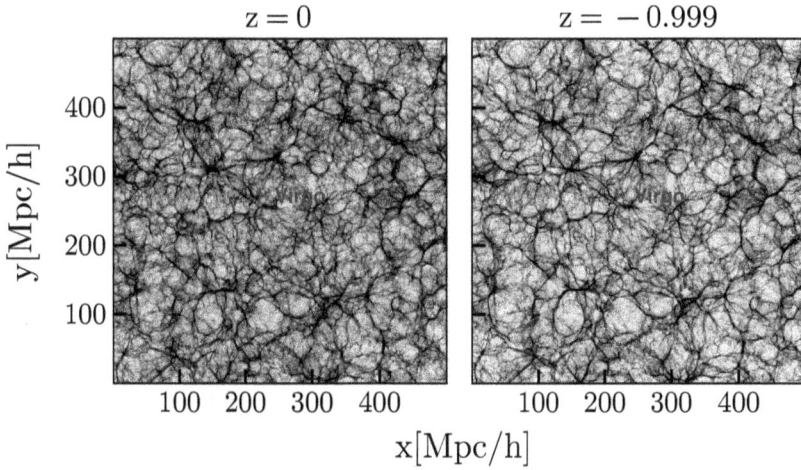

Fig. 9.8. The cosmic web in the forward simulation (dark matter component) within a slice of $10\ h^{-1}$ Mpc at the present time (left) and in the far future (right). The position of the Virgo cluster as the nearest cluster is marked (Seidel *et al.*, 2024).

whole supercluster at the early epoch $z = 120$, while the right panel shows the collapsed supercluster core region at $z = 0$.

Next Seidel *et al.* (2024) investigated the future of bound structures. Figure 9.8 shows a slice through this Clairvoyant simulation at the present time and in the far future. The figure shows that the shape of the cosmic web with superclusters remains in the far future similar to the structure at the present epoch (in comoving coordinates). In other words, the evolution of the cosmic web is already almost progressed to its final morphological state at the present time. It should be noted here that this standstill occurs only in comoving space, and in real space, the frozen structures are drifting apart.

Further, Seidel *et al.* (2024) studied the evolution of the local halo mass and accretion histories for the six supercluster regions, applying Clairvoyant forward simulations. There is little to no late time evolution in comoving space at the largest scales. Collapsing overdensities separate causally into gravitational "islands" with flows between those regions halted by the rapid expansion pulling theses islands apart in real space. For similar reasons, the voids continue to empty out after the present time, but there is also a significant amount of matter that is left behind in the voids. The suppression of structure

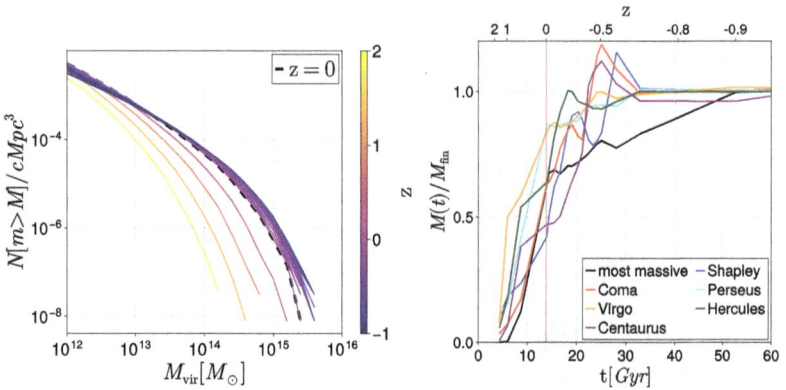

Fig. 9.9. Left: Evolution of the local halo mass function far into the future. Each line represents the halo mass function at the redshift indicated by the line colour The present time is the dashed purple line. Right: Late-time accretion histories for the six supercluster regions relative to their respective final halo masses (Seidel *et al.*, 2024).

formation is evident in the evolution of halo masses. The left panel of Fig. 9.9 shows the halo mass in the Clairvoyant simulation and its evolution with cosmic time. The halo mass function evolves to some degree after $z = 0$, it barely reaches objects with masses of $10^{16} M_\odot$, and the evolution is slowed down. Authors note that the largest structures assemble after the present time, but the evolution at the group and small cluster mass region is completely halted. This reflects the assembly of the supercluster structures that are dynamically still very young at the present time. Beyond 60 Gyr, the evolution of large-scale structures is stopped completely in comoving space.

The right panel of Fig. 9.9 shows the mass evolution of the haloes associated with the six superclusters in the Clairvoyant simulation. Additionally, we show the evolution of the most massive halo in the simulation volume overall (black line). Two classes of superclusters can be identified with reference to their mass growth history: (i) those that have already assembled their mass fully at the present time (Perseus and Virgo) and (ii) superclusters that undergo significant mass accretion and major mergers in the future before freezing out. The most prominent example of late growing superclusters is Shapley: the most massive supercluster experiences two major mergers in the future, associated with the infalls of A3560 and A3571.

A common peculiarity of these late-time mergers is that the virial mass of the haloes actually decreases after reaching a peak after the secondary cluster has fallen in. For the late-time mergers, the mass loss is generally not recovered through reaccretion of the lost material. This is illustrated by the trajectory of Hercules in Fig. 9.9: due to the final major merger happening early enough ($t \approx 18\,\text{Gyr}$), the supercluster is able to recover nearly all of its peak mass after an initial post-merger mass loss. In contrast, the mass lost by the last major mergers in the Coma supercluster is lost to the Hubble flow and is never recovered.

Chen *et al.* (2024) compiled a supercluster catalogue using Chon *et al.*'s (2015) physically motivated definition that superclusters should survive the accelerating expanding Universe and collapse in the future. Authors used the Hyper Suprime-Cam Subaru Strategic Program (hereafter the HSC survey) to find distant superclusters at redshifts $z = 0.5-1.0$. The survey was made with the Hyper Suprime-Cam camera installed at the 8.2-m Subaru telescope in Hawaii. The survey covers a sky area of about 1027 square degrees and has a limiting magnitude of $i \approx 26$. The supercluster catalogue consists of 673 objects with multiplicity from 2 to 5, based on photometric redshifts, checked by numerical simulations of the evolution to distant future up to scaling factor $a = 15$ ($a = 1$ is the present epoch). Properties of high-density cores (HDCs) of superclusters are similar to HDCs identified by Einasto *et al.* (2022b) in the BOSS Great Wall. Given the similarity between these characteristics, Chen *et al.* (2024) concluded that the supercluster candidates they have identified essentially correspond to future collapsed regions of HDCs.

9.2 Some geometrical properties of the cosmic web

Usually, the cosmic web is considered in terms of nodes, walls, filaments and voids, and the geometry of the location of galaxies is not discussed. However, the Local Universe is host to numerous peculiarities that possibly can challenge the current canonical cosmological model. Such peculiarities are observed on various scales from haloes to superclusters of galaxies. In the following sections, we describe two of these peculiarities: the quasi-regularity of the cosmic web and the presence of supercluster walls.

9.2.1   *Regularity of the cosmic web*

ARIEL: Would it not be interesting if the universe was periodic? Like most of our simulations.

RHEA: Periodic boundary conditions are technical tools to fit infinity into the computer. The real universe is, of course, nothing like that.

CALLISTUS: I'd take this seriously for a second.

RHEA: Of course, you would.

CALLISTUS: The universe could have a non-trivial topology. If that topology were a torus, it would be periodic.

RHEA: Cosmic microwave background measurements severely constrain non-trivial topologies below the horizon scale.

CALLISTUS: I remember back in the day, there was a lot of excitement, even controversy, about pencil beam surveys. Some interpreted them as a sign that the universe might be periodic with a scale of approximately $\simeq 120\,h^{-1}$ Mpc.

ARIEL: What is a pencil beam survey?

RHEA: This is ancient history: in the late 1980s, taking spectra of many galaxies in a wide-field survey was still too costly. There were no multi-object spectrographs like PFS on Subaru today. Astronomers instead measured redshifts in a tiny angular window, resulting in an essentially 1D survey.
         [*Callistus shows Fig. 9.10.*]

CALLISTUS: A clever idea. The result is this periodogram, essentially a 1D correlation function.

ARIEL: I remember how 2D correlation functions lose a lot of information with respect to the full 3D measurement. Isn't that the case for a 1D correlation function?

RHEA: Spot on. We have to be careful when interpreting lower dimensional correlation functions.

CALLISTUS: Especially in these legacy pencil beam surveys with a variable selection function. I believe the original paper never claimed that the universe was periodic, but that was the simplest picture the media could project.

RHEA: With hindsight, they probably discovered baryonic oscillations.

ARIEL: Photon baryon plasma carries baryons from potential wells with the sound speed approximately $1/\sqrt{(3)}$ of the speed of light. After decoupling and recombination, photons drop baryons at the distance sound waves could travel until then, the sound horizon, about $110\,h^{-1}$ Mpc. This scale imprints on the CMB and the large-scale structure of the universe... But wait, they measured something slightly larger!

RHEA: Even if you believe the error bars, the pencil beam measurements are within 2σ of the sound horizon scale. I think BAO is the most likely explanation.

CALLISTUS: And if you observe a rare wave, the largest random fluctuation, it will never be perfectly parallel with the line of sight. Thus, you will likely observe a larger scale than the actual sound horizon.

RHEA: Moreover, non-linearities modulate the BAO scale: larger densities get slightly closer due to gravitational attraction, while low densities experience the opposite effect.[1]

ARIEL: It just strikes me that they might have discovered the BAO features a decade earlier than others!

CALLISTUS: Nimis maturus.[2] You have to be ahead of everybody else... just a little bit, otherwise people will not understand your discovery.

RHEA: That's your life story, isn't it?

CALLISTUS: Let's not get into this now. It's too late.

ARIEL: Please continue! My TikTok channel desperately needs new gossip!

Already first pictures of the wedges shown in Fig. 2.18 show that there is some regularity in the distribution of rich superclusters. This issue became topical after the discovery by Broadhurst *et al.* (1990) that the distribution of high-density regions of galaxies is quasi-regular or periodic. Broadhurst *et al.* found such a regularity in the direction of the North and South Galactic poles, see Fig. 9.10.

Ryabinkov & Kaminker (2021, 2024) searched for quasi-periodical structures at moderate cosmological redshifts using SDSS DR7 and DR12 data on the luminous red galaxies and clusters of galaxies with redshifts $0.16 \leq z \leq 0.47$. Authors found that the strongest periodicity signal is observed in two nearly opposite directions: $\alpha = 170°$, $\delta = 29°$, and $\alpha = 346°$, $\delta = -29°$ with peaks at wavelengths $116 \pm 10 \ h^{-1}$ Mpc and $130 \pm 9 \ h^{-1}$ Mpc, respectively.

Einasto *et al.* (1994b, 1997b, 2001) compiled catalogues of superclusters formed by Abell clusters. She found that the median diameter of voids defined by clusters in superclusters is $120 \ h^{-1}$ Mpc. These values characterise the scale of cells of the Universe and the mean sizes of voids between superclusters. Further study by Einasto *et al.*

---

[1] Neyrinck *et al.* (2018).
[2] Too early.

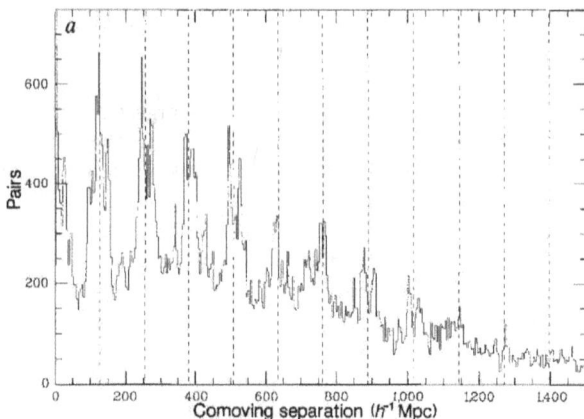

Fig. 9.10. One-dimensional pair-count correlation for all the redshift data available in the SGP-NGP axis. The dashed line indicates scales as multiples of $128\,h^{-1}$ Mpc (Broadhurst *et al.*, 1990). Used with permission of *Nature* from Broadhurst *et al.* (1990). Permission conveyed through Copyright Clearence Center, Inc.

(1997a) confirmed that high-density regions marked by rich clusters form a quasi-regular lattice of scale $\approx 120\ h^{-1}$ Mpc.

The quasi-regularity in the distribution of rich clusters and superclusters has a characteristic scale $\approx 120\ h^{-1}$ Mpc. This scale is close to the scale of baryonic oscillations, $109\ h^{-1}$ Mpc. Thus, a natural question arises: Are these phenomena related? Einasto *et al.* (2016) discussed the origin of shell-like structures and came to the conclusion that they are different from BAO shells. The radii of the possible shells are larger than expected for a BAO shell ($\approx 120-130\ h^{-1}$ Mpc versus $\approx 109\ h^{-1}$ Mpc). Detected shell structures are determined by very rich galaxy clusters and superclusters and have the density enhancement approximately equal to the central cluster, one of such clusters is A1795. However, the morphology of supercluster shells is different from the morphology of baryon cells — BAO shells are barely detected in the distribution of galaxies (Arnalte-Mur *et al.*, 2012). In contrast, the walls of the shell connected with cluster A1795 are formed by clusters. Together they form a density enhancement tens of times higher (stronger) and approximately 100 times more massive than the central cluster, A1795. Also, the physics of these phenomena is different. The large-scale distribution of clusters and superclusters is given by the initial density perturbation field generated during or after the inflation period of the evolution of the *dark matter*-dominated Universe. Baryonic acoustic oscillations are generated by sound waves in the *baryonic matter* before recombination.

To understand the formation of the 128 h^{-1} Mpc periodicity, Suhhonenko *et al.* (2011) studied the formation of a characteristic scale in the distribution of galaxies and clusters of galaxies. Authors used a series of toy models where amplitudes of density perturbations above a cutoff scale λ_{cut} were put to zero. The cutoff scale was varied from 8 to 768 h^{-1} Mpc, using simulation cubes of box sixes 100, 256 and 768 h^{-1} Mpc. The analysis showed that the characteristic scale of the cosmic web in models is determined by the cutoff wavelength λ_{cut}. But this regularity is valid only to models with cutoff wavelength $\lambda_{cut} \leq 128$ h^{-1} Mpc. Models with higher cutoff wavelength have the same structure as the model with cutoff wavelength $\lambda_{cut} = 128$ h^{-1} Mpc, i.e. largest waves do not influence the scale of the web. These large waves were outside of the horizon for most of the time and had no influence to the evolution of the cosmic web. These waves entered the horizon at the transition from the radiation-dominated stage to the matter-dominated one, which occurred at $z = z_{eq} \approx 3200$, see Fig. 10.1 in Chapter 10. Suhhonenko *et al.* (2011) concluded that the cosmic web with filamentary superclusters and voids is formed by the combined action of all perturbations up to the scale ≈ 120 h^{-1} Mpc. The largest perturbations in this range determine the scale of the supercluster-void network. Perturbations of the largest scales > 120 h^{-1} Mpc modulate the richness of galaxy systems from clusters to superclusters, and make voids emptier, but do not influence the scale of the web.

9.2.2   *Supercluster walls*

Another regularity was detected in the distribution of nearby Abell clusters and superclusters, as shown in Fig. 9.11. As discussed by Einasto & Miller (1983) and Einasto *et al.* (1983), this distribution suggests the presence of two local voids, the Northern Local Void and the Southern Local Void, each about 100 h^{-1} Mpc in diameter. Nearby Abell clusters and superclusters, as shown in Fig. 9.11, form a 100 Mpc supercluster wall in the Local Universe. The wall consists of Virgo, Coma, Perseus–Pisces (PP), Lynx–Ursa Major, Hydra–Centaurus and Pavo–Corona Australes superclusters between the Northern and the Southern Local Voids; its diameter is about

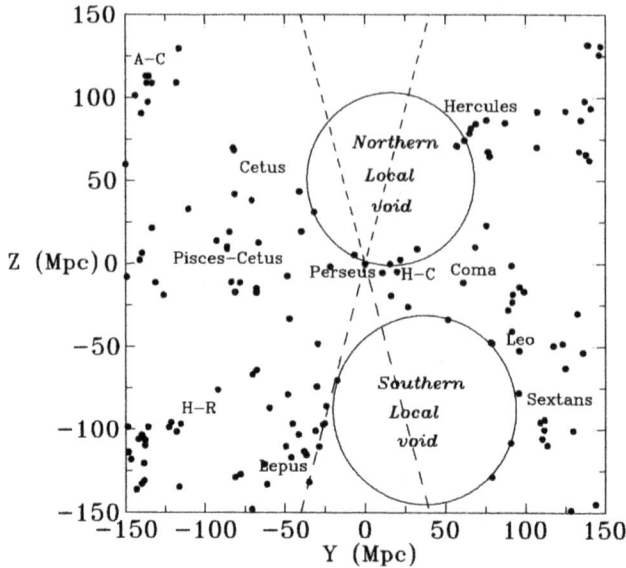

Fig. 9.11. The distribution of Abell clusters in rectangular supergalactic coordinates in the interval $X = -75 \ldots 50 \ h^{-1}$ Mpc. The zone of avoidance is shown by dashed lines, outer contours of Local Voids by circles. Supercluster names are marked. H–R, H–C and A–C are for the Horologium–Reticulum, Hydra–Centaurus and Aquarius–Capricornus superclusters, respectively (Einasto, 2024).

$125 \ h^{-1}$ Mpc, and thickness about $25 \ h^{-1}$ Mpc. Superclusters located at distances $75 - 125 \ h^{-1}$ Mpc (Hercules, Ursa Majoris–Leo, and several Southern superclusters) form sidewalls of Local Voids. The Hercules and Ursa Majoris–Leo superclusters form a wall between the Northern Local Void and the Bootes Void; the PP supercluster is located in a wall between the Northern Local Void and a void beyond that supercluster, as seen in Fig. 2.17.

The concentration of nearby Abell clusters close to the supergalactic plane was confirmed by Tully & Fisher (1987) and Shaver (1991), see the discussion by Jim Peebles (2022a, 2022b, 2023). Brent Tully and Peter Shaver also discovered independently local voids. The Northern Local Void is often called Tully's Void. The Local Supercluster plane is not the only huge planar structure discovered in the cosmic web. Einasto *et al.* (1997b) reported the discovery of another huge plane of superclusters, perpendicular to the Local Supercluster plane, named as the Dominant supercluster plane. The richest superclusters in the Dominant Supercluster plane are the Horologium–Reticulum, the Sculptor, the Hercules, the Bootes, the Virgo–Coma and other superclusters from Southern and Northern sky. As these planes cross each other, some superclusters (like the Hercules supercluster) are members of both planes. Jim Peebles (2022a, 2023, 2025) considered the concentration of rich clusters in walls as an anomaly in physical cosmology.

9.3 Structure and evolution of the ensemble of superclusters and voids

RHEA: Brrr. It's getting colder every day.

ARIEL: And reading about the Cold Spot[3] of the Cosmic Microwave background makes me feel even colder.

RHEA: How much colder? About $70\,\mu K$?

ARIEL: Just about. Maybe more. It looks mysterious and cold in the pictures. The imprint of a supervoid! And winter is coming. With that, the end of the semester, finals, my comprehensive –

RHEA: So you're looking for a distraction.

CALLISTUS: A Cold Spot question is highly likely on your comp. What do you know about it so far?

ARIEL: We have seen that the CMB is a 2.7 K black body radiation with tiny temperature fluctuations at the 10 s of μK level. We understand the CMB from inflationary initial conditions pretty well, except for a few unexplained anomalies we can't quite explain.

RHEA: And maybe we don't even need to, since these anomalies, the low quadrupole, the "Axis of Evil", the North–South asymmetry, and some special regions like the Loop and the Cold Spot are all pretty low significance.

ARIEL: Wow, the names resonate.

RHEA: They could be statistical flukes: when we analyse a sufficiently complex system with a wide range of statistical tools, eventually, we'll find something. However, counterintuitively, each trial lowers the significance of the result.

CALLISTUS: Thus, if you find something in the grey area that contradicts your theory, you can make more independent measurements with different methods. Each time you find nothing, the significance of your original result will be lower. Evanesco.[4] You can make it disappear.

---

[3]Cruz *et al.* (2005).
[4]Vanish.

ARIEL: Harry Potter spells always work because they are in Latin. But why would one measurement lower the significance of another independent one?

RHEA: Let's say I find a one in 20 or 2σ event. Barely interesting, but maybe it's worth contemplating a little bit. But if I add that I have done 20 independent numerical experiments, you know I have a 2σ event among them with probability 1. That makes my result completely insignificant. This is the *look elsewhere effect...* and people would not use it to make something disappear.

CALLISTUS: Not deliberately. Computers make it too easy to add a massive amount of measurements without establishing that they are truly independent. Then there is unconscious bias –

ARIEL: OK. What's the significance of the Cold Spot?

CALLISTUS: Depends on the precise handling of the look elsewhere effect, but it is in the few per cent range. To me, that's a potentially interesting hint worth looking into.

RHEA: Or it could be a fluke, which is still the prevailing view for all anomalies. And I agree.

CALLISTUS: It's changing as we speak RE: the Cold Spot.

ARIEL: The discovery of the Eridanus void resulted from a search motivated by the Cold Spot.[5] Motivation is hard to express statistically, but it must mean something.

CALLISTUS: The Eridanus supervoid is a rare structure aligned with another rare fluctuation in the sky. For the sake of simplicity, let's assume they are both so rare that there is only one each in the full sky, and they match with 1° accuracy. Since the sky is about 40,000 square degrees, the likelihood that two random structures overlap is 40,000 times smaller than one causing the other.

ARIEL: How could a supervoid imprint the Cold Spot? I read about exotic ideas for alternative explanations, such as topological defects, non-Gaussian fluctuations, or even a window to another universe.

CALLISTUS: Usually, the most reasonable explanation prevails.

RHEA: Which would be a statistical fluke.

CALLISTUS: It would be the Integrated Sachs–Wolfe effect. When CMB photons cross a large decaying potential of a superstructure, they imprint on the CMB. Imagine your car in neutral, speeding through a valley. You enter with your initial velocity and speed up at the bottom, but neglecting friction, you will exit with your initial velocity at the other end. Except, if you exit at a lower level than entering, you gain energy. This exactly happens when dark energy causes potentials to decay while crossing; you have to climb less to come out of the potential

---

[5]Szapudi *et al.* (2015); Finelli *et al.* (2016).

well. As a result, photons will heat up when crossing super-clusters and marking them as hot spots. Conversely, voids will show up as cold spots.

ARIEL: And a supervoid really close to us will imprint as the Cold Spot.

RHEA: Slow down, there is a paper claiming[6] that the Eridanus void can only produce about a quarter of the required temperature drop.

CALLISTUS: When you stack more distant superstructures, you get about 4–5 times the ΛCDM prediction. Applying this to the Eridanus voids explains Cold Spot. The ISW puzzle,[7] along with the H_0 and S_8 tensions, suggests that our concordance model needs modifications to describe the expansion history of our universe.

ARIEL: What are the alternatives?

RHEA: There are a bunch of theories, including dark photons and early dark energy, but at the moment, none of them has as much credence as the concordance model with simple vacuum energy. The latest DESI results[8] suggest a changing dark energy equation of state.

CALLISTUS: My favourites are emerging curvature theories.

RHEA: Of course, they are.

CALLISTUS: In Timescape[9] and AvERA,[10] negative curvature emerges from non-linear evolution. The latter even predicted the anomalous sign change of the ISW effect.[11]

RHEA: At redshifts larger than $z \gtrsim 1$, supervoids appear to imprint hot spots. If this result holds up, it would be extraordinary. No simple variants or additions to the concordance model would be consistent. Call it what you want, but at the moment, the significance is still moderate. Colour me slightly interested, but I'm holding out for more data.

CALLISTUS: Right. We'll get back to this in a few years.

ARIEL: The statistical argument for the Eridanus supervoid causing the Cold Spot is independent of the mechanism. The consistency with the ISW puzzle is just the icing on the cake.

RHEA: A cake that tastes bitter to me...

CALLISTUS: The universe is what it is. It might not be perfect.

ARIEL: Kind of like the weather around here.

RHEA: What a rainy ending to a perfect day.

---

[6]Mackenzie *et al.* (2017).

[7]Szapudi (2025).

[8]DESI Collaboration *et al.* (2025b).

[9]Wiltshire (2024).

[10]Rácz *et al.* (2017).

[11]Kovács *et al.* (2022).

In this section, we discuss the structure and evolution of the whole ensemble of superclusters using density and velocity fields. To identify structures in the density field, it is necessary to define a density threshold to separate high-density regions (superclusters) from low-density regions (voids). There is no natural value of the threshold density. Liivamägi *et al.* (2012) used for supercluster search two methods, one with a fixed density threshold and the other with an adaptive density threshold, depending on the distribution of galaxies in the particular region.

9.3.1 *Structure and evolution of the ensemble of superclusters from density field*

To investigate the evolution of the ensemble of superclusters, we use ΛCDM simulations described in Chapter 4, following Einasto *et al.* (2019b, 2021b). We extracted DM particle data for epochs $z = 30$, 10, 3, 1, 0, and calculated density fields with the B_3 spline, using smoothing scale 8 h^{-1} Mpc. To characterise the properties of the ensemble of superclusters, we apply percolation analysis, as described in Chapter 5. The percolation analysis consists of several steps: finding overdensity regions (clusters as potential superclusters) in the density field, calculation of parameters of potential superclusters, and finding the supercluster with the largest volume for a given density threshold. As traditional in the percolation analysis, overdensity regions are called clusters and underdensity regions voids (Stauffer, 1979).

We scan the density field in the range of threshold densities from $D_t = 0.1$ to $D_t = 10$ in mean density units. This range covers all densities of practical interest, since in low-density regions, the minimal density is ≈ 0.1, and the density threshold to find conventional superclusters is $D_t \approx 5$ (Liivamägi *et al.*, 2012). We mark all cells with density values equal to or above the threshold D_t as filled regions, and all cells below this threshold as empty regions. To find clusters, we scan all filled cells: two cells of the same type are considered as neighbours (friends) and members of the cluster if they have a common sidewall. Every cell can have at most six cells as neighbours. Members of clusters are selected using a Friend-of-Friend (FoF) algorithm: the friend of my friend is my friend.

For each cluster, we calculate diameters (lengths) along coordinate axes, Δx, Δy, Δz, geometrical diameters (lengths), $L_g = \sqrt{(\Delta x)^2 + (\Delta y)^2 + (\Delta z)^2}$, and geometrical volumes, V_g, defined as the volume in space where the density is equal or greater than the threshold density D_t and total masses (or luminosities), $\mathcal{L}$, the mass (luminosity) inside the density contour D_t of the cluster, in units of the mean density of the sample. All distances are in comoving units. We also calculate total volume of overdensity regions, equal to the sum of volumes of all clusters, $V_C = \sum V_g$, and the respective total filling factor:

$$F_f = N_f/N_{\text{cells}} = V_C/V_0, \tag{9.1}$$

where N_f is the number of filled (over-density) simulation cells and V_0 is the volume of the sample.

As additional parameters, we use fitness volumes and diameters. We define the fitness volume of the supercluster, V_f, proportional to its geometrical volume, V_g, divided by the total filling factor, F_f. The sum of fitness volumes is equal to the volume of the sample. In this way, fitness volumes give information on the volumes of supercluster BoAs around superclusters. These functions characterise general geometrical properties of the ensemble of superclusters in the cosmic web and allow us to select the proper threshold density to compile the actual supercluster catalogue.

Figure 9.12 shows geometrical length functions, L_g, and numbers of clusters, N, for cosmic epochs $z = 30, 10, 3, 1, 0$. Threshold

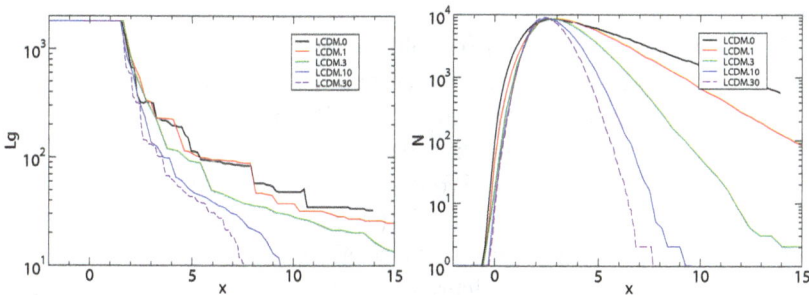

Fig. 9.12. Left: the evolution of geometrical length functions of clusters; Right: the evolution of number functions of clusters. As arguments of percolation functions, we use the reduced threshold density, $x = (D_t - 1)/\sigma$ (Einasto et al., 2019b).

densities are given in units of the dispersion of the density contrast:

$$x = (D_t - 1)/\sigma. \qquad (9.2)$$

We call these threshold densities 'reduced'; the use of reduced threshold densities is convenient to follow the evolution of the density field. In Chapter 5, we investigated the influence of the smoothing length to percolation functions, applying usual threshold densities D_t, see Fig. 5.3.

At small threshold densities, $D_t \leq 1$ ($x \leq 0$), there exists one percolating cluster, extending over the whole volume of the computational box. The percolation threshold density, $P = D_t$, is defined as follows: for $D_t \leq P$, there exists one and only one percolating cluster; for $D_t > P$, there are no percolating clusters (Stauffer, 1979). As we see, the reduced percolation threshold density of all epochs is almost identical, $x_P \approx 1.5$. In the reduced threshold density range, $x \leq 1.5$, geometrical diameters of clusters are equal to the diameter of the box, $L_g = \sqrt{3}\, L_0$.

When we increase the threshold density, then at $x \approx 0$, there appear additional clusters, and the number of clusters N starts to increase rapidly. At percolating threshold, $x \approx 1.5$, geometrical diameters of largest clusters, L_g, start to decrease: the large percolating cluster splits to smaller clusters. At $D_t = D_{\max} \approx 2.7$ ($x_{\max} \approx 2.5$), the number of clusters reaches a maximum, $N_{\max} \approx 8300$. At this threshold, density clusters are still complexes of large overdensity regions, connected by filaments to form systems of diameters $L_g \approx 300\ h^{-1}$ Mpc, i.e. largest overdensity regions are actually complexes of superclusters.

When we increase the threshold density more, then the number of clusters starts to decrease, since smallest clusters have maximal densities lower than the threshold density and disappear from the cluster sample. With further increase in the density threshold, geometrical diameters decrease. We denote the threshold density to find superclusters in our samples as D_t (x_t in reduced threshold density units). At threshold density D_t, the total filling factor of high-density regions is $F_f \approx 0.01$. The decrease in the number of clusters with increasing threshold continues until only central regions of clusters have densities higher than the threshold density.

Figure 9.13 shows cumulative distributions of geometrical diameters and luminosities of superclusters. Data are given for simulation

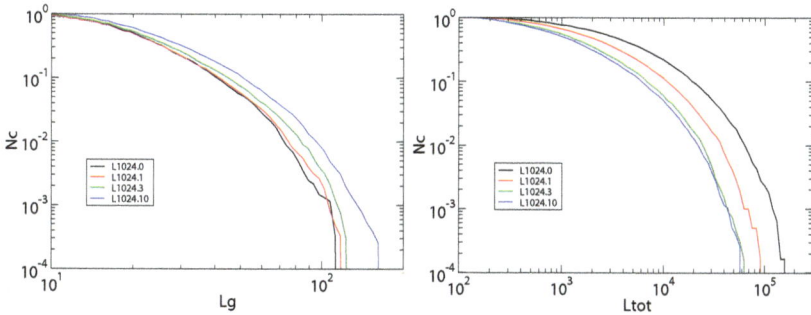

Fig. 9.13. Left: cumulative distribution of supercluster geometrical diameters, L_g; Right: total luminosities, $\mathcal{L}$, of L1024 models at different evolution epochs up to $z = 10$ (Einasto *et al.*, 2019b).

epochs up to $z = 10$. As we see from Fig. 9.13, geometrical diameters at early epochs are larger than at the present epoch (in comoving coordinates) approximately by a factor of 2. This means that in comoving coordinates, superclusters shrink during the evolution.

The right panel of Fig. 9.13 shows that masses of superclusters increase during the evolution approximately by a factor of 3. This result is in good agreement with all simulations of the growth of the cosmic web. The skeleton of the web with supercluster embryos forms already at early epoch. Superclusters grow by the infall of matter from low-density regions towards early forming knots and filaments, forming early superclusters.

Now, we discuss the evolution of percolation parameters of the ensemble of superclusters in more detail. Figure 9.14 shows the evolution of fitness lengths of largest clusters, spatial density of numbers of clusters, and filling factor of all high-density regions. We see that fitness diameters depend on the smoothing scale used to select clusters. This property is expected, since smoothing highlights properties of the cosmic web on different scales, as discussed in Chapter 5. Top left and right panels of Fig. 9.14 show the change of two percolation parameters of supercluster ensembles with cosmic epoch z: the fitness lengths of the largest cluster, $L_f(z)$, and the number of clusters, $N_{\max}(z)$. In this figure, numbers are actually spatial densities of clusters reduced to the volume of the sample of size $L_0 = 1\,\mathrm{Gpc}$ (in comoving units). Here, we use, in addition to the L1024 model, two models of smaller size, L512 and L256, and two

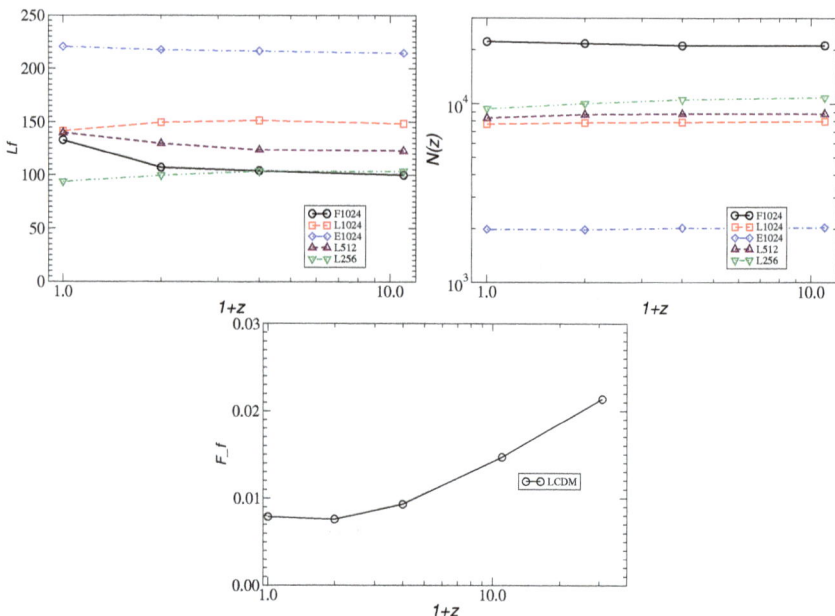

Fig. 9.14. Left top: the evolution of fitness lengths of largest clusters with epoch, $L_f(z)$ (in comoving units); Right top: the evolution of the spatial density of numbers of clusters with epoch, $N(z)$, per cubic cell of size $L_0 = 1\,\mathrm{Gpc}$; Bottom: the change of the filling factor of all high-density regions, $F_f(z)$, Model designations are described in text (Einasto *et al.*, 2019b).

models of size $L_0 = 1024\,h^{-1}$ Mpc with different smoothing scale, the model F1024 with smoothing scale $4\,h^{-1}$ Mpc, and the model E1024 with smoothing scale $16\,h^{-1}$ Mpc. The model F1024 selects smaller clusters (overdensity regions) than the L1024 model, thus numbers of clusters are about three times higher than in the L1024 model. These clusters are actually high-density cores of traditional super-clusters. The model E1024 has lower density contrast than L1024 and F1024 models. Numbers of clusters are about four times smaller than in models of the L1024 series, see Fig. 9.14. Fitness lengths of largest clusters of the E1024 series are about two times larger than fitness lengths of largest clusters of the L1024 series. These clusters are mostly complexes of traditional superclusters. An example is the Sloan Great Wall, which is a complex of three conventional superclusters (Liivamägi *et al.*, 2012). The smoothing length $R_B = 16\,h^{-1}$ Mpc was used by Liivamägi *et al.* (2012) to select superclusters from the Luminous Red Giant (LRG) sample of the SDSS survey. LRG superclusters, found with the adaptive threshold density, are approximately two times larger than superclusters of the SDSS main galaxy sample. Fitness diameters of largest clusters are determined by just one cluster, the largest one, and have a larger scatter than numbers of clusters, which are sums of all clusters. The main message from the analysis of fitness lengths and numbers is as follows: both parameters are approximately constant during evolution.

The bottom panel of Fig. 9.14 gives the change of the filling factor, F_f, of models during the evolution. This filling factor was used to calculate fitness volumes of superclusters, V_f. The total filling factor of superclusters decreases in comoving units during the evolution from redshift $z = 30$ to $z = 0$ about three times from $F_f = 0.022$ to $F_f = 0.0079$. This decrease of the filling factor was calculated using the density fields smoothed with 8 h^{-1} Mpc kernel. Actually, there exist voids inside superclusters, as seen from high-resolution density fields. Thus, the actual filling factor of high-density regions is even smaller.

To understand better the evolution of the cosmic web on super-cluster scale, we show in Fig. 9.15 the visual appearance of density fields of models L1024 at different epochs: in the left panel at the early epoch $z = 10$, in the middle panel at epoch $z = 3$, and in the

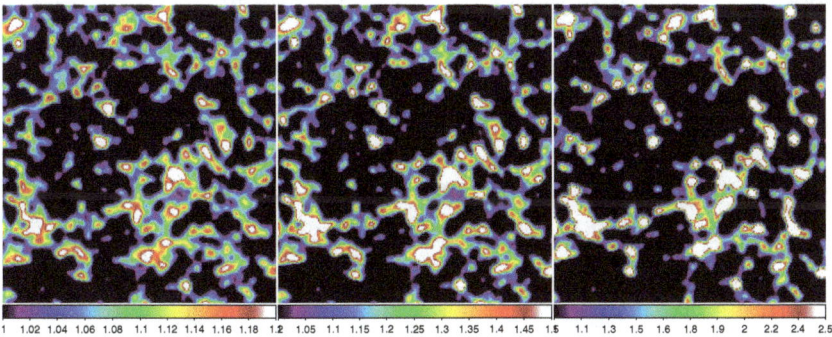

1 1.02 1.04 1.06 1.08 1.1 1.12 1.14 1.16 1.18 1.2   1.05 1.1 1.15 1.2 1.25 1.3 1.35 1.4 1.45 1.5   1.1 1.3 1.5 1.6 1.8 1.9   2   2.2 2.4 2.5

Fig. 9.15. Density fields of L1024.10, L1024.3 and L1024.0 models found with smoothing kernel of radius 8 h^{-1} Mpc. Left panel corresponds to the epoch $z = 10$ and middle panel to epoch $z = 3$, right panel to the present epoch $z = 0$. Cross-sections are shown for central 512×512 h^{-1} Mpc sections in a 2 h^{-1} Mpc thick layer, densities are expressed in linear scale. Adapted from Einasto *et al.* (2019b).

right panel at the present epoch $z = 0$, all smoothed with 8 h^{-1} Mpc comoving scale. This figure is complementary to Fig. 4.3, which shows the evolution of the model L256 from epoch $z = 30$ to the present epoch $z = 0$, and to Fig. 6.5 of TNG100 simulations from $z = 5$ to $z = 0$. The evolution of density fields can be followed by comparison of panels. This comparison suggests that supercluster-type structural elements of the cosmic web are present already at very early epochs. Of course, there are differences on small scales, but main supercluster-type elements of the web are seen at similar locations at all epochs. Basic visible changes are the increase in the density contrast: distributions of densities at epochs $z = 10$ and $z = 3$ are very similar, and only the amplitude of density perturbations has increased. This means that in this time interval, the evolution is near to a linear growth, as shown in Chapter 4, see Fig. 4.17. On later epochs, the non-linearity of the evolution is dominant. The flow of small-scale structural elements towards large ones, and the contraction of superclusters are more visible.

Elements of the cosmic web evolve with time. Physical clusters of galaxies grow by merging of smaller clusters and by infall of non-clustered matter, filaments merge, and voids became emptier. Superclusters also change, their sizes shrink in comoving coordinates, and masses grow by infall and merging. Similar general visual appearance of the density fields at very early and present epochs suggests that supercluster embryos were created very early, much earlier than the epoch $z = 30$ used in our calculations. This result is not surprising, already Kofman & Shandarin (1988) demonstrated that the whole present-day structure is seen in the initial fluctuation distribution, see also Fig. 4.1.

9.3.2 *Structure and evolution of ensemble of superclusters and voids from velocity data*

The distribution of matter can be alternatively and independently determined from surveys of the distribution of galaxies and from the motions of galaxies. If the distances of galaxies are measured, then their peculiar radial velocities can be calculated, which can be translated to true 3D peculiar velocities. The peculiar radial velocities are departures from the cosmic expansion and are collected in CosmicFlows data by Tully *et al.* (2013, 2016, 2023). To use peculiar radial velocities in an optimal way, Wiener filtering of data is used. This allows us to find BoAs, which are separated from each other by surfaces, where on the one side the smoothed velocity flow is directed to one supercluster and on the other side to an another supercluster. As noted in Chapter 8, in this way, the whole volume of the Universe is divided into supercluster cells of dynamical influence or BoAs, see Dupuy *et al.* (2019, 2020) and Dupuy & Courtois (2023).

9.3.3 *Method and first results*

The Wiener filtering method was originally proposed by Norbert Wiener to reconstruct noisy data in an optimal way, and examples are filtering of photographs blurred with noise and defocusing. In cosmology, this technique was developed by Zaroubi *et al.* (1995). The Wiener filtering has the purpose of reconstructing the large-scale structure of the Universe from noisy, sparse, and incomplete data. The problem is solved by assuming statistical prior information of the underlying field one is trying to measure. Large-scale structure is assumed to develop from gravitational instabilities out of primordial random Gaussian fluctuations. The developing density and velocity fields retain their Gaussian properties as long as the growth is in the linear regime; this evolution can be described by the gravitational perturbation theory. There is a double advantage in using peculiar velocities: first, they are highly linear and correlated on large scales; second, they are excellent tracers of the underlying gravitational field as they account for both the baryonic matter and the dark matter. For mathematical details of the Wiener filtering technique, see Zaroubi *et al.* (1995).

One of first applications of the Wiener reconstruction technique to velocity data was done by Tully *et al.* (2014). Authors suggested that one could employ peculiar velocities to define large systems of galaxies as non-virialised objects that will dissipate with the expansion. Regions of space which show gravitationally induced coherent inward motions, can be called "basins of attraction" (BoA), see also Chapter 8. BoAs are defined as locations where peculiar velocity flows diverge, as water does at watershed divides. They contain all galaxies and particles whose flow lines converge at a given attractor, the local minimum of the gravitational potential. This definition also allows one to identify regions that are dual to BoA — basins of repulsion

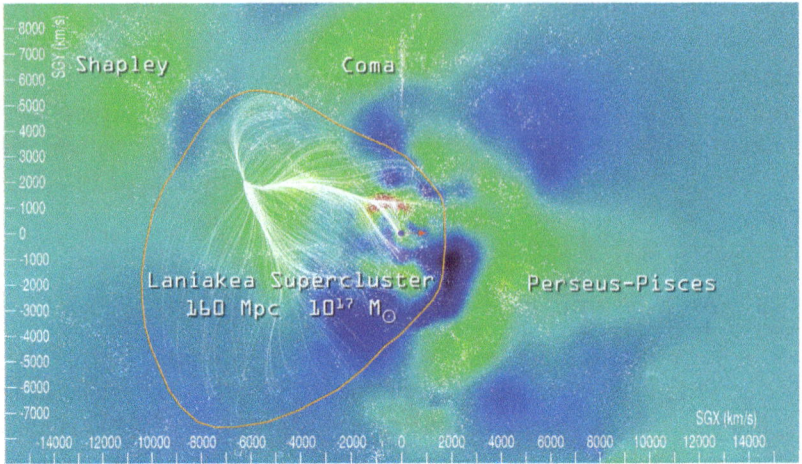

Fig. 9.16. A slice of the Laniakea supercluster in the supergalactic equatorial plane. Shaded contours represent density values within the equatorial slice with red at high densities and blue in voids. Individual galaxies from a redshift catalog are given as white dots. Velocity flow streams within the Laniakea BoA are shown in white. The orange contour encloses the outer limits of these streams. This domain has a extent of ∼12,000 km/s (∼160 Mpc diameter) and encloses ∼$10^{17} M_\odot$ (Tully *et al.*, 2014). Used with permission of *Nature* from Tully *et al.* (2014). Permission conveyed through Copyright Clearence Center, Inc.

(BoR), namely volumes of space with gravitationally coherent outward motions. Tully *et al.* (2014) defined our local BoA — Laniakea, see Fig. 9.16.

9.3.4   *Structure and evolution of the ensemble of basins of attractions and repulsions*

Dupuy *et al.* (2020) used SmallMultiDark simulations by Klypin *et al.* (2016) to segment the universe into dynamically coherent basins applying various smoothing lengths to velocity data. This simulation was performed in a box of size 400 h^{-1} Mpc. Basins were searched using three parameters, in addition to the smoothing length, and the integration step along streamlines the maximum streamlines length. At optimal search parameters, the number of basins converged to 647. Taking into account the size of the simulation box, these numbers are in fairly good agreement with the number of SDSS superclusters. As shown by Einasto *et al.* (2019b, 2021b), superclusters are high-density regions inside BoAs, and fill only a small fraction of BoAs

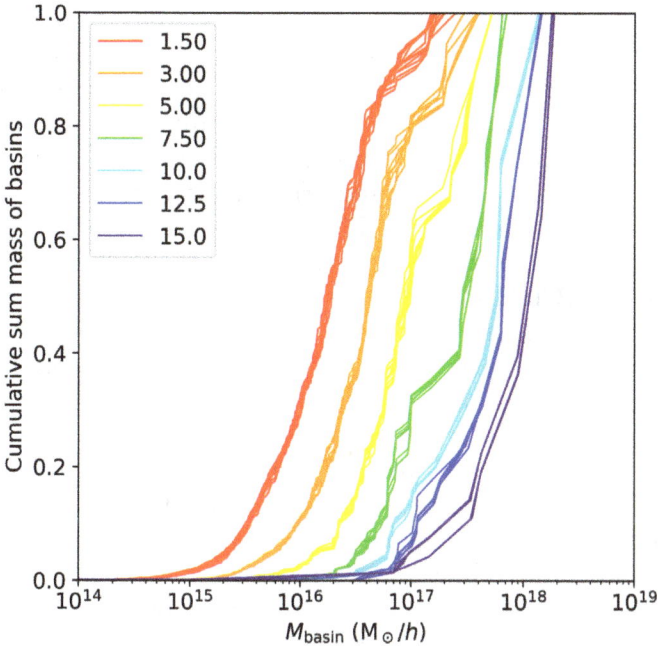

Fig. 9.17. Cumulative sum of the mass of basins (normalised by the total mass) as a function of the mass of the BoA segmented. The colour code represents the smoothing scale rs considered. The solid lines of the same colour correspond to velocity fields with the same smoothing radius but different redshifts (Dupuy *et al.*, 2020).

space, about 0.8% at the present epoch. The most massive ΛCDM model superclusters have at the present epoch masses, $M \approx 10^{16} M_\odot$ (Einasto *et al.*, 2021b). During the evolution, masses of superclusters increase about three times by the infall of surrounding matter inside BoAs to superclusters. In contrast, masses of basins remain approximately constant during the evolution. Dupuy *et al.* (2020) showed that most massive supercluster basins have at all epochs masses, $M \approx 2 \times 10^{17} M_\odot$, when the optimal smoothing radius 1.5 h^{-1} Mpc is used. The cumulative sum of masses of basins is shown in Fig. 9.17. This figure shows the dependence of BoA masses on smoothing scale. Very large smoothing increases BoA masses up to 100 times and decreases the number of BoAs to a few entities.

Dupuy *et al.* (2020) studied the changes of basins during the cosmic evolution. Authors found that the visual appearance as well as the number of basins are very close to each other at all redshifts. The comparison of the evolution of superclusters and BoAs shows that numbers of superclusters and BoAs remain almost constant during the evolution. But more interesting are differences in the evolution. Masses of BoAs remain constant during the evolution, whereas masses of superclusters increase during the evolution by a factor of about 3, see the right panel of Fig. 9.13.

The second important difference is in masses themselves. The most massive ΛCDM superclusters have at the present epoch masses, $M \approx 10^{16} M_\odot$. Figure 9.17 shows that most massive supercluster basins have at all epochs masses, $M \approx 2 \times 10^{17} M_\odot$. Such difference is expected. Volumes of superclusters at the early epoch are about 50 times smaller than volumes of BoAs; at the present epoch, this difference has increased to about 140 times. The difference of masses of superclusters and BoAs at the present epoch is only about 20 times. This means that regions of BoAs outside superclusters have much lower densities than inside superclusters. Since masses of BoAs remain constant during the evolution, the growth of supercluster masses can be explained by the infall of surrounding matter inside BoAs to superclusters. The exchange of matter between neighbouring BoAs is minimal because the velocity flows within BoAs is directed inwards.

9.4   Concluding remarks

ARIEL: My friend invited me to her group's unblinding party.

RHEA: Something happens when everybody finds out.

ARIEL: What happens? I'm clueless.

CALLISTUS: Caeci caecos ducentes.[12] Basically, if the result is good, everybody is happy. If it's not, then they reblind and redo.

RHEA: Woah, you're clearly not a fan.

ARIEL: I still don't know what he's not a fan of.

RHEA: Let's take a step back. Do you know about confirmation bias?

ARIEL: I know about unconscious bias: your subconscious preconceptions and biases influence your decisions and judgment regarding certain groups.

---

[12] Blind leading the blind.

CALLISTUS: Let's imagine you analyse a complex dataset. You filter and calibrate your data and apply some statistical techniques. After a lot of work, you finish your analysis and condense the results into a few numbers. What do you do?

ARIEL: I compare my numbers with previous experiments or observations.

CALLISTUS: What happens if you disagree?

ARIEL: That would be an extraordinary claim requiring extraordinary evidence. First, I would have to double check everything I did. Maybe I find a bug in my code.

RHEA: And you'd continue doing that until you finally agree?

ARIEL: That's a trap. I'm more likely to stop looking for errors when I agree with previous work.

RHEA: Yeah, you did everything right. Yet, even with the best intentions, you might be biased towards previous findings. Even if you don't harbour stereotypes in your subconscious.

CALLISTUS: Or attached to ΛCDM. Or MOND.

RHEA: I let that slide.

ARIEL: That's terrible. I mean confirmation bias. What can we do?

RHEA: The best you can do is blinding. You can try to hide the critical final results of an analysis until all important decisions have been made.

CALLISTUS: Let's say you measure cosmological parameters. Your final matter density Ω must be fairly close to the previous Planck results. So, a small sub-group applies a secret blinding transformation to your data that shifts the final results. Nobody else knows what exactly they did, and nobody from the rest of the team can compare with Planck while you're doing the work.

ARIEL: I see. Once I'm happy with my analysis, we throw a party and invite all my friends.

RHEA: To witness what happens. Unblinding is undoing the blinding transformations. If your results are consistent with previous ones, you can be sure it's not because of confirmation bias.

ARIEL: Clever and fun. I can't wait for the party!

RHEA: It's pretty standard for large enough groups where you can isolate a small blinding sub-group while the rest performs the analyses.

CALLISTUS: Right. What happens if you find an Ω that's way out there?

ARIEL: Ooops. If we did everything right, that should not happen.

RHEA: It rarely does.

ARIEL: But does it?

RHEA: Well...

CALLISTUS: It. Does. Sometimes. Then. What do you do?

ARIEL: I guess we're in a pickle. We'd need to start from scratch, ideally throwing away our data up to that point.

CALLISTUS: Throwing away years of work by many people, costing a lot of time and money. Realistically, people must go back, reblind and find the bug in their code.

ARIEL: Before unblinding. Again. Thus, they use information from previous results to find bugs, which is a recipe for...

CALLISTUS: Confirmation bias.

RHEA: This almost never happens. People know the stakes are high, and they avoid such costly and embarrassing mistakes.

CALLISTUS: But even if the theoretical possibility of reblinding is there –

RHEA: I grudgingly admit: confirmation bias cannot be fully eliminated.

ARIEL: I'm late. And now I'm worried about my friend. I hope it will go well. Thanks to you, statistical trickery like "look elsewhere" and "confirmation bias" triggers me.

RHEA: We'll put a trigger warning – in your thesis.

The definition of superclusters was discussed by Chon *et al.* (2015). To obtain a physically well-motivated definition of superclusters and following Dünner *et al.* (2006), Chon *et al.* (2015) proposed to select superclusters with an overdensity criterion that selects only those objects that will collapse in the future. For the presently accepted cosmological parameters, this minimum density is 2.36 times the critical density of the Universe. To avoid conflict with previously defined superclusters, Chon *et al.* (2015) suggested that those superclusters that will survive the accelerating cosmic expansion and collapse in the future be called *"superstes-clusters"*, where *"superstes"* means survivor in Latin. Authors found that the infall region of the Local Supercluster is limited to distance about 7.2 Mpc. Within this distance, the local overdensity is about 2.6. Authors found that the Laniakea supercluster as defined by Tully *et al.* (2014) is not an overdensity region, and it splits to several subunits in the future.

The density field method allows us to select superclusters. The velocity field method allows us to find supercluster BoAs, but not superclusters themselves. Thus, these methods are complementary.

Available data suggest that embryos of galaxies and superclusters were created by high peaks of the initial field. The initial velocity field around peaks is almost laminar. Pichon *et al.* (2011) and Dubois *et al.* (2012, 2014) showed that a significant fraction of the cold gas falls along filaments nearly radially to the centres of high redshift rare massive haloes, see Fig. 8.2. This process increases the mass of the central halo rapidly. We may conclude that, depending on the height of the initial density peak, in this way embryos of galaxies, clusters of galaxies, and superclusters were created. However, the further evolution of superclusters differs from the evolution of galaxies and clusters of galaxies. Galaxies and ordinary clusters of galaxies are local attractors and collect additional matter from their local environment. Superclusters are global attractors and collect matter from a much larger environment inside their BoAs.

The filamentary character of the cosmic web can be described using the skeleton, the 3D analogue of ridges in a mountainous landscape (Pichon *et al.*, 2010). Peaks of the cosmic web are connected by filaments. The number of filaments, connecting the clusters with other clusters, can be called connectivity for global connections (including bifurcation points), and multiplicity for local connections (Codis *et al.*, 2018). Kraljic *et al.* (2020) investigated the connectivity of the SDSS sample of galaxies. Authors first determined the skeleton of the SDSS sample traced by the DisPerSE algorithm by Sousbie (2011). Then authors calculated the connectivity of all clusters. They found that the connectivity of clusters of the SDSS sample has a peak at 3, and the multiplicity (local connectivity) has a peak at 2. Both parameters depend on the mass of the cluster. The mean connectivity of massive SDSS clusters is 4, and the multiplicity of most massive clusters is 6.

These results have a simple explanation. Low and medium mass clusters lie inside filaments and thus have the multiplicity 2 (connection is from both sides of the cluster inside the filament). Clusters move together with the filaments in the large potential well of superclusters. The simultaneous movement of clusters with their surrounding filament follows from the simple fact that the filamentary character of the cosmic web is preserved at the present epoch. If clusters would have large peculiar velocities with respect to surrounding filaments, then during the evolution, the filamentary character of the web would be destroyed. The laminar character of the velocity field is explained by the presence of the DE, as suggested already by Sandage *et al.* (2010). Very rich clusters are central clusters of superclusters and are connected with other structures by many filaments. This was demonstrated already by Tully & Fisher (1978) for the Virgo supercluster, by Jõeveer *et al.* (1977, 1978) for the Perseus–Pisces supercluster, and by Einasto *et al.* (2020b) for the A2142 supercluster.

Central clusters of these superclusters lie at minima of potential wells created by respective superclusters. They are fed by filaments from several sides and are suitable locations for cluster merging — small clusters fall to the central cluster along filaments surrounding the central cluster. The pattern of the cosmic web suggests that the high connectivity can be used as a signature for the presence of the central cluster of a supercluster.

Chapter 10

ΛCDM and Beyond

The ΛCDM model, or "Lambda Cold Dark Matter", has been the standard cosmological framework since the late 1990s, serving as the leading paradigm for over a quarter of a century. It rose to prominence following a series of pivotal measurements:

- **Supernova measurements:** In 1998, teams from the Supernova Cosmology Project (Perlmutter *et al.*, 1999) and the High-z Supernova Search (Riess *et al.*, 1998) observed the accelerated expansion of the Universe using Type Ia supernovae, strongly suggesting the presence of dark energy (DE) (cosmological constant, Λ).
- **Cosmic microwave background:** Measurements of CMB fluctuations by the MAXIMA (Hanany *et al.*, 2000) and BOOMERanG (de Bernardis *et al.*, 2000) experiments in 2000 provided crucial evidence. Specifically, the location of the first acoustic peak in the CMB angular power spectrum confirmed the spatial flatness of the Universe, aligning with predictions from inflationary models.
- **Large-scale structure surveys:** Galaxy redshift surveys like Las Campanas (Shectman *et al.*, 1996), 2dF (Peacock *et al.*, 2001), and SDSS (York *et al.*, 2000) mapped the distribution of matter on large scales, finding evidence for low matter density, which in combination with flatness requirements resulted in further support for the ΛCDM framework.

The model gained further credibility in 2003 with NASA's WMAP satellite, which provided precision measurements for several of its key parameters (Spergel *et al.*, 2003). ESA's Planck mission later refined these measurements, improving parameter constraints while reaffirming ΛCDM as the standard model of cosmology (Planck Collaboration *et al.*, 2014a, 2020).

Here, we give a brief overview of the ΛCDM model, outline its key assumptions, and introduce its parameters. Finally, we discuss some of its shortcomings.

10.1   ΛCDM

[*Callistus sits at his desk, staring at a plot on his screen. Ariel leans forward eagerly, while Rhea sips her coffee.*]

CALLISTUS: We all know the standard model of cosmology – ΛCDM – fits the data remarkably well. But tell me, what does that actually mean?

ARIEL: It means that our observations of the cosmic microwave background, galaxy clustering, and supernovae align with a universe dominated by cold dark matter (CDM) and a cosmological constant, right?

RHEA: Yes, but that's just a fit. It doesn't mean we actually understand what dark energy or dark matter are.

CALLISTUS: Precisely. And yet, how often do we hear statements like "the universe will expand exponentially forever, approaching a de Sitter state"? What do you think about that claim?

ARIEL: Isn't that just the prediction from ΛCDM? If dark energy remains a constant vacuum energy, then wouldn't an eternal de Sitter future be inevitable?

RHEA: That's an if, though. We have no idea whether dark energy is truly a cosmological constant. For all we know, it could be evolving over time. Or it might not even exist as a separate component at all.

CALLISTUS: Excellent. We should be cautious when extrapolating into the far future based on a model whose key components – dark matter and dark energy – are still complete mysteries. So, what are the alternatives?

ARIEL: You mean like quintessence? A dynamical field instead of a cosmological constant?

RHEA: Or modifications to gravity, where what we call dark energy is just an effect of large-scale deviations from general relativity.

CALLISTUS: Right. And if any of these alternatives turn out to be true, then our predictions for the long-term fate of the universe could change dramatically. Maybe the expansion slows down. Maybe dark energy decays away. Maybe the universe even collapses in a "big crunch". The truth is, we don't know.

ARIEL: So, every time I see a popular article saying "the universe will end in a cold, dark void", I should be skeptical?

RHEA: Very. That conclusion only follows if ΛCDM is exactly right forever. But since we don't even understand its fundamental ingredients, assuming that is a huge leap.

CALLISTUS: Precisely. Until we understand the nature of the dark sector, we should remain humble about what we claim to know about the universe's ultimate fate. Our best-fit model may work for now, but the story it tells is far from complete.

[Ariel leans back, thoughtful. Rhea takes another sip of her coffee, smirking.]

RHEA: So... what you're saying is that we need another big grant to figure all this out?

CALLISTUS: Now, you're thinking like a real cosmologist.

As common among many of the cosmological models, the construction of ΛCDM also starts with two key assumptions: (i) the validity of *Cosmological Principle* stating that the Universe is homogeneous and isotropic when viewed on large scales, i.e. looking the same at every point and in every direction, (ii) at large enough scales, only gravitational interactions matter, which are assumed to follow equations of Einstein's *general relativity*.

The assumption of homogeneity and isotropy dictates the form for the metric, which can be expressed as

$$ds^2 = -c^2 dt^2 + a(t)^2 \left[\frac{dr^2}{1 - kr^2} + r^2 \left(d\theta^2 + \sin^2 \theta \, d\phi^2 \right) \right].$$

This is the Friedmann–Lemaître–Robertson–Walker metric in spherical coordinates. Here, $a(t)$ is the scale factor and k determines the spatial curvature ($k = 0$ for flat, $k > 0$ for closed, $k < 0$ for open geometry).

Having specified a highly symmetric form for the metric, Einstein's equations simplify a lot. The evolution equation for the scale factor — the first Friedmann equation, which arises from the time–time component of Einstein's field equations — can be given as

$$\left(\frac{\dot{a}}{a} \right)^2 \equiv H^2(a) = H_0^2 \left[\Omega_r a^{-4} + \Omega_m a^{-3} + \Omega_k a^{-2} + \Omega_\Lambda \right] \equiv H_0^2 E^2(a),$$

$$\Omega_k \equiv -\frac{k}{H_0^2 a^2} = 1 - \Omega_r - \Omega_m - \Omega_\Lambda.$$

where, $H(a)$ is the Hubble function, H_0 the Hubble constant today, and the scale factor has been normalised such that $a(t = t_0) \equiv 1$

(then $a = 1/(1+z)$, z–redshift). This is the Friedmann equation for the ΛCDM and the density parameters Ω_x for all the relevant components — radiation (r), matter (m) and vacuum energy (Λ) are defined as

$$\Omega_x \equiv \frac{\rho_{x,0}}{\rho_{\text{crit},0}}, \quad x \in \{r, m, \Lambda\}, \quad \rho_{\text{crit},0} \equiv \frac{3H_0^2}{8\pi G} \quad \text{/critical density/,}$$

where the subscript 0 denotes the value taken today $(t = t_0)$. From the above relations, the density parameters for different evolutionary stages can be given as[1]

$$\Omega_r(a) = \frac{\Omega_r a^{-4}}{E^2(a)}, \quad \Omega_m(a) = \frac{\Omega_m a^{-3}}{E^2(a)}, \quad \Omega_\Lambda(a) = \frac{\Omega_\Lambda}{E^2(a)}.$$

These are plotted in Fig. 10.1.

Observations, in accordance with inflationary predictions, have shown spatial curvature to be consistent with a flat universe, $\Omega_k = 0.001 \pm 0.002$ (Planck Collaboration *et al.*, 2020), i.e. all the

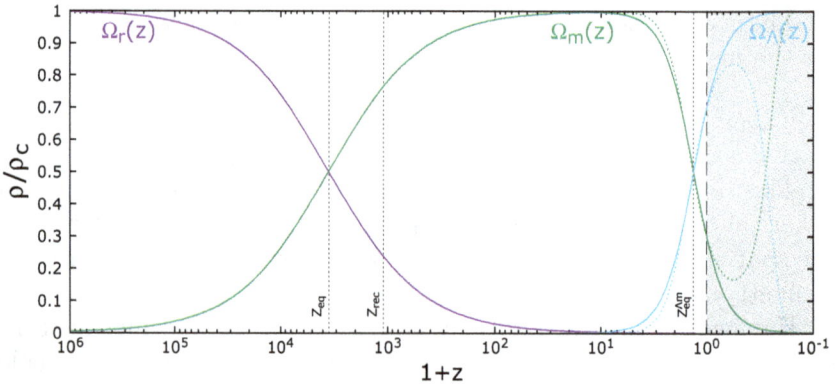

Fig. 10.1. The evolution of radiation ($\Omega_R(z)$), matter ($\Omega_m(z)$) and DE ($\Omega_\Lambda(z)$) densities shown as a function of redshift z. The vertical dashed line corresponds to the present moment, and gray shaded region represents the future. Solid lines show standard ΛCDM, dotted lines represent the model, where Λ is replaced by decaying DE, as suggested by recent DESI measurements (Adame *et al.*, 2025).

---

[1]The plain notation Ω_x without extras is implicitly assumed to represent today's value, and for different evolutionary stages, one explicitly writes $\Omega_x(a)$ (or $\Omega_x(z)$).

density components (r, m, Λ) at all times add up to the critical density.

In the following sections, we briefly look at radiation, matter and DE components of ΛCDM. Also, if not stated otherwise, $\Omega_k = 0$ is implicitly assumed.

10.1.1 Radiation

In standard cosmology at redshifts $z \ll 10^9$, i.e. when typical energies are significantly below the electron rest mass, there are only two noticeable contributors to the radiation density Ω_r: photons (γ) and neutrinos (ν), i.e. $\Omega_r = \Omega_\gamma + \Omega_\nu$. Since the CMB is a black body with a very precisely measured temperature[2] (Fixsen, 2009)

$$T_\gamma = 2.72548 \pm 0.00057 \,\text{K},$$

the photon contribution to Ω_r is also well known[3]:

$$\Omega_\gamma \simeq 2.47 \times 10^{-5} h^{-2}.$$

The temperature of the neutrino background (under instantaneous decoupling approximation) is given as

$$T_\nu = \left(\frac{4}{11}\right)^{1/3} T_\gamma,$$

i.e., $T_\nu \simeq 1.95$ K. Since the energy density $(\propto T^4)$ for fermions has an extra factor of 7/8 and both photons and neutrinos have two spin states, the density parameter for neutrinos can be expressed as

$$\Omega_\nu = \frac{7}{8} \left(\frac{4}{11}\right)^{4/3} N_{\text{eff}} \cdot \Omega_\gamma,$$

where N_{eff} is the effective number of neutrinos. In standard model, $N_{\text{eff}} = 3.044$ (Akita & Yamaguchi, 2020), which differs from the

---

[2]With relative error of only ~ 0.2‰, this is the most precisely known cosmological parameter. The other parameter known with almost similar precision, ~ 0.3‰, is the angular size of the sound horizon measured from the CMB maps.

[3]Here, h is the reduced Hubble constant, $h \equiv H_0/(100 \,\text{km/s/Mpc})$, $h \simeq 0.7$.

whole number due to several approximations used (e.g., instantaneous decoupling) to derive the above formulae. Observational bounds from the Planck satellite data in combination with low-redshift baryonic acoustic oscillation (BAO) measurements give $N_{\text{eff}} = 2.99 \pm 0.17$ (Planck Collaboration *et al.*, 2020), i.e. there is not much room for extra relativistic degrees of freedom beyond the standard model ones. With $N_{\text{eff}} \simeq 3$, $\Omega_\nu \simeq 0.69 \cdot \Omega_\gamma$ and $\Omega_r \simeq 1.69 \cdot \Omega_\gamma$.

The evolution of radiation density parameter $\Omega_r(z)$ is shown in Fig. 10.1. Thus, at high enough redshifts, the universe was dominated by relativistic degrees of freedom. To comply with observational abundance of light elements, radiation domination has to start at least as early as the epoch of Big Bang nucleosynthesis (BBN), which occurred at energies around 0.1 MeV i.e. at redshifts $z \sim 5 \times 10^8$ or at times when the universe was around 1 minute old. From BBN, one gets bounds on N_{eff} which are somewhat weaker (but still consistent with standard model predictions) than the above CMB+BAO values: $N_{\text{eff}} = 3.0 \pm 0.5$ (Pettini & Cooke, 2012) (or $\Delta N_{\text{eff}} \leq 1$ at 95% confidence level (Mangano & Serpico, 2011)).

10.1.2 *Matter*

The matter in ΛCDM has two components: visible (ordinary) or baryonic[4] matter and dark matter, i.e. $\Omega_m = \Omega_b + \Omega_{dm}$. Planck CMB data in combination with low-redshift BAO measurements from large galaxy redshift surveys have provided the following values[5] (Planck Collaboration *et al.*, 2020):

$$\Omega_m h^2 = 0.14240 \pm 0.00087, \quad \Omega_b h^2 = 0.02242 \pm 0.00014,$$

$$h = 0.6766 \pm 0.0042.$$

Thus, one can see that the total matter density is approximately 30% of the critical density and that there is around five times more dark than ordinary matter, i.e.:

$$\Omega_m \simeq 0.3, \quad \Omega_b \simeq 0.05, \quad \Omega_{dm} \simeq 0.25.$$

---

[4]Baryonic, since almost all of the mass of stable standard model particles is in the form of baryons (protons and neutrons).

[5]Note that all of these values are known with precision $\sim 0.6\%$.

Although the standard model of particle physics provides an exceptionally accurate description of ordinary matter, one major question remains unanswered: Why is there so much more matter than antimatter in the Universe? This unresolved issue is known as the baryon asymmetry problem, and the theoretical models developed to address it are referred to as baryogenesis models.

Any viable baryogenesis model must satisfy certain core criteria known as the Sakharov conditions (Sakharov, 1967):

(1) baryon number (B) violation;
(2) C and CP violation;
(3) loss of thermal equilibrium.

The need for the first condition is obvious. In the standard model of particle physics, B is violated in non-perturbative sphaleron processes.[6] However, having B-violating processes is far from being sufficient — it turns out that one also requires C and CP violation. On top of that, one also needs out-of-equilibrium conditions, which are naturally provided by the expanding background. This last requirement is qualitatively easy to understand, since in thermal equilibrium, the contribution of any reaction which could potentially create baryon asymmetry is exactly balanced by the corresponding reverse reaction, leading to zero net effect. Although in the standard model of particle physics and cosmology, all of the above three conditions can be met, it turns out (and generally believed) that there is not sufficient CP-violation available for achieving successful baryogenesis. Thus, one needs to seek for solutions outside the standard model. For a nice review on baryogenesis, see Cline (2006).

The evolution of matter density parameter $\Omega_m(z)$ is shown in Fig. 10.1. Considering the above-given numbers, the redshift below which matter starts to dominate over radiation, $1 + z_{eq} = \Omega_m/\Omega_r$, is approximately $z_{eq} \sim 3500$. This is the moment when the fluctuations in DM component can start to grow efficiently. Since the cosmic gas at that time is highly ionised (i.e., feels photon pressure), the fluctuations in baryonic component cannot grow and have to wait for another factor of ~ 3 of cosmic expansion (until redshift $z_{rec} \sim 1100$) for the universe to cool enough to recombine.

---

[6]Sphalerons can also be used to convert lepton asymmetry into baryon asymmetry — this way potentially achieving "baryogenesis via leptogenesis".

Although the origin of most of the DM is currently unknown, some minor fraction of it is provided by the known stuff — the standard model neutrinos. Neutrino oscillation measurements dictate that only one out of three standard model neutrinos can possibly be massless, while the other two have masses significant enough to make them non-relativistic, and thus contribute to Ω_m, today. Depending on the mass hierarchy, normal or inverted, the sum of masses is bounded from below by $\sum m_\nu \gtrsim 0.06$ eV or $\sum m_\nu \gtrsim 0.12$ eV, respectively. The upper bound (95% CL) from the Planck CMB + BAO measurements $\sum m_\nu \lesssim 0.12$ eV (Planck Collaboration *et al.*, 2020) or particularly from the more recent DESI (Dark Energy Spectroscopic Instrument) measurements $\sum m_\nu \lesssim 0.071$ eV (DESI Collaboration *et al.*, 2025b) seems to rule out inverted hierarchy and settle the absolute mass scale around masses 0.06–0.07 eV. It is interesting that for such masses, the total mass density of neutrinos ($\Omega_\nu \sim 0.0015$) is approximately half of the total contribution given by the stars ($\Omega_{stars} \sim 0.0027$ (Fukugita & Peebles, 2004)) and can explain a small fraction, $\sim$0.5%, of DM.

10.1.3 *Vacuum energy*

In the ΛCDM model, the observed low matter density and the flatness of the Universe are reconciled by including a significant contribution from vacuum energy given by

$$\Omega_\Lambda \simeq 0.7.$$

The vacuum energy is the simplest model for an energy density component that remains uniform in space and constant in time, i.e., it does not directly participate in structure formation (affecting only the background dynamics) and does not dilute with the universe's expansion, thereby driving accelerated expansion. However, vacuum energy suffers from enormous fine-tuning problems. Even though it has many contributions — each naturally orders of magnitude larger than the observed value — they somehow miraculously add up to such a small value. This raises the question of why it is so small yet nonzero, with one possible explanation being anthropic reasoning: only in a universe with such a small Λ could structures like galaxies — and ultimately observers — form. This reasoning was notably developed by Steven Weinberg in his 1987 paper (Weinberg, 1987), i.e. even before the measurement of accelerated expansion by

Type Ia supernovae. Despite this, there is an ongoing search for a deeper mechanism or symmetry that could explain this apparent fine-tuning.

The evolution of the density parameter $\Omega_\Lambda(z)$ is plotted in Fig. 10.1. The rightmost vertical dotted line marks the redshift where the vacuum energy density overtakes matter, which can be expressed as

$$z_{\text{eq}}^{\Lambda m} = (\Omega_\Lambda/\Omega_m)^{1/3} - 1 \simeq 1/3.$$

However, the accelerated expansion — see the lower panel of Fig. 10.2 — starts at a slightly earlier time

$$z_{\text{acc}} = (2\Omega_\Lambda/\Omega_m)^{1/3} - 1 \simeq 2/3.$$

In ΛCDM cosmology, the future expansion will be determined by the vacuum energy and the Hubble parameter will asymptote to a constant value of $H_0\sqrt{\Omega_\Lambda} \sim 60$ km/s/Mpc, leading to a limiting de Sitter universe with exponentially growing scale factor $a(t) \propto e^{t/t_\Lambda}$, where the e-folding time $t_\Lambda \equiv 1/(H_0\sqrt{\Omega_\Lambda}) \sim 17$ Gyr. Interestingly, in the standard ΛCDM cosmology, the universe exhibits de Sitter behaviour both in its earliest moments, during the inflationary epoch driven by the nearly constant potential energy of a scalar field, and in its far future. While the ΛCDM model predicts a de Sitter future for the universe, we will see that this scenario is highly uncertain, particularly in light of some recent observational findings.

ARIEL: It does bother me sometimes that we have only one universe.

RHEA: The 1.

ARIEL: My friends in physics can repeat their experiments many times. They have a lab, and they set things up the way they like.

CALLISTUS: The whole universe is our lab.

RHEA: It's the biggest experiment. Literally.

ARIEL: But we are stuck with this universe. And it might not even be the best of all possible universes.

CALLISTUS: Probably not. For a starter, it has Λ.

RHEA: What's your problem with Einstein's biggest blunder? I guess you never know –

CALLISTUS: If the quadrupole moment of the CMB just happens to be low or it's a sign... of new physics.

RHEA: If one thing had been different.

ARIEL: We'd have a completely different universe. That's exactly the feeling I can't cope with.

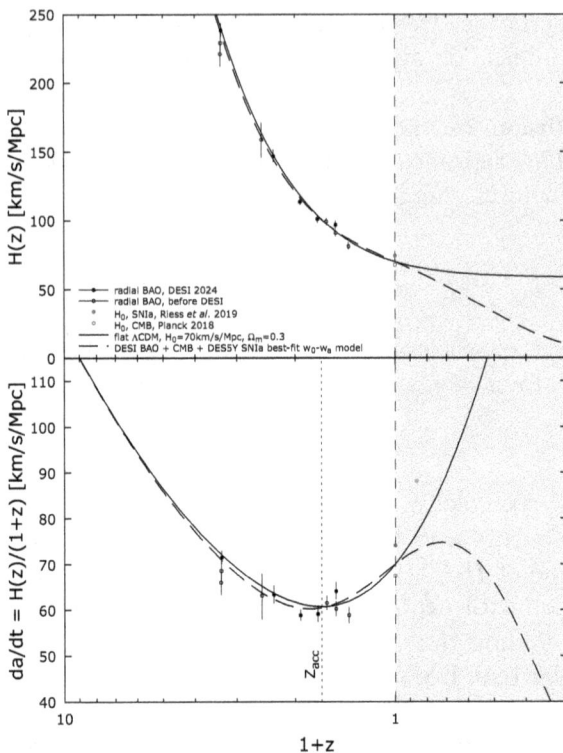

Fig. 10.2. The low-redshift evolution of the Hubble parameter (upper panel) and time derivative of the scale factor (lower panel). The vertical dashed line corresponds to the present moment, gray shaded region represents the future. Hubble parameter values from radial BAO measurements according to year one DESI data (Adame *et al.*, 2025) and from older radial BAO measurements taken from Alam *et al.* (2017); Blomqvist *et al.* (2019); de Sainte Agathe *et al.* (2019); Zarrouk *et al.* (2018) (in order of increasing redshift) are shown with filled and empty black circles, respectively. Local H_0 measurement using Type Ia supernovae (Riess *et al.*, 2022) and H_0 inferred from the Planck CMB data assuming flat ΛCDM (Planck Collaboration *et al.*, 2020) are shown with filled and empty red circles (only best-fit values are displayed). The plotted model curves — solid for standard ΛCDM and dashed for DESI best-fit w_0–w_a DE model — were required to pass through the $H_0 = 70\,\text{km/s/Mpc}$ point. The vertical dotted line in the lower panel indicates the transition from decelerated to accelerated expansion in the ΛCDM model, occurring at $z_\text{acc} \simeq 2/3$.

RHEA: Bayesian statistics will save the day!

ARIEL: Pff. I'm puzzled why we cosmologists worship a trivial formula: the prior times the likelihood divided by the evidence is the posterior. As if giving a name for each term in an equation would turn it into magic.

RHEA: This humble formula describes how observations update our knowledge. That's the closest to magic I know.

ARIEL: All my experimentalist friends form a frequentist clique. It's not fun to be a Bayesian around them. Microagression. Snide comments about "believers of the Bayesian cult".

RHEA: No! Frequentist statistics hide implicit priors which obscure interpretation. The joke is on them.

ARIEL: Thanks. I'll remember that.

CALLISTUS: A fun fact: the Bayesian posterior quantifies your state of mind. When I say $\Omega_c h^2 = 0.12 \pm 0.001$ according to Planck, it expresses my degree of belief about the value of a cosmological parameter based on the latest data.

RHEA: Frequentist confidence limits quantify the long-run frequency of a measurement. This is irrelevant to cosmology, given that we have only one universe.

CALLISTUS: Although you can apply frequentist inference to separate parts of the universe or to simulations. Still, the epistemic approach of Bayesian statistics is more natural for us.

ARIEL: That's cool! A probability obtained through Bayes' formula thus updates my degree of belief about possible values of cosmological parameters. But what if the underlying model is completely wrong?

RHEA: Bayesian statistics can test between different hypotheses: you analyse the data assuming different models and compare the results. The simplest choice is the one with the highest posterior.

CALLISTUS: And there are approximate recipes to choose between hypotheses, such as the Bayesian Information Criterion and the Akaike Information Criterion.

RHEA: They are generalisations of the elementary χ^2 with extra attention to the degrees of freedom: a model with fewer parameters fitting the data equally well would be preferable.

ARIEL: Occam shakes his razor in exuberant agreement. The one universe welcomes cosmologists in the Bayesian camp while my experimentalist friends slide down on a slippery, frequentist slope.

CALLISTUS: That's the spirit. Gradus ad infirmitatem.[7]

---

[7]Steps towards weakness.

ARIEL: I finally installed an AI translator on my phone. I will totally rub it in with Latin insults the next time I see them frequentists.

RHEA: Remember, friends are friends. We have to accept them even if they are recovering frequentists. Instead of trading insults, you can convince the heretics with the virtues of the all-powerful Markov Chain Monte Carlo or MCMC method.

CALLISTUS: A Markov Chain samples the posterior distribution quasi-randomly, with each sample depending on the previous one. It's super efficient near the maximum likelihood, equivalent to importance sampling for estimating parameters and their covariances.

ARIEL: But could it get stuck at a local maximum like other stochastic samplers?

RHEA: We usually start several independent chains to avoid that. In this day and age, standard values for cosmological parameters are obtained from data in a Bayesian context with MCMC (or similar ones, like Hamiltonian) samplers.

CALLISTUS: Moreover, the MCMC chains provide a straightforward way of combining datasets. The only downside is the large set of assumptions and priors that we tend to forget about.

RHEA: Even though they are listed in the papers.

CALLISTUS: Usually in the appendix, or the previous version of the paper.

ARIEL: One universe. Bayesian statistics. Markov Chains. They are all opportunities for trick questions for my comp... If I don't pass, please search the Internet for a sad DJ named MCMC Ariel.

CALLISTUS: I don't think the world is ready for that.

RHEA: I'd prefer to search for Dr. Ariel. In the a few years from now.

ARIEL: Dr. Ariel in one universe, and MCMC Ariel in another one. Which one is ours?

10.1.4 *Minimal set of ΛCDM parameters*

For the sake of completeness, we list here the set of standard ΛCDM parameters. On top of the three previously given parameters, $\Omega_m h^2$, $\Omega_b h^2$ and h, which described matter content and background expansion, there are two additional parameters describing initial fluctuations. As the initial fluctuations, assumed to be generated during the very early inflationary epoch, are compatible of being Gaussian in nature, their full statistical information is completely given by the two-point correlator, e.g., by power spectrum in Fourier space. Moreover, out of several possible fluctuation modes, only the adiabatic scalar mode dominates the evolution, the initial power spectrum for which is well approximated by a simple power law. This brings in two parameters: scalar amplitude A_s and spectral index n_s.

To describe the observable CMB fluctuations, one more parameter needs to be included: the optical depth to reionisation, τ, which gives the probability for the freely propagating CMB photons to scatter off of low-redshift electrons released by the reionisation process. Thus, depending on whether CMB data is included or not, the minimal set consists of five (six) parameters:

$$\Omega_m h^2, \Omega_b h^2, h, A_s, n_s, (\tau).$$

Quite obviously, this minimal set, or in general ΛCDM framework, can be extended, e.g., by allowing extra radiation components (N_{eff}), non-flat models (Ω_k), other than CDM components, or DE different from cosmological constant. Also, the assumption of general relativity can be relaxed and models of modified gravity investigated. Although these extended studies have been regularly performed, the ΛCDM with its underlying assumptions has still mostly managed to survive. However, during the last decade, some cracks have started to emerge, the severest of which will be described briefly.

10.1.5 *Problems with ΛCDM*

Recently, Di Valentino *et al.* (2025) published "The CosmoVerse White Paper", addressing observational tensions in cosmology. In the following, we provide short descriptions of the problems with ΛCDM cosmology:

- H_0 **tension:** There is currently more than 5σ discrepancy between the locally determined Hubble parameter from cepheid-calibrated Type Ia supernovae ($H_0 = 73.04 \pm 1.04$ km/s/Mpc (Riess *et al.*, 2022)) and the value inferred from Planck CMB and low-redshift BAO data under the assumption of flat ΛCDM ($H_0 = 67.66 \pm 0.42$ km/s/Mpc (Planck Collaboration *et al.*, 2020). This represents the most significant challenge faced by the ΛCDM model to date.

  For the overlapping redshift range of supernovae (SN) and BAO measurements, and assuming the validity of the distance–duality relation (which is expected to hold under a broad class of physical models as long as photon number is conserved and space–time is described by a metric theory), one can directly compare luminosity distances (from SN) and angular diameter distances (from

BAO). A constant offset between the datasets is observed, see, e.g., Poulin *et al.* (2024), suggesting a calibration issue, possibly involving either the supernova absolute magnitudes or the sound horizon scale.

The H_0 value derived from cepheid-calibrated SNIa has been supported by other distance ladder methods, such as measurements based on tip-of-the-red-giant-branch (TRGB) stars and asymptotic giant branch (AGB) stars (Riess *et al.*, 2024). However, according to other studies (Freedman *et al.*, 2024), differences persist between results obtained using cepheids and TRGB or AGB stars, highlighting calibration challenges and methodological variations. Upcoming data from the James Webb Space Telescope (JWST) are expected to provide improved precision and may help reconcile these discrepancies.

Assuming no issues with the calibration of SNIa, attention shifts to modifications of the sound horizon. Among the proposed solutions, early DE is currently one of the most promising candidates (Kamionkowski & Riess, 2023). Another possibility involves modifications to recombination dynamics, such as those induced by primordial magnetic fields (Jedamzik & Pogosian, 2020). A comprehensive overview of the approaches discussed in the literature for addressing the Hubble tension can be found in the review by Schöneberg *et al.* (2022).

However, it is important to note that no solution has yet emerged as entirely convincing. One also has to keep in mind that, while true in standard cosmology, the assumption that local measurements reliably approximate the global value of H_0 may itself require re-evaluation.

- σ_8 **tension:** The σ_8[8] tension refers to the discrepancy between the amplitude of matter density fluctuations as inferred from different cosmological probes. This tension was first noted with the Planck CMB measurements (Planck Collaboration *et al.*, 2014b), which suggested a higher σ_8 from the primary fluctuations, while secondary contributions from galaxy clusters indicated a lower value. The higher σ_8 value would imply roughly twice the number of

---

[8]The σ_8 parameter represents the variance of density fluctuations in spheres of comoving radius $8\,h^{-1}$ Mpc, a scale that approximately corresponds to the region where galaxy clusters typically accumulate matter during their formation.

galaxy clusters than were actually observed. This discrepancy has persisted in several subsequent measurements, though it remains relatively mild, typically amounting to no more than a $(2-3)\sigma$ difference. In particular, galaxy weak lensing measurements at low redshifts ($z \lesssim 0.5$–1) suggest weaker matter clustering than expected from the primary CMB data. However, CMB lensing, which probes the matter distribution at higher redshifts and larger scales, is consistent with the primary CMB fluctuations. Additionally, many large-scale structure measurements, such as clustering power spectra, are in good agreement with the CMB predictions (Ghirardini *et al.*, 2024; Papageorgiou *et al.*, 2024). Potential explanations for this discrepancy include uncertainties in baryonic feedback effects or possible modifications to the standard ΛCDM framework, such as changes to the properties of DM or DE.

- **ISW tension:** Decaying potential wells and peaks associated with superclusters and voids lead to the heating and cooling of photons, respectively, as they cross them. The resulting integrated Sachs–Wolfe (ISW) effect (Sachs & Wolfe, 1967) is considered a smoking gun of DE since in an Einstein–de Sitter universe (EdS), the effect would be negligible at the linear level.[9] Thus, stacking superstructures on the CMB should result in hot and cold spots for superclusters and supervoids. The emergence of the $\gtrsim 4\sigma$ ISW tension (Granett *et al.*, 2008) was the original clue that the expansion history deviates from the simple ΛCDM paradigm. Stacking superstructures has since confirmed the ISW levels 4–5× above predictions at low redshifts, and recently, a sign change was detected above redshifts of $z \gtrsim 1$ (Kovács *et al.*, 2022; Szapudi, 2025, and references therein). The latter is consistent with predictions of an emergent curvature model, the Average Expansion Rate Approximation (AvERA, Rácz *et al.*, 2017). A caveat is that CMB-LSS cross-correlations and stacking of smaller structures are mostly consistent with the concordance model, although with large errors. The largest superstructure, the Eridanus void, is the most likely cause of the CMB Cold Spot (Finelli *et al.*, 2016; Szapudi *et al.*, 2015) and thus corroborates the ISW tension at the same level as stacking measurements.

---

[9]The smaller non-linear Rees & Sciama (1968) effect exists even in an EdS universe.

- **Hints for dynamical dark energy:** In spring of 2024, the DESI collaboration announced evidence for dynamical DE by combining their own BAO measurements with CMB and SNIa data (Adame *et al.*, 2025). Depending on the specific SNIa dataset used, the evidence ranged from $\sim 2\sigma$ up to $\sim 4\sigma$. Their analysis assumed a simple linear parametrisation for the DE equation of state:

$$w_{\rm de}(a) = w_0 + (1-a)w_a,$$

with the data preferring values $w_0 > -1$ and $w_a < 0$. For this parametrisation, in the above-given ΛCDM Friedmann equation, the $\Omega_\Lambda = 1 - \Omega_m - \Omega_r$ term should be replaced by

$$(1 - \Omega_m - \Omega_r)e^{-3w_a(1-a)}a^{-3(1+w_0+w_a)},$$

and in the case of a generic equation of state $w_{\rm de}(a)$ by

$$(1 - \Omega_m - \Omega_r)\exp\left[3\int_a^1 \frac{da'}{a'}[1 + w_{\rm de}(a')]\right].$$

For example, when using the SNIa dataset from the Dark energy survey year five data release (Abbott *et al.*, 2024), the following values for w_0 and w_a were obtained:

$$w_0 = -0.727 \pm 0.067, \quad w_a = -1.05^{+0.31}_{-0.27},$$

and the model was favoured over ΛCDM by 3.9σ. It is evident that with these parameter values, the DE density is decaying – in stark contrast with the constant density behaviour of ΛCDM. A comparison of this model with ΛCDM is shown in Figs. 10.1 and 10.2. As can be seen, it is fair to say that, with our current understanding of DE, we cannot make any definite statements about the future of the Universe.

With the above values of w_0 and w_a, it is clear that at redshifts $z \gtrsim 0.35$, this linearly parametrised DE model suggests phantom behaviour, i.e., the DE equation of state parameter $w_{\rm de}$ drops below -1. This behaviour signals a violation of the null energy condition, which cannot occur under the usual assumption of scalar fields within the framework of general relativity. However, modified gravity models or scalar field models with non-standard kinetic terms (*k*-essence) can yield $w_{\rm de} < -1$, though they often suffer from pathologies (ghosts, superluminal modes, etc.).

That said, physically motivated scalar field (quintessence) models can fit the data just as well as this simple DE toy model with a linearly evolving equation of state (Gialamas *et al.*, 2025). As a result, the apparent indication of phantom behaviour is likely an artefact of this simplistic parametrisation. Furthermore, the best-fitting quintessence models predict an even steeper decline in DE density at low redshifts compared to the above w_0–w_a case. It also turns out the preference for dynamical DE seems to be primarily driven by low-redshift SNIa data. When supernovae with redshifts below ~0.1 are excluded from the analysis, concordance with ΛCDM is restored (Gialamas *et al.*, 2025).

Although the hints for evolving DE are still tentative, they provide an exciting direction for future research. As DESI has only released year one data, with more to come, the evidence for or against dynamical DE may become clearer, potentially reshaping our fundamental understanding of the Universe.

These four problems hint at a modification of the expansion and growth history of the concordance cosmological model. However, the significance of these deviations remains under intense scrutiny, with the Hubble tension standing out as the most severe among them. Could these tensions collectively point towards a unified solution requiring only minimal adjustments to ΛCDM? Perhaps, but such an expectation might be overly optimistic. After all, ΛCDM, while remarkably efficient in describing a wide range of observational data, is an exceedingly simple framework. For instance, beyond a few basic parameters, we lack a deeper understanding of the fundamental nature of DM or DE. What guarantees that the dark sector should itself be simple?

In the following, we review models of the dark sector. Our discussion of DE will be very brief, reflecting the current state of knowledge on the topic. In contrast, we will delve into DM in slightly greater detail.

10.2 Dark energy

In the broadest sense, DE can be modelled either as an additional component (with an equation of state $w_{\mathrm{de}} \simeq -1$ in the universal energy–momentum tensor or as a modification to general relativity (GR), i.e., either modifying right-hand (sources) or left-hand (geometric) sides of Einstein's equations. The other possibility could be due to our lack of understanding of the feedback caused by the non-linear evolution of the density perturbations or the breakup of cosmological principle without giving up Einstein's GR or adding extra energy–momentum components. Here, we enlist and discuss very briefly these options. For a thorough treatment of the topic, see textbooks by Amendola & Tsujikawa (2010) and Wang (2010).

- **Constant vacuum energy (cosmological constant):** Cosmological constant, or a vacuum energy which is constant in space and time, is the simplest option for the DE. Indeed, this is the component in concordance ΛCDM which provides the majority of the energy density of the low-redshift universe and drives the cosmic acceleration. Quite unavoidably, the vacuum energy density gets contributions from the zero-point energies of all the quantum fields along with condensates arising from various spontaneous symmetry breakings (e.g., Higgs and QCD condensates), the absolute values of each being orders of magnitude larger than the observationally needed one. Thus, all these numbers need somehow to miraculously cancel each other, leaving only a miniscule positive residual. Supersymmetry, in principle, helps to cancel out zero-point energies, as bosonic and fermionic fields provide equal but opposite in sign contributions. In reality, this does not help much, as supersymmetry, when it exists, needs to be broken in our Universe, and also, the problems with the above-mentioned condensates remain. As commonly done, these major fine-tuning problems are often swept under the carpet by assuming the existence of some yet-unknown symmetry which sets the vacuum energy to zero value (or assumes that the vacuum energy somehow does not gravitate) and addresses the observed DE via different mechanisms. Still one has to keep in mind that all of these DE alternatives (to be discussed in the following) typically leave the problem of vacuum energy unresolved. In ΛCDM, the smallness of the observed vacuum energy is often explained through anthropic arguments, which suggest that its value must lie within a narrow range to permit the formation of galaxies and, consequently, the emergence of observers like us.

Detailed reviews on vacuum energy and cosmological constant problem have been written by Weinberg (1989), Carroll *et al.* (1992), Peebles & Ratra (2003), Padmanabhan (2003).

- **Dynamical field:** The next most widely studied possibility is that DE originates from a dynamical field with only minimal coupling to gravity. Due to its simplicity and compatibility with the cosmological principle, this field is almost always assumed to be a scalar field. For a scalar field ϕ with a potential $V(\phi)$, the equation of state parameter w is given by

$$w = \frac{\frac{1}{2}\dot{\phi}^2 - V(\phi)}{\frac{1}{2}\dot{\phi}^2 + V(\phi)}.$$

Thus, if the field is slowly rolling in its potential, i.e., when $\dot{\phi}^2 \ll V(\phi)$, it closely emulates the observed behaviour $w_{\mathrm{de}} \simeq -1$. In general, the above expression for w allows values in the range $-1 < w < 1$.[10]

These scalar field models are commonly referred to as *quintessence* models. Depending on the behaviour of the DE equation of state $w_{\mathrm{de}}(z)$, these models are often classified as either "freezing" or "thawing". In the freezing case, $w_{\mathrm{de}}(z)$ approaches a value of -1 today, whereas in the thawing case, it moves away from -1. As mentioned previously, the latest BAO measurements from the DESI survey, in combination with CMB and SNIa data, seem to suggest DE that is currently decaying and thus preferring thawing models.

Although theoretically less well motivated, the data can also be equally well fitted with models that go *phantom*, i.e., $w_{\mathrm{de}} < -1$, over a range of redshifts. Such behaviour can be achieved in scalar field models with non-canonical kinetic terms, known as kinetic quintessence or *k-essence* models.

Compared to the cosmological constant case, dynamical field models can naturally include the so-called *tracker models*, which address the cosmic coincidence problem: Why, in the present universe, are the DM and DE densities so comparable in size?

---

[10] However, if the potential is allowed to be somewhat negative so that the total energy density in the denominator approaches zero, w can become arbitrarily large.

Additionally, models with DM and DE interactions can be constructed to help alleviate the Hubble tension.

Dynamical DE fields necessarily imply accompanying fluctuations. In the case of quintessence, however, since the effective sound speed of DE is equal to the speed of light, $c_s = c$, these fluctuations are only relevant on very large scales comparable to the Hubble length, c/H.

For k-essence, there are no such restrictions on c_s, and the speed can be arranged to be substantially smaller, $c_s \ll c$, resulting in DE fluctuations at significantly smaller scales, c_s/H. Although these fluctuations could have potentially interesting consequences, current observational data — so far providing only hints about the temporal behaviour of DE — still leaves this possibility open.

An extended discussion of the dynamical field models, along with relevant references, can be found in standard texts, e.g., Amendola & Tsujikawa (2010).

- **Modified gravity:** Instead of introducing an additional energy–momentum tensor component, another plausible scenario is that DE arises due to modifications of GR at large scales. According to Lovelock's theorem, GR is unique in being the only possible theory in four-dimensional space–time derivable from a local gravitational action that contains only second derivatives of the space–time metric and no additional field content. Therefore, any modification of GR must abandon at least one of these assumptions, with perhaps the most straightforward approach being the introduction of additional fields. A simple example is provided by scalar–tensor theories, where a scalar field is added to the metric.

The number of other possibilities, along with variations within different subclasses explored in the literature, is impressively large and will not be discussed further here. Extended discussions on modified gravity, along with references to the original papers, can be found in reviews by Sotiriou & Faraoni (2010), De Felice & Tsujikawa (2010), Clifton *et al.* (2012), Nojiri *et al.* (2017), Burrage & Sakstein (2018), and Heisenberg (2019).

Despite the wide range of theoretical possibilities, the class of phenomenologically viable models is significantly constrained. For example, these models must satisfy the following conditions:

o Preserve the success of GR in Solar System precision tests.
o Be consistent with LIGO/Virgo/KAGRA gravitational wave (GW) observations of black hole merger signals.
o Satisfy precision tests of the equivalence principle.
o Comply with bounds on the speed of GW propagation ($-3 \times 10^{-15} < (c_{\rm GW} - c)/c < 7 \times 10^{-16}$) as inferred from the neutron star (NS–NS) merger event GW170817 (Abbott et al., 2017).

Since modifications to gravity, invoked to explain DE, alter its behaviour on large scales, they must remain compatible with GR on small scales, where it is well tested. This is often achieved through non-linear *screening mechanisms* (e.g., Vainshtein, chameleon, symmetron), which suppress deviations from GR in high-density environments while allowing modifications to emerge on cosmological scales.

- **Other DE mimickers: backreaction and void models:** Finally, we mention a couple of alternatives to the popular DE models discussed above — specifically, those that neither modify GR nor introduce extra fields into the energy–momentum tensor. These models are motivated by the observation that accelerated expansion becomes significant around the time when cosmic large-scale structure develops strong non-linearities. They propose that the observed effects of DE could arise from the influence of these inhomogeneities on the large-scale dynamics of space–time, thereby naturally addressing the DE coincidence problem. Since these models achieve accelerated expansion through entirely different mechanisms, they are also free from the DE energy scale (or cosmological constant) problem, i.e., the question of why the energy scale of DE ($\sim$meV) is so vastly smaller than other fundamental energy scales.

One such approach suggests that within GR, we may not have properly accounted for how non-linear structures backreact on the background evolution. In particular, due to the non-linearity of the Einstein equations, averaging over an inhomogeneous matter distribution before solving for the geometry (the common approach) is not equivalent to solving the full inhomogeneous equations first and then averaging. As a result, the Friedmann equations acquire additional *backreaction* terms that could, in principle, drive cosmic acceleration. Although research on this topic is ongoing, it is widely believed that the level of achievable backreaction is significantly below what is required to explain the observed acceleration.

Another class of models suggests that the apparent accelerated expansion could be a consequence of living in an underdense void region. These void models include those based on a single inhomogeneous Lemaître–Tolman–Bondi (LTB) metric as well as *Swiss-cheese* models, where underdense voids (modelled by LTB solutions) are embedded in an FLRW background (the "cheese" component). The former clearly violates the cosmological principle, while the latter preserves statistical homogeneity on large scales.

To mimic cosmic acceleration, the void must be significantly underdense (at least $\sim 30\%$ below the cosmic mean) and extend to redshifts $z \sim 0.3$–0.5. If the observer is not located exactly at the void centre, one would expect to detect anisotropies in the expansion rate, while the underdensity itself should also produce a potentially measurable local *integrated Sachs–Wolfe (ISW) effect*. In the case of Swiss-cheese models, ISW distortions of the CMB due to voids should appear in multiple directions across the sky. Interestingly, some anomalously large ISW detections towards void regions (Granett *et al.*, 2008; Szapudi *et al.*, 2015) provide slight observational hints in favour of such scenarios.

Unravelling the nature of DE requires large, dedicated cosmological surveys to map observables sensitive to both the background evolution and the growth of fluctuations on this expanding background. As of winter 2025, two major ongoing projects have DE as a central scientific focus: the ground-based DESI survey and ESA's M2 space mission, Euclid. It remains to be seen whether the intriguing hints of time-varying DE from DESI's first-year data will be confirmed in future releases. Results from DESI's three-year data and Euclid's initial quick data release are both expected in spring 2025. Thus, in terms of observational progress in DE research, we are undoubtedly living in very exciting times.

10.3 Dark matter

Throughout the years, we have gained undeniable evidence for the existence of DM, starting from the largest scales, as probed by the CMB, down to supercluster, cluster and galactic scales. The evidence comes only via its gravitational effects. Beyond general requirements that DM should be cold and should not interact or interact very feebly with visible sector — facts known by now for several decades — we still do not know much about it, e.g., how massive it is, how was it produced and in what interactions (beyond gravitational) it participates in. Quite remarkably, although its interactions with ordinary matter are highly constrained, its self-interactions could be quite sizeable.

The list of other requirements for successful DM models are nicely summarised in Taoso *et al.* (2008). These limits eliminate several possible DM contenders, however the remaining possibilities are still vast, i.e. we still do not have any clue what DM could be. To illustrate this for a single fundamental parameter describing DM — its mass — in Fig. 10.3, we show a range of possible values discussed in the contemporary literature.

As seen, it can range for almost 90 orders of magnitude: starting from the ultralight or fuzzy DM (FDM) with 10^{-22} eV up to solar mass ($1\,M_\odot \sim 10^{57}$ GeV) or heavier primordial black holes (PBHs). Due to various PBH constraints, this range gets somewhat smaller, as only asteroid mass PBHs 10^{17}–10^{21}g can constitute the entirety of DM. This still leaves us with more than 70 orders of magnitude!

In Fig. 10.3, Planck mass, $M_{\text{Pl}} \sim 10^{-5}$ g, separates DM in the form of fundamental particles from the composite DM. DM lighter than $\mathcal{O}(0.1$–$1)$ keV cannot be fermionic, which is due to phase-space arguments — Pauli exclusion principle forbids fermions to be packed down to the observed dwarf galaxy-sized clumps. In addition, in order not to act as hot dark matter, and excessively wash out cosmic structures, these light-end DM candidates must also be produced non-thermally, i.e. there cannot be times when they where in thermal equilibrium with the standard model (SM) thermal bath.

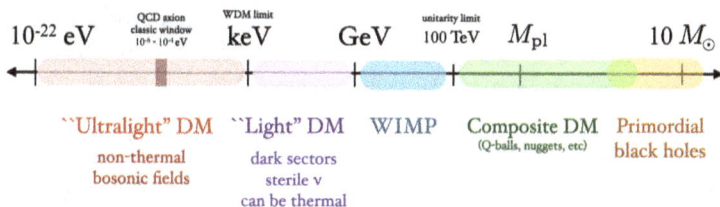

Fig. 10.3. Wide choice for DM mass (not to scale), adapted from Lin (2019).

Being compatible with observational bounds from large-scale structure (LSS) and CMB measurements, all the candidates in Fig. 10.3 behave as CDM at those well-probed scales while their differences start to appear typically at sub-galactic scales. CDM (Blumenthal *et al.*, 1982; Bond *et al.*, 1982; Peebles, 1982a) has been a standard paradigm for DM for many decades, but more recent studies have started to reveal some small-scale problems which include the following:

(1) **Core-cusp problem:** Central profiles of the Milky Way (MW) dwarf satellites are inferred to be shallower than the ones obtained from CDM-only N-body simulations.

(2) **Missing satellites problem:** Number of MW satellites are much smaller compared to CDM predictions.

(3) **Too-big-to-fail problem:** Many of the MW satellites predicted by CDM models are so massive that there is no way they could have not formed stars.

(4) **Galaxy diversity problem:** CDM models do not naturally predict as diverse range of rotation curve inner slopes as seen in observational data.

(5) **Planes of satellites problem:** The presence of planes of satellites as observed in MW and Andromeda galaxies are surprising in the context of CDM models.

Detailed discussions about small-scale CDM problems are extensively covered in reviews by Weinberg *et al.* (2015) and Bullock & Boylan-Kolchin (2017).

Many of the above problems can potentially be overcome by adjusting the treatments for the sub-grid baryonic physics, which is not appropriately resolved even in the highest-resolution cosmological hydrodynamic simulations. Still, these potential shortcomings have initiated vigorous activity in searches for DM models that go beyond vanilla CDM. In the following, along with weakly interacting particles (WIMPs) — a standard CDM candidate — we briefly cover some of the widely discussed alternatives.

- **Weakly interacting massive particles:** WIMP is by far the most extensively studied DM candidate. Its popularity owes to a simple fact that a $\mathcal{O}(100)\,\text{GeV}$ DM particle with an electroweak-scale self-annihilation cross-section, having been once in thermal equilibrium with visible sector and subsequently decoupling via the freeze-out mechanism, provides a correct cosmic abundance for the DM — the coincidence known as "WIMP miracle". In general, the WIMP class is significantly widened by abandoning the above electroweak assumptions, but still requiring large enough interaction cross-sections for the particles to be in equilibrium with thermal bath at early times. This broader class of WIMPs can have masses in the range $\sim 1\,\mathrm{GeV}$–$100\,\mathrm{TeV}$, where the upper bound is dictated by the unitarity limit — the cross-sections are bounded from above by the perturbative unitarity requirement.

The most well-known WIMP arises in supersymmetric extensions of the SM as the lightest supersymmetric particle (Ellis *et al.*, 1984; Jungman *et al.*, 1996) — typically neutralino. Another example is provided by the lightest Kaluza–Klein particle in models with universal extra dimensions (Cheng *et al.*, 2002; Servant & Tait, 2003).

Among the DM candidates, WIMPs have enjoyed the most elaborate search program, schematically depicted in Fig. 10.4, which consists of the following: (i) direct detection: search for DM-induced nuclear recoils in deep underground labs, (ii) indirect detection: search for DM annihilation or decay signals in cosmic radiation/particle fields, and (iii) searches in particle colliders.

Over the years, the constraints on WIMP models have become tighter and tighter. Every now and then, some positive hints for the existence of WIMPs have been obtained, but unfortunately these have not withstood the test of time. Due to these reasons, the focus of the community has largely shifted towards other not so thoroughly investigated DM candidates. This does not mean that

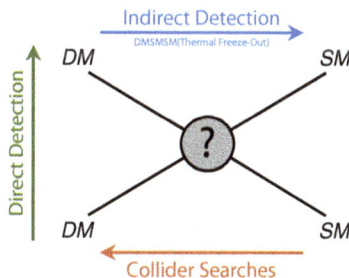

Fig. 10.4. Well-established WIMP search program.

WIMPs have completely lost their attractiveness, just that there are more opportunities elsewhere.

- **Sterile neutrinos and warm dark matter:** Alternatives in the form of WDM have been investigated in order to address some of the above-listed CDM shortcomings. In comparison to CDM, faster moving particles of WDM result in extra level of random motions which smooth out structure on small scales. Typical example of WDM is provided by a sterile neutrino (Dodelson & Widrow, 1994; Dolgov & Hansen, 2002; Shi & Fuller, 1999) with mass in the range $\mathcal{O}(1\text{--}10)$ keV. Sterile neutrinos being right-handed particles do not participate in weak interactions. Having only gravitational interactions, they provide a perfect DM candidate. In comparison to supersymmetric extensions of the SM which double the number of particles, the inclusion of sterile neutrino(s) looks like a significantly milder way for extending the SM. Sterile neutrinos have also been used in several other contexts, e.g., giving natural explanation for the smallness of SM neutrino masses or participating in processes which generated matter–antimatter asymmetry in the early universe (Abazajian, 2017). However, in these cases, their masses are orders of magnitude above the keV scale needed for them to act as WDM.

  WDM has tight small-scale bounds derived from the quasar Lyman-alpha forest measurements. It turns out that for the masses still available, $m_{\mathrm{WDM}} > 5.3\,\mathrm{keV}$ (Iršič *et al.*, 2017), the WDM acts almost like CDM at scales of relevance and thus possibly shares the same small-scale problems as CDM.

- **Ultralight dark matter: axions, axion-like particles, fuzzy dark matter:** In this category, the most important example is given by the QCD axion field which was initially introduced to resolve the strong CP-problem (Peccei & Quinn, 1977) — the fact that the strong interaction, without immediate reasons, somehow achieves to respect the CP-symmetry. It was soon realised that axions could also act as DM (Abbott & Sikivie, 1983; Dine & Fischler, 1983; Preskill *et al.*, 1983): non-thermally produced QCD axions with masses typically in the range $\sim 10^{-6}\text{--}10^{-4}$ eV provide a perfect example of CDM. In terms of structure formation, in particular with an aim to address CDM shortcomings, the situation gets interesting once one allows significantly smaller masses, e.g., down to $\sim\!10^{-22}$ eV as then the corresponding de Broglie wavelength reaches kiloparsec scale. This type of ultralight DM is often called Fuzzy dark matter (FDM) (Hu *et al.*, 2000; Hui *et al.*, 2017). As scalar fields are ubiquitous in beyond SM theories, FDM could be in the form of axion-like particles or moduli fields. Compared to CDM, FDM due to its quantum pressure effects leads to a strong suppression in number density of small mass haloes. The central parts of the haloes are also significantly modified due to the development of coherent solitonic cores.

- **Primordial black holes:** PBHs (Hawking, 1971; Zeldovich & Novikov, 1967) are among the oldest proposed DM candidates (Chapline, 1975; Meszaros, 1975). The research activity in this field has seen two major activity periods. The first one was coinciding with MACHO (Massive Compact Halo Object) search projects starting at early 1990s (Alcock *et al.*, 1993; Paczynski, 1986). The second wave, starting from 2016, was initiated (Bird *et al.*, 2016) by the detection of gravitational waves from black hole (BH) mergers with $\sim 30 \, \mathrm{M_\odot}$ masses by the LIGO and VIRGO interferometers (Abbott *et al.*, 2016), which at the time seemed suspiciously heavy to be formed by standard stellar evolution. However, later on, several astrophysical paths to forming so heavy BHs were recognised, still the observed excess of merger events at around this $\sim 30 \, \mathrm{M_\odot}$ scale over the much broader spectrum of masses still needs a good explanation, one of which could be PBHs. Ongoing LIGO/Virgo/KAGRA and future gravitational wave measurements will help shed light on possible PBH contribution. The decisive difference between astrophysical BH (ABH) and PBH populations arises due to differences in their event rate redshift dependences, with the ABH mergers following star formation which, contrary to the PBHs, is expected to fade out at high redshifts. The initial investigation of LIGO/Virgo data having both ABH and PBH populations was carried out in Hütsi *et al.* (2021).

  PBHs with much larger masses — $\mathcal{O}(10^5) \, \mathrm{M_\odot}$ and above — could provide seeds for supermassive black hole population, which are notoriously hard to grow in time in standard ΛCDM cosmology. They could also facilitate early galaxy formation hinted by the recent JWST measurements (Hütsi *et al.*, 2023).

  Due to various observational constraints (accretion, structure formation or other dynamical constraints), it is important to point out that so massive PBHs — $\mathcal{O}(10)$ or $\mathcal{O}(10^5) \, \mathrm{M_\odot}$ and above — can only form a minor fraction of DM. At very low PBH masses, $< \mathcal{O}(10^{16}\text{–}10^{17})$g, PBHs as DM is strongly constrained by the BH evaporation bounds, whereas the microlensing bounds rule out PBH DM above masses $\mathcal{O}(10^{21})$g. Thus, the only window where PBHs can provide entirety of DM falls in the asteroid mass range of $10^{17}\text{–}10^{21}$ g.

The detailed discussion on physics and observational status of PBHs are nicely covered by recent reviews (Carr & Kuhnel, 2020; Escrivà *et al.*, 2022; Green & Kavanagh, 2021; Villanueva-Domingo *et al.*, 2021).

- **Self-interacting dark matter:** It would be unfair if we do not mention self-interacting dark matter (SIDM) as a possible solution to small-scale CDM troubles. In fact, the DM self-interactions could be quite significant: the constraints on self-interaction cross-section derived from the Bullet Cluster observations, $\lesssim 1\,\mathrm{cm}^2/\mathrm{g}$, are of the same order to nuclear cross-sections. Detailed overview about SIDM as an attractive alternative to CDM can be found in, e.g., Adhikari *et al.* (2022).

ARIEL: Panta Kykloutai!

RHEA: Panta Kykloutai!

CALLISTUS: Wait, next thing I see you have a secret handshake?
> [*Rhea and Ariel execute a secret handshake.*]

CALLISTUS: Why do I feel excluded?

ARIEL: You can join us!

CALLISTUS: I understand Panta Rhei. Everything moves. And it's my favourite progressive rock band. Is Panta Kykloutai an example of what Gen Z inexplicably calls music?

ARIEL: Let's fly over the generational abyss between our musical tastes and go back to Heraclitus of Ephesus.[11]

RHEA: He would have liked Abyss. It's my favourite album.

RHEA: Heraclitus looked at the universe–
> [*Callistus looks at his phone.*]

CALLISTUS: He said Panta Rhei, not Panta Kykloutai. According to chat-GPT, it means–

ARIEL: Everything rotates!

RHEA: Exactly! Think about it, the Earth!

ARIEL: The Moon!

CALLISTUS: The Sun and the planets!

RHEA: Our solar system, stars, planets, and solar systems in our Galaxy!

---

[11]c. 535 BC–c. 475 BC.

ARIEL: Black holes! Nobody has seen a Schwartzschild black hole outside a GR textbook! Kerr it is, baby!

CALLISTUS: Our Galaxy!

ARIEL: All galaxies!

RHEA: Galaxy clusters!

[*Beat.*]

CALLISTUS: So what?

ARIEL: What about the universe?

CALLISTUS: Rotating? I hate to throw cold water on your fire, but the CMB constraints are extremely tight on that.[12] The universe is unlikely to be rotating.

RHEA: Party pooper. It's so unlike you to be against an out-of-the-box idea.

CALLISTUS: It's unlike you to take this seriously! This is sooo not ΛCDM! Not even MOND!

ARIEL: General Relativity was motivated by Mach's principle stating that distant stars somehow define local rotation. But ultimately, GR states exactly the opposite: our local inertial frame has nothing to do with distant objects. When you reject off-hand that the universe rotates, you implicitly endorse Mach against Einstein.

RHEA: And you don't wanna do that. Although you could test Mach's principle with high-precision local astrometry.[13]

ARIEL: There are rotating metrics that escape the CMB limits[14] based on imprinting of vector perturbations.

CALLISTUS: Congrats! You're ready for your comp and beyond! I bet you're motivated by this new paper[15] that's on my reading list but didn't get to. What does it say? Panta Kykloutai!

ARIEL: They observe that everything rotates, so they ask the question of how rotation would affect the expansion history of the universe.

CALLISTUS: I see. The universe would expand slightly faster due to rotation, which relates to the Hubble puzzle.

RHEA: Not only that, they find that the maximum allowed rotation speed –

ARIEL: which is exquisitely tiny, one rotation in 500 Gy. Otherwise, you would have closed time-like loops, aka, speeds faster than the speed of light within the horizon.

RHEA: The maximum allowed speed appears to solve the Hubble puzzle.

ARIEL: And here is the rub: most black holes rotate with the maximum allowed speed. Since one way to think about cosmology is that

---

[12]Saadeh *et al.* (2016).

[13]Szapudi (2021).

[14]Obukhov (2000).

[15]Endre Szigeti *et al.* (2025).

we are in an inside-out black hole ... I'll let you put two and two together.

CALLISTUS: Right. Except you need a slight anisotropy in the expansion.

RHEA: Yes. There is no clear evidence for that, but some hints are interesting.[16]

ARIEL: It's convincing. It's exciting! I wanna work on this!

CALLISTUS: It's certainly amusing. Might be worth more studies ... Why not? Panta Kykloutai!

RHEA AND ARIEL: Panta Kykloutai! [*The three of them perform the secret handshake.*]

To conclude, over the years, a lot of DM models have been explored, some of them eliminated or ever more tightly constrained, many — particularly the ones with phenomenology close to CDM — still surviving. Except for PBHs, for which ongoing and future gravitational wave experiments will play a decisive role, or SIDM with dark-sector interactions only, for the other, the above-discussed DM candidates' essential progress relies heavily on our luck in obtaining signals from outside the gravitational sector. Considering the vastness of the space of possible manifestations, we are by no means guaranteed to succeed.

Over the years, several hints have been obtained. Unfortunately, most of them have not survived, e.g., the above-discussed Fermi LAT 130 GeV line (Weniger, 2012) or 750 GeV diphoton excess at LHC (ATLAS collaboration, 2015; CMS collaboration, 2015), both could have been signals from WIMP DM. A 3.5 keV X-ray line (Bulbul *et al.*, 2014) — result of a possible 7 keV sterile neutrino decay — has shared a similar faith. Currently, the only signal still possibly surviving is a Galactic centre gamma-ray excess at $\mathcal{O}(1)$ GeV (Hooper & Goodenough, 2011), which might be due to $\mathcal{O}(10\text{--}100)$ GeV WIMP annihilation or perhaps more realistically due to our lack of knowledge regarding the astrophysics of our Galactic centre — the signal could be due to population of pulsars (Abazajian, 2011).

The investigation of the nature of DM is actively continuing. The emphasis on WIMPs has somewhat declined with a main focus on alternatives. Especially active are investigations related to PBHs and FDM.

---

[16]Migkas (2025), Rameez (2025).

10.4 Concluding comments

It is remarkable that such a simple construct as ΛCDM has withstood the test of time, consistently fitting a wide range of observational data. However, despite its success, ΛCDM remains an effective model, leaving room for diverse physical realisations that could produce similar observational outcomes. The real challenge lies in our lack of knowledge about the dark sector, which, so far, is revealed to us only through gravitational evidence. To unravel its nature, we must continue with dedicated observational efforts — cosmological surveys, DM searches, and gravitational wave experiments — and hope for new evidences (beyond gravity) to lead us forward.

Encouragingly, recent hints from CMB polarisation measurements suggest that propagation of electromagnetic waves at cosmological scales might not respect parity symmetry, with left- and right-handed polarisation modes propagating differently — an effect known as cosmic birefringence (Komatsu, 2022; Minami & Komatsu, 2020). This signal, detected at $\sim 9\sigma$ level (or $\sim 3\sigma$ when accounting for instrumental uncertainties), could point to the presence of an axion-like field permeating the Universe. It is interesting that such fields (ubiquitously appearing in string theories) could not only explain cosmic birefringence but also constitute DM and potentially drive cosmic acceleration. These intriguing prospects offer hope that ongoing observations may some day provide the breakthroughs needed to unlock the mysteries of the dark sector.

Epilogue

As in all sciences, there are numerous open problems in astrophysics and cosmology to be solved. Most of these problems concern the structure and evolution of various astronomical bodies from planets and stars to galaxies and the Universe as a whole. These problems were discussed in the previous chapters of this book. But there exist several global problems not solved yet. One of these problems is the physics of the birth of the Universe.

By the end of the 20th century, it was clear that our Universe was formed about 13 billion years ago by Big Bang. What preceded this event was not clear. Also, it was not clear why the Universe has properties which allow the birth of life. These questions were discussed by Martin Rees (2000), see also Rees & Kolb (2000), who asked thought-provoking questions in *Just Six Numbers: The Deep Forces That Shape the Universe*. Rees describes these forces as follows:

- The first number $\mathcal{N}$ measures the strength of the electrical forces that hold atoms together, divided by the force of gravity between them. This number is $\mathcal{N} \approx 10^{36}$ — the ratio of the strength of electrical forces that hold atoms together to the force of gravity; If $\mathcal{N}$ had a few less zeros, only a short-lived miniature universe could exist: no creatures could grow larger than insects, and there would be no time for biological evolution.

- The second number is $\epsilon = 0.007$ — the effectivity of the nuclear burning of hydrogen to helium. It defines how firmly atomic nuclei are bound together and how all the atoms were made. Its value controls the power from the Sun and how stars transmute hydrogen into all the atoms of the periodic table. If ϵ were 0.006 or 0.008, we could not exist.

- The third cosmic number $\Omega_m = 0.3$ — the amount of matter in the Universe, in units of the critical density. Ω_m measures the amount of material in our Universe: galaxies, diffuse gas, and "dark matter". It tells us the relative importance of gravity and expansion energy in the Universe. If this ratio were too high relative to a particular "critical" value, the Universe would have collapsed long ago; had it been too low, no galaxies or stars would have formed. The initial expansion speed seems to have been finely tuned.

- The fourth number, $\Omega_\Lambda = 0.7$, is new force — a cosmic "antigravity". It controls the expansion of our Universe and is destined to become ever more dominant over gravity and other forces as our Universe becomes ever darker and emptier. Fortunately, for us, Ω_Λ is very small. Otherwise, its effect would have stopped galaxies and stars from forming, and cosmic evolution would have been stifled before it could even begin.

- The seeds for all cosmic structures — stars, galaxies and clusters of galaxies — were all imprinted in the Big Bang. The fabric of our Universe depends on one number: $Q \approx 10^{-5}$ — the ratio of the energy needed to disperse large structures (superclusters) to their internal rest mass energy (mc^2). If Q were smaller, the Universe would be inert and structureless; if Q were much larger, it would be a violent place, in which no stars or solar systems could survive.

- The sixth crucial number is the number of spatial dimensions in our world, D, and equals three. Life couldn't exist if D were two or four. Time is a fourth dimension, but distinctively different from the others in that it has a built-in arrow: we "move" only towards the future.

Martin Rees (2000) explains that if these numbers were a bit different from their actual values, our Universe in such form as we know it would be impossible. The question is: Why do these numbers have exactly the values needed for the existence of our Universe and the life in it, including ourselves?

The other open problem is the birth of the Universe and its early inflationary phase of evolution. The classical version of the inflation model, described in previous chapters, can be called "old". It provides a simple solution to several crucial problems, but does not answer some other fundamental questions: Why does the Universe have elementary particle parameters as they are? What happened before inflation started? To answer these questions, Andrei Linde (1982, 1983, 2002a, 2002b) suggested a scenario of chaotic inflation. Linde (2002a) explains the problems which led him to his new inflation scenario as follows:

Most of the parameters of elementary particles look more like a collection of random numbers than a unique manifestation of some hidden harmony of Nature. For example, the mass of the electron is 3 orders of magnitude smaller than the mass of the proton, which is 2 orders of magnitude smaller than the mass of the W-boson, which is 17 orders of magnitude smaller than the Planck mass M_p. Meanwhile, it was pointed out long ago that a minor change (by a factor of two or three) in the mass of the electron, the fine-structure constant, the strong-interaction constant, or the gravitational constant would lead to a universe in which life as we know it could never have arisen. These facts, as well as a number of other observations, lie at the foundation of the so-called anthropic principle. According to this principle, we observe the universe to be as it is because only in such a universe could observers like ourselves exist... One can consider different universes with different laws of physics in each of them. This does not necessarily require introduction of quantum cosmology. It is sufficient to consider an extended action represented by a sum of all possible actions of all possible theories in all possible universes. One may call this structure a "multiverse."

According to the inflationary theory, all elementary particles in the universe emerged after the end of inflation in a process called reheating (Kofman *et al.*, 1994b). The most far-reaching prediction of the theory of inflationary universe is the suggestion of a model of a self-reproducing inflationary universe consisting of different parts (Linde, 1982; Linde *et al.*, 1994). These parts are exponentially large and uniform because of inflation. Therefore, each of these parts looks like a separate mini-universe, or pocket universe, independent of what happens in other parts of the universe.

Inhabitants of each of these parts might think that the universe everywhere looks the same. However, different pocket universes may have different laws of low-energy physics operating in each of them. Thus, our world, instead of being a single spherically symmetric expanding balloon, becomes a huge fractal, an inflationary multiverse consisting of many different pocket universes with different properties. This provided a simple interpretation of the cosmological anthropic principle (Linde, 2005).

Linde & Vanchurin (2010) investigated the problem: How many universes are in the multiverse? Authors argued that the evolution and properties of all important macroscopic features of the universe can be traced back to the two main ingredients:

(1) the properties of our vacuum state, represented by one of the many vacua in the landscape;
(2) the properties of the slow-roll inflation and the large-scale perturbations of metric and physical fields produced at that stage.

The goal was to understand how many different locally Friedmann universes constitute the multiverse, which, as a whole, looks like a very inhomogeneous and anisotropic non-Friedmann eternally growing fractal. Authors found that the total number of such universes may exceed $10^{10^{10^7}}$.

RHEA: Wait, are we now drinking in the office?
  [*Prof. Callistus pours champagne into three flutes.*]
ARIEL: During office hours?
CALLISTUS: Right. Only behind closed doors. My office is a safe space, so nobody will know.
ARIEL: Wow, Professor, I'm not the only one learning here. Pour me a glass.
RHEA: Although the initial effects of alcohol in low doses are relaxation and exuberance that quickly gives way to... whatever, we do have a reason to celebrate. [*Rhea smiles.*]
CALLISTUS: We have many reasons.
ARIEL: The semester ended.
CALLISTUS: Finis coronat opus.[1] And I heard somebody passed their comprehensive exam with flying colours!
RHEA: Yay! And they also got an A+ from a tough professor who never gives an A+.
CALLISTUS: There is a first for everything. To Ariel. Looks like the long arch of our universe bends towards Dr Ariel, away from MCMC Ariel.
ARIEL: That's like one in $10^{10^{10^7}}$ universes... and both titles sound great!

---

[1]The ending crowns the work (Ovid).

CALLISTUS: Speaking of titles, here's to our new Tenured Associate Professor, Rhea!

RHEA: Thank you!

CALLISTUS: Don't thank me; the vote was unanimous! To Rhea!

RHEA: And last but not least... our JWST proposal is accepted!

CALLISTUS: Woohoo! So many good things never happen in real life. I'm either dreaming, or somebody is writing this story.

RHEA: Now, if only we could solve all the ΛCDM tensions!

CALLISTUS: All of them? Now, you're dreaming.

ARIEL: The H_0 tension, whereby the universe expands a bit faster than it should.

RHEA: The S_8 tension: fluctuations grow slightly slower than they should.

CALLISTUS: And my favourite: the ISW puzzle. The imprint of superstructures on the CMB is about 4–5× larger than it should be at low redshift and changes sign at redshifts above $z \gtrsim 1$.

ARIEL: Let's not forget the emerging evidence for dynamic dark energy.

CALLISTUS: Can we pinpoint where all these tensions come from?

ARIEL: The CMB constrains the Hubble constant and the level of fluctuations at high redshifts, $z \simeq 1000$. Then our theoretical model crunches through the expansion history of the universe to predict the present value of cosmological parameters.

RHEA: Unlike local measurements, which are more direct and less dependent on the details of the ΛCDM model.

CALLISTUS: The high redshift indirect measurements and direct measurements at low redshift should agree.

ARIEL: But they don't. However, the ISW tension doesn't need such a comparison.

CALLISTUS: Right. The ISW imprint depends on the derivative of the growth function at a particular redshift; it still hints at the same problem with the assumed expansion history as the other tensions.

RHEA: So, all we need to figure out is the correct expansion history, and all tensions are solved.

ARIEL: I'm a little bit lost as to why this is so important. $H_0 = 68$ or 72, should I care?

RHEA: These are 4–6σ tensions. That's the definition of big. Statistically speaking.

CALLISTUS: Assuming we fully understand calibration uncertainties.

RHEA: Pesky calibration problems. You can always explain away anything. But we are optimists, so calibration worries will not stop us from thinking about real solutions.

ARIEL: To new physics!

CALLISTUS: There are already perhaps too many solutions.[2] Just their basic classes are exhausting to recount.

---

[2] Abdalla *et al.* (2022), Valentino *et al.* (2021).

RHEA: None of them are perfect. The infamous Hubble Olympics, open to all theorists working on early dark energy,

CALLISTUS: late dark energy,

ARIEL: models with extra degrees of freedom like sterile neutrinos or dark photons,

RHEA: decaying or interacting dark matter,

CALLISTUS: modified gravity,

RHEA: inflationary models,

ARIEL: modified recombination,

CALLISTUS: and several alternative proposals like emerging curvature, of which we looked at a few.

ARIEL: I think there are dozens of contenders in each of these categories. Why so many?

CALLISTUS: There are infinitely many ways to tweak the expansion and hence growth history.

RHEA: It boils down to the fact that we don't have the faintest clue what dark matter and dark energy are.

CALLISTUS: Right. Sometimes, I think these are just words to cover our ignorance.

ARIEL: Maybe we have to take a step back and think about it from a different perspective. Figure out what dark energy really is. And dark matter. I'm sure the puzzle pieces will lock together when we look at them from the right angle.

RHEA: To put that puzzle together, we'll need more data with special attention to calibration.

CALLISTUS: It will take an effort from all of us, observers taking and calibrating data meticulously and theorists trying to make sense of it. And funding.

RHEA: This is where our new JSWT data will be crucial.

ARIEL: Can I help reduce it?

CALLISTUS: I thought you'd never ask.

RHEA: We'll figure this out. Together.

ARIEL: So... is now a good time to do a PhD in Cosmology?

ALL: Cheers to that!

References

Aaronson, M., Huchra, J., Mould, J., Schechter, P. L., & Tully, R. B. 1982, The velocity field in the Local Supercluster, *ApJ*, 258, 64.

Abazajian, K. N. 2011, The Consistency of Fermi-LAT observations of the Galactic Center with a millisecond pulsar population in the central stellar cluster, *JCAP*, 03, 010.

Abazajian, K. N. 2017, Sterile neutrinos in cosmology, *Phys. Rept.*, 711–712, 1.

Abbott, B. P. *et al.* 2016, Observation of gravitational waves from a binary black hole merger, *Phys. Rev. Lett.*, 116, 061102.

Abbott, B. P. *et al.* 2017, Gravitational waves and gamma-rays from a binary neutron star merger: *GW170817 and GRB 170817A*, *ApJ*, 848, L13.

Abbott, L. F. & Sikivie, P. 1983, A cosmological bound on the invisible axion, *Phys. Lett. B*, 120, 133.

Abbott, T. M. C. *et al.* 2024, The dark energy survey: Cosmology results with $\sim$ 1500 new high-redshift type ia supernovae using the full 5 yr data set, *ApJ*, 973, L14.

Abdalla, E., Abellán, G. F., Aboubrahim, A., *et al.* 2022, Cosmology intertwined: A review of the particle physics, astrophysics, and cosmology associated with the cosmological tensions and anomalies, *J. High Energy Astrophys.*, 34, 49.

Abel, T., Hahn, O., & Kaehler, R. 2012, Tracing the dark matter sheet in phase space, *MNRAS*, 427, 61.

Abell, G. O. 1958, The distribution of rich clusters of galaxies, *ApJS*, 3, 211.

Abell, G. O., Corwin, Jr., H. G., & Olowin, R. P. 1989, A catalog of rich clusters of galaxies, *ApJS*, 70, 1.

Abraham, R. G., Tanvir, N. R., Santiago, B. X., *et al.* 1996, Galaxy morphology to I = 25 mag in the Hubble Deep Field, *MNRAS*, 279, L47.

Adame, A. G., Aguilar, J., Ahlen, S., *et al.* 2025, DESI 2024 VI: Cosmological constraints from the measurements of baryon acoustic oscillations, *J. Cosmology Astropart. Phys.*, 2025, 021.

Adhikari, S., Banerjee, A., Boddy, K. K., *et al.* 2022, Astrophysical tests of dark matter self-interactions, arXiv:2207.10638.

Akita, K. & Yamaguchi, M. 2020, A precision calculation of relic neutrino decoupling, *JCAP*, 08, 012.

Alam, S. *et al.* 2017, The clustering of galaxies in the completed SDSS-III Baryon Oscillation Spectroscopic Survey: cosmological analysis of the DR12 galaxy sample, *MNRAS*, 470, 2617.

Albrecht, A. & Magueijo, J. 1999, Time varying speed of light as a solution to cosmological puzzles, *Phys. Rev. D*, 59, 043516.

Alcock, C. *et al.* 1993, Possible gravitational microlensing of a star in the large magellanic cloud, *Nature*, 365, 621.

Alpher, R. A., Herman, R., & Gamow, G. A. 1948, Thermonuclear reactions in the expanding universe, *Physical Review*, 74, 1198.

Ambartsumian, V. A. 1961, Instability phenomena in systems of galaxies, *AJ*, 66, 536.

Amendola, L. & Tsujikawa, S. 2010, *Dark Energy: Theory and Observations* (Cambridge University Press).

André, P., Men'shchikov, A., Bontemps, S., *et al.* 2010, From filamentary clouds to prestellar cores to the stellar IMF: Initial highlights from the Herschel Gould Belt Survey, *A&A*, 518, L102.

Anglés-Alcázar, D., Faucher-Giguère, C.-A., Kereš, D., *et al.* 2017, The cosmic baryon cycle and galaxy mass assembly in the FIRE simulations, *MNRAS*, 470, 4698.

Angulo, R. E., Hahn, O., Ludlow, A. D., & Bonoli, S. 2017, Earth-mass haloes and the emergence of NFW density profiles, *MNRAS*, 471, 4687.

Aragón-Calvo, M. A., Jones, B. J. T., van de Weygaert, R., & van der Hulst, J. M. 2007, The multiscale morphology filter: Identifying and extracting spatial patterns in the galaxy distribution, *A&A*, 474, 315.

Araya-Melo, P. A., Reisenegger, A., Meza, A., *et al.* 2009, Future evolution of bound superclusters in an accelerating Universe, *MNRAS*, 399, 97.

Arnalte-Mur, P., Labatie, A., Clerc, N., *et al.* 2012, Wavelet analysis of baryon acoustic structures in the galaxy distribution, *A&A*, 542, A34.

Arnold, V. I., Shandarin, S. F., & Zeldovich, I. B. 1982, The large scale structure of the universe I. General properties. One-and two-dimensional models, *Geophys. Astrophys. Fluid Dyn.*, 20, 111.

Arzoumanian, D., André, P., Könyves, V., *et al.* 2019, Characterizing the properties of nearby molecular filaments observed with Herschel, *A&A*, 621, A42.

ATLAS Collaboration. 2015, *ATLAS-CONF-2015-081*.

Baade, W. 1951, Galaxies – Present day problems, Publications of Michigan Observatory, 10, 7.

Baade, W. & Zwicky, F. 1934, Remarks on super-novae and cosmic rays, *Phys. Rev.*, 46, 76.

Babcock, H. W. 1939, The rotation of the Andromeda nebula, *Lick Observ. Bull.*, 19, 41.

Bacchini, C., Fraternali, F., Iorio, G., & Pezzulli, G. 2019, Volumetric star formation laws of disc galaxies, *A&A*, 622, A64.

Ballesteros-Paredes, J., André, P., Hennebelle, P., *et al.* 2020, From diffuse gas to dense molecular cloud cores, *Space Sci. Rev.*, 216, 76.

Barrow, J. D. & O'Toole, C. 2001, Spatial variations of fundamental constants, *MNRAS*, 322, 585.

Bastian, N. & Lardo, C. 2018, Multiple stellar populations in globular clusters, *ARA&A*, 56, 83.

Behroozi, P. S., Wechsler, R. H., & Conroy, C. 2013, The average star formation histories of galaxies in dark matter halos from $z = 0$-8, *ApJ*, 770, 57.

Bell, E. F., Monachesi, A., Harmsen, B., *et al.* 2017, Galaxies grow their bulges and black holes in diverse ways, *ApJ*, 837, L8.

Bellstedt, S., Robotham, A. S. G., Driver, S. P., *et al.* 2021, Galaxy and mass assembly (GAMA): The inferred mass-metallicity relation from $z = 0$ to 3.5 via forensic SED fitting, *MNRAS*, 503, 3309.

Benson, A. J. & Bower, R. 2011, Accretion shocks and cold filaments in galaxy formation, *MNRAS*, 410, 2653.

Benson, A. J., Cole, S., Frenk, C. S., Baugh, C. M., & Lacey, C. G. 2000, The nature of galaxy bias and clustering, *MNRAS*, 311, 793.

Bernardeau, F. 1994, Skewness and kurtosis in large-scale cosmic fields, *ApJ*, 433, 1.

Bernardeau, F., Colombi, S., Gaztañaga, E., & Scoccimarro, R. 2002, Large-scale structure of the Universe and cosmological perturbation theory, *Phys. Rep.*, 367, 1.

Bernardeau, F. & Kofman, L. 1995, Properties of the cosmological density distribution function, *ApJ*, 443, 479.

Bertin, E. & Clusel, M. 2006, Generalized extreme value statistics and sum of correlated variables, *J. Phys. A Math. Gen.*, 39, 7607.

Bertschinger, E. 1995, COSMICS: Cosmological initial conditions and microwave anisotropy codes, ArXiv:astro-ph/9506070.

Bethe, H. A. 1939, Energy production in stars, *Phys. Rev.*, 55, 434.

Bidaran, B., La Barbera, F., Pasquali, A., *et al.* 2022, On the accretion of a new group of galaxies on to Virgo - II. The effect of pre-processing on the stellar population content of dEs, *MNRAS*, 515, 4622.

Bidaran, B., La Barbera, F., Pasquali, A., *et al.* 2023, On the accretion of a new group of galaxies onto Virgo - III. The stellar population radial gradients of dEs, *MNRAS*, 525, 4329.

Bidaran, B., Pasquali, A., Lisker, T., *et al.* 2020, On the accretion of a new group of galaxies on to Virgo: I. Internal kinematics of nine in-falling dEs, *MNRAS*, 497, 1904.

Bidaran, B., Pérez, I., Sánchez-Menguiano, L., *et al.* 2025, The puzzle of isolated and quenched dwarf galaxies in cosmic voids, *A&A*, 693, L16.

Bignone, L. A., Helmi, A., & Tissera, P. B. 2019, A Gaia-Enceladus analog in the EAGLE simulation: Insights into the early evolution of the Milky Way, *ApJ*, 883, L5.

Binggeli, B., Tammann, G. A., & Sandage, A. 1987, Studies of the virgo cluster. VI. Morphological and kinematical structure of the virgo cluster, *AJ*, 94, 251.

Binney, J. & Tremaine, S. 2008, *Galactic Dynamics: Second Edition* (Princeton University Press).

Bird, S., Cholis, I., Muñoz, J. B., *et al.* 2016, Did LIGO detect dark matter?, *Phys. Rev. Lett.*, 116, 201301.

Birnboim, Y. & Dekel, A. 2003, Virial shocks in galactic haloes?, *MNRAS*, 345, 349.

Blanton, M. R. & Roweis, S. 2007, K-corrections and filter transformations in the ultraviolet, optical, and near-infrared, *AJ*, 133, 734.

Blomqvist, M. *et al.* 2019, Baryon acoustic oscillations from the cross-correlation of Lyα absorption and quasars in eBOSS DR14, *A&A*, 629, A86.

Blumenthal, G. R., Faber, S. M., Primack, J. R., & Rees, M. J. 1984, Formation of galaxies and large-scale structure with cold dark matter, *Nature*, 311, 517.

Blumenthal, G. R., Pagels, H., & Primack, J. R. 1982, Galaxy formation by dissipationless particles heavier than neutrinos, *Nature*, 299, 37.

Boizelle, B. D., Walsh, J. L., Barth, A. J., *et al.* 2021, Black hole mass measurements of radio galaxies NGC 315 and NGC 4261 Using ALMA CO Observations, *ApJ*, 908, 19.

Bond, J. R., Kofman, L., & Pogosyan, D. 1996, How filaments of galaxies are woven into the cosmic web, *Nature*, 380, 603.

Bond, J. R., Szalay, A. S., & Turner, M. S. 1982, Formation of galaxies in a gravitino-dominated universe, *Phys. Rev. Lett.*, 48, 1636.

Bondi, H. & Gold, T. 1948, The steady-state theory of the expanding universe, *MNRAS*, 108, 252.

Bosma, A. 1978, The distribution and kinematics of neutral hydrogen in spiral galaxies of various morphological types, PhD thesis, Groningen University.

Bournaud, F. & Combes, F. 2003, Formation of polar ring galaxies, *A&A*, 401, 817.

Boylan-Kolchin, M., Springel, V., White, S. D. M., Jenkins, A., & Lemson, G. 2009, Resolving cosmic structure formation with the Millennium-II Simulation, *MNRAS*, 398, 1150.

Bradač, M., Clowe, D., Gonzalez, A. H., *et al.* 2006, Strong and weak lensing united. III. Measuring the mass distribution of the merging galaxy cluster 1ES 0657-558, *ApJ*, 652, 937.

Broadhurst, T. J., Ellis, R. S., Koo, D. C., & Szalay, A. S. 1990, Large-scale distribution of galaxies at the Galactic poles, *Nature*, 343, 726.

Bromm, V. & Loeb, A. 2003, The formation of the first low-mass stars from gas with low carbon and oxygen abundances, *Nature*, 425, 812.

Brooks, A. M., Governato, F., Quinn, T., Brook, C. B., & Wadsley, J. 2009, The role of cold flows in the assembly of galaxy disks, *ApJ*, 694, 396.

Brüns, C., Kerp, J., Kalberla, P. M. W., & Mebold, U. 2000, The head-tail structure of high-velocity clouds. A survey of the northern sky, *A&A*, 357, 120.

Bruzual, G. & Charlot, S. 2003, Stellar population synthesis at the resolution of 2003, *MNRAS*, 344, 1000.

Buitrago, F., Trujillo, I., Conselice, C. J., *et al.* 2008, Size evolution of the most massive galaxies at $1.7 < z < 3$ from GOODS NICMOS survey imaging, *ApJ*, 687, L61.

Bulbul, E., Markevitch, M., Foster, A., *et al.* 2014, Detection of an unidentified emission line in the stacked X-ray spectrum of galaxy clusters, *ApJ*, 789, 13.

Bullock, J. S. & Boylan-Kolchin, M. 2017, Small-scale challenges to the ΛCDM paradigm, *ARA&A*, 55, 343.

Bunn, E. F. & Hogg, D. W. 2009, The kinematic origin of the cosmological redshift, *Am. J. Phys.*, 77, 688.

Burbidge, E. M., Burbidge, G. R., Fowler, W. A., & Hoyle, F. 1957, Synthesis of the elements in stars, *Rev. Mod. Phys.*, 29, 547.

Burkert, A. 1995, The structure of dark matter halos in dwarf galaxies, *ApJ*, 447, L25.

Burrage, C. & Sakstein, J. 2018, Tests of chameleon gravity, *Living Rev. Relativ.*, 21, 1.

Burrows, A., Radice, D., Vartanyan, D., *et al.* 2020, The overarching framework of core-collapse supernova explosions as revealed by 3D FORNAX simulations, *MNRAS*, 491, 2715.

Burrows, A. & Vartanyan, D. 2021, Core-collapse supernova explosion theory, *Nature*, 589, 29.

Cappellari, M., Emsellem, E., Krajnović, D., *et al.* 2011a, The ATLAS^{3D} project - I. A volume-limited sample of 260 nearby early-type galaxies: Science goals and selection criteria, *MNRAS*, 413, 813.

Cappellari, M., Emsellem, E., Krajnović, D., *et al.* 2011b, The ATLAS^{3D} project - VII. A new look at the morphology of nearby galaxies: The kinematic morphology-density relation, *MNRAS*, 416, 1680.

Cappellari, M., McDermid, R. M., Alatalo, K., *et al.* 2013, The ATLAS^{3D} project - XX. Mass-size and mass-σ distributions of early-type galaxies: Bulge fraction drives kinematics, mass-to-light ratio, molecular gas fraction and stellar initial mass function, *MNRAS*, 432, 1862.

Carr, B. & Kuhnel, F. 2020, Primordial black holes as dark matter: Recent developments, *Ann. Rev. Nucl. Part. Sci.*, 70, 355.

Carroll, S. M., Press, W. H., & Turner, E. L. 1992, The cosmological constant, *ARA&A*, 30, 499.

Carron, J. & Szapudi, I. 2013, Optimal non-linear transformations for large-scale structure statistics, *MNRAS*, 434, 2961.

Cautun, M., van de Weygaert, R., Jones, B. J. T., & Frenk, C. S. 2014, Evolution of the cosmic web, *MNRAS*, 441, 2923.

Cen, R. 2003, The Universe was reionized twice, *ApJ*, 591, 12.

Cen, R. & Chisari, N. E. 2011, Star formation feedback and metal-enrichment history of the intergalactic medium, *ApJ*, 731, 11.

Cen, R. & Ostriker, J. P. 1992, Galaxy formation and physical bias, *ApJ*, 399, L113.

Centrella, J. & Melott, A. L. 1983, Three-dimensional simulation of large-scale structure in the universe, *Nature*, 305, 196.

Chabrier, G. 2005, The initial mass function: From Salpeter 1955 to 2005, in *Astrophysics and Space Science Library*, Vol. 327, *The Initial Mass Function 50 Years Later*, ed. E. Corbelli, F. Palla, & H. Zinnecker, 41.

Chand, H., Srianand, R., Petitjean, P., & Aracil, B. 2004, Probing the cosmological variation of the fine-structure constant: Results based on VLT-UVES sample, *A&A*, 417, 853.

Chandrasekhar, S. 1931, The maximum mass of ideal white dwarfs, *ApJ*, 74, 81.

Chantavat, T., Chongchitnan, S., & Silk, J. 2023, The most massive Population III stars, *MNRAS*, 522, 3256.

Chapline, G. F. 1975, Cosmological effects of primordial black holes, *Nature*, 253, 251.

Chen, T.-C., Lin, Y.-T., Schive, H.-Y., *et al.* 2024, A systematic search of distant superclusters with the subaru hyper suprime-cam survey, *ApJ*, 975, 200.

Cheng, H.-C., Feng, J. L., & Matchev, K. T. 2002, Kaluza-Klein dark matter, *Phys. Rev. Lett.*, 89, 211301.

Chernin, A. D. 1981, The rest mass of primordial neutrinos, and gravitational instability in the hot universe, *AZh*, 58, 25.

Chon, G., Böhringer, H., & Zaroubi, S. 2015, On the definition of superclusters, *A&A*, 575, L14.

Christopher, D. & Andrew, J. S. 2016. Information content of the non-linear power spectrum: The effect of boat-coupling to the large scales, *MNRAS*, 371(3), 1205–1215.

Clark, P. C., Glover, S. C. O., Klessen, R. S., & Bromm, V. 2011a, Gravitational fragmentation in turbulent primordial gas and the initial mass function of population III stars, *ApJ*, 727, 110.

Clark, P. C., Glover, S. C. O., Smith, R. J., *et al.* 2011b, The formation and fragmentation of disks around primordial protostars, *Science*, 331, 1040.

Clifton, T., Ferreira, P. G., Padilla, A., & Skordis, C. 2012, Modified gravity and cosmology, *Phys. Rep.*, 513, 1.

Cline, J. M. 2006, Baryogenesis, in *Les Houches Summer School - Session 86: Particle Physics and Cosmology: The Fabric of Spacetime.* eprint arXiv:hep-ph/0609145.

Clowe, D., Bradač, M., Gonzalez, A. H., *et al.* 2006a, A direct empirical proof of the existence of dark matter, *ApJ*, 648, L109.

Clowe, D., Gonzalez, A., & Markevitch, M. 2004, Weak-lensing mass reconstruction of the interacting cluster 1E 0657-558: Direct evidence for the existence of dark matter, *ApJ*, 604, 596.

Clowe, D., Schneider, P., Aragón-Salamanca, A., *et al.* 2006b, Weak lensing mass reconstructions of the ESO distant cluster survey, *A&A*, 451, 395.

CMS Collaboration. 2015, CMS-PAS-EXO-15-004.

Codis, S., Pogosyan, D., & Pichon, C. 2018, On the connectivity of the cosmic web: Theory and implications for cosmology and galaxy formation, *MNRAS*, 479, 973.

Coles, P. & Jones, B. 1991, A lognormal model for the cosmological mass distribution, *MNRAS*, 248, 1.

Connolly, A. J., Scranton, R., Johnston, D., *et al.* 2002, The angular correlation function of galaxies from early sloan digital sky survey data, *ApJ*, 579, 42.

Conselice, C. J. 2003, The relationship between stellar light distributions of galaxies and their formation histories, *ApJS*, 147, 1.

Conselice, C. J. 2014, The evolution of galaxy structure over cosmic time, *ARA&A*, 52, 291.

Conselice, C. J., Mortlock, A., Bluck, A. F. L., Grützbauch, R., & Duncan, K. 2013, Gas accretion as a dominant formation mode in massive galaxies from the GOODS NICMOS Survey, *MNRAS*, 430, 1051.

Conselice, C. J., Rajgor, S., & Myers, R. 2008, The structures of distant galaxies - I. Galaxy structures and the merger rate to z ~3 in the Hubble Ultra-Deep Field, *MNRAS*, 386, 909.

Croton, D. J. 2013, Damn you, little h! (Or, Real-world applications of the Hubble constant using observed and simulated data), *PASA*, 30, e052.

Croton, D. J., Springel, V., White, S. D. M., *et al.* 2006, The many lives of active galactic nuclei: Cooling flows, black holes and the luminosities and colours of galaxies, *MNRAS*, 365, 11.

Cruz, M., Martínez-González, E., Vielva, P., & Cayón, L. 2005, Detection of a non-Gaussian spot in WMAP, *MNRAS*, 356, 29.

Cyburt, R. H., Fields, B. D., & Olive, K. A. 2003, Primordial nucleosynthesis in light of WMAP, *Phys. Lett. B*, 567, 227.

Davis, B. L., Graham, A. W., & Cameron, E. 2019a, Black hole mass scaling relations for spiral galaxies. I. M_{BH}-$M_{*,sph}$, *ApJ*, 873, 85.

Davis, B. L., Graham, A. W., & Combes, F. 2019b, A consistent set of empirical scaling relations for spiral galaxies: The (v_{max}, M_{oM})-(σ_0, M_{BH}, ϕ) relations, *ApJ*, 877, 64.

Davis, M., Efstathiou, G., Frenk, C. S., & White, S. D. M. 1985, The evolution of large-scale structure in a universe dominated by cold dark matter, *ApJ*, 292, 371.

Davis, M., Geller, M. J., & Huchra, J. 1978, The local mean mass density of the universe: New methods for studying galaxy clustering, *ApJ*, 221, 1.

Davis, M. & Peebles, P. J. E. 1983, A survey of galaxy redshifts. V. The two-point position and velocity correlations, *ApJ*, 267, 465.

de Bernardis, P. *et al.* 2000, A flat Universe from high resolution maps of the cosmic microwave background radiation, *Nature*, 404, 955.

De Bianchi, S. & Szapudi, I. 2025, Achronotopic interpretation of quantum mechanics: Quantum objects and their measurement in emergent space–time scenarios, *Found. Phys.*, 55, 4.

De Felice, A. & Tsujikawa, S. 2010, $f(R)$ *Theor.*, *Living Rev. Relativ.*, 13, 3.

de Lapparent, V., Geller, M. J., & Huchra, J. P. 1986, A slice of the universe, *ApJ*, 302, L1.

de Sainte Agathe, V. *et al.* 2019, Baryon acoustic oscillations at $z = 2.34$ from the correlations of Lyα absorption in eBOSS DR14, *A&A*, 629, A85.

de Sitter, W. 1917, Einstein's theory of gravitation and its astronomical consequences. Third paper, *MNRAS*, 78, 3.

de Swart, J.-G. 2024, Five decades of missing matter, *Phys. Today*, 77, 24.

de Vaucouleurs, G. 1948, Recherches sur les nebuleuses extragalactiques, *Annales d'Astrophysique*, 11, 247.

de Vaucouleurs, G. 1953, Evidence for a local super-galaxy, *AJ*, 58, 30.

de Vaucouleurs, G. 1958, The Local supercluster of galaxies, *Nature*, 182, 1478.

de Vaucouleurs, G. 1970, The case for a hierarchical cosmology, *Science*, 167, 1203.

de Vaucouleurs, G. 1978, The extragalactic distance scale. IV - Distances of nearest groups and field galaxies from secondary indicators, *ApJ*, 224, 710.

de Vaucouleurs, G. 1982, The extragalactic distance scale and the Hubble constant, *The Observatory*, 102, 178.

de Vaucouleurs, G. & Corwin, H. G. J. 1985, S Andromedae 1885: A centennial review, *ApJ*, 295, 287.

de Zeeuw, P. T., Bureau, M., Emsellem, E., *et al.* 2002, The SAURON project – II. Sample and early results, *MNRAS*, 329, 513.

Dekel, A. 1998, Galaxy biasing: Nonlinear, stochastic and measurable, in wide field surveys in cosmology, ed. S. Colombi, Y. Mellier, & B. Raban, Vol. 14, 47.

Dekel, A., Birnboim, Y., Engel, G., *et al.* 2009, Cold streams in early massive hot haloes as the main mode of galaxy formation, *Nature*, 457, 451.

Dekel, A. & Silk, J. 1986, The origin of dwarf galaxies, cold dark matter, and biased galaxy formation, *ApJ*, 303, 39.

DESI Collaboration, Adame, A. G., Aguilar, J., *et al.* 2025a, DESI 2024 VII: Cosmological constraints from the full-shape modeling of clustering measurements, *JCAP*, 07, 28.

DESI Collaboration, Karim, M. A., Aguilar, J., *et al.* 2025b, DESI DR2 results II: Measurements of baryon acoustic oscillations and cosmological constraints, arXiv e-prints, arXiv:2503.14738.

Desjacques, V., Jeong, D., & Schmidt, F. 2018, Large-scale galaxy bias, *Phys. Rep.*, 733, 1.

Di Matteo, P., Haywood, M., Lehnert, M. D., *et al.* 2019, The Milky Way has no in-situ halo other than the heated thick disc. Composition of the stellar halo and age-dating the last significant merger with Gaia DR2 and APOGEE, *A&A*, 632, A4.

Di Matteo, T., Springel, V., & Hernquist, L. 2005, Energy input from quasars regulates the growth and activity of black holes and their host galaxies, *Nature*, 433, 604.

Di Valentino, E., Levi Said, J., Riess, A. *et al.* 2025. The cosmoverse white paper: Addressing observational tensions in cosmology with systematics and fundamental physics. arXiv e-prints, arXiv:2504.01669.

Dine, M. & Fischler, W. 1983, The not so harmless axion, *Phys. Lett. B*, 120, 137.

Diwanji, P., Walker, S. A., & Mirakhor, M. S. 2024, A rare, strong shock front in the merging cluster SPT-CLJ 2031-4037, *ApJ*, 969, 115.

Dodelson, S. & Schmidt, F. 2020, *Modern Cosmology* (Academic Press).

Dodelson, S. & Widrow, L. M. 1994, Sterile-neutrinos as dark matter, *Phys. Rev. Lett.*, 72, 17.

Dolag, K., Sorce, J. G., Pilipenko, S., *et al.* 2023, Simulating the LOcal Web (SLOW). I. Anomalies in the local density field, *A&A*, 677, A169.

Dolgov, A. D. & Hansen, S. H. 2002, Massive sterile neutrinos as warm dark matter, *Astropart. Phys.*, 16, 339.

Doroshkevich, A. G., Kotok, E. V., Poliudov, A. N., *et al.* 1980, Two-dimensional simulation of the gravitational system dynamics and formation of the large-scale structure of the universe, *MNRAS*, 192, 321.

Doroshkevich, A. G., Shandarin, S. F., & Zeldovich, I. B. 1982, Three-dimensional structure of the universe and regions devoid of galaxies, *Comments Astrophysics*, 9, 265.

Doroshkevich, A. G., Sunyaev, R. A., & Zeldovich, I. B. 1974, The formation of galaxies in Friedmannian universes, in IAU Symposium, Vol. 63, Confrontation of Cosmological Theories with Observational Data, ed. M. S. Longair, 213.

Doumler, T., Courtois, H., Gottlöber, S., & Hoffman, Y. 2013, Reconstructing cosmological initial conditions from galaxy peculiar velocities – II. The effect of observational errors, *MNRAS*, 430, 902.

Duan, Q., Conselice, C. J., Li, Q., *et al.* 2025, Galaxy mergers in the epoch of reionization I: A JWST study of pair fractions, merger rates, and stellar mass accretion rates at $z = 4.5 - 11.5$, *MNRAS*, 540, 774.

Dubois, Y., Peirani, S., Pichon, C., *et al.* 2016, The HORIZON-AGN simulation: Morphological diversity of galaxies promoted by AGN feedback, *MNRAS*, 463, 3948.

Dubois, Y., Pichon, C., Haehnelt, M., *et al.* 2012, Feeding compact bulges and supermassive black holes with low angular momentum cosmic gas at high redshift, *MNRAS*, 423, 3616.

Dubois, Y., Pichon, C., Welker, C., *et al.* 2014, Dancing in the dark: Galactic properties trace spin swings along the cosmic web, *MNRAS*, 444, 1453.

Dünner, R., Araya, P. A., Meza, A., & Reisenegger, A. 2006, The limits of bound structures in the accelerating universe, *MNRAS*, 366, 803.

Dupuy, A. & Courtois, H. M. 2023, Dynamic cosmography of the local universe: Laniakea and five more watershed superclusters, *A&A*, 678, A176.

Dupuy, A., Courtois, H. M., Dupont, F., *et al.* 2019, Partitioning the Universe into gravitational basins using the cosmic velocity field, *MNRAS*, 489, L1.

Dupuy, A., Courtois, H. M., Libeskind, N. I., & Guinet, D. 2020, Segmenting the universe into dynamically coherent basins, *MNRAS*, 493, 3513.

Eddington, A. S. 1914, *Stellar Movement and the Structure of the Universe* (London: McMillan and Co.).

Eddington, A. S. 1924, On the relation between the masses and luminosities of the stars, *MNRAS*, 84, 308.

Eddington, A. S. 1926, *The Internal Constitution of the Stars* (Cambridge University Press).

Eggen, O. J., Lynden-Bell, D., & Sandage, A. R. 1962, Evidence from the motions of old stars that the Galaxy collapsed, *ApJ*, 136, 748.

Einasto, J. 1965, *On the construction of a composite model for the Galaxy and on the determination of the system of Galactic parameters, Trudy Astrophys. Inst. Alma-Ata* (Tartu Astr. Obs. Teated 17), 5, 87.

Einasto, J. 1969, On galactic descriptive functions, *Astron. Nachr.*, 291, 97.

Einasto, J. 1972, Structure and evolution of regular galaxies, PhD thesis, *Tartu Observatory*.

Einasto, J. 2024, *Dark Matter and Cosmic Web Story*, 2nd edn. (Singapore: World Scientific).

Einasto, J., Einasto, M., Gottloeber, S., *et al.* 1997a, A 120 Mpc periodicity in the three-dimensional distribution of galaxy superclusters, *Nature*, 385, 139.

Einasto, J., Einasto, M., Tago, E., *et al.* 1999, Steps toward the power spectrum of matter. II. The biasing correction with sigma_8 normalization, *ApJ*, 519, 456.

Einasto, J., Hütsi, G., & Einasto, M. 2021a, Correlation functions in 2D and 3D as descriptors of the cosmic web, *A&A*, 652, A152.

Einasto, J., Hütsi, G., Einasto, M., *et al.* 2003, Clusters and superclusters in the sloan digital sky survey, *A&A*, 405, 425.

Einasto, J., Hütsi, G., Kuutma, T., & Einasto, M. 2020a, Correlation function: Biasing and fractal properties of the cosmic web, *A&A*, 640, A47.

Einasto, J., Hütsi, G., Liivamägi, L. J., *et al.* 2023a, Evolution of matter and galaxy clustering in cosmological hydrodynamical simulations, *MNRAS*, 523, 4693.

Einasto, J., Hütsi, G., Suhhonenko, I., Liivamägi, L. J., & Einasto, M. 2021b, Evolution of superclusters and supercluster cocoons in various cosmologies, *A&A*, 647, A17.

Einasto, J., Jõeveer, M., Kaasik, A., & Vennik, J. 1976a, The dynamics of aggregates of galaxies as related to their main galaxies, *A&A*, 53, 35.

Einasto, J., Jõeveer, M., Kaasik, A., & Vennik, J. 1976b, The missing mass around galaxies, in *Stars and Galaxies from Observational Points of View*, ed. E. K. Kharadze, 431.

Einasto, J., Jõeveer, M., & Saar, E. 1980a, Structure of superclusters and supercluster formation, *MNRAS*, 193, 353.

Einasto, J., Jõeveer, M., & Saar, E. 1980b, Superclusters and galaxy formation, *Nature*, 283, 47.

Einasto, J., Kaasik, A., & Saar, E. 1974a, Dynamic evidence on massive coronas of galaxies, *Nature*, 250, 309.

Einasto, J., Klypin, A., Hütsi, G., Liivamägi, L.-J., & Einasto, M. 2021c, Evolution of skewness and kurtosis of cosmic density fields, *A&A*, 652, A94.

Einasto, J., Klypin, A., & Shandarin, S. 1983, Structure of neighboring superclusters - A quantitative analysis, in *IAU Symposium*, Vol. 104, *Early Evolution of the Universe and its Present Structure*, ed. G. O. Abell & G. Chincarini, 265.

Einasto, J., Liivamägi, L. J., & Einasto, M. 2023b, The time evolution of bias, *MNRAS*, 518, 2164.

Einasto, J., Liivamägi, L. J., Suhhonenko, I., & Einasto, M. 2019a, The biasing phenomenon, *A&A*, 630, A62.

Einasto, J. & Miller, R. H. 1983, Neighboring superclusters and their environs, in *IAU Symposium*, Vol. 104, *Early Evolution of the Universe and Its Present Structure*, ed. G. O. Abell & G. Chincarini, 405.

Einasto, J. & Saar, E. 1987, Spatial distribution of galaxies – Biased galaxy formation, supercluster-void topology, and isolated galaxies, in *IAU Symposium*, Vol. 124, *Observational Cosmology*, ed. A. Hewitt, G. Burbidge, & L. Z. Fang, 349.

Einasto, J., Saar, E., Einasto, M., Freudling, W., & Gramann, M. 1994a, The fraction of matter in voids, *ApJ*, 429, 465.

Einasto, J., Saar, E., Kaasik, A., & Chernin, A. D. 1974b, Missing mass around galaxies – Morphological evidence, *Nature*, 252, 111.

Einasto, J., Saar, E., & Klypin, A. A. 1986, Structure of superclusters and supercluster formation. V – Spatial correlation and voids, *MNRAS*, 219, 457.

Einasto, J., Suhhonenko, I., Liivamägi, L. J., & Einasto, M. 2018, Extended percolation analysis of the cosmic web, *A&A*, 616, A141.

Einasto, J., Suhhonenko, I., Liivamägi, L. J., & Einasto, M. 2019b, Evolution of superclusters in the cosmic web, *A&A*, 623, A97.

Einasto, M., Deshev, B., Tenjes, P., *et al.* 2020b, Multiscale cosmic web detachments, connectivity, and preprocessing in the supercluster SCl A2142 cocoon, *A&A*, 641, A172.

Einasto, M., Einasto, J., Tago, E., Dalton, G. B., & Andernach, H. 1994b, The structure of the universe traced by rich clusters of galaxies, *MNRAS*, 269, 301.

Einasto, M., Einasto, J., Tago, E., Müller, V., & Andernach, H. 2001, Optical and X-Ray clusters as tracers of the supercluster-void network. I. Superclusters of abell and X-Ray clusters, *AJ*, 122, 2222.

Einasto, M., Einasto, J., Tenjes, P., *et al.* 2024, Galaxy groups and clusters and their brightest galaxies within the cosmic web, *A&A*, 681, A91.

Einasto, M., Heinämäki, P., Liivamägi, L. J., *et al.* 2016, Shell-like structures in our cosmic neighbourhood, *A&A*, 587, A116.

Einasto, M., Kipper, R., Tenjes, P., *et al.* 2022a, Death at watersheds: Galaxy quenching in low-density environments, *A&A*, 668, A69.

Einasto, M., Kipper, R., Tenjes, P., *et al.* 2021d, The Corona Borealis supercluster: Connectivity, collapse, and evolution, *A&A*, 649, A51.

Einasto, M., Tago, E., Jaaniste, J., Einasto, J., & Andernach, H. 1997b, The supercluster-void network I. The supercluster catalogue and large-scale distribution, *A&AS*, 123, 119.

Einasto, M., Tenjes, P., Gramann, M., *et al.* 2022b, The evolution of high-density cores of the BOSS Great Wall superclusters, *A&A*, 666, A52.

Einstein, A. 1916, Die Grundlage der allgemeinen Relativitätstheorie, *Annalen der Physik*, 354, 769.

Einstein, A. & de Sitter, W. 1932, On the relation between the expansion and the mean density of the universe, *Proc. Nat. Acad. Sci.*, 18, 213.

Eisenhauer, F., Genzel, R., Alexander, T., *et al.* 2005, SINFONI in the galactic center: Young stars and infrared flares in the central light-month, *ApJ*, 628, 246.

Elia, D., Molinari, S., Schisano, E., *et al.* 2022, The star formation rate of the Milky Way as seen by Herschel, *ApJ*, 941, 162.

Ellis, J. R., Hagelin, J. S., Nanopoulos, D. V., Olive, K. A., & Srednicki, M. 1984, Supersymmetric relics from the big bang, *Nucl. Phys. B*, 238, 453.

Ellis, R. S. 2022, *When Galaxies Were Born. The Quest for Cosmic Dawn* (Cambridge University Press).

Endre Szigeti, B., Szapudi, I., Ferenc Barna, I., & Gábor Barnaföldi, G. 2025, Can rotation solve the Hubble puzzle?, arXiv e-prints, arXiv:2503.13525.

Escrivà, A., Kuhnel, F., & Tada, Y. 2022, Primordial black holes, arXiv:2211.05767.

Event Horizon Telescope Collaboration, Akiyama, K., Alberdi, A., *et al.* 2022, First sagittarius A* event horizon telescope results. I. The shadow of the supermassive black hole in the center of the Milky Way, *ApJ*, 930, L12.

Event Horizon Telescope Collaboration, Akiyama, K., Alberdi, A., *et al.* 2019, First M87 event horizon telescope results. IV. Imaging the central supermassive black hole, *ApJ*, 875, L4.

Faber, S. M. & Jackson, R. E. 1976, Velocity dispersions and mass-to-light ratios for elliptical galaxies, *ApJ*, 204, 668.

Falck, B. L., Neyrinck, M. C., & Szalay, A. S. 2012, ORIGAMI: Delineating halos using phase-space folds, *ApJ*, 754, 126.

Fall, S. M. & Efstathiou, G. 1980, Formation and rotation of disc galaxies with haloes., *MNRAS*, 193, 189.

Faucher-Giguère, C.-A. & Oh, S. P. 2023, Key physical processes in the circumgalactic medium, *ARA&A*, 61, 131.

Feldbrugge, J., van de Weygaert, R., Hidding, J., & Feldbrugge, J. 2018, Caustic skeleton & cosmic web, *J. Cosmology Astropart. Phys.*, 2018, 027.

Feldbrugge, J., van Engelen, M., van de Weygaert, R., Pranav, P., & Vegter, G. 2019, Stochastic homology of gaussian vs. non-gaussian random fields: graphs towards betti numbers and persistence diagrams, *J. Cosmology Astropart. Phys.*, 2019, 052.

Feldbrugge, J., Yan, Y., & van de Weygaert, R. 2023, Statistics of tidal and deformation eigenvalue fields in the primordial Gaussian matter distribution: The two-dimensional case, *MNRAS*, 526, 5031.

Ferreras, I. 2019, *Fundamentals of Galaxy Dynamics, Formation and Evolution* (UCL Press, London).

Finelli, F., García-Bellido, J., Kovács, A., Paci, F., & Szapudi, I. 2016, Supervoids in the WISE-2MASS catalogue imprinting cold spots in the cosmic microwave background, *MNRAS*, 455, 1246.

Finkelstein, S. L. & Bagley, M. B. 2022, On the coevolution of the AGN and star-forming galaxy ultraviolet luminosity functions at $3 < z < 9$, *ApJ*, 938, 25.

Finkelstein, S. L., Leung, G. C. K., Bagley, M. B., *et al.* 2024, The complete CEERS early universe galaxy sample: A surprisingly slow evolution of the space density of bright galaxies at $z{\sim}8.5$–14.5, *ApJ*, 969, L2.

Fixsen, D. J. 2009, The temperature of the cosmic microwave background, *ApJ*, 707, 916.

Freedman, W. L. 2021, Measurements of the Hubble constant: Tensions in perspective, *ApJ*, 919, 16.

Freedman, W. L., Madore, B. F., Gibson, B. K., *et al.* 2001, Final results from the Hubble Space Telescope key project to measure the Hubble constant, *ApJ*, 553, 47.

Freedman, W. L., Madore, B. F., Jang, I. S., *et al.* 2025, Status report on the Chicago-Carnegie Hubble Program (CCHP): Three independent astrophysical determinations of the Hubble Constant using the James Webb Space Telescope, 2025, *ApJ*, 985, 203.

Freeman, K. C. 1970, On the disks of spiral and S0 galaxies, *ApJ*, 160, 811.

Freeman, P. E., Izbicki, R., Lee, A. B., *et al.* 2013, New image statistics for detecting disturbed galaxy morphologies at high redshift, *MNRAS*, 434, 282.

Friedmann, A. 1922, Über die Krümmung des Raumes, *Z. Phys.*, 10, 377.

Friedmann, A. 1924, Über die Möglichkeit einer Welt mit konstanter negativer Krümmung des Raumes, *Z. Phys.*, 21, 326.

Fukugita, M. & Peebles, P. J. E. 2004, The Cosmic energy inventory, *ApJ*, 616, 643.

Gaia Collaboration, Antoja, T., McMillan, P. J., *et al.* 2021, Gaia early data release 3. The galactic anticentre, *A&A*, 649, A8.

Gaia Collaboration, Katz, D., Antoja, T., *et al.* 2018, Gaia data release 2. Mapping the Milky Way disc kinematics, *A&A*, 616, A11.

Gamow, G. & Schoenberg, M. 1941, Neutrino theory of stellar collapse, *Phys. Rev.*, 59, 539.

Gao, L., Springel, V., & White, S. D. M. 2005a, The age dependence of halo clustering, *MNRAS*, 363, L66.

Gao, L., White, S. D. M., Jenkins, A., Frenk, C. S., & Springel, V. 2005b, Early structure in ΛCDM, *MNRAS*, 363, 379.

Gebhardt, K., Bender, R., Bower, G., *et al.* 2000, A relationship between nuclear black hole mass and galaxy velocity dispersion, *ApJ*, 539, L13.

Geller, M. J. & Huchra, J. P. 1989, Mapping the universe, *Science*, 246, 897.

Gentile, F., Casey, C. M., Akins, H. B., *et al.* 2024, Not-so-little red dots: Two massive and dusty starbursts at z ∼ 5–7 Pushing the limits of star formation discovered by JWST in the COSMOS-web survey, *ApJ*, 973, L2.

Genzel, R., Eisenhauer, F., & Gillessen, S. 2010, The galactic center massive black hole and nuclear star cluster, *Rev. Modern Physics*, 82, 3121.

Ghez, A. M., Salim, S., Weinberg, N. N., *et al.* 2008, Measuring distance and properties of the Milky Way's central supermassive black hole with stellar orbits, *ApJ*, 689, 1044.

Ghirardini, V., Bulbul, E., Artis, E., *et al.* 2024, The SRG/eROSITA all-sky survey: Cosmology constraints from cluster abundances in the western Galactic hemisphere, *A&A*, 689, A298.

Gialamas, I. D., Hütsi, G., Kannike, K., *et al.* 2025, Interpreting DESI 2024 BAO: late-time dynamical dark energy or a local effect?, 2025, *Phys. Rev. D*, 111, d3540.

Gnat, O. & Sternberg, A. 2007, Time-dependent ionization in radiatively cooling gas, *ApJS*, 168, 213.

Gnedin, N. Y. & Madau, P. 2022, Modeling cosmic reionization, *Living Rev. in Comput. Astrophys.*, 8, 3.

Gott, J. R., Miller, J., Thuan, T. X., *et al.* 1989, The topology of large-scale structure. III. Analysis of observations, *ApJ*, 340, 625.

Gott, J. R., Weinberg, D. H., & Melott, A. L. 1987, A quantitative approach to the topology of large-scale structure, *ApJ*, 319, 1.

Gott, J. R. 2016, *The Cosmic Web: Mysterious Architecture of the Universe* (Princeton University Press).

Gott, J. R., Dickinson, M., & Melott, A. L. 1986, The sponge-like topology of large-scale structure in the universe, *ApJ*, 306, 341.

Graham, A. W. 2016, Galaxy bulges and their massive black holes: A review, in *Astrophysics and Space Science Library*, Vol. 418, *Galactic Bulges*, ed. E. Laurikainen, R. Peletier, & D. Gadotti, 263.

Gramann, M. & Einasto, J. 1992, The power spectrum in nearby superclusters, *MNRAS*, 254, 453.

Gramann, M., Einasto, M., Heinämäki, P., *et al.* 2015, Characteristic density contrasts in the evolution of superclusters. The case of A2142 supercluster, *A&A*, 581, A135.

Granett, B. R., Neyrinck, M. C., & Szapudi, I. 2008, An imprint of superstructures on the microwave background due to the integrated sachs-wolfe effect, *ApJ*, 683, L99.

Graziani, R., Courtois, H. M., Lavaux, G., *et al.* 2019, The peculiar velocity field up to z=0.05 by forward-modelling Cosmicflows-3 data, *MNRAS*, 488, 5438.

Green, A. M. & Kavanagh, B. J. 2021, Primordial Black Holes as a dark matter candidate, *J. Phys. G*, 48, 043001.

Greene, J. E., Labbe, I., Goulding, A. D., *et al.* 2024, UNCOVER spectroscopy confirms the surprising ubiquity of active galactic nuclei in red sources at $z > 5$, *ApJ*, 964, 39.

Gregory, S. A. & Thompson, L. A. 1978, The Coma/A1367 supercluster and its environs, *ApJ*, 222, 784.

Gregory, S. A., Thompson, L. A., & Tifft, W. G. 1981, The Perseus supercluster, *ApJ*, 243, 411.

Gunn, J. E. & Tinsley, B. M. 1975, An accelerating universe, *Nature*, 257, 454.

Guth, A. H. 1981, Inflationary universe: A possible solution to the horizon and flatness problems, *Phys. Rev. D*, 23, 347.

Hanany, S. *et al.* 2000, MAXIMA-1: A Measurement of the cosmic microwave background anisotropy on angular scales of 10 arcminutes to 5 degrees, *ApJ*, 545, L5.

Harikane, Y., Nakajima, K., Ouchi, M., *et al.* 2024, Pure spectroscopic constraints on UV luminosity functions and cosmic star formation history from 25 galaxies at $z_{spec} = 8.61$-13.20 confirmed with JWST/NIRSpec, *ApJ*, 960, 56.

Harris, W. E. 1991, Globular cluster systems in galaxies beyond the Local Group, *ARA&A*, 29, 543.

Harris, W. E. & van den Bergh, S. 1981, Globular clusters in galaxies beyond the Local Group, I. New cluster systems in selected northern ellipticals, *AJ*, 86, 1627.

Hartwig, E. 1885, S And, *Astron. Nachr.*, 112, 355.

Hartwig, T., Magg, M., Chen, L.-H., *et al.* 2022, Public release of A-SLOTH: Ancient stars and local observables by tracing halos, *ApJ*, 936, 45.

Haud, U. & Einasto, J. 1989, Galactic models with massive corona – Part two – Galaxy, *A&A*, 223, 95.

Hawking, S. 1971, Gravitationally collapsed objects of very low mass, *MNRAS*, 152, 75.

Haywood, M., Lehnert, M. D., Di Matteo, P., *et al.* 2016, When the Milky Way turned off the lights: APOGEE provides evidence of star formation quenching in our Galaxy, *A&A*, 589, A66.

Heckman, T. M. & Best, P. N. 2014, The coevolution of galaxies and supermassive black holes: Insights from surveys of the contemporary universe, *ARA&A*, 52, 589.

Heiles, C. & Troland, T. H. 2005, The millennium Arecibo 21 centimeter absorption-line survey. IV. Statistics of magnetic field, column density, and turbulence, *ApJ*, 624, 773.

Heisenberg, L. 2019, A systematic approach to generalisations of general relativity and their cosmological implications, *Phys. Rep.*, 796, 1.

Helmi, A. 2020, Streams, substructures, and the early history of the Milky Way, *ARA&A*, 58, 205.

Helmi, A., Babusiaux, C., Koppelman, H. H., *et al.* 2018, The merger that led to the formation of the Milky Way's inner stellar halo and thick disk, *Nature*, 563, 85.

Helmi, A. & White, S. D. M. 1999, Building up the stellar halo of the Galaxy, *MNRAS*, 307, 495.

Hennebelle, P. & Grudić, M. Y. 2024, The physical origin of the stellar initial mass function, *ARA&A*, 62, 63.

Hernández-Aguayo, C., Springel, V., Pakmor, R., *et al.* 2023, The MillenniumTNG Project: High-precision predictions for matter clustering and halo statistics, *MNRAS*, 524, 2556.

Hidding, J. 2010, PhD thesis, University of Groningen, Netherlands.

Hidding, J., Shandarin, S. F., & van de Weygaert, R. 2014, The Zel'dovich approximation: Key to understanding cosmic web complexity, *MNRAS*, 437, 3442.

Hockney, R. W. & Eastwood, J. W. 1981, *Computer Simulation Using Particles* (McGraw-Hill, New York).

Hoffman, Y., Pomarède, D., Tully, R. B., & Courtois, H. M. 2017, The dipole repeller, *Nature Astronomy*, 1, 0036.

Hoffman, Y. & Ribak, E. 1991, Constrained realizations of Gaussian fields: A simple algorithm, *ApJ*, 380, L5.

Hofstadter, D. R. 1999, *Gödel, Escher, Bach: An Eternal Golden Braid* (Basic Books. Inc.).

Hooper, D. & Goodenough, L. 2011, Dark matter annihilation in the galactic center as seen by the Fermi Gamma Ray Space Telescope, *Phys. Lett. B*, 697, 412.

Hopkins, P. F., Lauer, T. R., Cox, T. J., Hernquist, L., & Kormendy, J. 2009, Dissipation and extra light in galactic nuclei. III. "Core" Ellipticals and "Missing" Light, *ApJS*, 181, 486.

Hoyle, F. 1948, A new model for the expanding universe, *MNRAS*, 108, 372.

Hoyle, F. & Fowler, W. A. 1960, Nucleosynthesis in Supernovae, *ApJ*, 132, 565.

Hoyle, F. & Schwarzschild, M. 1955, On the evolution of type II stars, *ApJS*, 2, 1.

Hu, W., Barkana, R., & Gruzinov, A. 2000, Cold and fuzzy dark matter, *Phys. Rev. Lett.*, 85, 1158.

Hubble, E. 1929a, A relation between distance and radial velocity among extra-galactic nebulae, *Proc. Nat. Acad. Sci.*, 15, 168.

Hubble, E. 1934, The distribution of extra-galactic nebulae, *ApJ*, 79, 8.

Hubble, E. & Humason, M. L. 1931, The velocity-distance relation among extra-galactic nebulae, *ApJ*, 74, 43.

Hubble, E. P. 1925, NGC 6822, a remote stellar system, *ApJ*, 62, 409.

Hubble, E. P. 1926, A spiral nebula as a stellar system: Messier 33, *ApJ*, 63, 236.

Hubble, E. P. 1929b, A spiral nebula as a stellar system, Messier 31, *ApJ*, 69, 103.

Hui, L., Ostriker, J. P., Tremaine, S., & Witten, E. 2017, Ultralight scalars as cosmological dark matter, *Phys. Rev. D*, 95, 043541.

Hütsi, G., Raidal, M., Urrutia, J., Vaskonen, V., & Veermäe, H. 2023, Did JWST observe imprints of axion miniclusters or primordial black holes?, *Phys. Rev. D*, 107, 043502.

Hütsi, G., Raidal, M., Vaskonen, V., & Veermäe, H. 2021, Two populations of LIGO-Virgo black holes, *JCAP*, 03, 068.

Iben, Icko, J. 1991, Single and binary star evolution, *ApJS*, 76, 55.

Inayoshi, K., Visbal, E., & Haiman, Z. 2020, The assembly of the first massive black holes, *ARA&A*, 58, 27.

Iršič, V. *et al.* 2017, New constraints on the free-streaming of warm dark matter from intermediate and small scale Lyman-α forest data, *Phys. Rev. D*, 96, 023522.

Ivleva, A., Remus, R.-S., Valenzuela, L. M., & Dolag, K. 2024, Merge and strip: Dark matter-free dwarf galaxies in clusters can be formed by galaxy mergers, *A&A*, 687, A105.

Jõeveer, M. 1972, An attempt to estimate the Galactic mass density in the vicinity of the Sun, *Tartu Astr. Obs. Teated*, 37, 3.

Jõeveer, M. 1974, Ages of delta cephei stars and the galactic mass density near the Sun, *Tartu Astr. Obs. Teated*, 46, 35.

Jõeveer, M. & Einasto, J. 1978, Has the universe the cell structure, in *IAU Symposium*, Vol. 79, *Large Scale Structures in the Universe*, ed. M. S. Longair & J. Einasto, 241.

Jõeveer, M., Einasto, J., & Tago, E. 1977, The cell structure of the universe, *Tartu Astr. Obs. Preprint*, 3.

Jõeveer, M., Einasto, J., & Tago, E. 1978, Spatial distribution of galaxies and of clusters of galaxies in the southern galactic hemisphere, *MNRAS*, 185, 357.

Janka, H.-T., Melson, T., & Summa, A. 2016, Physics of core-collapse supernovae in three dimensions: A sneak preview, *Ann. Rev. Nucl. Part. Sci.*, 66, 341.

Jeans, J. H. 1922, The motions of stars in a Kapteyn universe, *MNRAS*, 82, 122.

Jedamzik, K. & Pogosian, L. 2020, Relieving the Hubble tension with primordial magnetic fields, *Phys. Rev. Lett.*, 125, 181302.

Joachimi, B. & Taylor, A. N. 2011, Forecasts of non-gaussian parameter spaces using Box-Cox transformations, *MNRAS*, 416, 1010.

Johnson, J. A. 2019, Populating the periodic table: Nucleosynthesis of the elements, *Science*, 363, 474.

Jungman, G., Kamionkowski, M., & Griest, K. 1996, Supersymmetric dark matter, *Phys. Rep.*, 267, 195.

Juszkiewicz, R., Bouchet, F. R., & Colombi, S. 1993, Skewness induced by gravity, *ApJ*, 412, L9.

Kahn, F. D. & Woltjer, L. 1959, Intergalactic matter and the Galaxy, *ApJ*, 130, 705.

Kaiser, N. 1984, On the spatial correlations of Abell clusters, *ApJ*, 284, L9.

Kaiser, N. 1987, Clustering in real space and in redshift space, *MNRAS*, 227, 1.

Kalirai, J. S., Beaton, R. L., Geha, M. C., *et al.* 2010, The *SPLASH* survey: Internal kinematics, chemical abundances, and masses of the andromeda I, II, III, VII, X, and XIV dwarf spheroidal galaxies, *ApJ*, 711, 671.

Kamionkowski, M. & Riess, A. G. 2023, The Hubble tension and early dark energy, *Ann. Rev. Nucl. Part. Sci.*, 73, 153.

Kannan, R., Garaldi, E., Smith, A., *et al.* 2022, Introducing the THESAN project: Radiation-magnetohydrodynamic simulations of the epoch of reionization, *MNRAS*, 511, 4005.

Kannan, R., Springel, V., Hernquist, L., *et al.* 2023, The MillenniumTNG project: the galaxy population at z $\geq$ 8, *MNRAS*, 524, 2594.

Karachentsev, I. D. 1966, The virial mass-luminosity ratio and the instability of different galactic systems, *Astrofizika*, 2, 81.

Karachentsev, I. D. & Nasonova, O. G. 2010, The observed infall of galaxies towards the virgo cluster, *MNRAS*, 405, 1075.

Kasen, D., Metzger, B., Barnes, J., Quataert, E., & Ramirez-Ruiz, E. 2017, Origin of the heavy elements in binary neutron-star mergers from a gravitational-wave event, *Nature*, 551, 80.

Kashibadze, O. G., Karachentsev, I. D., & Karachentseva, V. E. 2020, Structure and kinematics of the Virgo cluster of galaxies, *A&A*, 635, A135.

Kennicutt, Robert C., J. 1998, The global schmidt law in star-forming galaxies, *ApJ*, 498, 541.

Kereš, D., Katz, N., Weinberg, D. H., & Davé, R. 2005, How do galaxies get their gas?, *MNRAS*, 363, 2.

Kerr, F. J. & Lynden-Bell, D. 1986, Review of galactic constants, *MNRAS*, 221, 1023.

Kiang, T. 1967, On the clustering of rich clusters of galaxies, *MNRAS*, 135, 1.

Kleinmann, R. 2016, Similarities between basic mechanisms of cosmic and biologic systems, *Int. J. Phys. Sci.*, 11, 1.

Klessen, R. S. & Glover, S. C. O. 2023, The first stars: Formation, properties, and impact, *ARA&A*, 61, 65.

Klypin, A., Hoffman, Y., Kravtsov, A. V., & Gottlöber, S. 2003, Constrained simulations of the real universe: The Local supercluster, *ApJ*, 596, 19.

Klypin, A., Kravtsov, A. V., Valenzuela, O., & Prada, F. 1999, Where are the missing galactic satellites?, *ApJ*, 522, 82.

Klypin, A., Prada, F., Betancort-Rijo, J., & Albareti, F. D. 2018, Density distribution of the cosmological matter field, *MNRAS*, 481, 4588.

Klypin, A., Yepes, G., Gottlöber, S., Prada, F., & Heß, S. 2016, MultiDark simulations: The story of dark matter halo concentrations and density profiles, *MNRAS*, 457, 4340.

Klypin, A. A. & Shandarin, S. F. 1983, Three-dimensional numerical model of the formation of large-scale structure in the Universe, *MNRAS*, 204, 891.

Kobayashi, C., Mandel, I., Belczynski, K., *et al.* 2023, Can neutron star mergers alone explain the r-process enrichment of the Milky Way?, *ApJ*, 943, L12.

Kofman, L., Bertschinger, E., Gelb, J. M., Nusser, A., & Dekel, A. 1994a, Evolution of one-point distributions from gaussian initial fluctuations, *ApJ*, 420, 44.

Kofman, L., Linde, A., & Starobinsky, A. A. 1994b, Reheating after inflation, *Phys. Rev. Lett.*, 73, 3195.

Kofman, L., Pogosian, D., & Shandarin, S. 1990, Structure of the universe in the two-dimensional model of adhesion, *MNRAS*, 242, 200.

Kofman, L., Pogosyan, D., Shandarin, S. F., & Melott, A. L. 1992, Coherent structures in the universe and the adhesion model, *ApJ*, 393, 437.

Kofman, L. A. & Shandarin, S. F. 1988, Theory of adhesion for the large-scale structure of the universe, *Nature*, 334, 129.

Kofman, L. A. & Starobinskii, A. A. 1985, Effect of the cosmological constant on largescale anisotropies in the microwave background, *Sov. Astron. Lett.*, 11, 271.

Kofman, L. A., Starobinskii, A. A., Shandarin, S. F., & Einasto, J. E. 1986, Second Tartu cosmology seminar – Missing mass largescale structure microwave background, *Sov. Astron.*, 30, 729.

Kokorev, V., Caputi, K. I., Greene, J. E., *et al.* 2024, A census of photometrically selected little red dots at $4 < z < 9$ in JWST blank fields, *ApJ*, 968, 38.

Kolb, E. W. & Turner, M. S. 1990, *The Early Universe*, Vol. 69 (Westview Press).

Komatsu, E. 2022, New physics from the polarized light of the cosmic microwave background, *Nature Rev. Phys.*, 4, 452.

Komatsu, E., Dunkley, J., Nolta, M. R., *et al.* 2009, Five-year Wilkinson Microwave Anisotropy Probe observations: Cosmological interpretation, *ApJS*, 180, 330.

Komatsu, E., Smith, K. M., Dunkley, J., *et al.* 2011, Seven-year Wilkinson Microwave Anisotropy Probe (WMAP) observations: Cosmological interpretation, *ApJS*, 192, 18.

Kormendy, J. 2013, *Secular Evolution of Galaxies*, ed. J. Falcón-Barroso & J. H. Knapen (Cambridge University Press), 1.

Kormendy, J. & Bender, R. 2019, Structural analogs of the Milky Way galaxy: Stellar populations in the boxy bulges of NGC 4565 and NGC 5746, *ApJ*, 872, 106.

Kormendy, J., Bender, R., & Cornell, M. E. 2011, Supermassive black holes do not correlate with galaxy disks or pseudobulges, *Nature*, 469, 374.

Kormendy, J., Fisher, D. B., Cornell, M. E., & Bender, R. 2009, Structure and formation of elliptical and spheroidal galaxies, *ApJS*, 182, 216.

Kormendy, J. & Ho, L. C. 2013, Coevolution (Or Not) of supermassive black holes and host galaxies, *ARA&A*, 51, 511.

Kormendy, J. & Kennicutt, R. C., Jr. 2004, Secular evolution and the formation of pseudobulges in disk galaxies, *ARA&A*, 42, 603.

Kovács, A., Jeffrey, N., Gatti, M., *et al.* 2022, The DES view of the Eridanus supervoid and the CMB cold spot, *MNRAS*, 510, 216.

Kraljic, K., Pichon, C., Codis, S., *et al.* 2020, The impact of the connectivity of the cosmic web on the physical properties of galaxies at its nodes, *MNRAS*, 491, 4294.

Kroupa, P. 2002, The initial mass function of stars: Evidence for uniformity in variable systems, *Science*, 295, 82.

Kroupa, P. & Weidner, C. 2003, Galactic-field initial mass functions of massive stars, *ApJ*, 598, 1076.

Kruijssen, J. M. D. 2015, Globular clusters as the relics of regular star formation in 'normal' high-redshift galaxies, *MNRAS*, 454, 1658.

Kruijssen, J. M. D., Pfeffer, J. L., Chevance, M., *et al.* 2020, Kraken reveals itself - The merger history of the Milky Way reconstructed with the E-MOSAICS simulations, *MNRAS*, 498, 2472.

Kruijssen, J. M. D., Pfeffer, J. L., Crain, R. A., & Bastian, N. 2019a, The E-MOSAICS project: Tracing galaxy formation and assembly with the age-metallicity distribution of globular clusters, *MNRAS*, 486, 3134.

Kruijssen, J. M. D., Pfeffer, J. L., Reina-Campos, M., Crain, R. A., & Bastian, N. 2019b, The formation and assembly history of the Milky Way revealed by its globular cluster population, *MNRAS*, 486, 3180.

Kruijssen, J. M. D., Schruba, A., Chevance, M., *et al.* 2019c, Fast and inefficient star formation due to short-lived molecular clouds and rapid feedback, *Nature*, 569, 519.

Kuzmin, G. 1952, On the distribution of mass in the galaxy, *Tartu Astr. Obs. Publ.*, 32, 211.

Kuzmin, G. 1953, The third integral of motions of stars and the dynamics of the stationar galaxy. I, *Tartu Astr. Obs. Publ.*, 32, 332.

Kuzmin, G. G. 2022, Etudes on the dynamics of stellar systems, arXiv e-prints, arXiv:2201.04136.

Lahav, O., Bridle, S. L., Percival, W. J., *et al.* 2002, The 2dF galaxy redshift survey: The amplitudes of fluctuations in the 2dFGRS and the CMB, and implications for galaxy biasing, *MNRAS*, 333, 961.

Lemaître, G. 1927, Un Univers homogène de masse constante et de rayon croissant rendant compte de la vitesse radiale des nébuleuses extra-galactiques, *Ann. Soc. Sci. Bruxelles*, 47, 49.

Leroy, A. K., Walter, F., Brinks, E., *et al.* 2008, The star formation efficiency in nearby galaxies: Measuring where gas forms stars effectively, *AJ*, 136, 2782.

Levan, A. J., Gompertz, B. P., Salafia, O. S., *et al.* 2024, Heavy-element production in a compact object merger observed by JWST, *Nature*, 626, 737.

Libeskind, N. I., van de Weygaert, R., Cautun, M., *et al.* 2018, Tracing the cosmic web, *MNRAS*, 473, 1195.

Liddle, A. R. & Lyth, D. H. 2000, *Cosmological Inflation and Large-Scale Structure* (Cambridge University Press).

Liivamägi, L. J., Tempel, E., & Saar, E. 2012, SDSS DR7 superclusters. The catalogues, *A&A*, 539, A80.

Lin, T. 2019, Dark matter models and direct detection, *PoS*, 333, 009.

Lindblad, B. 1927, On the state of motion in the galactic system, *MNRAS*, 87, 553.

Linde, A. 2002a, Inflation, quantum cosmology and the anthropic principle, ArXiv:hep-th/0211048.

Linde, A. 2002b, Inflationary theory versus ekpyrotic/cyclic scenario, arXiv:astro-ph/0205259v1.

Linde, A. 2005, Particle physics and inflationary cosmology, arXiv e-prints. hep-th/0503203, hep.

Linde, A., Linde, D., & Mezhlumian, A. 1994, From the Big Bang theory to the theory of a stationary universe, *Phys. Rev. D*, 49, 1783.

Linde, A. & Vanchurin, V. 2010, How many universes are in the multiverse?, *Phys. Rev. D*, 81, 083525.

Linde, A. D. 1982, A new inflationary universe scenario: A possible solution of the horizon, flatness, homogeneity, isotropy and primordial monopole problems, *Phys. Lett. B*, 108, 389.

Linde, A. D. 1983, Chaotic inflation, *Phys. Lett. B*, 129, 177.

Loeb, A. & Furlanetto, S. R. 2013, *The First Galaxies in the Universe* (Princeton University Press).

Longair, M. S. 2006, *The Cosmic Century: A History of Astrophysics and Cosmology* (Cambridge University Press).

Longair, M. S. 2023, *Galaxy Formation* (Cambridge University Press).

Lotz, J. M., Davis, M., Faber, S. M., et al. 2008, The evolution of galaxy mergers and morphology at $z < 1.2$ in the extended groth strip, *ApJ*, 672, 177.

Lundmark, K. 1924, The determination of the curvature of space-time in de Sitter's world, *MNRAS*, 84, 747.

Luparello, H., Lares, M., Lambas, D. G., & Padilla, N. 2011, Future virialized structures: An analysis of superstructures in the SDSS-DR7, *MNRAS*, 415, 964.

Lynden-Bell, D. 1967, Statistical mechanics of violent relaxation in stellar systems, *MNRAS*, 136, 101.

Lynden-Bell, D. 1969, Galactic nuclei as collapsed old quasars, *Nature*, 223, 690.

Lynden-Bell, D. 1976, Dwarf galaxies and globular clusters in high velocity hydrogen streams, *MNRAS*, 174, 695.

Macciò, A. V., Moore, B., & Stadel, J. 2006, The origin of polar ring galaxies: Evidence for galaxy formation by cold accretion, *ApJ*, 636, L25.

Mackenzie, R., Shanks, T., Bremer, M. N., et al. 2017, Evidence against a super-void causing the CMB Cold Spot, *MNRAS*, 470, 2328.

Mackereth, J. T., Schiavon, R. P., Pfeffer, J., et al. 2019, The origin of accreted stellar halo populations in the Milky Way using APOGEE, gaia, and the EAGLE simulations, *MNRAS*, 482, 3426.

Mackey, D., Lewis, G. F., Brewer, B. J., et al. 2019, Two major accretion epochs in M31 from two distinct populations of globular clusters, *Nature*, 574, 69.

Maddox, S. J., Efstathiou, G., Sutherland, W. J., & Loveday, J. 1990, Galaxy correlations on large scales, *MNRAS*, 242, 43.

Majewski, S. R., Schiavon, R. P., Frinchaboy, P. M., *et al.* 2017, The Apache Point observatory galactic evolution experiment (APOGEE), *AJ*, 154, 94.

Mandelbrot, B. B. 1982, *The Fractal Geometry of Nature* (San Francisco, CA: Freeman).

Mandelbrot, B. B. 1986, Multifractals and fractals, *Phys. Today*, 39, 11.

Mangano, G. & Serpico, P. D. 2011, A robust upper limit on N_{eff} from BBN, circa 2011, *Phys. Lett. B*, 701, 296.

Mann, R. G., Peacock, J. A., & Heavens, A. F. 1998, Eulerian bias and the galaxy density field, *MNRAS*, 293, 209.

Mather, J. C., Fixsen, D. J., Shafer, R. A., Mosier, C., & Wilkinson, D. T. 1999, Calibrator design for the COBE Far-Infrared Absolute Spectrophotometer (FIRAS), *ApJ*, 512, 511.

Marasco, A., Cresci, G., Posti, L., *et al.* 2021, A universal relation between the properties of supermassive black holes, galaxies, and dark matter haloes, *MNRAS*, 507, 4274.

Margutti, R. & Chornock, R. 2021, First multimessenger observations of a neutron star merger, *ARA&A*, 59, 155.

Markevitch, M., Gonzalez, A. H., David, L., *et al.* 2002, A textbook example of a bow shock in the merging galaxy cluster 1E 0657-56, *ApJ*, 567, L27.

Martinez, V. J. & Jones, B. J. T. 1990, Why the universe is not a fractal, *MNRAS*, 242, 517.

Martínez, V. J. & Saar, E. 2002, *Statistics of the Galaxy Distribution* (Chapman & Hall/CRC).

Materne, J. & Tammann, G. A. 1976, On the stability of groups of galaxies and the question of hidden matter, in *Stars and Galaxies from Observational Points of View*, ed. E. K. Kharadze, 455.

Mathewson, D. S., Cleary, M. N., & Murray, J. D. 1974, The Magellanic stream, *ApJ*, 190, 291.

Mathewson, D. S. & Schwarz, M. P. 1976, The origin of the Magellanic stream, *MNRAS*, 176, 47P.

Mathewson, D. S., Schwarz, M. P., & Murray, J. D. 1977, The Magellanic stream: The turbulent wake of the Magellanic Clouds in the halo of the Galaxy., *ApJ*, 217, L5.

Matthee, J., Naidu, R. P., Brammer, G., *et al.* 2024, Little red dots: An abundant population of faint active galactic nuclei at $z \sim 5$ revealed by the EIGER and FRESCO JWST Surveys, *ApJ*, 963, 129.

McCall, M. L. 2014, A council of giants, *MNRAS*, 440, 405.

Melott, A. L., Einasto, J., Saar, E., *et al.* 1983, Cluster analysis of the nonlinear evolution of large-scale structure in an axion/gravitino/photino-dominated universe, *Phys. Rev. Lett.*, 51, 935.

Merten, J., Coe, D., Dupke, R., *et al.* 2011, Creation of cosmic structure in the complex galaxy cluster merger Abell 2744, *MNRAS*, 417, 333.

Merton, R. K. 1961, Singletons and multiples in scientific discovery, in *Sociology of Science*, Vol. 105, *Proceedings of the American Philosophical Society*, 470.

Meszaros, P. 1975, Primeval black holes and galaxy formation, *A&A*, 38, 5.

Migkas, K. 2025, Galaxy clusters as probes of cosmic isotropy, *Philos. Trans. R. Soc. Lond. A*, 383, 20240030.

Milgrom, M. & Bekenstein, J. 1987, The modified Newtonian dynamics as an alternative to hidden matter, in *IAU Symposium*, Vol. 117, *Dark Matter in the Universe*, ed. J. Kormendy & G. R. Knapp, 319.

Minami, Y. & Komatsu, E. 2020, New extraction of the cosmic birefringence from the Planck 2018 polarization data, *Phys. Rev. Lett.*, 125, 221301.

Moore, B., Quinn, T., Governato, F., Stadel, J., & Lake, G. 1999, Cold collapse and the core catastrophe, *MNRAS*, 310, 1147.

Mor, R., Robin, A. C., Figueras, F., & Antoja, T. 2018, BGM FASt: Besançon galaxy model for big data. Simultaneous inference of the IMF, SFH, and density in the solar neighbourhood, *A&A*, 620, A79.

Mor, R., Robin, A. C., Figueras, F., & Lemasle, B. 2017, Constraining the thin disc initial mass function using Galactic classical Cepheids, *A&A*, 599, A17.

Mortlock, A., Conselice, C. J., Hartley, W. G., *et al.* 2013, The redshift and mass dependence on the formation of the Hubble sequence at $z > 1$ from CANDELS/UDS, *MNRAS*, 433, 1185.

Mowla, L. A., van Dokkum, P., Brammer, G. B., *et al.* 2019, COSMOS-DASH: The evolution of the galaxy size-mass relation since z $\sim$ 3 from new Wide-field WFC3 imaging combined with CANDELS/3D-HST, *ApJ*, 880, 57.

Nandal, D., Farrell, E., Buldgen, G., Meynet, G., & Ekström, S. 2024, The evolution and impact of $\sim$3000 $M_\odot$ stars in the early universe, *A&A*, 685, A159.

Navarro, J. F., Frenk, C. S., & White, S. D. M. 1996, The structure of cold dark matter halos, *ApJ*, 462, 563.

Navarro, J. F., Frenk, C. S., & White, S. D. M. 1997, A universal density profile from hierarchical clustering, *ApJ*, 490, 493.

Nelson, D., Genel, S., Pillepich, A., *et al.* 2016, Zooming in on accretion – I. The structure of halo gas, *MNRAS*, 460, 2881.

Neyrinck, M. C., Szapudi, I., McCullagh, N., *et al.* 2018, Density-dependent clustering – I. Pullingback the curtains on motions of the BAO peak, *MNRAS*, 478, 2495.

Neyrinck, M. C., Szapudi, I., & Szalay, A. S. 2009, Rejuvenating the matter power spectrum: Restoring information with a logarithmic density mapping, *ApJ*, 698, L90.

Nipoti, C. 2017, On the origin of Sérsic profiles of galaxies and Einasto profiles of dark-matter halos, in *IAU Symposium*, Vol. 321, *Formation and Evolution of Galaxy Outskirts*, ed. A. Gil de Paz, J. H. Knapen, & J. C. Lee, 87.

Nojiri, S., Odintsov, S. D., & Oikonomou, V. K. 2017, Modified gravity theories on a nutshell: Inflation, bounce and late-time evolution, *Phys. Rep.*, 692, 1.

Norberg, P., Baugh, C. M., Hawkins, E., *et al.* 2001, The 2dF Galaxy Redshift Survey: Luminosity dependence of galaxy clustering, *MNRAS*, 328, 64.

Obukhov, Y. N. 2000, On physical foundations and observational effects of cosmic rotation, in *Colloquium on Cosmic Rotation*, ed. M. Scherfner, T. Chrobok, & M. Shefaat, 23.

Olive, K. A., Steigman, G., & Walker, T. P. 2000, Primordial nucleosynthesis: theory and observations, *Phys. Rep.*, 333, 389.

Oort, J. H. 1927, Observational evidence confirming Lindblad's hypothesis of a rotation of the galactic system, *Bull. Astron. Inst. Netherlands*, 3, 275.

Oort, J. H. 1928, Dynamics of the Galactic system in the vicinity of the Sun, *Bull. Astron. Inst. Netherlands*, 4, 269.

Oort, J. H. 1940, Some problems concerning the structure and dynamics of the galactic system and the elliptical nebulae NGC 3115 and 4494, *ApJ*, 91, 273.

Öpik, E. 1921, Probable distance of the great Andromeda nebula and the nature of spiral nebulae, *Mirovedenie*, 10, 12.

Öpik, E. 1922a, An estimate of the distance of the Andromeda Nebula, *ApJ*, 55, 406.

Öpik, E. 1922b, Notes on stellar statistics and stellar evolution, *Publications of the Tartu Astr. Obs.*, 25, 1.

Öpik, E. 1923, On the luminosity-curve of components of double stars, *Tartu Astr. Obs. Publ.*, 25, 1.

Öpik, E. 1924, Statistical studies of double stars: On the distribution of relative luminosities and distances of double stars in the harvard revised photometry north of declination -31 deg, *Tartu Astr. Obs. Publ.*, 25, 1.

Öpik, E. 1938, Stellar structure, source of energy, and evolution, *Tartu Astr. Obs. Publ.*, 30, 1.

Ostriker, J. P. & Peebles, P. J. E. 1973, A numerical study of the stability of flattened galaxies: or, can cold galaxies survive?, *ApJ*, 186, 467.

Ostriker, J. P., Peebles, P. J. E., & Yahil, A. 1974, The size and mass of galaxies, and the mass of the universe, *ApJ*, 193, L1.

Paczynski, B. 1986, Gravitational microlensing by the galactic halo, *ApJ*, 304, 1.

Padmanabhan, T. 2003, Cosmological constant-the weight of the vacuum, *Phys. Rep.*, 380, 235.

Pagel, B. E. J. 1997, *Nucleosynthesis and Chemical Evolution of Galaxies* (Cambridge University Press).

Papageorgiou, A., Plionis, M., Basilakos, S., & Abdullah, M. H. 2024, The cluster mass function and the σ_8 tension, *MNRAS*, 527, 5559.

Park, C., Choi, Y.-Y., Vogeley, M. S., *et al.* 2005, Topology analysis of the Sloan Digital Sky Survey. I. Scale and luminosity dependence, *ApJ*, 633, 11.

Park, C., Lee, J., Kim, J., *et al.* 2022, Formation and morphology of the first galaxies in the cosmic morning, *ApJ*, 937, 15.

Pawlowski, M. S. & Kroupa, P. 2020, The Milky Way's disc of classical satellite galaxies in light of Gaia DR2, *MNRAS*, 491, 3042.

Peacock, J. A. 1999, *Cosmological Physics* (Cambridge University Press).

Peacock, J. A., Cole, S., Norberg, P., *et al.* 2001, A measurement of the cosmological mass density from clustering in the 2dF galaxy redshift survey, *Nature*, 410, 169.

Pease, F. G. 1918, The rotation and radial velocity of the central part of the Andromeda nebula, *Proc. Nat. Acad. Sci.*, 4, 21.

Peccei, R. D. & Quinn, H. R. 1977, CP conservation in the presence of instantons, *Phys. Rev. Lett.*, 38, 1440.

Peebles, P. J. E. 1973, Statistical analysis of catalogs of extragalactic objects. I. Theory, *ApJ*, 185, 413.

Peebles, P. J. E. 1974, Statistical analysis of catalogs of extragalactic objects. IV. Cross-correlation of the abell and shane-wirtanen catalogs, *ApJS*, 28, 37.

Peebles, P. J. E. 1976, A cosmic virial theorem, *Ap&SS*, 45, 3.

Peebles, P. J. E. 1980, *The Large-Scale Structure of the Universe*, Princeton Series in Physics (Princeton University Press).

Peebles, P. J. E. 1982a, Large-scale background temperature and mass fluctuations due to scale-invariant primeval perturbations, *ApJ*, 263, L1.

Peebles, P. J. E. 1982b, Primeval adiabatic perturbations - Effect of massive neutrinos, *ApJ*, 258, 415.

Peebles, P. J. E. 1984, Tests of cosmological models constrained by inflation, *ApJ*, 284, 439.

Peebles, P. J. E. 1989, The fractal galaxy distribution, *Physica D*, 38, 273.

Peebles, P. J. E. 1998, The standard cosmological model, astro-ph/9806201.

Peebles, P. J. E. 2001, The galaxy and mass N-Point correlation functions: A blast from the past, in *Astronomical Society of the Pacific Conference Series*, Vol. 252, *Historical Development of Modern Cosmology*, ed. V. J. Martínez, V. Trimble, & M. J. Pons-Bordería, 201.

Peebles, P. J. E. 2002, Nineteenth and twentieth century clouds over the twenty-first century virtual observatory, ArXiv:astro-ph/0209403.

Peebles, P. J. E. 2020, *Cosmology's Century: An Inside History of our Modern Understanding of the Universe* (Princeton University Press).

Peebles, P. J. E. 2022a, Anomalies in physical cosmology, *Ann. Phys.*, 447, 169159.

Peebles, P. J. E. 2022b, The extended Local Supercluster, *MNRAS*, 511, 5093.

Peebles, P. J. E. 2022c, *The Whole Truth. A Cosmologist's Reflections on the Search for Objective Reality* (Princeton University Press).

Peebles, P. J. E. 2023, Flat patterns in cosmic structure, *MNRAS*, 526, 4490.

Peebles, P. J. E. 2024, The physicists philosophy of physics, arXiv e-prints, arXiv:2401.16506.

Peebles, P. J. E. 2025, Status of the ΛCDM theory: Supporting evidence and anomalies, *Philosophical Transactions of the Royal Society of London Series A*, 383, 20240021.

Peebles, P. J. E. & Groth, E. J. 1975, Statistical analysis of catalogs of extragalactic objects. V - Three-point correlation function for the galaxy distribution in the Zwicky catalog, *ApJ*, 196, 1.

Peebles, P. J. E., Page, Lyman A., J., & Partridge, R. B. 2009, *Finding the Big Bang* (Cambridge University Press).

Peebles, P. J. E. & Ratra, B. 2003, The cosmological constant and dark energy, *Rev. Mod. Phys.*, 75, 559.

Peebles, P. J. E. & Yu, J. T. 1970, Primeval adiabatic perturbation in an expanding universe, *ApJ*, 162, 815.

Perlmutter, S., Aldering, G., Goldhaber, G., *et al.* 1999, Measurements of Omega and Lambda from 42 high-redshift supernovae, *ApJ*, 517, 565.

Perlmutter, S. & Schmidt, B. P. 2003, in *Supernovae and Gamma-Ray Bursters*, ed. K. Weiler, Vol. 598 (Lecture Notes in Physics), 195–217.

Pettini, M. & Cooke, R. 2012, A new, precise measurement of the primordial abundance of Deuterium, *MNRAS*, 425, 2477.

Pichon, C., Gay, C., Pogosyan, D., *et al.* 2010, The skeleton: Connecting large scale structures to Galaxy formation, in *American Institute of Physics Conference Series*, Vol. 1241, *American Institute of Physics Conference Series*, ed. J.-M. Alimi & A. Fuözfa, 1108.

Pichon, C., Pogosyan, D., Kimm, T., *et al.* 2011, Rigging dark haloes: Why is hierarchical galaxy formation consistent with the inside-out build-up of thin discs?, *MNRAS*, 418, 2493.

Pietronero, L. 1987, The fractal structure of the universe: Correlations of galaxies and clusters and the average mass density, *Physica A Statistical Mechanics and its Applications*, 144, 257.

Pillepich, A., Nelson, D., Hernquist, L., *et al.* 2018, First results from the IllustrisTNG simulations: The stellar mass content of groups and clusters of galaxies, *MNRAS*, 475, 648.

Planck Collaboration, Ade, P. A. R., Aghanim, N., *et al.* 2014a, Planck 2013 results. XVI. Cosmological parameters, *A&A*, 571, A16.

Planck Collaboration, Ade, P. A. R., Aghanim, N., *et al.* 2014b, Planck 2013 results. XX. Cosmology from Sunyaev-Zeldovich cluster counts, *A&A*, 571, A20.

Planck Collaboration, Aghanim, N., Akrami, Y., *et al.* 2020, Planck 2018 results. VI. Cosmological parameters, *A&A*, 641, A6.

Platen, E., van de Weygaert, R., & Jones, B. J. T. 2007, A cosmic watershed: The WVF void detection technique, *MNRAS*, 380, 551.

Pol, N., McLaughlin, M., & Lorimer, D. R. 2020, An updated galactic double neutron star merger rate based on radio pulsar populations, *Res. Notes Am. Astron. Soci.*, 4, 22.

Pomarède, D., Tully, R. B., Hoffman, Y., & Courtois, H. M. 2015, The arrowhead mini-supercluster of galaxies, *ApJ*, 812, 17.

Poulin, V., Smith, T. L., Calderón, R., & Simon, T. 2024, On the implications of the 'cosmic calibration tension' beyond H_0 and the synergy between early- and late-time new physics, arXiv e-prints 2407.18292.

Preskill, J., Wise, M. B., & Wilczek, F. 1983, Cosmology of the invisible axion, *Phys. Lett. B*, 120, 127.

Press, W. H. & Schechter, P. 1974, Formation of galaxies and clusters of galaxies by self-similar gravitational condensation, *ApJ*, 187, 425.

Primack, J. R. & Blumenthal, G. R. 1984, What is the dark matter? – Implications for galaxy formation and particle physics, in *NATO ASIC Proc. 117: Formation and Evolution of Galaxies and Large Structures in the Universe*, ed. J. Audouze & J. Tran Thanh Van, 163.

Prole, L. R., Clark, P. C., Klessen, R. S., & Glover, S. C. O. 2022, Fragmentation-induced starvation in Population III star formation: A resolution study, *MNRAS*, 510, 4019.

Putman, M. E., Peek, J. E. G., & Joung, M. R. 2012, Gaseous galaxy halos, *ARA&A*, 50, 491.

Rácz, G., Dobos, L., Beck, R., Szapudi, I., & Csabai, I. 2017, Concordance cosmology without dark energy, *MNRAS*, 469, L1.

Rácz, G., Szapudi, I., Csabai, I., & Dobos, L. 2018, Compactified cosmological simulations of the infinite universe, *MNRAS*, 477, 1949.

Rácz, G., Szapudi, I., Csabai, I., & Dobos, L. 2021, The anisotropy of the power spectrum in periodic cosmological simulations, *MNRAS*, 503, 5638.

Rameez, M. 2025, Anisotropy in the cosmic acceleration inferred from supernovae, *Philoso. Trans. R. Soc. Lond. A*, 383, 20240032.

Rees, M. 2000, *Just Six Numbers: The Deep Forces That Shape the Universe* (Basic Books).

Rees, M. & Kolb, E. W. 2000, Just six numbers: The deep forces that shape the universe, *Phys. Today*, 53, 67.

Rees, M. J. 1977, Cosmology and Galaxy formation, in *Evolution of Galaxies and Stellar Populations*, ed. B. M. Tinsley & R. B. Larson, 339.

Rees, M. J. 1984a, Black hole models for active Galactic nuclei, *ARA&A*, 22, 471.

Rees, M. J. 1984b, Is the universe flat?, *J. Astro. Astron.*, 5, 331.

Rees, M. J. & Sciama, D. W. 1968, Large-scale density inhomogeneities in the Universe, *Nature*, 217, 511.

Repp, A. & Szapudi, I. 2018, Precision prediction for the cosmological density distribution, *MNRAS*, 473, 3598.

Repp, A. & Szapudi, I. 2019a, A gravitational ising model for the statistical bias of galaxies, arXiv e-prints, arXiv:1904.05048.

Repp, A. & Szapudi, I. 2019b, Empirical validation of the ising galaxy bias model, arXiv e-prints, arXiv:1912.05557.

Repp, A. & Szapudi, I. 2022, Indicator power spectra: Surgical excision of non-linearities and covariance matrices for counts in cells, *MNRAS*, 509, 586.

Riess, A. G., Filippenko, A. V., Challis, P., *et al.* 1998, Observational evidence from supernovae for an accelerating universe and a cosmological constant, *AJ*, 116, 1009.

Riess, A. G., Scolnic, D., Anand, G. S., *et al.* 2024, JWST validates HST distance measurements: Selection of supernova subsample explains differences in JWST estimates of local H_0, *ApJ*, 977, 120.

Riess, A. G., Yuan, W., Macri, L. M., *et al.* 2022, A comprehensive measurement of the local value of the Hubble Constant with $1\,\mathrm{km\,s^{-1}\,Mpc^{-1}}$ uncertainty from the Hubble Space Telescope and the SH0ES team, *ApJ*, 934, L7.

Rimes C. D., Hamilton A. J. S., 2005, Information content of the non-linear matter power spectrum, *MNRAS*, 360, L82.

Roberts, M. S. 1966, A high-resolution 21-CM hydrogen-line survey of the Andromeda nebula, *ApJ*, 144, 639.

Robertson, B., Johnson, B. D., Tacchella, S., *et al.* 2024, Earliest galaxies in the JADES origins field: Luminosity function and cosmic star formation rate density 300 Myr after the Big Bang, *ApJ*, 970, 31.

Robertson, B. E. 2022, Galaxy formation and reionization: Key unknowns and expected breakthroughs by the James Webb Space Telescope, *ARA&A*, 60, 121.

Robertson, B. E., Tacchella, S., Johnson, B. D., *et al.* 2023, Identification and properties of intense star-forming galaxies at redshifts $z \geq 10$, *Nat. Astron.*, 7, 611.

Romanowsky, A. J., Strader, J., Brodie, J. P., *et al.* 2012, The ongoing assembly of a central cluster galaxy: Phase-space substructures in the halo of M87, *ApJ*, 748, 29.

Röpke, F. K., Hillebrandt, W., Schmidt, W., *et al.* 2007, A three-dimensional deflagration model for type ia supernovae compared with observations, *ApJ*, 668, 1132.

Rubin, K. H. R., Prochaska, J. X., Koo, D. C., & Phillips, A. C. 2012, The direct detection of cool, metal-enriched gas accretion onto galaxies at $z \sim 0.5$, *ApJ*, 747, L26.

Rubin, V. C., Ford, Jr., W. K., & Rubin, J. S. 1973, A curious distribution of radial velocities of SC i galaxies with $14.0 < M < 15.0$, *ApJ*, 183, L111.

Rubin, V. C. & Ford, W. K. J. 1970, Rotation of the Andromeda Nebula from a spectroscopic survey of emission regions, *ApJ*, 159, 379.

Rubin, V. C., Ford, W. K. J., & Thonnard, N. 1980, Rotational properties of 21 SC galaxies with a large range of luminosities and radii, from NGC 4605 R = 4kpc to UGC 2885 R = 122 kpc, *ApJ*, 238, 471.

Ryabinkov, A. I. & Kaminker, A. D. 2021, Traces of anisotropic quasi-regular structure in the SDSS Data, *Universe*, 7, 289.

Ryabinkov, A. I. & Kaminker, A. D. 2024, Search for a possible quasi-periodic structure based on data of the SDSS DR12 LOWZ, *MNRAS*, 527, 1813.

Ryden, B. 2017, *Introduction to Cosmology* (Cambridge University Press).

Ryden, B. S. & Pogge, R. W. 2021, *Interstellar and intergalactic medium* (Cambridge University Press).

Saadeh, D., Feeney, S. M., Pontzen, A., Peiris, H. V., & McEwen, J. D. 2016, How isotropic is the universe?, *Phys. Rev. Lett.*, 117.

Sachs, R. K. & Wolfe, A. M. 1967, Perturbations of a cosmological model and angular variations of the microwave background, *ApJ*, 147, 73.

Sakai, S., Giovanelli, R., & Wegner, G. 1994, Distribution of galaxies around Abell 262 and the NGC 383 and NGC 506 groups, *AJ*, 108, 33.

Sakharov, A. D. 1967, Violation of CP invariance, C asymmetry, and baryon asymmetry of the universe, *Pisma Zh. Eksp. Teor. Fiz.*, 5, 32.

Salim, S. & Narayanan, D. 2020, The dust attenuation law in galaxies, *ARA&A*, 58, 529.

Salpeter, E. E. 1955, The luminosity function and stellar evolution, *ApJ*, 121, 161.

Sandage, A. 1961, The ability of the 200-inch telescope to discriminate between selected world models, *ApJ*, 133, 355.

Sandage, A. 1970, Main-sequence photometry, color-magnitude diagrams, and ages for the globular clusters M3, M13, M15, and M92, *ApJ*, 162, 841.

Sandage, A. & Eggen, O. J. 1969, Isochrones, ages, curves of evolutionary deviation, and the composite C-M diagram for old galactic clusters, *ApJ*, 158, 685.

Sandage, A., Reindl, B., & Tammann, G. A. 2010, The linearity of the cosmic expansion field from 300 to 30,000 km s^{-1} and the bulk motion of the Local supercluster with respect to the cosmic microwave background, *ApJ*, 714, 1441.

Sandage, A. & Tammann, G. A. 1975, Steps toward the Hubble constant. V – The Hubble Constant from nearby galaxies and the regularity of the local velocity field, *ApJ*, 196, 313.

Sandage, A. & Tammann, G. A. 1976, Steps toward the Hubble constant. VII – Distances to NGC 2403, M101, and the Virgo cluster using 21 centimeter line widths compared with optical methods: The global value of H sub 0, *ApJ*, 210, 7.

Sandage, A., Tammann, G. A., Saha, A., *et al.* 2006, The Hubble Constant: A summary of the Hubble Space Telescope program for the luminosity calibration of type Ia supernovae by means of cepheids, *ApJ*, 653, 843.

Sanders, R. H. 2010, *The Dark Matter Problem: A Historical Perspective* (Cambridge University Press).

Sankhyayan, S., Bagchi, J., Tempel, E., *et al.* 2023, Identification of superclusters and their properties in the Sloan Digital Sky Survey using the WHL cluster catalog, *ApJ*, 958, 62.

Sarkar, A., Chakraborty, P., Vogelsberger, M., *et al.* 2025, Unveiling the cosmic chemistry: Revisiting the mass–metallicity relation with JWST/NIRSpec at $4 < z < 10$, *ApJ*, 978, 136.

Savino, A., Weisz, D. R., Dolphin, A. E., *et al.* 2025, The Hubble Space Telescope survey of M31 satellite galaxies. IV. Survey overview and lifetime star formation histories, *ApJ*, 979, 205.

Savino, A., Weisz, D. R., Skillman, E. D., *et al.* 2023, The Hubble Space Telescope survey of M31 satellite galaxies. II. The star formation histories of ultrafaint dwarf galaxies, *ApJ*, 956, 86.

Savino, A., Weisz, D. R., Skillman, E. D., *et al.* 2022, The Hubble Space Telescope survey of M31 satellite galaxies. I. RR lyrae-based distances and refined 3D geometric structure, *ApJ*, 938, 101.

Savorgnan, G. A. D., Graham, A. W., Marconi, A., & Sani, E. 2016, Supermassive black holes and their host spheroids. II. The red and blue sequence in the M_{BH}-$M_{*,sph}$ diagram, *ApJ*, 817, 21.

Sawala, T., Cautun, M., Frenk, C., *et al.* 2023, The Milky Way's plane of satellites is consistent with ΛCDM, *Nat. Astron.*, 7, 481.

Sawala, T., Delhomelle, J., Deason, A. J., *et al.* 2024, Apocalypse When? no certainty of a Milky Way – Andromeda collision, arXiv e-prints, arXiv:2408.00064.

Sawala, T., Frenk, C. S., Fattahi, A., *et al.* 2016, The APOSTLE simulations: Solutions to the local group's cosmic puzzles, *MNRAS*, 457, 1931.

Sawala, T., Pihajoki, P., Johansson, P. H., *et al.* 2017, Shaken and stirred: The Milky Way's dark substructures, *MNRAS*, 467, 4383.

Scalo, J. M. 1986, The stellar initial mass function, *Fund. Cosmic Phys.*, 11, 1.

Schaap, W. E. 2007, DTFE: The Delaunay Tessellation Field Estimator, PhD thesis, University of Groningen, Netherlands.

Schaap, W. E. & van de Weygaert, R. 2000, Continuous fields and discrete samples: Reconstruction through Delaunay tessellations, *A&A*, 363, L29.

Schaye, J., Crain, R. A., Bower, R. G., *et al.* 2015, The EAGLE project: Simulating the evolution and assembly of galaxies and their environments, *MNRAS*, 446, 521.

Schechter, P. 1976, An analytic expression for the luminosity function for galaxies, *ApJ*, 203, 297.

Schlafly, E. F. & Finkbeiner, D. P. 2011, Measuring reddening with Sloan Digital Sky Survey stellar spectra and recalibrating SFD, *ApJ*, 737, 103.

Schlegel, D. J., Finkbeiner, D. P., & Davis, M. 1998, Maps of dust infrared emission for use in estimation of reddening and cosmic microwave background radiation foregrounds, *ApJ*, 500, 525.

Schmidt, M. 1959, The rate of star formation, *ApJ*, 129, 243.

Schöneberg, N., Franco Abellán, G., Pérez Sánchez, A., *et al.* 2022, The H0 Olympics: A fair ranking of proposed models, *Phys. Rept.*, 984, 1.

Schwarzschild, M., Rabinowitz, I., & Härm, R. 1953, Inhomogeneous stellar models. III. models with partially degenerate isothermal cores, *ApJ*, 118, 326.

Schwarzschild, M. & Spitzer, L. 1953, On the evolution of stars and chemical elements in the early phases of a galaxy, *The Observatory*, 73, 77.

Schweizer, F., Whitmore, B. C., & Rubin, V. C. 1983, Colliding and merging galaxies. II. SO galaxies with polar rings., *AJ*, 88, 909.

Seidel, B. A., Dolag, K., Remus, R. S., *et al.* 2024, SLOW IV: Not all that is close will merge in the end. Superclusters and their Lagrangian collapse regions, arXiv e-prints, arXiv:2412.08708.

Seldner, M., Siebers, B., Groth, E. J., & Peebles, P. J. E. 1977, New reduction of the Lick catalog of galaxies, *AJ*, 82, 249.

Sersic, J. L. 1968, *Atlas de galaxias australes* (Cordoba, Argentina: Observatorio Astronomico).

Servant, G. & Tait, T. M. P. 2003, Is the lightest Kaluza-Klein particle a viable dark matter candidate?, *Nucl. Phys. B*, 650, 391.

Shandarin, S., Habib, S., & Heitmann, K. 2012, Cosmic web, multistream flows, and tessellations, *Phys. Rev. D*, 85, 083005.

Shandarin, S. F. 2011, The multi-stream flows and the dynamics of the cosmic web, *J. Cosmology Astropart. Phys.*, 2011, 015.

Shandarin, S. F. & Klypin, A. A. 1984, Rich galaxy clusters may result from largescale motions inside superclusters, *Soviet Ast.*, 28, 491.

Shandarin, S. F. & Zeldovich, I. B. 1983, Topology of the large-scale structure of the universe, *Comments on Astrophys.*, 10, 33.

Shandarin, S. F. & Zeldovich, I. B. 1984, Topological mapping properties of collisionless potential and vortex motion, *Phys. Rev. Lett.*, 52, 1488.

Shandarin, S. F. & Zeldovich, Y. B. 1989, The large-scale structure of the universe: Turbulence, intermittency, structures in a self-gravitating medium, *Rev. Mod. Phys.*, 61, 185.

Shane, C. & Wirtanen, C. 1967, The distribution of galaxies, *Publ. Lick Obs.*, 22.

Shapley, H. & Ames, A. 1932, *A Survey of the External Galaxies Brighter than the Thirteenth Magnitude* (Annals of Harvard College Observatory, Cambridge, MA.: Astronomical Observatory of Harvard College, 1932).

Sharma, P., McCourt, M., Quataert, E., & Parrish, I. J. 2012, Thermal instability and the feedback regulation of hot haloes in clusters, groups and galaxies, *MNRAS*, 420, 3174.

Shaver, P. A. 1991, Radio surveys and large scale structure, *Australian Journal of Physics*, 44, 759.

Shectman, S. A., Landy, S. D., Oemler, A., *et al.* 1996, The Las Campanas redshift survey, *ApJ*, 470, 172.

Shen, J. & Gebhardt, K. 2010, The supermassive black hole and dark matter halo of NGC 4649 (M60), *ApJ*, 711, 484.

Shi, X.-D. & Fuller, G. M. 1999, A New dark matter candidate: Nonthermal sterile neutrinos, *Phys. Rev. Lett.*, 82, 2832.

Simon, D. A., Cappellari, M., & Hartke, J. 2024, Supermassive black hole mass in the massive elliptical galaxy M87 from integral-field stellar dynamics using OASIS and MUSE with adaptive optics: Assessing systematic uncertainties, *MNRAS*, 527, 2341.

Smartt, S. J., Chen, T. W., Jerkstrand, A., *et al.* 2017, A kilonova as the electromagnetic counterpart to a gravitational-wave source, *Nature*, 551, 75.

Smith, M. D., Bureau, M., Davis, T. A., *et al.* 2021, WISDOM project – VI. Exploring the relation between supermassive black hole mass and galaxy rotation with molecular gas, *MNRAS*, 500, 1933.

Sofue, Y. 2017, Rotation and mass in the Milky Way and spiral galaxies, *PASJ*, 69, R1.

Sofue, Y. & Rubin, V. 2001, Rotation curves of spiral galaxies, *ARA&A*, 39, 137.

Soneira, R. M. & Peebles, P. J. E. 1978, A computer model universe – Simulation of the nature of the galaxy distribution in the Lick catalog, *AJ*, 83, 845.

Sorce, J. G., Gottlöber, S., Hoffman, Y., & Yepes, G. 2016a, How did the virgo cluster form?, *MNRAS*, 460, 2015.

Sorce, J. G., Gottlöber, S., Yepes, G., *et al.* 2016b, Cosmicflows constrained local universE simulations, *MNRAS*, 455, 2078.

Sotiriou, T. P. & Faraoni, V. 2010, f(R) theories of gravity, *Rev. Mod. Phys.*, 82, 451.

Sousbie, T. 2011, The persistent cosmic web and its filamentary structure – I. Theory and implementation, *MNRAS*, 414, 350.

Spergel, D. N., Verde, L., Peiris, H. V., *et al.* 2003, First-year Wilkinson Microwave Anisotropy Probe (WMAP) observations: Determination of cosmological parameters, *ApJS*, 148, 175.

Springel, V., Di Matteo, T., & Hernquist, L. 2005a, Black holes in galaxy mergers: The formation of red elliptical galaxies, *ApJ*, 620, L79.

Springel, V. & Farrar, G. R. 2007, The speed of the 'bullet' in the merging galaxy cluster 1E0657-56, *MNRAS*, 380, 911.

Springel, V., Frenk, C. S., & White, S. D. M. 2006, The large-scale structure of the Universe, *Nature*, 440, 1137.

Springel, V., Pakmor, R., Pillepich, A., *et al.* 2018, First results from the IllustrisTNG simulations: Matter and galaxy clustering, *MNRAS*, 475, 676.

Springel, V., White, S. D. M., Jenkins, A., *et al.* 2005b, Simulations of the formation, evolution and clustering of galaxies and quasars, *Nature*, 435, 629.

Stacy, A., McKee, C. F., Lee, A. T., Klein, R. I., & Li, P. S. 2022, Magnetic fields in the formation of the first stars – II. Results, *MNRAS*, 511, 5042.

Stark, D. P. 2016, Galaxies in the first billion years after the Big Bang, *ARA&A*, 54, 761.

Starobinsky, A. A. 1980, A new type of isotropic cosmological models without singularity, *Phys. Lett. B*, 91, 99.

Starobinsky, A. A. 1982, Dynamics of phase transition in the new inflationary universe scenario and generation of perturbations, *Phys. Lett. B*, 117, 175.

Starobinsky, A. A. 1985, Multicomponent de Sitter (inflationary) stages and the generation of perturbations, *ZhETF Pis ma Redaktsiiu*, 42, 124.

Stauffer, D. 1979, Scaling theory of percolation clusters, *Phys. Rep.*, 54, 1.

Strader, J., Romanowsky, A. J., Brodie, J. P., *et al.* 2011, Wide-field Precision Kinematics of the M87 Globular Cluster System, *ApJS*, 197, 33.

Strömberg, G. 1924, The asymmetry in stellar motions and the existence of a velocity-restriction in space, *ApJ*, 59, 228.

Suhhonenko, I., Einasto, J., Liivamägi, L. J., *et al.* 2011, The cosmic web for density perturbations of various scales, *A&A*, 531, A149+.

Szalay, A. S. & Marx, G. 1976, Neutrino rest mass from cosmology, *A&A*, 49, 437.

Szapudi, I. 2009, *Introduction to Higher Order Spatial Statistics in Cosmology*, Vol. 665 (Springer), 457–492.

Szapudi, I. 2021, Constraining Mach's principle with high precision astrometry, arXiv e-prints, arXiv:2105.05337.

Szapudi, I. 2025, The ISW puzzle, *Philos. Trans. R. Soc. Lond. A*, 383, 20240026.

Szapudi, I., Kovács, A., Granett, B. R., *et al.* 2015, Detection of a supervoid aligned with the cold spot of the cosmic microwave background, *MNRAS*, 450, 288.

Szapudi, I., Pan, J., Prunet, S., & Budavári, T. 2005, Fast edge-corrected measurement of the two-point correlation function and the power spectrum, *ApJ*, 631, L1.

Szapudi, I., Prunet, S., & Colombi, S. 2001, Fast analysis of inhomogenous megapixel cosmic microwave background maps, *ApJ*, 561, L11.

Talia, M., Cimatti, A., Mignoli, M., *et al.* 2014, Listening to galaxies tuning at $z = 2.5 - 3.0$: The first strikes of the Hubble fork, *A&A*, 562, A113.

Tamm, A., Tempel, E., Tenjes, P., Tihhonova, O., & Tuvikene, T. 2012, Stellar mass map and dark matter distribution in M 31, *A&A*, 546, A4.

Taoso, M., Bertone, G., & Masiero, A. 2008, Dark matter candidates: A ten-point test, *JCAP*, 03, 022.

Tarenghi, M., Tifft, W. G., Chincarini, G., Rood, H. J., & Thompson, L. A. 1978, The structure of the Hercules supercluster, in IAU Symposium, Vol. 79, *Large Scale Structures in the Universe*, ed. M. S. Longair & J. Einasto, 263.

Taylor, R. P., Micolich, A. P., & Jonas, D. 1999, Fractal analysis of Pollock's drip paintings, *Nature*, 399, 422.

Teerikorpi, P., Chernin, A. D., & Baryshev, Y. V. 2005, The quiescent Hubble flow, local dark energy tests, and pairwise velocity dispersion in a $\Omega = 1$ universe, *A&A*, 440, 791.

Tegmark, M., Blanton, M. R., Strauss, M. A., et al. 2004, The three-dimensional power spectrum of galaxies from the Sloan Digital Sky Survey, *ApJ*, 606, 702.

Tegmark, M. & Bromley, B. C. 1999, Observational evidence for stochastic biasing, *ApJ*, 518, L69.

Tegmark, M. & Peebles, P. J. E. 1998, The time evolution of bias, *ApJ*, 500, L79.

Tempel, E., Einasto, J., Einasto, M., Saar, E., & Tago, E. 2009, Anatomy of luminosity functions: The 2dFGRS example, *A&A*, 495, 37.

Tempel, E., Stoica, R. S., Martínez, V. J., et al. 2014a, Detecting filamentary pattern in the cosmic web: A catalogue of filaments for the SDSS, *MNRAS*, 438, 3465.

Tempel, E., Tamm, A., Gramann, M., et al. 2014b, Flux- and volume-limited groups/clusters for the SDSS galaxies: Catalogues and mass estimation, *A&A*, 566, A1.

Tempel, E., Tamm, A., & Tenjes, P. 2010, Dust-corrected surface photometry of M 31 from Spitzer far-infrared observations, *A&A*, 509, A91.

Tenjes, P., Haud, U., & Einasto, J. 1994, Galactic models with massive coronae IV. The Andromeda galaxy, M 31, *A&A*, 286, 753.

Thom, C., Tumlinson, J., Werk, J. K., et al. 2012, Not dead Yet: Cool circumgalactic gas in the halos of early-type galaxies, *ApJ*, 758, L41.

Thompson, L. A. 2021, *The Discovery of Cosmic Voids* (Cambridge University Press).

Tifft, W. G. & Gregory, S. A. 1978, Observations of the large scale distribution of galaxies, in *IAU Symposium*, Vol. 79, *Large Scale Structures in the Universe*, ed. M. S. Longair & J. Einasto, 267.

Tinsley, B. M. 1968, Evolution of the stars and gas in galaxies, *ApJ*, 151, 547.

Toomre, A. 1964, On the gravitational stability of a disk of stars, *ApJ*, 139, 1217.

Toomre, A. 1977, Mergers and some consequences, in *Evolution of Galaxies and Stellar Populations*, ed. B. M. Tinsley & R. B. G. Larson, D. Campbell, 401.

Toomre, A. & Toomre, J. 1972, Galactic bridges and tails, *ApJ*, 178, 623.

Toyouchi, D., Inayoshi, K., Li, W., Haiman, Z., & Kuiper, R. 2023, Radiative feedback on supermassive star formation: The massive end of the Population III initial mass function, *MNRAS*, 518, 1601.

Trujillo, I., Conselice, C. J., Bundy, K., et al. 2007, Strong size evolution of the most massive galaxies since z $\sim$ 2, *MNRAS*, 382, 109.

Tully, R. B. 1982, The Local supercluster, *ApJ*, 257, 389.

Tully, R. B. 2015, Galaxy groups, *AJ*, 149, 54.

Tully, R. B., Courtois, H., Hoffman, Y., & Pomarède, D. 2014, The Laniakea supercluster of galaxies, *Nature*, 513, 71.

Tully, R. B., Courtois, H. M., Dolphin, A. E., *et al.* 2013, Cosmicflows-2: The data, *AJ*, 146, 86.

Tully, R. B., Courtois, H. M., & Sorce, J. G. 2016, Cosmicflows-3, *AJ*, 152, 50.

Tully, R. B. & Fisher, J. R. 1978, A tour of the Local Supercluster, in *IAU Symposium*, Vol. 79, *Large Scale Structures in the Universe*, ed. M. S. Longair & J. Einasto, 214.

Tully, R. B. & Fisher, J. R. 1987, *Atlas of Nearby Galaxies* (Cambridge University Press).

Tully, R. B., Kourkchi, E., Courtois, H. M., *et al.* 2023, Cosmicflows-4, *ApJ*, 944, 94.

Tully, R. B., Pomarède, D., Graziani, R., *et al.* 2019, Cosmicflows-3: Cosmography of the local void, *ApJ*, 880, 24.

Tumlinson, J., Peeples, M. S., & Werk, J. K. 2017, The circumgalactic medium, *ARA&A*, 55, 389.

Turner, M. S., Steigman, G., & Krauss, L. M. 1984, Flatness of the universe - Reconciling theoretical prejudices with observational data, *Phys. Rev. Lett.*, 52, 2090.

Turok, N., ed. 1997, *Critical Dialogues in Cosmology* (World Scientific).

Uzan, J.-P. 2003, The fundamental constants and their variation: Observational and theoretical status, *Rev. Mod. Phys.*, 75, 403.

Valade, A., Libeskind, N. I., Pomarède, D., *et al.* 2024, Identification of basins of attraction in the Local Universe, *Nature Astron.*, 8, 1610.

Valentino, F., Daddi, E., Puglisi, A., *et al.* 2021, The effect of active galactic nuclei on the cold interstellar medium in distant star-forming galaxies, *A&A*, 654, A165.

van Albada, T. S. 1982, Dissipationless galaxy formation and the R to the 1/4-power law, *MNRAS*, 201, 939.

van de Weygaert, R. 2002, Froth across the universe dynamics and stochastic geometry of the cosmic foam, ArXiv:astro-ph/0206427.

van den Bergh, S. 1961a, The halo phase of galactic evolution, *PASP*, 73, 135.

van den Bergh, S. 1961b, The stability of clusters of galaxies, *AJ*, 66, 566.

van den Bergh, S. 1962, Are clusters of galaxies stable?, *AJ*, 67, 285.

van den Bergh, S. 1976, A new classification system for galaxies, *ApJ*, 206, 883.

van den Bergh, S. 1992, The Hubble parameter, *PASP*, 104, 861.

van den Bergh, S. 1994, The Hubble parameter revisited, *PASP*, 106, 1113.

van der Marel, R. P., Cretton, N., de Zeeuw, P. T., & Rix, H.-W. 1998, Improved evidence for a black hole in M32 from HST/FOS Spectra. II. Axisymmetric dynamical models, *ApJ*, 493, 613.

van der Marel, R. P., Fardal, M., Besla, G., *et al.* 2012, The M31 velocity vector. II. Radial orbit toward the Milky Way and implied Local Group mass, *ApJ*, 753, 8.

Vazza, F. & Feletti, A. 2020, The quantitative comparison between the neuronal network and the cosmic web, *Front. Phys.*, 8, 491.

Verde, L., Heavens, A. F., Percival, W. J., *et al.* 2002, The 2dF galaxy redshift survey: The bias of galaxies and the density of the Universe, *MNRAS*, 335, 432.

Villanueva-Domingo, P., Mena, O., & Palomares-Ruiz, S. 2021, A brief review on primordial black holes as dark matter, *Front. Astron. Space Sci.*, 8, 87.

Vogeley, M. S., Park, C., Geller, M. J., Huchra, J. P., & Gott, III, J. R. 1994, Topological analysis of the CfA redshift survey, *ApJ*, 420, 525.

Vogelsberger, M., Genel, S., Springel, V., *et al.* 2014, Introducing the Illustris Project: Simulating the coevolution of dark and visible matter in the Universe, *MNRAS*, 444, 1518.

Wang, J., Bose, S., Frenk, C. S., *et al.* 2020, Universal structure of dark matter haloes over a mass range of 20 orders of magnitude, *Nature*, 585, 39.

Wang, X., Neyrinck, M., Szapudi, I., *et al.* 2011, Perturbation theory of the cosmological log-density field, *ApJ*, 735, 32.

Wang, Y. 2010, *Dark Energy* (Wiley-VCH).

Wang, Y., Brunner, R. J., & Dolence, J. C. 2013, The SDSS galaxy angular two-point correlation function, *MNRAS*, 432, 1961.

Weinberg, D. H., Bullock, J. S., Governato, F., Kuzio de Naray, R., & Peter, A. H. G. 2015, Cold dark matter: Controversies on small scales, *Proc. Nat. Acad. Sci.*, 112, 12249.

Weinberg, S. 1987, Anthropic bound on the cosmological constant, *Phys. Rev. Lett.*, 59, 2607.

Weinberg, S. 1989, The cosmological constant problem, *Rev. Mod. Phys.*, 61, 1.

Weniger, C. 2012, A tentative gamma-ray line from dark matter annihilation at the Fermi large area telescope, *JCAP*, 08, 007.

White, S. D. M., Frenk, C. S., & Davis, M. 1983, Clustering in a neutrino-dominated universe, *ApJ*, 274, L1.

White, S. D. M., Frenk, C. S., Davis, M., & Efstathiou, G. 1987, Clusters, filaments, and voids in a universe dominated by cold dark matter, *ApJ*, 313, 505.

White, S. D. M. & Rees, M. J. 1978, Core condensation in heavy halos - A two-stage theory for galaxy formation and clustering, *MNRAS*, 183, 341.

Whitmore, B. C., Lucas, R. A., McElroy, D. B., *et al.* 1990, New observations and a photographic atlas of polar-ring galaxies, *AJ*, 100, 1489.

Wiltshire, D. L. 2024, Solution to the cosmological constant problem, arXiv e-prints, arXiv:2404.02129.

Wirtz, C. 1922, Einiges zur Statistik der Radialbewegungen von Spiralnebeln und Kugelsternhaufen, *Astron. Nach.*, 215, 349.

Wirtz, C. 1924, De Sitters Kosmologie und die Radialbewegungen der Spiralnebel, *Astron. Nach.*, 222, 21.

Wolk, M., Carron, J., & Szapudi, I. 2015, On the total cosmological information in galaxy clustering: An analytical approach, *MNRAS*, 454, 560.

Wood, R. A., Jones, C., Machacek, M. E., *et al.* 2017, The infall of the Virgo elliptical galaxy M60 toward M87 and the gaseous structures produced by Kelvin-Helmholtz instabilities, *ApJ*, 847, 79.

Wyithe, J. S. B. & Loeb, A. 2006, Suppression of dwarf galaxy formation by cosmic reionization, *Nature*, 441, 322.

Yan, Z., Jerabkova, T., & Kroupa, P. 2017, The optimally sampled galaxy-wide stellar initial mass function. Observational tests and the publicly available GalIMF code, *A&A*, 607, A126.

Yan, Z., Li, J., Kroupa, P., *et al.* 2024, The variation in the galaxy-wide initial mass function for low-mass stars: Modeling and observational insights, *ApJ*, 969, 95.

York, D. G. *et al.* 2000, The Sloan Digital Sky Survey: Technical summary, *Astron. J.*, 120, 1579.

Zahid, H. J., Geller, M. J., Kewley, L. J., *et al.* 2013, The chemical evolution of star-forming galaxies over the last 11 billion years, *ApJ*, 771, L19.

Zaroubi, S., Hoffman, Y., & Dekel, A. 1999, Wiener reconstruction of large-scale structure from peculiar velocities, *ApJ*, 520, 413.

Zaroubi, S., Hoffman, Y., Fisher, K. B., & Lahav, O. 1995, Wiener reconstruction of the large-scale structure, *ApJ*, 449, 446.

Zarrouk, P. *et al.* 2018, The clustering of the SDSS-IV extended baryon oscillation spectroscopic survey DR14 quasar sample: Measurement of the growth rate of structure from the anisotropic correlation function between redshift 0.8 and 2.2, *MNRAS*, 477, 1639.

Zehavi, I., Weinberg, D. H., Zheng, Z., *et al.* 2004, On departures from a power law in the galaxy correlation function, *ApJ*, 608, 16.

Zehavi, I., Zheng, Z., Weinberg, D. H., *et al.* 2011, Galaxy clustering in the completed sdss redshift survey: The dependence on color and luminosity, *ApJ*, 736, 59.

Zeldovich, I. B. & Novikov, I. D. 1975, Structure and evolution of the universe. (Nauka).

Zeldovich, Y. B. 1970, Gravitational instability: An approximate theory for large density perturbations., *A&A*, 5, 84.

Zeldovich, Y. B., Einasto, J., & Shandarin, S. F. 1982, Giant voids in the universe, *Nature*, 300, 407.

Zeldovich, Y. B. & Novikov, I. D. 1967, The hypothesis of cores retarded during expansion and the hot cosmological model, *Soviet Ast.*, 10, 602.

Zwicky, F. 1933, Die rotverschiebung von extragalaktischen nebeln, *Helvetica Physica Acta*, 6, 110.

Zwicky, F. 1937, On the masses of nebulae and of clusters of nebulae, *ApJ*, 86, 217.

Zwicky, F., Herzog, E., & Wild, P. 1968, *Catalogue of Galaxies and of Clusters of Galaxies* (Pasadena: California Institute of Technology (CIT), 1961–1968).

Index

A

A1795, 436
A2061, 414
A2062, 414
A2065, 414
A2089, 414
A2199, 360
A262, 59, 412, 430
A2744, 407
A347, 59, 412
A3560, 432
A3571, 432
A426, 59, 62, 430
acceleration potential, 33
accretion, 251–252, 256, 269, 274,
    284–294, 321, 324–326, 333, 336
active galactic nuclei, 318–319, 322,
    326–327, 338
adhesion model, 145–147, 221, 345
age of a universe, 95
amplitude of the initial power, 122
Andromeda, distance of, 383
Andromeda galaxy, 278, 295, 317,
    383, 390, 419–420
Andromeda, halo of, 390
Andromeda, model of, 384, 385
Andromeda, satellites of, 388–389,
    391
angular diameter distance, 98
angular power spectrum, 121

anti-biasing, 234
Apache Point Observatory Galactic
    Evolution Experiment (APOGEE),
    350, 379–380
APOGEE survey, 348
Aquila, 269
Atacama Large Millimeter Array
    (ALMA), 364
ATLAS survey, 392–395
attractor, 344–345
AWM7, 430
axial ratio, 25
axion-like particles, 482
axions, 56

B

B-polarisation, 122
B_3 spline, 151
backreaction models, 477
BAO shell, 436
baryogenesis, 113
baryon energy density, 122
baryon loading, 120
baryon number violation, 113
baryon to photon ratio, 112
baryonic acoustic oscillations (BAO),
    120, 129
baryonic density, 71
baryonic matter, 250, 254, 256, 260,
    275, 303, 306, 322

531

basins of attraction (BoA), 345, 413, 426–428, 449–452, 455

basins of repulsion (BoR), 427

BBKS approximation, 128

bias $1/F_c$ criterion, 223, 238, 240, 242

bias function, 225, 226, 228–229, 231, 233–234

bias parameter, evolution of, 224–227, 229, 234–243

biasing, 223–224, 229, 232

biasing, physical, 224

biasing, statistical, 223

big bang, 46, 48, 68–69, 78, 102, 301–302, 329

big bang nucleosynthesis, 78

big bang theory, 13

bispectrum, 131

black hole (BH), 301–302, 305, 312–327

Boltzmann equation, 29–30

break in the spectrum, 129

bulge, 26, 28, 45, 50, 57, 317–319, 321, 376, 384–386

C

3C129, 430

ΛCDM, 81, 457

ΛCDM model, 76, 96, 146, 148

ΛCDM parameters, 468

ΛCDM tensions, 469

ΛCDM universe, 145, 255

C,CP violation, 113

Canada–France–Hawaii Telescope (CFHT), 356, 392

caustics, 225

central density, 24

Chandra X-ray Observatory, 407

characteristic radius, 25

characteristic time, 39

chemical content, 250–251, 276, 280, 296, 298

chemical element, 301

chemical enrichment, 304

chemical equilibrium, 115

chemical evolution, 260, 272–273, 275, 295, 304, 306, 311

chemical potential, 115

circular velocity, 278

circumgalactic gas, 291

circumgalactic medium, 275–277, 285, 288–289, 292–294

classical gas, 94

closed universe, 88

Cloud in Cell (CIC), 154

cluster, bullet, 407

cluster, Coma, 49, 58, 360, 402

cluster, Draco, 419

cluster, Fornax, 419

cluster, Hydra, 360

cluster, nuclear star, 397

cluster, Perseus, 58, 62, 430, 432

cluster, Sculptor, 419

cluster, Ursa Minor, 419

cluster, Virgo, 317, 360, 394–398, 400, 403–405, 407, 424–426, 431

clustering, hierarchical, 141

clusters, 62, 65, 145, 169, 202, 214, 224, 256, 275–276, 317, 323, 330

clusters, colliding, 406

clusters, evolution of, 408

clusters, globular, 349, 399

clusters, internal structure of, 200, 204–205

CMB anisotropies, 119

CMB map, 123

CMB polarization, 124

cold dark matter (CDM), 56, 67, 80, 255

comoving coordinates, 125

comoving Hubble horizon, 104

Compton y-parameter, 124

concordance model, 96

consistency conditions, 108

continuity equations, 125

core, 45, 57, 384

correlation function, 64–65, 68, 186, 189–190, 195

correlation function, amplitude of, 199–202, 204

correlation function, angular, 17, 196–205, 223, 238
correlation function, evolution of, 231
correlation function, spatial, 188, 193, 196–197, 200, 203–205, 227–229, 231–232, 239
correlation length, 191, 193
Cosmic Background Explorer (COBE), 79
cosmic birefringence, 487
cosmic microwave background (CMB), 83, 117, 246, 261, 457
cosmic microwave temperature, 97
cosmic web, 47, 71, 83, 151, 153, 162, 192–193, 195, 214, 226
cosmic web, connectivity of, 178, 203
cosmic web, evolution of, 437
cosmic web, filamentary, 197
cosmic web, global properties of, 203
cosmic web, pattern of, 186, 201–202, 221
CosmicFlows, 424, 427–429
cosmological constant (constant vacuum energy), 123, 474
cosmological principle, 459
critical density, 47, 55, 460
cross-correlation, 123
current proper distance, 98
curvature term, 92, 96
curved geometry, 123
cyclic models, 102

D

2dF Galaxy Redshift Survey, 238, 242
dark ages, 251, 329
dark energy (DE), 71–73, 80, 83, 94, 472–473
dark halo, 254, 277–278, 286, 292–293
dark matter (DM), 47, 49, 51–56, 71, 83, 94, 146, 247, 250, 254, 256–258, 275, 277, 284–285, 321–322, 330, 338, 478–479
dark matter, fuzzy, 497, 482
dark matter, warm, 482
decaying mode, 127

deceleration parameter, 47, 71, 99
decoupling, 111
Delaunay Tesselation Field Estimator (DTFE), 155–158, 161, 163
density, 36
density contrast, 210–211
density distribution function, 210, 215
density field, 150–153, 175, 186, 198, 211, 214, 226–227
density field, angular, 202
density field, spatial, 202, 205
density gradient, 35–36
density limit, 149, 189, 194–195, 231, 233
density parameters, 460
density profile, 194–195, 255
density, reduced, 214
description function, 24
deuterium, 78, 247, 302
deuterium production, 115
diagram, color-magnitude, 76
diagram, Hertzsprung–Russell, 20–21, 41, 47, 351–352, 378
diagram, Hubble, 72–73, 77
diagram, Strömberg, 36–37, 54–55, 349
diagram, Toomre, 349–350
diagram, wedge, 58–61
disk, 26, 45, 57, 250–251, 256, 268, 270, 273, 275, 286, 290, 292–293, 319–320, 334, 377, 384–386
disk, thick, 351, 376
disk, thin, 376, 384
disk, young, 385–386
distance modulus, 101
dust, 261, 267–268, 280, 282, 311–312, 326, 338
dwarf galaxies, 329

E

η_b, 112
60 e-foldings, 107
Earth, 20
effective number of spin states, 110

Einstein–de Sitter universe, 471
Electroweak era, 112
elliptical galaxies, 254, 328
emergent space-time, 103
energy conservation, 93
energy distribution, 44
entropy density, 110
equation of state, 94
equations of motion, 125
ESO 2.2-m telescope, 364–365
ESO Very Large Telescope (VLT),
    395
Euclidean geometry, 98
Euler equations, 125
Eulerian position, 144
evolution, of galaxies, 39
evolution, stellar, 39
evolutionary diagrams, 215,
    217–219
evolutionary tracks, 215–217,
    219, 328

F

filaments, 59–60, 62, 65, 145, 164–166,
    169, 185, 202, 214, 224, 226–227,
    268–269, 271, 287–288, 290
filling factor, 180–185, 444, 447
flat disk, 28
flat universe, 98
flatness problem, 97, 103
Fornax, 360
Fourier transform, fast, 141, 197
fractal dimension function, 167–168,
    188–189, 193–195, 204–205
fractal properties, 192–193
fraction, of clustered matter, 223,
    226–227, 229, 237–238, 243
fraction, of unclustered matter, 240
fraction, of void matter, 239–241
free streaming, 119
freezing quintessence, 475
Friedmann equation, 92, 96, 459
Friedmann–Lemaitre–Robertson–
    Walker (FLRW) metric,
    87

G

Gaia mission, 348, 351, 377
Gaia–Enceladus, 350–352, 368–369
galactic disk, evolution of, 348
galactic parameters, 376
galactic plane, 37, 275, 312
galactic rotation constants, 33
galaxies, brightest cluster, 369, 398
galaxies, brightest group, 408–411
galaxies, cD, 356–357
galaxies, dwarf elliptical, 40, 349, 354
galaxies, early type, 392–395
galaxies, elliptical, 23, 38, 40, 44, 46,
    50, 62, 354
galaxies, evolution of, 22, 36, 38–40,
    43
galaxies, first ranked, 371–372
galaxies, formation of, 56
galaxies, halos of, 51
galaxies, isolated, 370–371, 375, 409
galaxies, masses of, 50
galaxies, models of, 23, 29, 31, 46
galaxies, polar ring, 418
galaxies, radio, 62
galaxies, S0, 393
galaxies, satellite, 369–372
galaxies, spheroidal, 354
galaxies, spiral, 38
Galaxy, 22–23, 27, 33, 35–38, 40, 50,
    58, 317
galaxy, polar ring, 419
galaxy, supergiant elliptical (cD), 61
gas, cooling, 251, 253, 261, 265–268,
    270, 276, 278–281, 286–288, 290,
    300, 323, 338
gas, heating, 265, 276, 278, 280, 282,
    286, 297, 338
gas, inflows, 275, 288, 290–292, 294,
    304, 306, 311, 323
gas, molecular, 267–271, 274, 282, 306
gas, outflows, 275, 288, 290–293, 304,
    306, 311, 323
gas, primordial, 248, 250–252, 256
general relativity, 459
giant molecular clouds, 360, 365

globular cluster, 23, 40–41, 44, 57, 250, 327

globular cluster, evolution of, 359–361, 363–364, 367–368

graceful exit, 109

gradient function, 188, 198, 200, 204,

grand unified theory, 112

gravitational clustering, 154

gravitational constant, 33, 246

gravitational instability, 137, 141

gravitational lensing, 123

gravitational potential, 24, 38, 143–144, 148, 254, 292, 345–346

gravitational waves, 299–300, 303, 477, 483

gravitinos, 56

great attractor, 424

growing mode, 127

H

H_0 tension, 469

halo, 45, 57

halo, evolution, 432

halo, stellar, 376–377, 384–386

haloes, 194

haloes, dark matter, 375, 385

haloes, internal structure of, 194–196, 200

harmonic oscillator, 108

He mass fraction, 116

helium, 113

Herschel Space Observatory, 268–269

hierarchical clustering, 57

Higgs field, 105

homogeneity, 88

horizon problem, 103

horizontal branch, 21

hot dark matter (HDM), 67, 145

Hubble constant, 13, 47, 90, 95

Hubble drag, 107

Hubble parameter, 46, 71, 90, 146

Hubble Space Telescope (HST), 73, 76, 317, 383, 391, 407

Hubble tension, 77

Hubby–Eberly Telescope, 359

hydrodynamical model, 29, 31

hydrogen, 78, 113, 251, 258, 262–263, 268, 279–280, 301–302, 329–330

hydrogen burning, 21

I

100-inch Mount Wilson telescope, 16

IAU Tallinn symposium, 58, 63

IC 3443, 399

indicator function, 132

inflation, 69, 71, 80, 83, 103, 246, 491

inflation ended, 106

information plateau, 130

initial amplitude, C_{10}, 122

initial conditions, 103–104

initial mass function, 39–40, 252, 271–274, 304–306, 324

instability, gravitational, 267, 270

instability, Kelvin–Helmholtz, 269

instability, Rayleigh–Taylor, 298–299

instability, thermal, 265–266

integrated Sachs–Wolfe (ISW) effect, 121, 471, 478

intergalactic medium, 264, 285, 288–289, 293–294, 329, 336

interstellar medium, 262, 265–267, 277, 302–303

IRAM 30-m Radio Telescope, 392

isochrones, 352

isotopes, 114

isotropy, 88

ISW tension, 471

J

James Webb Space Telescope (JWST), 77, 258

Jeans equations, 30–31

Jeans mass, 248–251, 270–271

K

k-essence, 475

Kaiser bias, 133

Kaluza–Klein particles, 481
kurtosis, cosmological, 211–212,
    215–221
kurtosis, mathematical, 211, 216–217,
    219–220
Kuzmin parameter, 38

L

Lagrange singularities, 68
Lagrangian position, 144
large-scale structure surveys, 457
late-time ISW, 123
Lemaître–Tolman–Bondi model, 478
Lick observatory, 17
Lick survey, 18
lightest supersymmetric particle, 481
LIGO/Virgo/KAGRA, 477, 483
linear bias, 129, 133
linear power spectrum, 128
little red dots, 326–327
local density fluctuations, 125
local group, 50, 58, 73, 75, 383, 404,
    420–422
local universe, 99
local web simulation, 429
log-power spectra, 131
Lovelock's theorem, 476
luminosity distance, 99
luminosity function, 43, 229–230, 259,
    261, 337–339, 369–372
luminosity function, evolution of, 373
luminosity limit, 231

M

2MASS redshift catalog, 425
M110, 388
M31, 15–16, 28, 35, 50, 57, 317, 383,
    421–422
M32, 388, 390, 392
M33, 16, 355–356, 390, 422
M49 (NGC 4472), 399–400
M59, 400
M60 (NGC 4649), 399, 403, 407
M87 (NGC 4486), 398–401, 403
Mach number, 283, 285

Magellan 6.5-m telescope, 406
Magellanic Cloud, Large, 295, 388,
    418, 422
Magellanic Cloud, Small, 388, 418
Magellanic, Greater, 418
magnetic fields, 252, 267, 270
magnitude, 101
magnitude limit, 232–233
mass fraction, 116
mass function, 26
mass-to-light function, 44
mass-to-light ratio, 25, 27–28, 40, 42,
    45–46, 311
massive neutrinos, 55, 67
matter density, 122, 462
matter-radiation equality, 112
mergers, 254–256, 299–304, 321, 323,
    333, 336
merging, 349, 351
merging of galaxies, 346
merging, dry, 347
merging, history of, 352–354
merging, wet, 347
Meszáros equation, 127
metallicity, 251, 264–265, 275, 278,
    286–287, 306–312
metric tensor, 87
Milky Way, 15–16, 254, 266, 270,
    275–276, 278, 290, 292, 295,
    300–301, 312–317, 348–352, 359,
    362, 368, 421–422, 426
Milky Way, evolution of, 369, 376
Milky Way, model of, 376
Milky Way, rotation of, 377
Milky Way, satellites of, 380,
    388–389, 391
Minkowski metric, 87
modified gravity, 476
Modified Newton dynamics (MOND),
    56
monopole problem, 104
multiplicity, 64
multiplicity distribution, 68
multiplicity function, 65–66
multiscale morphology filter (MMF),
    157

N

negative curvature, 97
neutrino, 55, 295–298
neutrino energy density, 117
neutron star, 297, 299–305
NEXUS filter, 160–161
NEXUS+ filter, 160–161
NGC 1052, 402
NGC 1275, 62
NGC 147, 388, 390
NGC 185, 390
NGC 187, 389
NGC 205, 389–390
NGC 2808, 361–362
NGC 300, 364–365
NGC 315, 412
NGC 3384, 392
NGC 383, 412
NGC 4365, 364, 392
NGC 4476, 399
NGC 4478, 399, 401
NGC 4482, 357
NGC 4486 (M87), 356–357
NGC 4486A, 399, 401
NGC 4486B, 399, 401
NGC 4565, 359
NGC 4736, 355–356
NGC 507, 412, 430
NGC 5746, 359
NGC 5813, 392
NGC 6822, 16
nodes, 164–165
non-equilibrium conditions, 113
non-linear bias, 133
nuclear statistical equilibrium, 115
nucleosynthesis, 47, 112–113,
    300–301
nucleus, 28, 45, 385–386

O

opacity parameter, τ, 122
open clusters, 41, 45
open universe, 88
Orion molecular cloud, 269

P

pancakes, 143
parameters, 107
particle density, 197
percolation, 64–65, 68, 180
percolation analysis, 178–179, 203,
    442
percolation function, 179–184,
    443–444
percolation parameter, 445
percolation threshold, 184, 444
perturbation theory, 134, 212,
    218–221
phantom dark energy, 475
phantom energy, 94
phase coupling, 174
photinos, 56
photoheating, 280, 282
photon temperature, 109
photon–baryon fluid, 119
photons, thermal, 109
Planck mass, 102
Planck space observatory, 79, 97, 261
Poisson equation, 33–34
population I, 78
population II, 78, 253–254, 263, 329
population III, 78, 252–254, 258–259,
    261–262, 294, 323, 329
population, clustered, 225–229, 231
population, non-clustered, 225, 231
positive curvature, 97
power spectrum, 122, 129
Press–Schechter theory, 143
primordial black holes, 479, 483
principal descriptive function, 23
probability distribution function
    (PDF), 132
profile, Burkert, 386, 388
profile, de Vaucouleurs, 25, 346
profile, density, 25
profile, Einasto, 24–25, 194, 255, 328,
    346, 356, 386–388
profile, exponential, 25
profile, Moore, 386

profile, NFW, 194, 255, 291–292, 328, 346, 380, 386, 388
profile, Sersic, 356–358

Q

QCD axion, 482
quantum fluctuations, 104
quantum gravity, 102
quark-hadron phase transition, 112
quasars, 264, 275, 280, 318, 326, 328
quintessence models, 475

R

radiation, 94
radiation density, 117, 461
recombination, 56, 112, 117, 247–249, 261–262, 266, 280, 329
red giant, 20
redshift, 88
region, future collapse, 413
region, turnaround, 413
region, zero gravity, 413
reheating, 106
reionization, 123, 261–263
relaxation, violent, 346
rotation curve, 28, 53, 57, 321
Russell hypothesis, 19

S

Sachs–Wolfe effect, 119
Sachs–Wolfe regime, 122
Saha equation, 115, 117
Sakharov conditions, 113, 463
scalar field, 105, 475
scalar modes, 108
screening mechanisms, 477
self-interacting dark matter, 484
Seyfert galaxy, 62
sheet, local, 422–423
Shell theorem, 91
shock wave, 256, 265–266, 269, 271, 282–286, 296–297, 302
shot noise, 130

Silk damping, 121
simulation (ΛCDM), 149, 152–153, 181, 184, 186, 189, 199, 210, 214–218
simulation, EAGLE, 150, 191, 352, 366–368, 379, 420–421
simulation, Illustris The Next Generation (TNG), 150, 226–227, 229–242, 257, 448
simulation, Millennium, 150, 159, 167, 191, 197–201, 254–256
simulation, Millennium II, 254–255
simulations, hydrodynamical, 149–150, 256, 267, 288, 347
skewness, cosmological, 211–212, 215–221
skewness, mathematical, 211–212, 216–217, 219–220
Sloan 2.5-m telescope, 379
Sloan Digital Sky Survey (SDSS), 169, 319–320, 334, 348
Sloan Digital Sky Survey Telescope, 391
slow-roll, 106–107
small-scale CDM problems, 480
smoothing, 150, 214–215, 344
smoothing length, 184–186, 217–218, 220
sound speed, 247–250, 270, 278, 282–283, 285, 287, 298
sound speed of DE, 476
space–time, 103
spatial density, 24, 306
spatially flat gauge, 108
spectral energy distribution, 384–385
spin states, 111
spiral galaxies, 320–321
spontaneous symmetry breaking, 112
stacking structure on the CMB, 123
standard candles, 101
standard rods, 101
standard ruler, 129
star formation, 27, 46, 54, 253, 257, 261, 263, 274, 287–288, 290, 292,

294, 302, 307–308, 310–312, 338, 352
star formation function, 43
star formation rate, 39, 253, 264, 305, 309, 337, 339, 365, 410
star formation, history of, 353
steady-state model, 13, 102
stellar evolution, 19, 303, 305, 308–309
stellar halo, 26, 28, 54, 327
stellar halo, evolution of, 348–349
stellar mass, 252, 257, 260, 264–265, 271, 273–274, 292, 305–307, 312, 318–320, 322, 327, 331–334, 336
stellar populations, 23, 308–309, 312, 328
stellar winds, 254, 259, 293, 302–303, 306, 322
stellar yield, 305, 307
sterile neutrinos, 482
stream, Draco–Ursa minor, 418
stream, Helmi, 349, 368–369
stream, Magellanic, 349, 418, 420
stream, Sagittarius, 349
stream, Sequioa, 350
stream, Thamnos, 350
streams, 347–348
strong CP-problem, 482
structure function, 188–190, 193, 201, 204
Subaru 8.2-m telescope Hawaii, 433
Sunyaev and Zeldovich (SZ) effect, 124
supercluster wall, 437
supercluster, A2142, 413–415, 455
supercluster, Apus, 428
supercluster, Aquarius–Capricornus, 438
supercluster, BoA, 443
supercluster, BOSS Great Wall, 413, 416, 433
supercluster, catalogue of, 435
supercluster, Centaurus, 430

supercluster, Coma, 58, 61, 428, 430, 433, 437
supercluster, Corona Borealis, 413, 414
supercluster, diameter of, 445
supercluster, Dominant plane of, 438
supercluster, evolution of, 431–433, 442, 445–448, 452, 454–455
supercluster, Hercules, 58, 428, 430, 433, 438
supercluster, Horologium–Reticulum, 438
supercluster, Hydra–Centaurus, 437–438
supercluster, Laniakea, 428, 450
supercluster, Lepus, 428
supercluster, Local, 58, 61, 146, 398, 422, 424
supercluster, Local plane of, 438
supercluster, Pavo–Corona Australes, 437
supercluster, Perseus-Pisces, 58–59, 61–62, 412, 424, 428, 430, 432, 437, 455
supercluster, regularity of distribution, 435–437
supercluster, Shapley, 147, 428, 432
supercluster, Sloan Great Wall, 171–172, 428, 446
supercluster, Ursa Majoris–Leo, 438
supercluster, Virgo, 432, 437, 455
superclusters, 60, 146, 169, 171, 185, 214, 224, 226, 345–346
superclusters, evolution of, 345, 412
supermassive black hole (SMBH), 313–319
supernova explosion, 22, 254, 293–295, 299, 301, 303, 322
supernova measurements, 457
supernova, core-collapse, 295–298, 302
supernova, explosion, 296, 300–302

supernova, thernonuclear, 295–296,
298, 303
supernova, type Ia, 72, 74,
296, 298
supervoid, Eridanus, 441
surface brightness, 26, 285, 318
surface density, 23–24, 291, 306
Swiss-cheese model, 478

T

σ_8 tension, 470
tensor modes, 108
tensor-to-scalar ratio, 122
tesselation, Delaunay, 155–156
tesselation, Voronoi, 155–156
thawing quintessence, 475
thermal equilibrium, 115
thick disk, 28
thickness, 197–202, 204, 226–227,
283
third integral of motion, 34
threshold density, 179–182, 184, 186,
442–444
Tip of the Red Giant Branch
(TRGB), 74
topology, 182, 186
topology, bubble, 180, 182
topology, meatball, 180, 182
topology, sponge, 180, 182
tracker models, 475
transfer function, 127
transition between gauges, 109
triple α reaction, 115
turbulence, 270, 296, 298

U

UGC2562, 430
UGC3358, 430
ultralight dark matter, 482
Universe, evolution of, 489–492
universes, 98

V

vacuum energy, 464
VCC0170, 395–396
velocity dispersion, 30, 32–37, 45, 54,
317, 319–320
velocity ellipsoid, 34
velocity, circular, 26
velocity, rotation, 31, 37, 45, 50, 376,
379
very large telescope, 315–316
violent relaxation, 38, 254–255
virial mass, 254
virial radius, 275, 286–288, 291–294
virial temperature, 250, 256, 277–278,
292, 323
voids, 59, 145, 164–166, 184–185, 202,
224
void models, 477
void, Bootes, 438
void, Local, 437–438

W

walls, 164–166, 268
wavelet, 169, 170–171, 174
weakly interacting massive particles,
480
Westerbork Radio Synthesis
Telescope, 392
white dwarf, 265, 295–296, 298–299,
301, 303, 305
Wiener filtering method, 424–425, 449
Wilkinson Microwave Anisotropy
Probe (WMAP), 79
William Herschel 4.2-m Telescope,
392

Z

Zeldovich approximation, 61, 67–68,
109, 143–148, 221, 224, 345, 424
Zeldovich scale-invariant spectrum,
109

www.ingramcontent.com/pod-product-compliance
Lightning Source LLC
Chambersburg PA
CBHW050534190326
41458CB00007B/1776